Fundamentals of Pulse and Digital Circuits

Third Edition

RONALD J. TOCCI

Monroe Community College

Charles E. Merrill Publishing Co.
A Bell & Howell Company
Columbus Toronto London Sydney

Merrill's International Series in Electrical and Electronics Technology

BATESON	*Introduction to Control System Technology, 2nd Edition, 8255–2*
BOYLESTAD	*Introductory Circuit Analysis, 4th Edition, 9938–2*
	Student Guide to Accompany Introductory Circuit Analysis, 4th Edition, 9856–4
BOYLESTAD/ KOUSOUROU	*Experiments in Circuit Analysis, 4th Edition, 9858–0*
NASHELSKY/ BOYLESTAD	*BASIC Applied to Circuit Analysis, 20161–6*
BREY	*Microprocessor/Hardware Interfacing and Applications, 20158–6*
FLOYD	*Digital Fundamentals, 2nd Edition, 9876–9*
	Electronic Devices, 20157–8
	Essentials of Electronic Devices, 20062–8
	Principles of Electric Circuits, 8081–9
	Electric Circuits, Electron Flow Version, 20037–7
STANLEY, B. H.	*Experiments in Electric Circuits, 9805–X*
GAONKAR	*Microprocessor Architecture, Programming and Applications With the 8085/8080A, 20159–4*
ROSENBLATT/ FRIEDMAN	*Direct and Alternating Current Machinery, 2nd Edition, 20160–8*
SCHWARTZ	*Survey of Electronics, 2nd Edition, 8554–3*
SEIDMAN/ WAINTRAUB	*Electronics: Devices, Discrete and Integrated Circuits, 8494–6*
STANLEY, W. D.	*Operational Amplifiers With Linear Integrated Circuits, 20090–3*
TOCCI	*Fundamentals of Electronic Devices, 3rd Edition 9887–4*
	Electronic Devices, 3rd Edition, Conventional Flow Version, 20063–6
	Fundamentals of Pulse and Digital Circuits, 3rd Edition, 20033–4
	Introduction to Electric Circuit Analysis, 2nd Edition, 20002–4
WARD	*Applied Digital Electronics, 9925–0*

Published by Charles E. Merrill Publishing Co.
A Bell & Howell Company
Columbus, Ohio 43216

This book was set in Times Roman and Korinna.
Production Editor: Jan Hall
Cover Design: Tony Faiola
Cover Photo: Western Electric

Library of Congress Catalog Card Number: 82–61752
International Standard Book Number: 0–675–20033–4
Printed in the United States of America
4 5 6 7 8 9 10—87 86

To Carrie Anne
The most wonderful Christmas
gift a family could have.

preface

Most two-year electronic technology programs are feeling the impact of the "digital revolution" and have responded by developing new courses on microprocessors, or by including microprocessors in existing courses. In many cases this response has resulted in the reduction or elimination of material that was previously deemed important enough to be part of the curriculum.

While the de-emphasized or deleted material sometimes consists of topics made obsolete by newer developments, it all too often includes areas still extremely important to the technician. Architects of electronics technology curricula must be careful not to eliminate topics and experiences that contribute to the development of circuit analysis and troubleshooting skills, and they should not assume that ICs have greatly decreased the need for such skills.

If anything, electronic systems are becoming more complex since they are designed to be more versatile, have more operating modes, and often contain many different circuit technologies and the necessary interfacing between them. A good technician has to know more about ICs than their input/output relationship, which means understanding their terminal electrical characteristics and their input and output circuitry. Of course, this circuitry is made up of discrete components.

Unfortunately, most textbooks and courses give only a cursory treatment of IC terminal characteristics with little regard for their importance not only in design, but in testing and troubleshooting. Even less attention is given to discrete circuits used for signal conditioning, signal generation, interfacing, etc., functions prevalent in most digital systems.

This edition is substantially revised and reorganized. It presents a balanced approach to pulse and digital circuits. It retains only the discrete circuit material that has not become obsolete and has significantly increased coverage of IC characteristics and applications.

The major reorganizational changes are as follows:

1. The old Chapter One (review of circuit theorems) has been deleted.
2. The old Chapter Four on *RL* and *RLC* circuits has been deleted.
3. The old Chapter Five on pulse transformers and delay lines has been deleted.
4. A new Chapter Four covers the switching characteristics of semiconductor devices.
5. A new Chapter Five presents the basic operation of operational amplifiers in the open-loop mode, and with negative and positive feedback.

6. A new Chapter Six covers signal conditioning circuits (clippers, limiters, level translators, buffers, comparators, and Schmitt triggers). Both discrete and IC circuits are presented.
7. Chapter Seven on logic gates and Boolean algebra has been expanded to include more illustrative examples and a section on interpreting logic schematics.
8. Chapter Eight on flip-flops has been updated to include the latest developments.
9. Chapter Nine on IC logic families has been updated to include Schottky TTL, SOS, and I^2L.
10. A new Chapter Ten addresses the problem of interfacing digital ICs to different logic families, and to other devices such as LEDs and high-voltage/high-current loads.
11. Chapter Eleven on signal generating circuits now includes Schmitt trigger oscillators, and more on the 555 timer. In addition, the complete circuit for an oscilloscope triggered sweep waveform is analyzed in detail.

The following are some of the distinctive features of this text, some of which are new to this edition:

1. The emphasis is on developing analytical skills that can be applied to a wide variety of circuits. This text will *not* supply sets of circuit design equations. I feel that these lessen the student's motivation for understanding the underlying concepts and often discourage any reasoning processes other than those used to choose the correct formulas.
2. A large number and wide variety of illustrative examples are used to enhance understanding and to show how analytical techniques are employed. Most of these examples do *not* merely plug numbers into formulas, but rather they show how to use basic concepts and analytical tools to solve challenging problems. Many of the examples demonstrate the types of reasoning to employ when troubleshooting a faulty circuit.
3. Numerous end-of-chapter problems are keyed to the appropriate sections in the text. Many of them are new for this edition and, like the examples, very few of these are plug-ins. Most of them are challenging problems designed to reinforce the concepts, show new applications, and present different aspects of the circuits covered in the text. A good number of the problems are troubleshooting exercises that present symptoms of a faulty circuit operation and require the student to determine the most likely cause of the malfunction.
4. Studying this text requires basic technical mathematics (algebra, trig, logs, etc.). No calculus is employed; however, the concepts of the derivative (rate of change) and the exponential form (ϵ^x) are fully explained as they are needed. A basic background in dc/ac circuits and semiconductor devices is also required, although many of these topics are reviewed throughout the text.

The material in this text has been used in a one-semester, sophomore-level course for several years. Recently, however, some topics from the text have been taught in the freshman electronics course, notably logic gates and Boolean algebra because they require very little math or electronics background.

I would like to express appreciation to my colleagues at Monroe Community College, who have offered numerous constructive suggestions; and to others who kindly gave of their time to help make this a better book:

John D. Meese, *Ohio Institute of Technology*
W. B. Oltman, *Atlas Electric Devices Co.*
Dr. Samuel Oppenheimer, *Broward Community College*
Homer L. Apple, *Guilford Technical Institute, North Carolina*
Angus Brown, *Wharton County Junior College, Texas*
Russell E. Clark, *Harrisburg Area Community College, Pennsylvania*
Dr. John H. Fuller, *Prairie View A&M University, Texas*
Herbert N. Hall, Jr., *Lakeland College, Ohio*
Paul Hardy, *Massasoit Community College, Massachusetts*
David M. Hata, *Portland Community College, Oregon*
Kirby Franklin, *Albuquerque Technical Vocational Institute*
Albert R. Maez, *New Mexico Highlands University*
E. W. McCulloch, *Rowan Technical College, North Carolina*
Charles G. Nelson, *California State University at Sacramento*
D. H. Peterson, *Northern Montana College, Havre, Montana*
Albert D. Robinson, *Washtenaw Community College, Ann Arbor, Michigan*
Terry R. Stanhope, *California Polytechnic State University at San Luis Obispo*
James P. Thompson, *Grand Rapids Junior College, Michigan*
Tom Waddoups, *Cochise College, Arizona*
John W. Walstrum, *Catonsville Community College, Maryland*

My deepest love and appreciation to my devoted wife, Cathy, who typed most of the manuscript in her almost nonexistent spare time. I know that the reader will find this edition significantly improved and worthy of its stated objectives.

contents

7 LOGIC GATES AND BOOLEAN ALGEBRA

8 FLIP-FLOPS

9 INTEGRATED CIRCUIT LOGIC FAMILY CHARACTERISTICS

1 Basic Concepts

1.1 Introduction

This book is about electronic circuits classified as pulse, digital, or switching circuits. These three common terms refer to circuits that share one fundamental characteristic: *their inputs and/or outputs switch rapidly from one level (value) to another.*

1.2 Switching Circuits

Figure 1–1 illustrates switching circuits for three separate cases. In (a) the input signal switches between 0 V and +10 V, while the output signal has a slowly varying (ramp) portion followed by a rapid jump from +10 V to 0 V. The circuit in (b) has a smoothly changing sinewave input, but its output switches levels rapidly. In (c), both input and output signals make fast transitions from one level to another.

All of these circuits are *switching* circuits because either their input signal, output signal, or both exhibit rapid transitions at some time. The circuit in (c) is a special type, where both input and output signals switch between two distinct levels or states. It is called a *digital circuit,* in which all inputs and outputs will nominally be at one of two distinct values; for example, 0 V and +5 V, as in Figure 1–1(c).

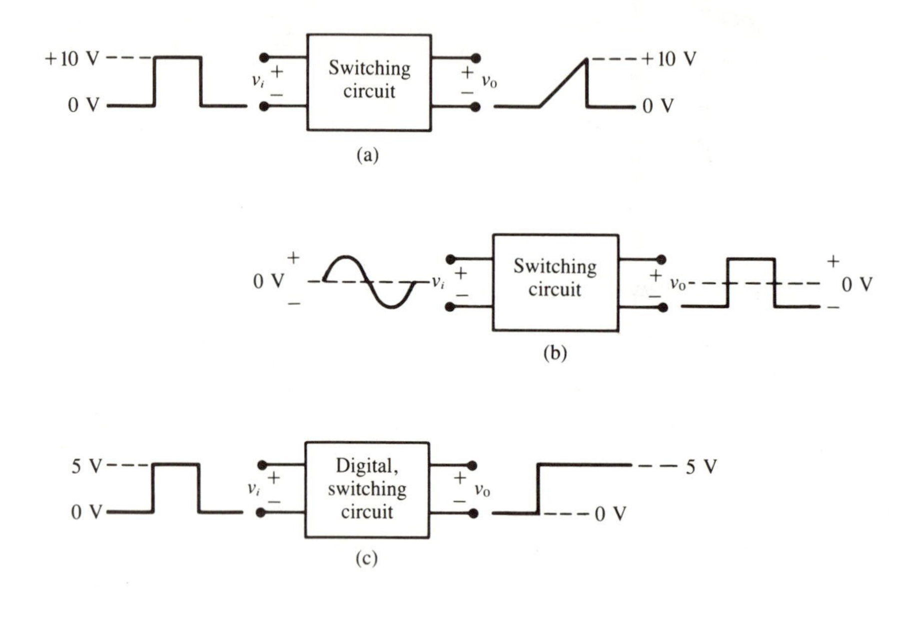

Figure 1–1 **Examples of switching circuits**

Linear circuits Circuits whose input and output signals change smoothly and continuously are classified as *linear,* or *analog,* circuits. An example would be an amplifier with a sinewave input and a sinewave output. Although linear circuits play an important role in the electronics industry, more and more electronic functions are being taken over by digital and switching circuits.

Digital integrated circuits The dominance of digital and switching circuits has been brought about by the rapid developments in integrated circuit fabrication technology. Digital circuits, because of their relative simplicity, reliability, and low-power consumption, are ideally suited for integration. We have seen digital integrated circuits (ICs) advance from several transistors per chip to thousands per chip. Digital switching circuits are the most widely used devices in modern electronic systems.

Despite the tremendous impact of ICs, discrete switching circuits are still important, especially in applications such as signal generation, signal conditioning, and interfacing. Moreover, the input and output circuitry of an IC is always visualized as discrete components for purposes of analysis. Thus, both ICs and discrete circuits should be included in any meaningful study of switching circuits. We will devote considerable time to both in the following chapters.

1.3 Ideal Pulse Signals

There are many different signals that occur in switching circuits, and many of them will be encountered in our ensuing study. Right now, however, we will introduce the two most common switching signals: the *step* waveform, and the *pulse* waveform.

Ideal step Figure 1–2(a) shows an *ideal* step voltage waveform in which the voltage is at the V_1 level for some time, and then jumps to the V_2 level. The step is ideal in that the transition from the V_1 level to the V_2 level is instantaneous, requiring no time to occur. It is a positive-going step because it jumps from one level to a more positive level. Note that V_1 and V_2 can be any values and of either polarity, as long as V_2 is more positive than V_1.

To compare, Figure 1–2(b) is an ideal negative-going step waveform that makes an instantaneous transition from V_2 to V_1. Again, V_2 is more positive than V_1.

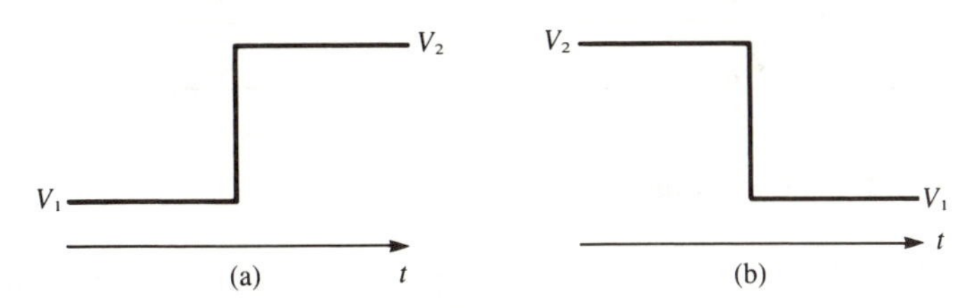

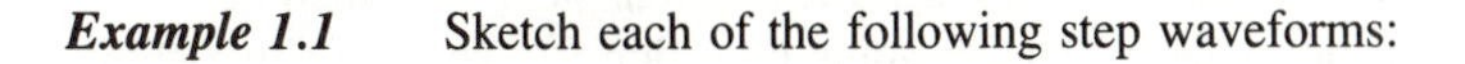
Figure 1–2 **Ideal step waveforms**

Example 1.1 Sketch each of the following step waveforms:

(a) positive-going step from −2 V to +6 V
(b) positive-going step from −6 V to −2 V
(c) negative-going step from +5 V to 0 V
(d) negative-going step from +4 V to −1 V

Solution: The results are drawn in Figure 1–3. Note that there is no correlation between the polarity of the two levels and the direction of the step.

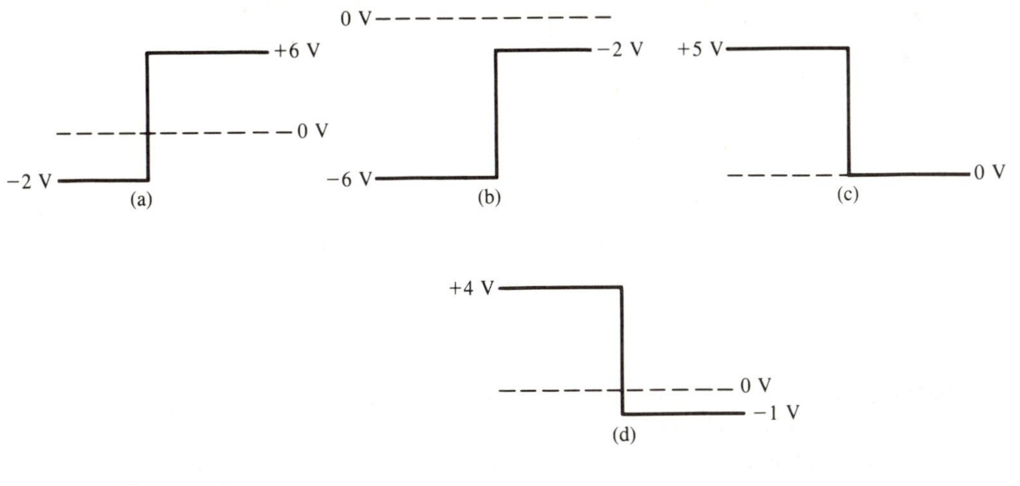

Figure 1–3

Ideal pulse Figure 1–4(a) shows an ideal voltage pulse that consists of two steps: the first from V_1 to V_2 at time t_1, and the second from V_2 to V_1 at time t_2. The V_1-to-V_2 transition is called the *rising edge* of the pulse; it is also referred to as the positive-going transition. Likewise, the V_2-to-V_1 transition is called the *falling edge,* or negative-going transition, of the pulse.

This pulse is ideal because its transitions are instantaneous. Also, the portion between transitions is perfectly flat (constant V_2 level). This flat portion is called the *pulse width,* or *pulse duration,* and is given the symbol, t_p.

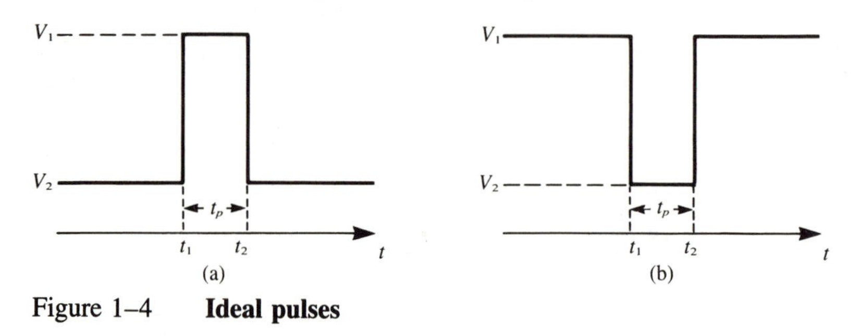

Figure 1–4 **Ideal pulses**

The pulse in Figure 1–4(a) is a positive-going pulse resting on a V_1 *baseline*. That is, the signal is normally at the V_1 level (baseline), except for a short time (t_p) during which it pulses in the positive direction to its *peak* value, V_2.

Likewise, the pulse in Figure 1–4(b) is a negative-going pulse from the V_1 baseline to its peak value, V_2. This pulse has a negative-going transition at t_1 and a positive-going transition at t_2.

The levels V_1 and V_2 can be any values of either polarity, although the most common baseline voltage is 0 V. Several different examples are shown in Figure 1–5.

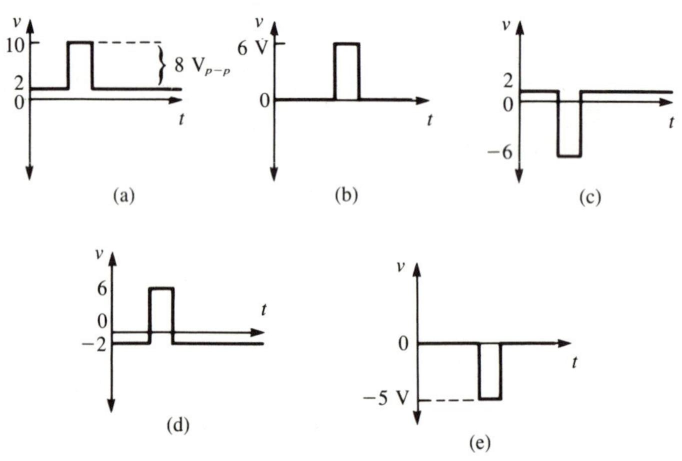

Figure 1–5

Pulse amplitude The amplitude of a pulse signal is best specified as its *peak-to-peak* (*p-p*) variation from baseline to its peak value. For example, the signal in Figure 1–5(a) has a +2 V baseline and a +10 V peak. The total change from its baseline to its peak is therefore 8 V. Thus, the amplitude of this pulse is expressed as 8 V *p-p*.

Example 1.2 Which of the pulses in Figure 1–5 is an 8 V *p-p* positive-going pulse on a −2 V baseline?

Solution: The pulse in Figure 1–5(d).

Example 1.3 Sketch a 5 V *p-p* negative-going pulse on a 0 V baseline.

Solution: The result should look like Figure 1–5(e).

Non-ideal pulses In practice it is impossible to produce step and pulse waveforms with instantaneous rising and falling edges, and perfectly flat pulse durations. The next chapter will examine some of the types of distortion that occur on practical pulse waveforms. However, in a lot of our subsequent circuit analysis we will use the ideal pulse signal in order to simplify the understanding of circuit operation. Once we have examined the circuit fully, we will then look at the effects of a practical pulse input.

1.4 Ideal Switching Devices

The most important component in any switching circuit is the *switching device*. This device can switch between two distinct operating states; for example, the conducting state and the non-conducting state. As we shall see in Chapter Four there are many semiconductor devices that can operate as switches. For now we will discuss the ideal switching device and the characteristics that make it ideal.

An ideal switch is symbolized in Figure 1–6. It has two terminals labelled x and y, and a moveable contact z. The position of the contact determines the electrical characteristic between x and y. In Figure 1–6(a), there is no electrical path between x and y. The switch is "open"; there is an open circuit between the switch terminals. This position will be referred to as the OFF state of the switch.

In Figure 1–6(b) the contact z has been moved so that there is a direct connection between x and y. The switch is "closed"; there is a short circuit between the switch terminals. This position will be called the ON state of the switch.

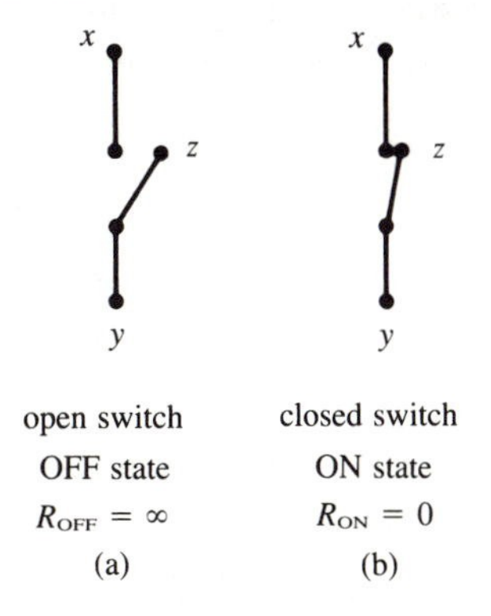

Figure 1–6 **Two states of an ideal switch**

Thus, we can say that an ideal switch has an *infinite* resistance between its terminals in the OFF state, and zero resistance in the ON state. That is $R_{OFF} = \infty$ and $R_{ON} = 0$. Currently, there are no semiconductor switching devices that can exactly match both of these resistance values, although some of them can have extremely high values of R_{OFF}. Mechanical switches and relay-controlled switch contacts come very close to the ideal characteristics of Figure 1–6(b), but their usefulness in modern switching circuits is limited by their relatively slow switching speed, contact bounce, and contact deterioration.

The ideal controlled switch Something has to control the state of a switching device; i. e., cause it to switch from one state to another. Semiconductor switching devices are controlled by either a voltage or current applied to a control terminal. Figure 1–7 symbolizes an ideal voltage-controlled switch. The terminal labelled c is the control input terminal. As the accompanying table shows, a high voltage (or current) at the control terminal causes the switch to go to the ON state, while a low voltage (or current) causes it to go to the OFF state.

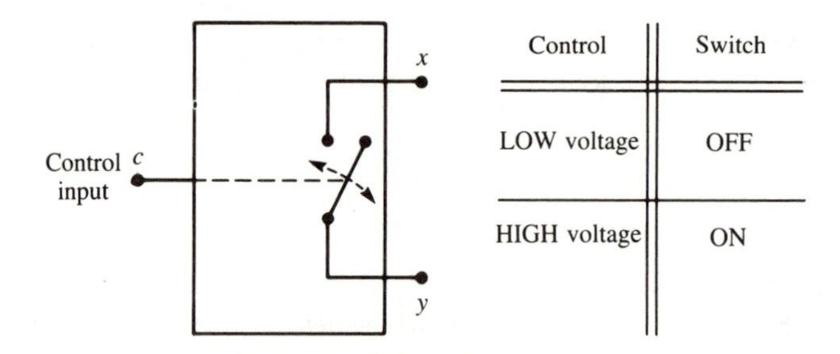

Control	Switch
LOW voltage	OFF
HIGH voltage	ON

Figure 1–7 **Ideal voltage-controlled switch**

Figure 1–8 shows a very common circuit arrangement that uses a voltage-controlled switch. The circuit output voltage is taken across the switch terminals. When the control voltage input, V_{IN}, is at 0 V, the switch is OFF so that an open circuit exists between the switch terminals. Thus, no current flows through the resistor, and the full +10 V source voltage appears across the switch terminals so that V_{OUT} = +10 V. When V_{IN} = +10 V, the switch is ON and there is a short circuit between the switch terminals. Thus, current will flow from the +10 V source through R and the shorted switch terminals to ground, and V_{OUT} will be 0 V.

The accompanying table in Figure 1–8 summarizes the voltages and currents for the two input voltage conditions. The following important points should be noted:

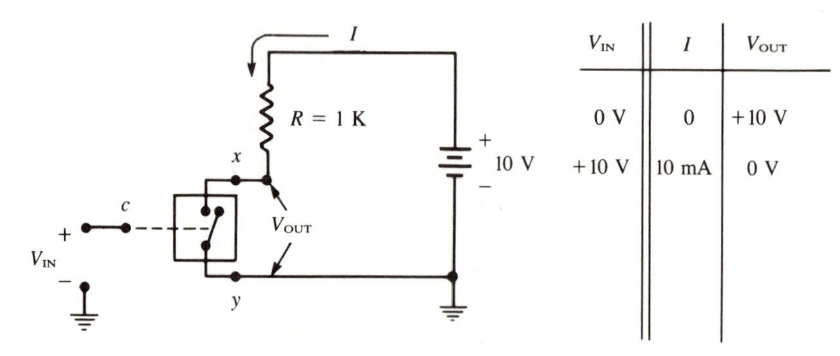

V_{IN}	I	V_{OUT}
0 V	0	+10 V
+10 V	10 mA	0 V

Figure 1–8 **Ideal controlled switch used as a voltage inverter**

(a) The output voltage is at a high level (+10 V) when the input voltage is at a low level (0 V), and vice-versa. For this reason, this circuit arrangement is called a *voltage level inverter,* or simply an *inverter.* The inverter is the basis for numerous switching and digital circuits, and we will encounter it very often in our subsequent study.

(b) The current flow through the switch is zero when the switch is OFF, but the voltage across the switch terminals, (V_{OUT}) is high (+10 V). On the other hand, the current is high (10 mA) when the switch is ON, but the voltage across the switch terminals is low (0 V). This point is a crucial one that often confuses students. Just remember that whenever a switching device is used in an inverter circuit, the current will be high when the output voltage is low, and the current will be low when the output is high.

Example 1.4 What would the output voltage of the inverter circuit look like when the input is a 10 V *p-p* positive-going pulse on a 0 V baseline?

Solution: Figure 1–9 shows the result. The inverter's output will be the *inverse* of the input, so that it will be a 10 V *p-p* negative-going pulse on a +10 V baseline. Note that the output pulse levels and transitions are always opposite to the input pulse transitions.

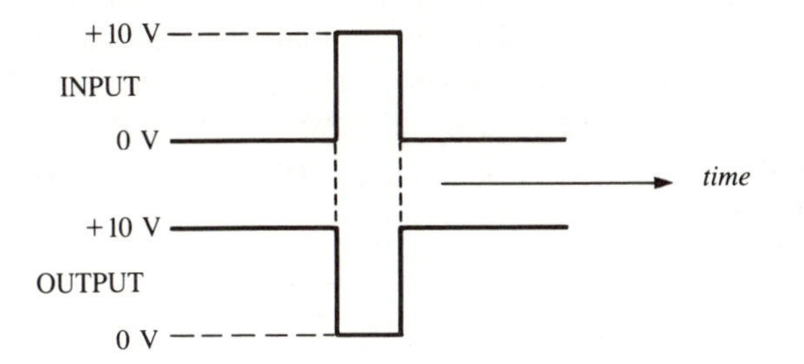

Figure 1–9 **A pulse applied to an inverter circuit produces an output pulse with opposite voltage levels and transitions**

Ideal switch response time The output waveform in Figure 1–9 indicates that the output and input pulse transitions occur at exactly the same time, and that they occur instantaneously. This situation can be true only if the controlled switch has a zero *switching response time*. An ideal switch has zero response time, but *all* practical switching devices require a certain amount of time to respond to the changes in the control input. The wide range of semiconductor switches have response times ranging from a couple of nanoseconds (ns) to a couple of milliseconds (ms). Figure 1–10 shows how the response of an inverter circuit might appear if it uses a *real* switching device. Note that the output transitions occur a short time after the input transitions, and they do not occur instantaneously (the transitions have a non-vertical slope).

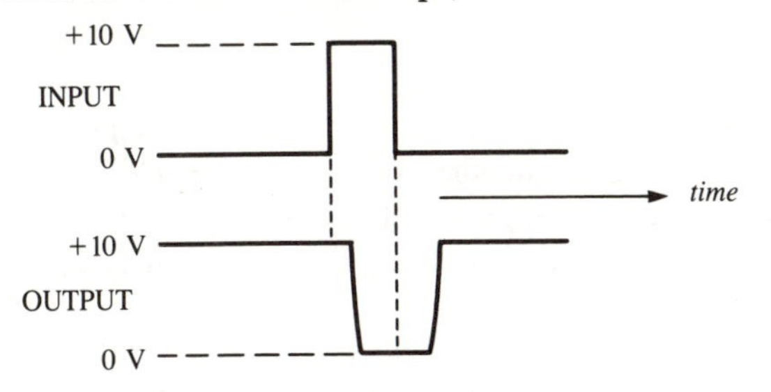

Figure 1–10 **Response of a non-ideal inverter shows effects of switch's response time**

Summary

An ideal switching device has two operating states, ON and OFF. In the ON state it has zero resistance between its terminals and will act like a perfect short circuit. In the OFF state, it is an open circuit and will conduct no current. An ideal switching device also will switch states instantaneously in response to a control input signal.

1.5 Current Direction

Throughout this text, we will use the *conventional* direction for current flow. That is, we will assume that current is the movement of *positive* charges. Although this assumption is contrary to what is known about electron movement in solids, the electronics industry has adopted this convention; our study will be easier if we use it, too.

Figure 1–11 shows how to determine this current direction. Since we are assuming that positive charges carry the current, and that the positive terminal of the voltage source will repel positive charges, the current will flow *away* from the positive terminal of the source. Likewise, the current will flow *toward* the negative terminal of the source since this terminal will attract positive charges.

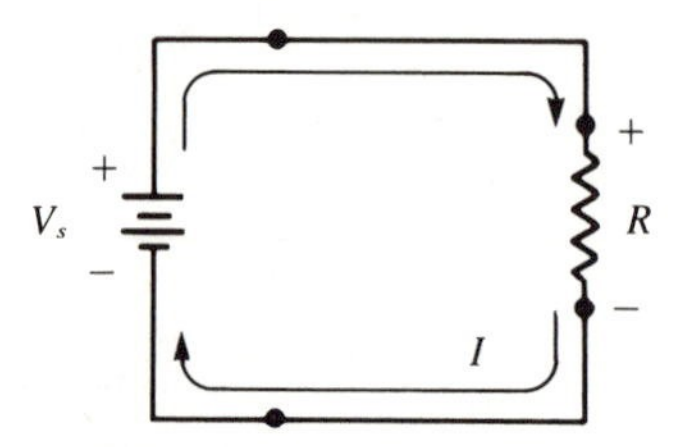

Figure 1–11 **Conventional current flows out of the source's + terminal and into its − terminal**

The flow of current, as we have defined it, always will be *into* the positive terminal of a resistance, and *out of* the negative terminal. Thus it will always be possible to determine the polarity of voltage across a resistive device if the direction of current is known, and vice-versa.

1.6 Voltage Notation

Whenever we refer to a voltage in a circuit, we are always talking about the voltage at one point in the circuit with respect to another point in the circuit. To avoid confusion, we will often use a double-subscript notation to define a voltage. For example, V_{xy} will always mean the voltage at point x with respect to point y. Likewise, V_{yx} means the voltage at point y with respect to point x.

If a single subscript is used, then the voltage refers to that terminal with respect to *ground*. For example, V_x means the voltage at point x with respect to ground.

Example 1.5 Refer to the circuit of Figure 1–12 and find the values of V_{xy} and V_{yx}.

Solution: The top terminal of R_1 is + with respect to its bottom terminal by 4 V. Thus, since point x is more positive than y, we have

$$V_{xy} = +4 \text{ V}$$

We can express this alternatively as

$$V_{yx} = -4 \text{ V}$$

The negative sign indicates that point y is negative with respect to x.

Note that $V_{yx} = -V_{xy}$. It is always true that reversing the subscripts changes the polarity of the voltage.

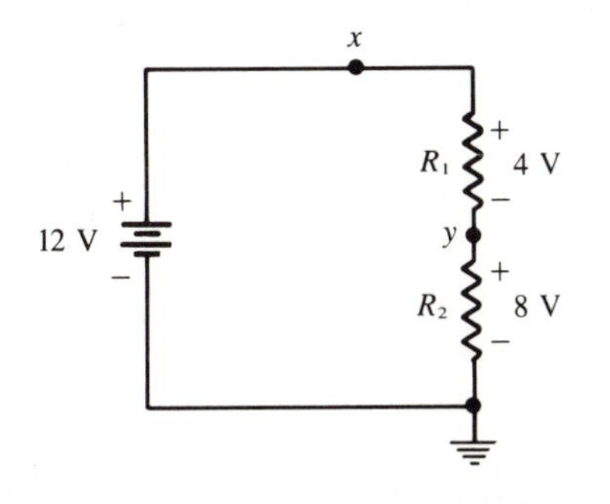

Figure 1–12

Example 1.6 Find the values of V_x and V_y in Figure 1–12.

Solution: Terminal x is 12 V more positive than ground since it is connected to the + source terminal. Thus,

$$V_x = +12 \text{ V}$$

Similarly, point y is +8 V with respect to ground, so that

$$V_y = +8 \text{ V}$$

Note from this last example that

$$V_x - V_y = +4 \text{ V}$$

which is exactly the value of voltage across R_1. In general, it is true that

$$V_x - V_y = V_{xy} \qquad \textbf{(1–1)}$$

This relationship states that the voltage between two points in a circuit is equal to the difference in the voltages from each point with respect to ground. This fact will be useful in some of our circuit analysis.

1.7 Representing dc Source Voltages

Most circuits use one or more dc source voltages, and very often several circuit branches are connected to the source terminals. In some cases, use of the dc source symbol (as in

Figure 1–12) becomes too cumbersome. For those cases, it is common to show only one terminal of the source, with the understanding that the other terminal is connected to the circuit ground point. To illustrate, the circuit of Figure 1–12 has been redrawn in Figure 1–13(a) using the simplified source representation. The terminal labelled "+12 V" is connected to the + terminal of a source. The negative terminal of the 12 V source is understood to be connected to ground.

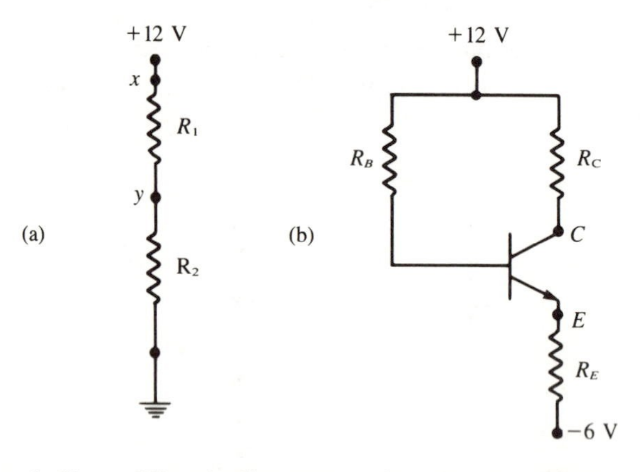

Figure 1–13 **Circuit diagrams using the simplified representation of dc source voltages**

Another illustration is given in Figure 1–13(b), where two dc sources are used. The terminal labelled "−6 V" is connected to the negative terminal of a 6 V source, and the + terminal of the 6 V source is understood to be connected to ground (along with the negative terminal of the 12 V source).

1.8 Time-Varying Currents and Voltages

Pure dc voltages and currents have constant values. These quantities will be symbolized by uppercase letters V and I respectively. Voltages and currents whose values are not constant, but change with time, are called time-varying voltages and currents. These quantities are symbolized throughout this text by lowercase v and i.

To illustrate, Figure 1–14 shows the two different cases of a pure dc voltage and a time-varying voltage signal. The dc voltage, V_{xy}, is the same value for all values of time, t. The signal, v_{xy}, has different values of voltage at different times.

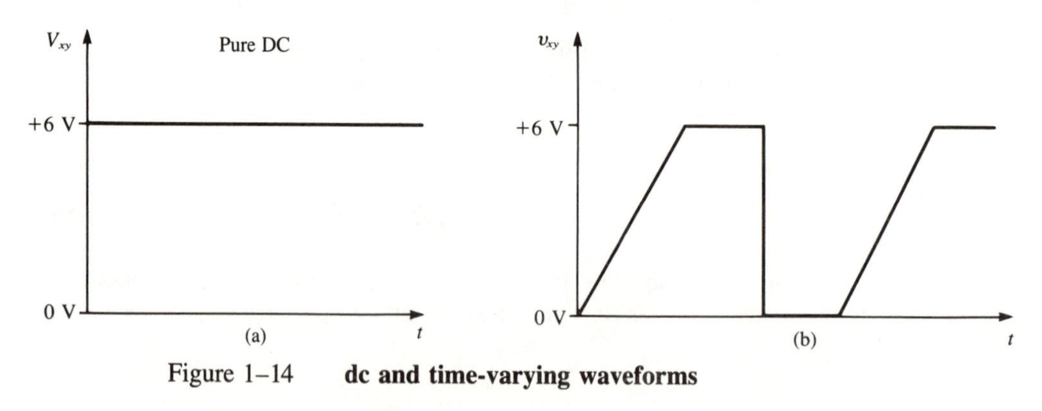

Figure 1–14 **dc and time-varying waveforms**

Questions

Sections 1.1–1.2

1.1 What is the common characteristic of all switching circuits?
1.2 What is the special characteristic of a digital circuit?
1.3 How do linear circuits differ from switching circuits?

Section 1.3

1.4 Sketch an ideal negative-going step that starts at +5 V and makes a 3 V transition at time t_1.
1.5 Sketch a positive-going step that makes a 6 V transition up to the +10 V level at time t_1.
1.6 Is the following statement *true* or *false*: A negative-going step always jumps from a positive level to a zero or negative level.
1.7 Which pulse in Figure 1–5 is a 6 V *p-p* positive-going pulse on a 0 V baseline?
1.8 Which pulse in Figure 1–5 is a 6 V *p-p* negative-going pulse on a +2 V baseline?
1.9 Sketch a 12 V *p-p* negative-going pulse on a +1 V baseline.
1.10 Sketch a 10 V *p-p* positive-going pulse on a −5 V baseline.
1.11 What is the *p-p* amplitude of a pulse signal that has a +2 V baseline and −16 V peak?
1.12 A pulse signal has an amplitude of 7 V *p-p*. What is its baseline voltage if its peak voltage is +4 V?

Section 1.4

1.13 What are the principle characteristics of an ideal switching device?
1.14 *True* or *false*: When an ideal switching device is in its OFF state, it can have a large voltage across its terminals.
1.15 *True* or *false*: In the ON state, an ideal switching device will have no current flowing through it.
1.16 Assume that the signal in Figure 1–15 is applied to the input of the inverter circuit of Figure 1–8, and the inverter output is applied to the input of a second inverter circuit. Sketch the output voltage of the second inverter.

+ 10 V

0 V

Figure 1–15 **Problem 1.16**

1.17 The circuit of Figure 1–16 uses two ideal voltage-controlled switches like the one in Figure 1–7. Determine the value of V_{OUT} for each of the possible sets of values of control voltage inputs.

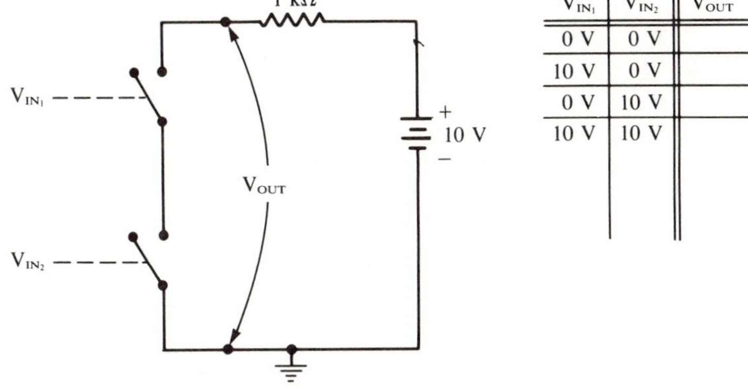

V_{IN_1}	V_{IN_2}	V_{OUT}
0 V	0 V	
10 V	0 V	
0 V	10 V	
10 V	10 V	

Figure 1–16 **Problem 1.17**

Sections 1.5–1.8

1.18 Determine the direction of conventional current flow in Figure 1–12.

1.19 Each circuit in Figure 1–17 has 2 mA of current flowing in it. Determine V_{xy}, V_{yx}, V V_y for each.

1.20 If $V_y = +50$ V and $V_{xy} = +22$ V in Figure 1–18, determine the value of V_x. Then det the direction of current through the resistors.

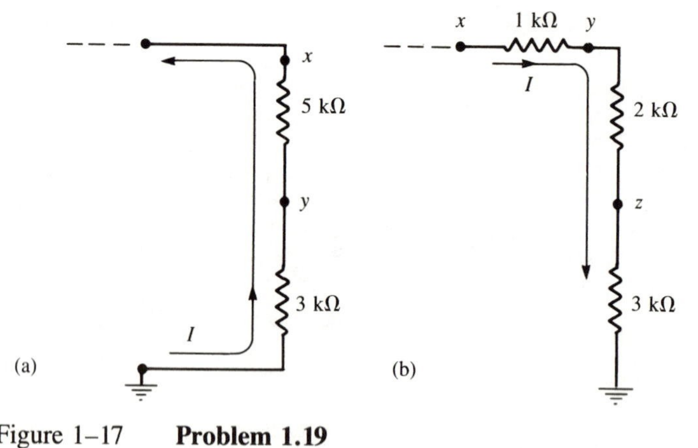

Figure 1–17 **Problem 1.19**

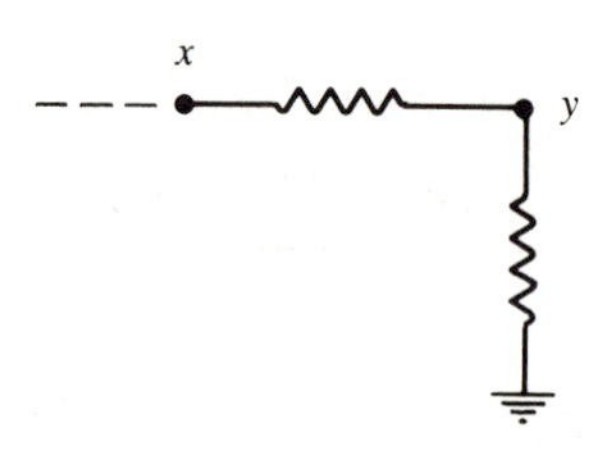

Figure 1–18 **Problem 1.20**

1.21 For the circuit of Figure 1–13(a), determine V_{xy} if $V_y = +3$ V.

1.22 For the circuit of Figure 1–13(b), the voltage across each resistor is 2 V (+ on top, – on bottom). Determine the values of V_C, V_E and V_{CE}.

1.23 How do we differentiate between dc and time-varying voltages and currents?

2

Pulse Waveform Analysis

2.1 Introduction

As stated in Chapter One, it is not possible to exactly generate ideal pulse waveforms. In this chapter we will examine the various types of non-ideal characteristics that are present, to varying degrees, on all practical pulses, and we will see how to quantitatively describe these characteristics. We will discuss the concept of *harmonic* content of a waveform and use it to predict the effects that various circuit types will have on the shape of a pulse waveform.

2.2 Types of Pulse Distortion

Figure 2–1 illustrates the most common types of distortion that appear on practical pulses. In (a) the pulse has non-zero transition times. These transition times are called the *rise time*, t_r, and *fall time*, t_f, for the positive-going and negative-going transitions respectively.

The pulse in (b) demonstrates the characteristic of *tilt* since the pulse peak is not perfectly flat. In (c) the pulse exhibits *overshoot* where the waveform voltage momentarily goes beyond the peak and baseline values during the transitions.

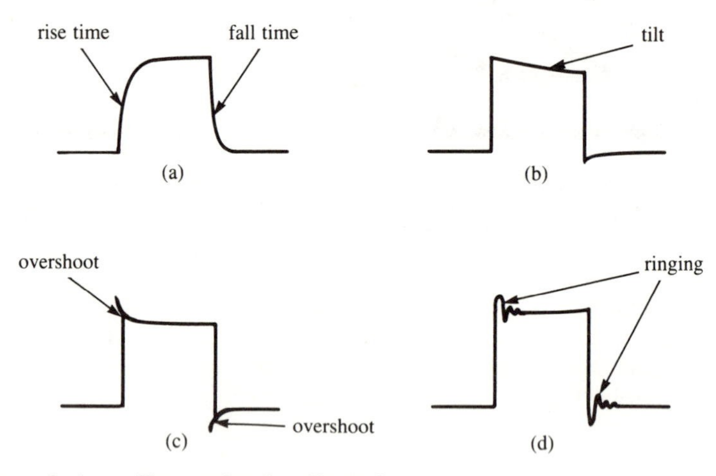

Figure 2–1 **Types of pulse distortion**

Of course, the various kinds of distortion can appear singly or in combination. An extreme case is shown in Figure 2–2, where a pulse waveform exhibits all of the previously described characteristics.

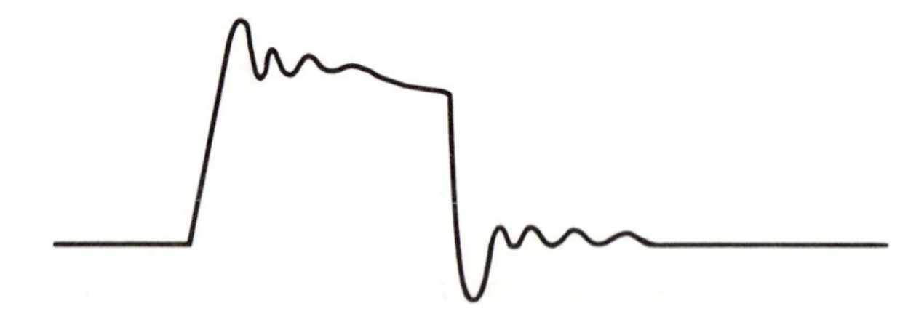

Figure 2–2 **Practical pulse with each type of distortion**

2.3 t_r, t_f, and t_p

Figure 2–3 shows a pulse with a peak value of V_1 on a 0 V baseline. The rise time, t_r, of this pulse is defined as the time interval between when the transition is 10 percent completed to when it is 90 percent completed. In other words, t_r is the time it takes for the voltage to rise from 0.1 V_1 to 0.9 V_1 during the positive-going transition.

Similarly, the fall time, t_f, is the time it takes the voltage to fall from 0.9 V_1 to 0.1 V_1 during the negative-going transition. In general, the values of t_r and t_f will not be the same for a given pulse, but will depend on how the pulse is being generated.

Because of the non-zero transition times, it is necessary to define the pulse duration, t_p, as the time interval between the 50 percent points on the pulse edges, as shown in the figure.

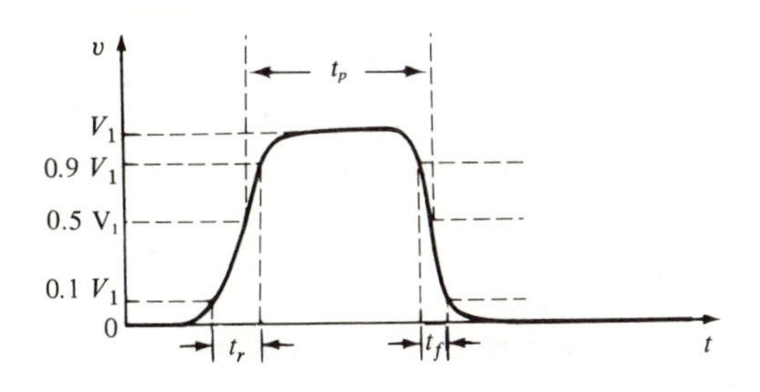

Figure 2–3 **A practical pulse**

Example 2.1 Figure 2–4 shows an oscilloscope trace of a practical pulse waveform. The horizontal and vertical scales are given. Determine t_r, t_f, and t_p.

Solution: First locate the 10 percent and 90 percent points on the rising edge, labelled a and c respectively. Then use the horizontal scale to measure the time interval between a and c. The result is

$$t_r \approx 3.4\ \mu s$$

Similarly, points d and f on the falling edge are used to determine

$$t_f \approx 3.2\ \mu s$$

The pulse duration is the time interval between points b and e. Thus,

$$t_p \approx 12\ \mu s$$

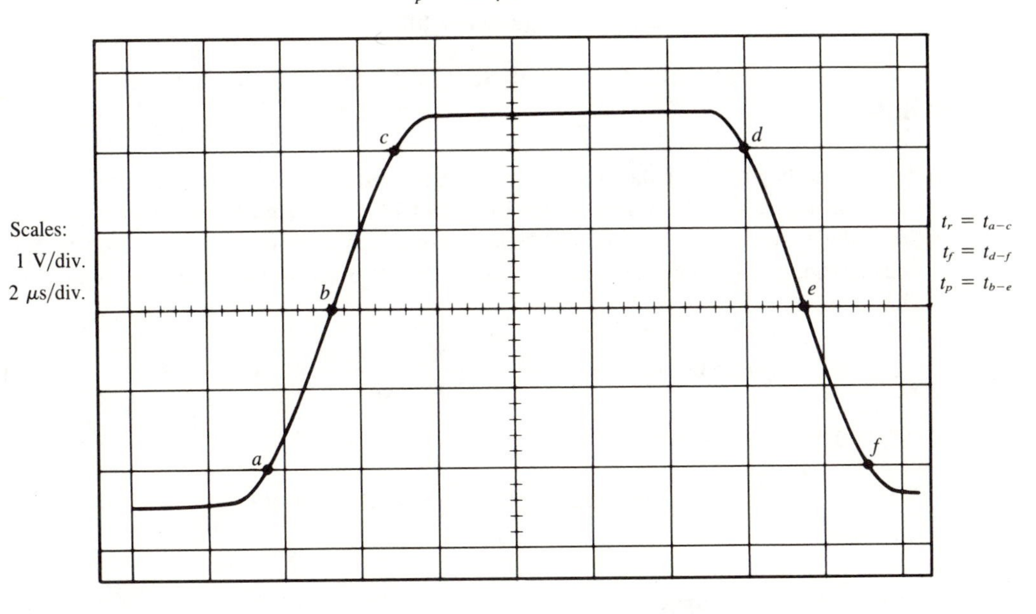

Figure 2–4 **Example 2.1**

2.4 Tilt

Figure 2–5 shows a pulse exhibiting significant tilt. Generally, a pulse that has tilt on its peak value will also have tilt on its baseline. The amount of tilt on the peak is simply the amount by which the voltage changes during the time when it should be ideally constant at the peak value. In this example, the voltage decreases from its peak value of 10 V to a value of 8 V. Thus, the tilt is 2 V.

Percentage tilt is defined as the ratio of the amount of tilt to the peak value, in percent. That is,

$$\text{percent tilt} = \frac{\text{amount of tilt}}{\text{peak value}} \times 100\% \tag{2–1}$$

For the pulse in Figure 2–5, we have

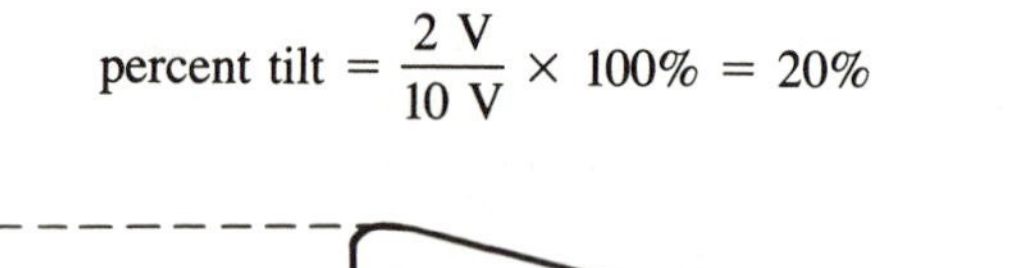

$$\text{percent tilt} = \frac{2\text{ V}}{10\text{ V}} \times 100\% = 20\%$$

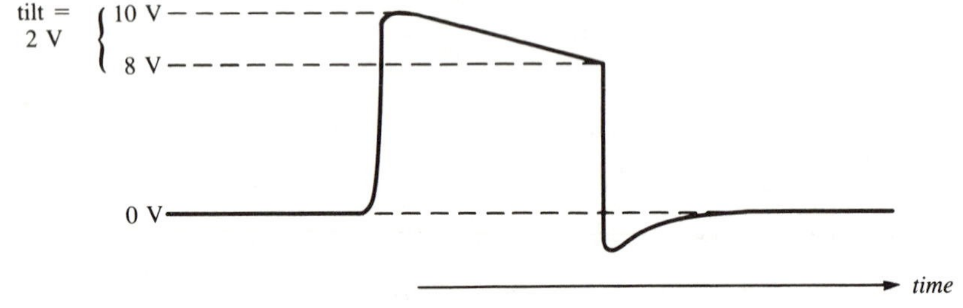

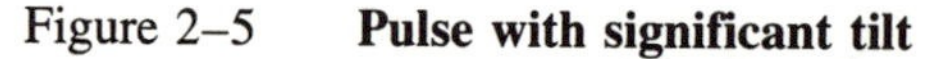

Figure 2–5 **Pulse with significant tilt**

2.5 Overshoot

Figure 2–6 shows a pulse that has overshoot on its positive-going transition. This positive transition jumps from the 0 V baseline to a peak of 13 V before the pulse levels off at 10 V. Since the transition exceeds the 10 V level by 3 V, the amount of overshoot is 3 V. The percentage overshoot is the ratio of the amount of overshoot to the pulse's final level. That is,

$$\text{percent overshoot} = \frac{\text{amount of overshoot}}{\text{peak level}} \times 100\% \tag{2–2}$$

For the pulse in Figure 2–6,

$$\text{percent overshoot} = \frac{3\text{ V}}{10\text{ V}} \times 100\% = 30\%$$

2.6 Periodic Pulse Waveforms

When a pulse is made to repeat itself at fixed periods of time, a periodic pulse waveform, or a *periodic train* of pulses, results. Figure 2–7 shows a train of rectangular pulses in

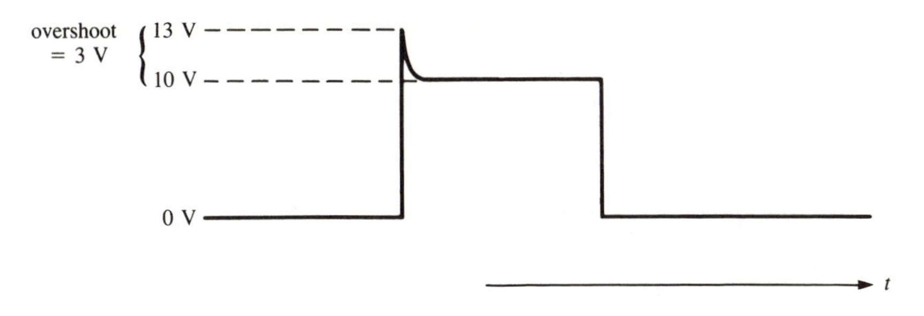

Figure 2–6 **Pulse with significant overshoot**

which the positive-going pulses have a duration of t_p and occur every T s. The parameter T is called the *period* of the pulse waveform because within each time interval (T) the behavior of the waveform is exactly the same.

The number of pulses that occurs within one second is called the *pulse repetition frequency,* which is abbreviated PRF. The PRF is related to the period T in the following manner:

$$\text{PRF} = \frac{1}{T} \qquad \textbf{(2–3)}$$

It has the units of *pulses per second,* abbreviated pps. For example, if T is 0.5 s, then the PRF would be 2 pps, or if T were 0.5 ms, the PRF would be 2000 pps (2 kpps).

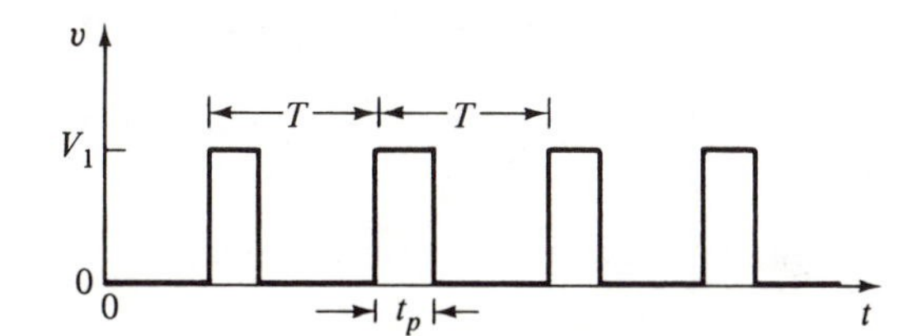

Figure 2–7 **Rectangular periodic ideal pulse train**

The *duty cycle* of a pulse train is defined as the ratio of the pulse duration t_p to the period T in percent.

$$\text{Duty cycle} = \frac{t_p}{T} \times 100\% \qquad \textbf{(2–4)}$$

In most pulse waveforms the duty cycle is 50 percent or less.

Example 2.2 Determine the PRF and duty cycle of the waveform in Figure 2–8.

Solution: The period T can be obtained by determining the time difference between successive leading edges. The value obtained in this case is T = 6 ms. Thus, PRF = 1/6 ms ≈ **167 pps.**

The pulse duration is equal to 2 ms, so the duty cycle is

$$\frac{2}{6} \times 100\% = \mathbf{33.3\%}$$

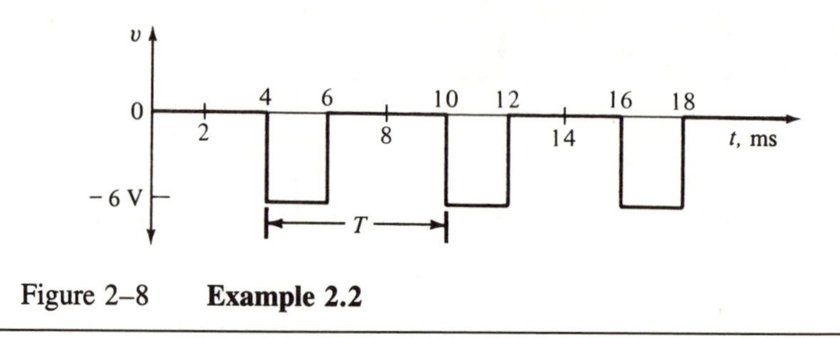

Figure 2–8 **Example 2.2**

Square waves A special type of pulse waveform occurs when the duty cycle is around 50 percent. In this case, the pulse width (t_p) is half the period (T), and we have what is called a *square wave*. Two types of square waves are illustrated in Figure 2–9. The first one is *unipolar* and extends in only one polarity. The other is *bipolar* and extends both positively and negatively.

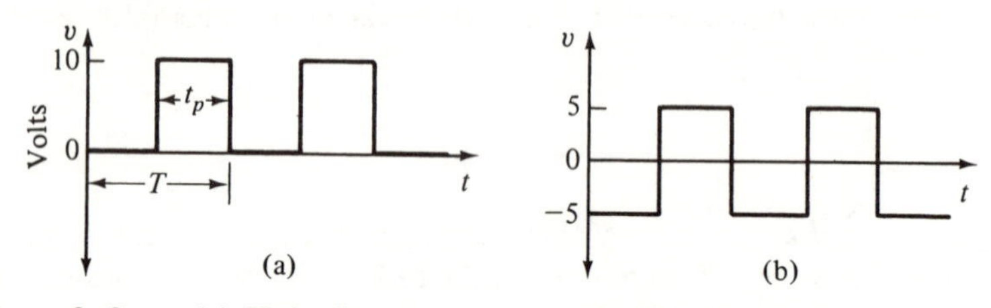

Figure 2–9 **(a) Unipolar square wave (b) Bipolar square wave**

Example 2.3 A certain square wave has a pulse duration of 2.5 μs. What is its repetition frequency?

Solution: Since its pulse duration is half of one period, $T = 5$ μs, and

$$f = 1/5 \text{ s} = \mathbf{200 \text{ kpps}}$$

Example 2.4 Determine the frequency and duty cycle of the waveform in Figure 2–10.

Solution: The period can be determined as the time between successive positive-going transitions. Thus, $T = 10$ μs. The pulse duration is the duration time during which the pulse is at its positive peak. Thus, $t_p = 2$ μs. Using these values, we have

$f = 1/10$ μs
$= \mathbf{100 \text{ kpps}}$

duty cycle $= 2/10$
$= \mathbf{20\%}$

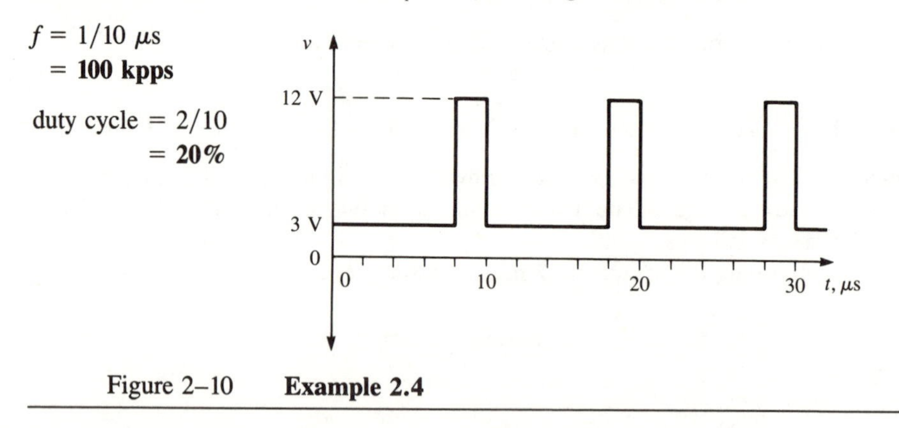

Figure 2–10 **Example 2.4**

2.7 Average Value (dc)

The average value of a waveform is the amount of dc the waveform contains. As we shall see shortly, every waveform is made up of many component parts, and dc may be one of them. Pure ac is a sinusoidal waveform that makes equal excursions above and below zero so that its average value is zero; that is, it contains no dc. Pure dc is constant at a given level, and so its average value is that level.

The average value of any rectangular pulse waveform can be found by using the following expression:

$$\text{Average value} = (\text{duty cycle}) \times (\text{peak voltage}) + (100\%\text{-duty cycle}) \times (\text{baseline voltage}) \quad \textbf{(2–5)}$$

This expression says that the average value is equal to the pulse's peak voltage multiplied by the percentage of the time that the pulse is at its peak (duty cycle), *plus* the baseline value multiplied by the percentage of the time the pulse is at its baseline (100 percent minus duty cycle).

Example 2.5 How much dc does the waveform of Figure 2–8 contain?

Solution: The baseline is 0 V, peak value is −6 V, and the duty cycle is 33.3 percent. Using Expression 2–5,

$$\begin{aligned}\text{Average value} &= (33.3\%) \times (-6\text{ V}) + (66.7\%) \times (0\text{ V}) \\ &= \mathbf{-2\ V}\end{aligned}$$

We could have guessed that this waveform would have a negative average value since it never goes positive.

Example 2.6 Determine the average value of the pulses in Figure 2–10.

Solution: Duty cycle = 20 percent; peak value = 12 V; baseline = 3 V. Thus,

$$\begin{aligned}\text{Average value} &= 20\% \times (12\text{ V}) + 80\% \times (3\text{ V}) \\ &= \mathbf{4.8\ V}\end{aligned}$$

Example 2.7 Determine the amount of dc in both square waves of Figure 2–9.

Solution: We could use Expression 2–5 again. However, you will find that for a square wave, the average value is always exactly halfway between the baseline and peak values. Thus, the square wave in (a) has an average value of + 5 V, and the one in (b) has an average value of zero. The last result means that this square wave contains no dc component.

Measuring average value Expression 2–5 can be used only for rectangular pulses. The average value of *any* waveform can be measured by using a *true* dc voltmeter. The dc voltmeter filters out all of the waveform's components except for the dc, and the meter indication is the waveform's dc, or average value.

An oscilloscope can also be used to measure the dc component of any waveform by using the dc/ac coupling switch on the scope's vertical amplifier. When dc coupling is used, the scope displays the waveform as it actually exists. When ac coupling is used, a coupling capacitor blocks the dc component of the waveform so that the scope displays the waveform *minus* its dc component.* By noting the vertical shifting of the waveform as you switch from dc to ac coupling, you can determine the waveform's average value.

Figure 2–11 illustrates this procedure. With the scope set at dc coupling, the display appears as shown in (a). With the scope switched to ac coupling, the display appears as shown in (b). The waveform using ac coupling is exactly the same shape as that using dc coupling, but it has shifted downward by *two* divisions, which represents 4 V. Since it shifted *downward* by 4 V when its dc component was removed, the waveform must have a dc average value of +4 V.

Example 2.8 If a waveform shifts *upward* by 1.7 V when the scope is switched from dc to ac coupling, what is the average value (dc) of the waveform?

Solution: Removing the waveform's dc component made it shift *upward*. Thus, its average value must be **−1.7 V,** since removing a negative value is like adding a positive value.

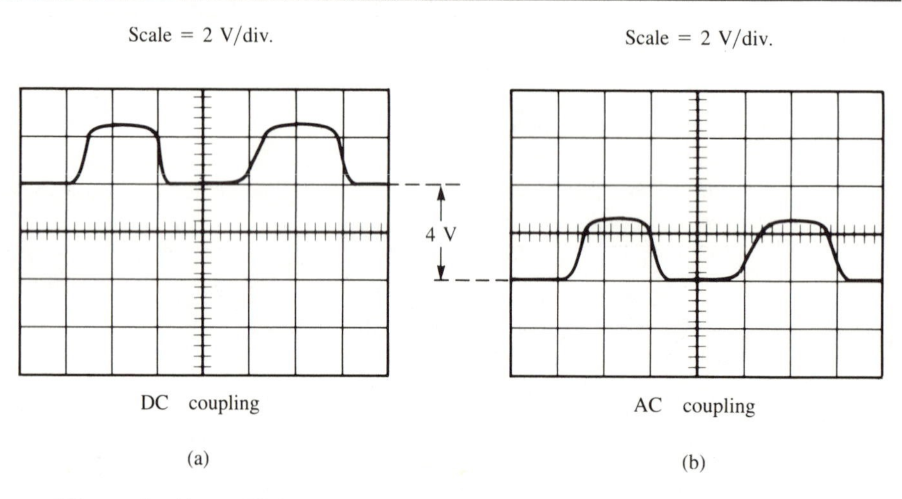

Figure 2–11 **Using scope input coupling to determine waveform dc component**

2.8 Harmonic Content of Periodic Waveforms

Every periodic waveform has a repetition frequency, also called the *fundamental frequency.* For example, the signal in Figure 2–10 has a repetition frequency of 100 kpps. Thus, the waveform has a fundamental frequency of 100 kHz.

*The scope might also cause some tilt when ac coupling is used. We will examine this in the next chapter.

It can be mathematically proven that every periodic waveform having a fundamental frequency (f_1) has these components:

(a) a dc component (which can be positive, negative, or zero)
(b) a sinusoidal component at a frequency equal to the fundamental frequency, f_1
(c) a sinusoidal component at a frequency of $2f_1$ (twice the fundamental), called the *second* harmonic
(d) a sinusoidal component at a frequency of $3f_1$, called the *third* harmonic
(e) sinusoidal components at $4f_1$ (*fourth* harmonic), $5f_1$ (*fifth* harmonic), and so on out to infinity

In other words, any periodic waveform can be broken down into a dc component, and sinusoidal components at the fundamental frequency and integer multiples of the fundamental frequency.

For example, the waveform of Figure 2–10 has a dc component of 4.8 V (calculated in Example 2.6) and sinusoidal components at frequencies of 100 kHz (fundamental), 200 kHz (second harmonic), 300 kHz (third harmonic), 400 kHz (fourth harmonic), and so on. In general, each of the fundamental and harmonic components will have a different amplitude. These amplitudes can be determined only by using a complex mathematical procedure called *Fourier Analysis,* although there are tables of values for most common waveforms. We will not be concerned with knowing the values of these harmonic amplitudes.

Consider the bipolar square wave shown in Figure 2–12. Obviously, this waveform does not have a dc component, since its average value is zero. It does, however, contain harmonics. Part (b) of the figure shows the fundamental, third, and fifth harmonics of this square wave. A perfect square wave (50 percent duty cycle) possesses only *odd* harmonics (i. e., first, third, fifth, seventh, etc.), so the even harmonics of a square wave have a *zero* amplitude.

Examination of Figure 2–12 reveals several interesting points. First, the amplitude of each successive harmonic decreases. The fundamental has an amplitude of 6.37 V-peak; the third harmonic has an amplitude of 2.12 V-peak; the fifth harmonic has an amplitude of 1.27 V-peak. These three harmonics are not the only harmonics. In fact, the ideal square wave has all of the *odd* harmonics out to infinity. However, the amplitudes of each successive harmonic get even smaller. For example, the 63rd harmonic of this square wave has an amplitude of approximately 0.1 V. Most common pulse waveforms have harmonics whose amplitudes behave in a similar manner.

Second, all of the square-wave harmonics are *in phase*. That is, at the beginning of each fundamental period (T), all of the harmonics are at zero and are increasing in the positive direction. In general, the harmonics of every nonsinusoidal waveform are *not* all in phase. The phase relationship among the harmonics depends upon the shape of the nonsinusoidal waveform.

Third, the peak amplitude of the fundamental is larger than the peak amplitude of the square wave itself. The fact that the square wave is made up of the *sum* of all its component harmonics resolves this apparent dilemma. It is not necessary to have all the harmonics to prove this. Figure 2–13 shows the result of adding the *fundamental* and *third* harmonics point by point. As can be seen, even using only the first two harmonics produces a resultant that begins to resemble the square wave. Notice that the resultant

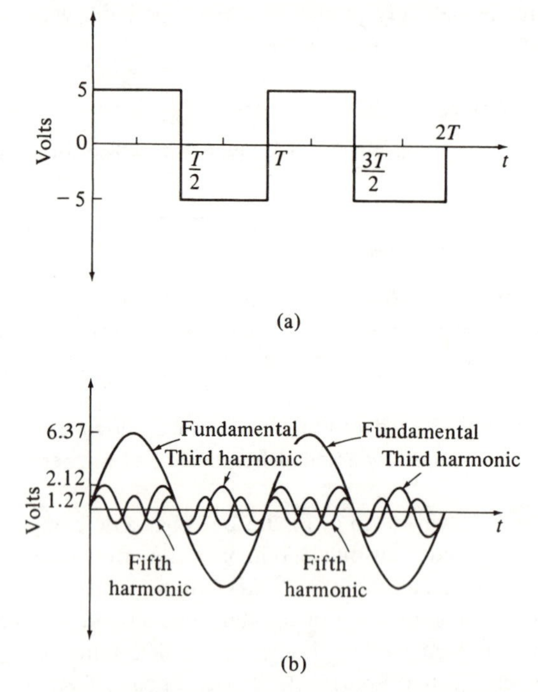

Figure 2–12 **(a) Ideal bipolar square wave (b) First three harmonics of the square wave**

has a dip that occurs at the point where the fundamental is at its positive peak and the third harmonic is at its negative peak. This indicates that even though the amplitude of the fundamental is 6.37 V, some of it is cancelled out by the negative portions of the third harmonic.

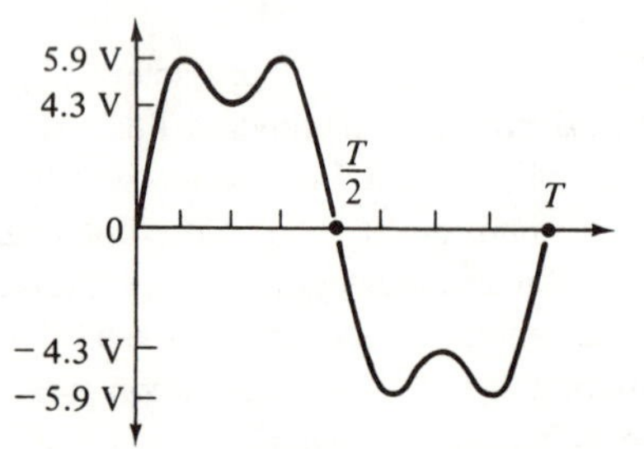

Figure 2–13 **Resultant produced by combining the first and third harmonics**

By including more and more of the harmonics, a resultant can be constructed that more closely resembles the square wave. Actually, an infinite number of harmonics have to be included to completely construct the original square wave. However, in practice it is necessary to consider only a limited number of harmonics, because beyond a certain point the higher harmonics contribute very little to the overall waveform (their amplitude is insignificant).

Example 2.9 Determine the frequency components of the waveform in Figure 2–14(a).

Solution: This waveform is really the same as the one in Figure 2–12(a), except that it is shifted upward by 5 V, so that it has a dc (average value) component of +5 V. Thus, this waveform has the same fundamental frequency and *odd* harmonic frequency components as the bipolar square wave of Figure 2–12, *plus* it has a dc component.

Note that the addition of a dc component does nothing to the harmonic content of a waveform since dc has no effect on the waveform shape.

Example 2.10 Determine the frequency components of the waveform of 2–14(b).

Solution: The period of this signal is 1 ms, so its fundamental frequency is 1 kHz. Thus, it has harmonics at 2 kHz, 3 kHz, 4 kHz, 5 kHz, and so on. The waveform also has a dc component since it is always positive.

This signal is not a rectangular pulse waveform; it is called a *ramp,* or *sweep,* waveform. It is very useful in many types of switching circuits, and we will encounter it later in the text.

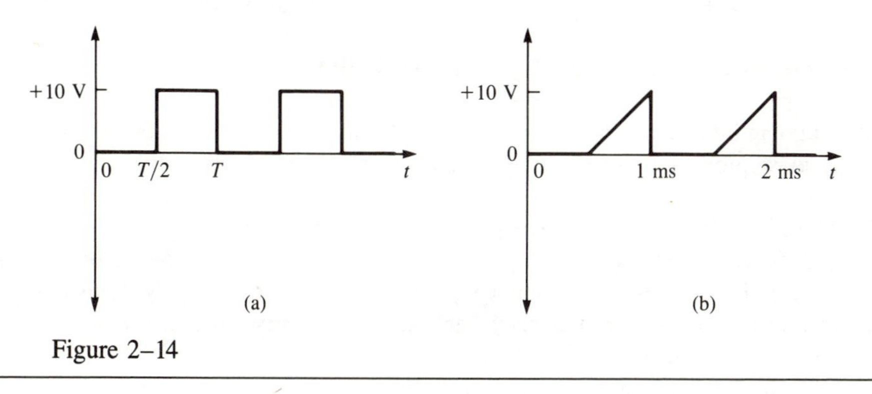

Figure 2–14

Frequency spectrum The frequency content of a signal can be conveniently represented by a diagram called a *frequency spectrum.* Figure 2–15 shows a portion of the frequency spectrum for the accompanying square wave. The horizontal axis represents frequency, and the vertical axis represents amplitude. The vertical arrows show the amplitude of each of the frequency components of the square wave (except for dc, which is normally not included in a frequency spectrum).

For example, the arrow at 1 kHz has a height of 6.37 V, indicating that the fundamental frequency component of this waveform has a peak amplitude of 6.37 V. There is no arrow at 2 kHz, since a square wave has no even harmonics. The arrow at 3 kHz (third harmonic) is 2.12 V, the one at 5 kHz is 1.27 V, and so on. Note that the amplitude of the harmonics decrease significantly with frequency. The 99th harmonic (not shown) has an amplitude of 0.064 V. The 999th harmonic has an amplitude of 0.006 V. This decrease of harmonic amplitude with frequency is typical of many simple waveforms. The effect is even more pronounced for *practical* pulse waveforms that have non-zero transition times.

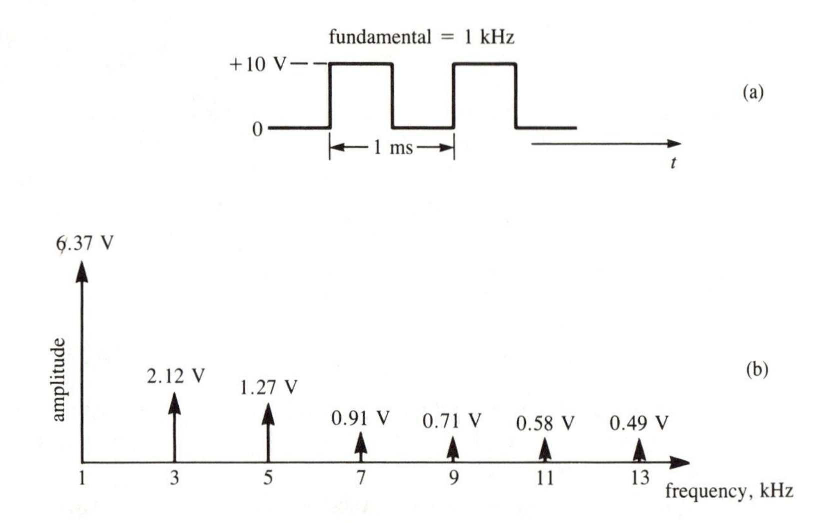

Figure 2–15 **Frequency spectrum of the ideal square wave shown above**

2.9 Relationship between Harmonic Content and Shape

Without going into a mathematical proof, we will state two general principles concerning the relationship between the shape of a pulse waveform and its harmonic content.

High frequency harmonics The pulse transition times are principally determined by its higher frequency harmonics. Any change in the relative amplitudes or phase of the harmonics at the high end of the frequency spectrum will effect the pulse transitions. On the other hand, the higher frequency harmonics have very little effect on the slow-changing, flat portions of a pulse waveform.

To illustrate this principle, Figure 2–16 shows two square waves with the same peak-to-peak amplitude and the same fundamental frequency. Thus, both waveforms have harmonics at the same frequencies. However, the high frequency harmonics of the square wave in (b) have lower amplitudes than the corresponding harmonics of the square wave in (a). The result is that the square wave in (b) has slower transition times than the square wave in (a).

This correspondence between a waveform's transition times and its high frequency harmonics should be easy to remember, because it makes sense that the fastest-changing portions of the pulse waveform (its rising and falling edges) should depend on the fastest-changing harmonics.

Theoretically, a pulse waveform has an infinite number of harmonics. But, as stated earlier, the amplitudes of the higher frequency harmonics become smaller and smaller. In fact, for any practical pulse waveforms, we can neglect all of the harmonics above a certain limit without any noticeable effect on the waveform transition times. That is, we can state that a pulse waveform has significant harmonics up to a frequency

$$f_H = \frac{1}{2\,t_t} \qquad \textbf{(2–6)}$$

where f_H is called the waveform's *highest significant frequency,* and t_t is the waveform's *shortest* transition time.

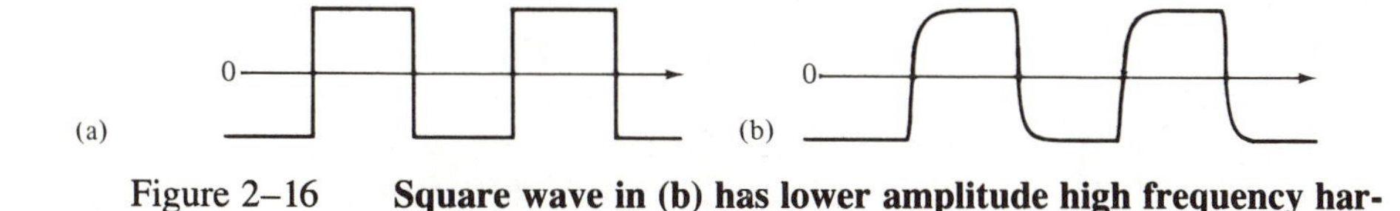

Figure 2–16 **Square wave in (b) has lower amplitude high frequency harmonics than the one in (a)**

Example 2.9 A pulse waveform has a rise time of 2 μs and a fall time of 1.5 μs. What is the highest significant frequency component of this waveform?

Solution: Since t_f is shorter than t_r, we use $t_t = t_f = 1.5\ \mu$s in Equation 2–6 to obtain

$$f_H = \frac{1}{2 \times 1.5\ \mu\text{s}} = \mathbf{333{,}333\ Hz}$$

When we neglect all of a waveform's harmonics above f_H, we are introducing only a very small error. The neglected harmonics, if they were included, would only reduce the transition times by a few percent. What we are saying, then, is that we can get a fairly accurate reproduction of the pulse waveform by summing all of its harmonic components up to f_H.

Example 2.10 Determine the *significant* frequency components of the pulse waveform in Figure 2–10 if it has $t_r = 0.1\ \mu$s and $t_f = 0.2\ \mu$s.

Solution: This waveform has a dc component as determined in Example 2.6, and it has a fundamental frequency of 100 kHz. The highest significant harmonic will be determined by using t_r in Equation 2–6.

$$f_H = \frac{1}{2 \times (0.1\ \mu\text{s})} = 5\ \text{MHz}$$

Thus, this waveform has *significant* components at the following frequencies:

dc, 100 kHz, 200 kHz, 300 kHz, 400 kHz---------5 MHz.

Note that this waveform has both odd and even harmonics since it is *not* a square wave.

This last example shows that a pulse waveform has sinusoidal components ranging from the fundamental frequency, on the low end, to the highest significant harmonic, f_H, on the high end. In other words, its useful frequency spectrum contains the fundamental and integer multiples of the fundamental up to f_H.

Lower frequency harmonics A pulse waveform's low-frequency harmonics contribute very little to the pulse transition times because these harmonics change slowly. They do, however, have a significant effect on the slow-changing, flat portions of the waveform. Any change in the amplitude or phase of the low frequency harmonics will cause distortion on the flat portions of the waveform.

Figure 2–17 illustrates this principle, where an ideal square wave is shown in (a) and three distorted square waves are shown in (b), (c), and (d). The waveform in (b) is a result of an increase in the amplitude of the fundamental component, while the waveform in (c) is a result of a decrease in the amplitude of the fundamental component. The waveform in (d) shows what happens when the amplitude of the fundamental is not changed, but its phase is shifted by a few degrees. This last result is important because it means that shifting the phase of a harmonic can cause waveform distortion even if the amplitude of the harmonic is unchanged.

Although the illustrations in Figure 2–17 show the effects of changing the fundamental component, similar results occur when other low-frequency harmonics have their amplitudes or phase changed. Whenever a pulse exhibits tilt on its flat portion, it is because its low-frequency harmonic content has been altered. Note in Figure 2–17 that the changes in the low frequency harmonics have very little effect on the waveform transitions. The t_r and t_f are essentially unchanged.

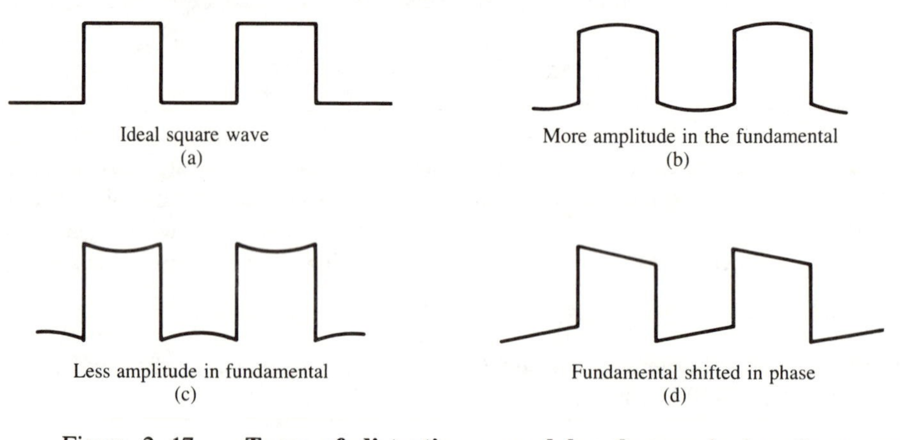

Figure 2–17 **Types of distortion caused by changes in low frequency harmonics**

2.10 Effect of Linear Networks on Pulse Waveforms

Now that we have reviewed some general principles, we can apply them to a study of how various linear circuits affect the harmonic content and, therefore, the shape of a pulse waveform.

Since a nonsinusoidal waveform consists of the sum of all its harmonics plus its dc component, it appears to be the same as a series combination of generators [shown in Figure 2–18(a)]. There is a dc source representing the dc component (average value) of the waveform, and there are ac sinusoidal generators, each representing one of the waveform harmonics. Each generator, in general, has a different phase and amplitude. The total series combination of the various sources is equivalent to the original waveform.

If the waveform is applied to a linear network, as in part (b) of Figure 2–18, it is possible to use the principle of *superposition* to determine the circuit output. That is, the circuit output for each individual generator (replacing the others by a short circuit) is determined and then all of the various outputs are added to get the total resultant output. Such a process would be prohibitively tedious even if just the harmonics up to f_H were included. Fortunately, the superposition principle can be used in some of the work with harmonics without actually applying it mathematically.

Very often the linear network has characteristics that vary with frequency, such as a simple *RC* or *RL* series circuit. The output of such a circuit when driven by a nonsinusoidal waveform will depend on the circuit's frequency response characteristic and on

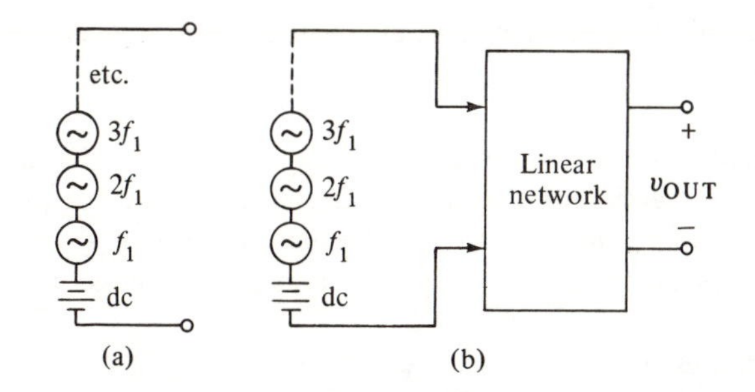

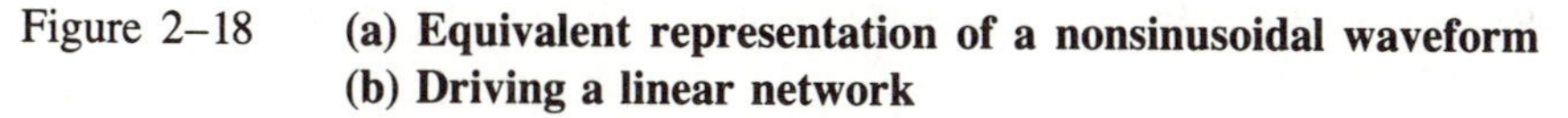

Figure 2–18 **(a) Equivalent representation of a nonsinusoidal waveform (b) Driving a linear network**

the harmonic content of the waveform. Before expanding on this concept, it will be useful to review three common network frequency response characteristics that occur in practice. Refer to Figure 2–19, where the frequency response characteristics of *low-pass, high-pass,* and *band-pass* networks are shown.

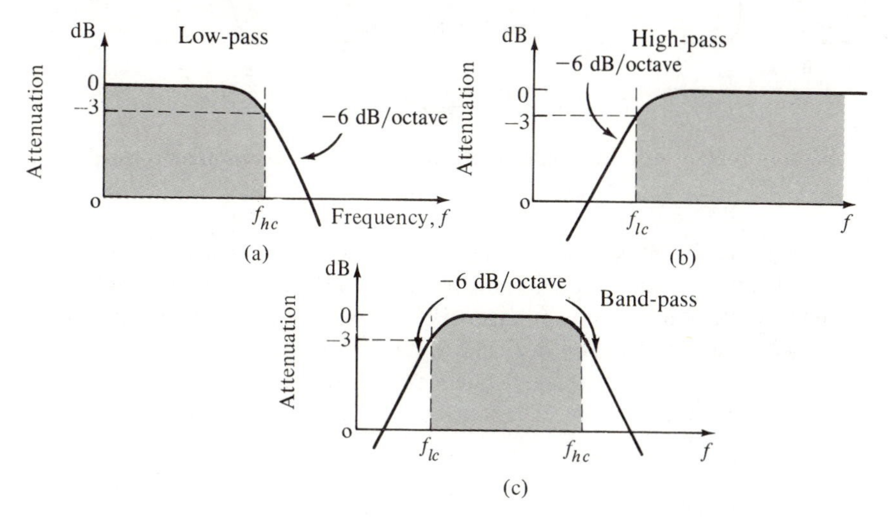

Figure 2–19 **Frequency response characteristics of (a) low-pass (b) high-pass and (c) band-pass networks. Shaded areas are pass-bands.**

The low-pass network passes frequencies within a pass-band that extends from 0 Hz (dc) to its *high frequency cutoff,* f_{hc}. At this frequency (f_{hc}), the low-pass attenuates a signal by a factor of 0.707, which is equal to −3 dB. For this reason, f_{hc} is also called the −3 dB frequency. The attenuation typically increases at a rate of −6 dB/octave for frequencies greater than f_{hc}, so signals at frequencies greater than f_{hc} will be increasingly attenuated as frequency increases.

The high-pass network passes all of the frequencies above its *low frequency cutoff,* f_{lc} (also called the −3 dB frequency). Signals at frequencies lower than f_{lc} will be increasingly attenuated (−6 dB/octave) as frequency decreases below f_{lc}.

The band-pass network has a pass-band extending from a low frequency cutoff, f_{lc}, to a high frequency cutoff, f_{hc}. Signals at these frequencies are attenuated by −3 dB.

Signals outside these limits are increasingly attenuated as their frequencies get further from the −3 dB points.

High frequency distortion When a pulse signal is applied to a low-pass circuit, some of its high frequency harmonics may be attenuated and shifted in phase as they pass through the circuit, while the signal's low frequency harmonics are unaffected. This occurrence will cause a distortion of the pulse's transitions; that is, an increase (slowing down) in t_r and t_f.

For example, assume that a pulse waveform with $t_r = t_f = 0.5\ \mu s$ is applied to a low-pass circuit with $f_{hc} = 250$ kHz (Figure 2–20). This waveform has significant harmonics up to

$$f_H = \frac{1}{2t_r} = 1 \text{ MHz}$$

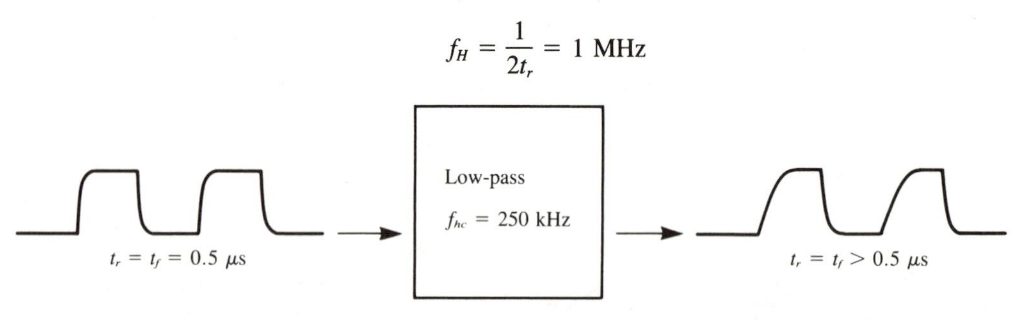

Figure 2–20 **A low-pass will slow down a waveform's transitions if $f_{hc} < 2f_H$**

Since f_{hc} is only 250 kHz, many of the significant waveform harmonics fall outside the pass-band of the low-pass circuit. These harmonics will be attenuated, thereby producing an output waveform with much longer rise and fall times than the input waveform.

To avoid this deterioration in t_r and t_f, the low-pass circuit should have a cutoff frequency greater than the highest significant harmonic of the pulse signal. A factor of two is sufficient to ensure an increase of less than 5 percent in the transition times. That is, when

$$f_{hc} \geqslant 2f_H \tag{2–7}$$

the low-pass will essentially pass all of the signal's significant high frequency harmonics and will have no significant effect on t_r and t_f. Of course, the low-pass will also pass all of the low frequency harmonics and the dc component.

Example 2.11 What low-pass circuit cutoff frequency is required in Figure 2–20 if the output is to have no high frequency distortion?

Solution: The input signal has $f_H = 1$ MHz. Thus, the low-pass circuit has to have $f_{hc} \geqslant 2$ MHz to produce an output with no distortion.

Low frequency distortion When a pulse signal is applied to a high-pass circuit, some of its low frequency harmonics can be attenuated and/or shifted in phase as they go through the circuit, thereby producing a distortion of the flat portions of the pulse. This distortion is in the form of tilt on the pulse peak. Figure 2–21 shows various degrees of tilt that can occur when the input in (a) is applied to a high-pass network.

The degree of tilt becomes more pronounced as the high-pass circuit's cutoff frequency, f_{lc}, is made larger, thereby causing more of the signal's harmonics to fall outside the circuit pass-band. It can be shown mathematically that a tilt of 10 percent will occur when

$$f_{lc} = \frac{1}{63t_p} \quad \text{(for 10\% tilt)} \qquad \textbf{(2–8)}$$

where t_p is the signal's pulse duration.

Thus, any f_{lc} lower than this value will produce a tilt of less than 10 percent. As a rule of thumb, we will say that the degree of tilt is acceptable whenever

$$f_{lc} < \frac{1}{63t_p} \qquad \textbf{(2–9)}$$

Of course, the lower the value of f_{lc}, the lower the degree of tilt.

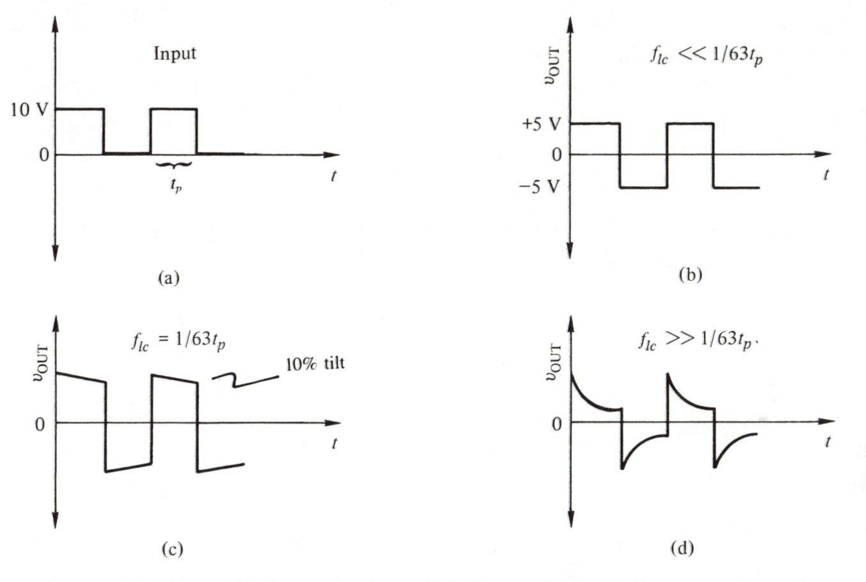

Figure 2–21 **Various degrees of tilt caused by high-pass network**

Example 2.12 A pulse waveform with a PRF of 1 kpps, and a 20 percent duty cycle is applied to a high-pass circuit with f_{lc}=200 Hz. Will the output have significant tilt?

Solution:

$$\begin{aligned} T &= 1/1 \text{ kpps} = 1 \text{ ms} \\ t_p &= \text{duty cycle} \times T \\ &= 20\% \times 1 \text{ ms} \\ &= 0.2 \text{ ms} \end{aligned}$$

For acceptable tilt, we want

$$f_{lc} < \frac{1}{63t_p} = 79.4 \text{ Hz}$$

Since the given high-pass circuit only has $f_{lc} = 200$ Hz, there *will* be significant output tilt.

This last example should elicit the following question: Why does the circuit cutoff frequency have to be so small (79.4 Hz) when the waveform fundamental frequency is 1 kHz? The reason is that even a few degrees of phase shift of the fundamental will cause tilt in the output. The f_{lc} has to be made much smaller than it would be if we were concerned only with preserving the amplitude of the fundamental. Satisfying Expression 2–9 will ensure that the phase shift of the fundamental will be minimal.

Example 2.13 What effects, other than tilt, can a high-pass circuit have on a pulse waveform?

Solution: Examination of Figure 2–21 shows that the input in (a) has a dc component of +5 V (recall how to obtain the average value of a square wave), but the various outputs have no dc component. This difference is because a high-pass circuit *always* attenuates dc no matter how low the value of f_{lc}.

The waveforms in Figure 2–21 also show that the high-pass circuit does not slow down the signal transitions since the high-pass circuit does not attenuate the signal's high frequency harmonics.

Band-pass distortion Since a band-pass circuit attenuates both the higher and lower frequencies, it combines the effects of high-pass and low-pass. As such, it will do the following:

(a) Always attenuate a signal's dc component.

(b) Slow down the signal transitions *unless* its high frequency cutoff, f_{hc}, is greater than $2f_H$.

(c) Produce tilt *unless* the circuit low frequency cutoff, f_{lc}, is less than $1/63t_p$.

The following examples will illustrate.

Example 2.14 Figure 2–22(a) shows a 1 kpps waveform with $t_r = t_f = 2\ \mu$s and a pulse duration of 250 ms. What will happen to this waveform if it is applied to a band-pass circuit with $f_{lc} = 100$ Hz and $f_{hc} = 100$ kHz?

Solution: The signal has harmonics up to $f_H = 1/2t_r = 250$ kHz. The band-pass f_{hc} is only 100 kHz, so it will produce high-frequency distortion. The signal has $t_p = 250$ ms so that

$$\frac{1}{63t_p} = 63.5 \text{ Hz}$$

However, the band-pass f_{lc} is 100 Hz, which is greater than 63.5 Hz. Thus, the circuit will produce low frequency distortion (tilt).

The resulting output waveform would look something like the waveform shown in Figure 2–22(b). Compared to the input in (a), it has slower transitions and it has tilt. Also, it has no dc component.

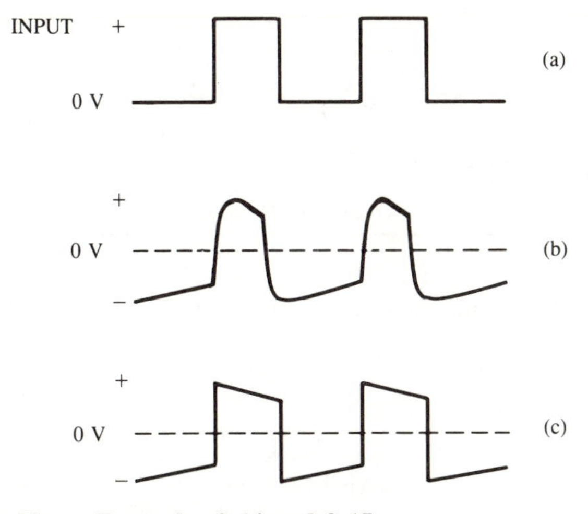

Figure 2–22 **Examples 2.14 and 2.15**

Example 2.15 What will happen to the output signal if the band-pass f_{hc} is increased to 1 MHz?

Solution: f_{hc} is now more than twice the value of f_H so that *no* high frequency distortion will occur. Figure 2–22(c) shows the result. There still is tilt and loss of dc.

2.11 Oscilloscope Effect on Pulse Measurements

When a signal is applied to the vertical amplifier input of an oscilloscope, it passes through several amplifier stages before it is finally displayed on the CRT screen. The complete oscilloscope circuitry from input to CRT has a low-pass characteristic. In other words, the oscilloscope acts like a low-pass circuit. When the manufacturer specifies the scope *bandwidth,* what actually is specified is the value of f_{hc} for the scope circuitry. For example, a scope with a 100 MHz bandwidth will act as a low-pass circuit with f_{hc}=100 MHz.

Since every scope has a high frequency cutoff, each can produce high frequency distortion of a pulse waveform. The degree of distortion will, of course, depend on the scope bandwidth and the pulse transition times. Whenever the scope f_{hc} does not satisfy

$$f_{hc} \geqslant 2f_H \qquad \textbf{(2–7)}$$

the signal displayed on the CRT will have significantly longer transition times than the signal applied to scope input.

In order for a scope to *accurately* display a pulse signal, it should have a bandwidth (f_{hc}) that satisfies Equation 2–7. Thus, you should have some idea of the signal rise and fall times if you measure them with an appropriate scope.

Example 2.16 What scope bandwidth is required to accurately display a pulse that has $t_r = t_f = 50$ ns?

Solution:

$$f_H = \frac{1}{2 \times 50\text{ ns}} = 10\text{ MHz}$$

Thus, the scope needs a bandwidth

$$f_{hc} \geqslant 2f_H = \mathbf{20\ MHz}$$

Example 2.17 What are the shortest transition times that can be accurately displayed on a 50 MHz scope?

Solution: We want

$$f_{hc} \geqslant 2f_H$$
$$50\text{ MHz} \geqslant 2f_H$$

This requires that f_H be less than or equal to 25 MHz

$$f_H = \frac{1}{2t_t} < 25\text{ MHz}$$

or

$$t_t \geqslant \frac{1}{50\text{ MHz}} = 20\text{ ns}$$

This means that transition times of **20 ns** or more will be accurately displayed on a 50 MHz scope.

Scope rise time Very often a manufacturer will specify the oscilloscope rise time instead of the oscilloscope bandwidth. They are related according to

$$t_r = \frac{0.35}{f_{hc}} \tag{2–9}$$

This is the rise time that would be displayed on the CRT if an *ideal* pulse were applied to the scope input. In other words, it is the rise time caused by the scope's internal circuitry.

When a signal with transition times much shorter than the scope's rise time is applied to the scope input, the displayed rise time will be approximately equal to the scope rise time and will bear no resemblance to the actual signal rise time. To obtain an accurate display of a signal's transitions, the scope rise time should be smaller than the signal rise time by at least a factor of three. That is, we want

$$t_r(\text{scope}) \leqslant 1/3t_r(\text{signal}) \tag{2–10}$$

if the scope is to have negligible effect on the signal rise time. As the next example will show, this criterion is essentially the same as the $f_{hc} \geqslant 2f_H$ criterion. Requiring a small scope t_r relative to the signal t_r is the same as requiring a large scope f_{hc} compared to the signal's f_H.

Example 2.18 Use the rise time criterion (2–10) to answer the question of Example 2.17.

Solution: The 50 MHz scope has a rise time given by Equation 2–9.

$$t_r(\text{scope}) = \frac{0.35}{50\text{ MHz}} = 7\text{ ns}$$

Thus, using 2–10, this scope can accurately display signals with rise times of 21 ns or more. This result is essentially the same as that of Example 2.17.

2.12 Non-periodic Pulses

So far we have been talking about periodic waveforms. Frequently the waveforms encountered in pulse circuitry are not repetitive at a given frequency. A good example of a non-periodic signal is an isolated single pulse that occurs only once over a relatively long time. Much of what we have said concerning periodic pulse waveforms is equally applicable to non-periodic pulses. There is one major difference: non-periodic signals do not have a fundamental frequency and equally spaced harmonics. Instead, they have a *continuous* band of sinusoidal components from dc to infinity. A practical, non-periodic pulse, however, has significant frequency components up to a frequency f_H.

Example 2.19 Assume that the pulse in Figure 2–4 is a single isolated pulse applied to a band-pass circuit. Determine the f_{lc} and f_{hc} limits for the circuit, if the pulse is to be passed without significant distortion.

Solution: The pulse parameters were determined in Example 2.1 as $t_r = 3.4\ \mu\text{s}$, $t_f = 3.2\ \mu\text{s}$ and $t_p = 9\ \mu\text{s}$. Using t_f, the pulse has significant frequency components up to

$$f_H = \frac{1}{2t_f} = 156.2\text{ kHz}$$

Thus, the band-pass should have

$$f_{hc} \geqslant 2f_H = \mathbf{312.4\ kHz}$$

Using t_p, the band-pass should have

$$f_{lc} \leqslant \frac{1}{63t_p} = \mathbf{1.763\ kHz}$$

Questions

Sections 2.1–2.5

2.1 Determine t_r, t_f, and t_p for the pulse shown in Figure 2–23.
2.2 Sketch pulses demonstrating the characteristics of overshoot, tilt, and ringing.
2.3 Draw a pulse waveform on a 0 V baseline, having a peak value of 12 V and 25 percent tilt.

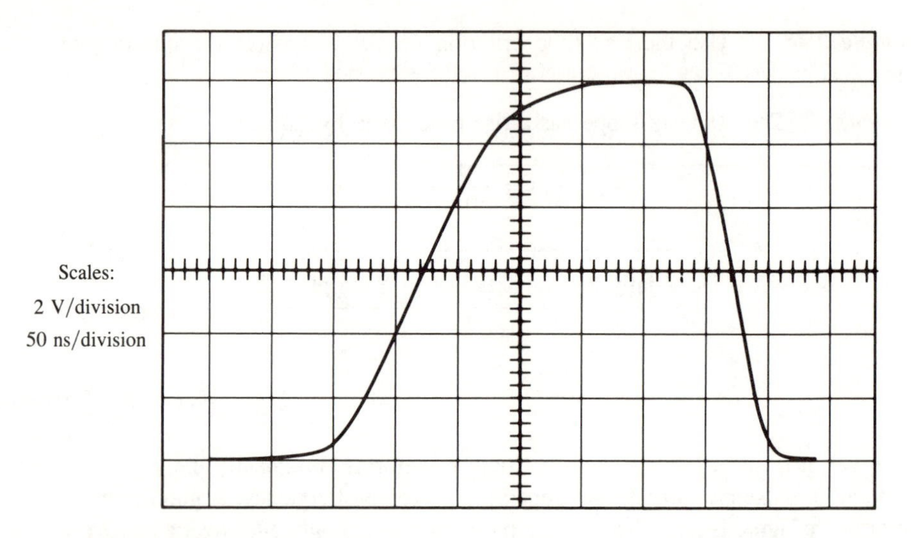

Figure 2–23

2.4 Modify the waveform of Figure 2–8 so that it has a 10 percent tilt.

2.5 Modify the waveform of Figure 2–8 so that it has a 20 percent overshoot.

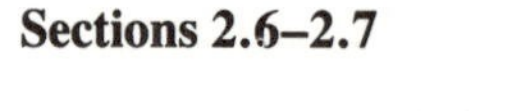

2.6 Determine the frequency, duty cycle, and average value of the waveform in Figure 2–24.

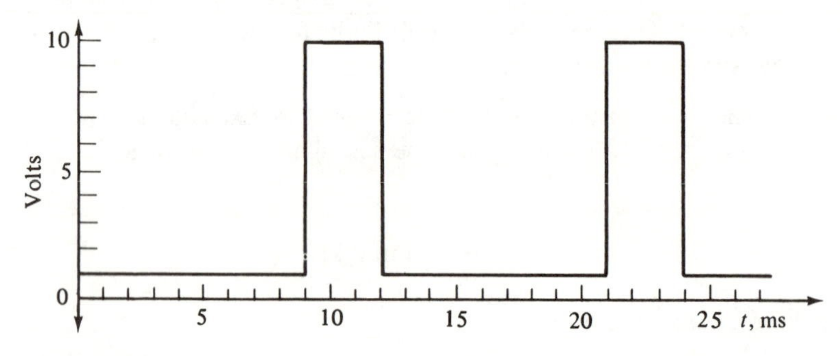

Figure 2–24

2.7 Indicate the effect (increase, decrease, or remain the same) of the following changes on the *duty cycle* and *average value* of the waveform considered in the preceding question:

(a) Leave frequency the same but decrease the pulse duration.

(b) Increase the pulse amplitude.

(c) Halve the frequency of the waveform and double the pulse duration.

2.8 Sketch a 24 V *p-p* bipolar square wave with a PRF of 2 MHz.

2.9 (a) Draw an ideal pulse waveform having the following characteristics:

PRF = 8 kHz　　　baseline = +2 V

duty-cycle = 40%　　　peak value = −10 V

(b) Determine t_p for this waveform.

(c) Determine the amount of dc that this waveform contains.

2.10 Suppose the signal of Problem 2.9 is displayed on an oscilloscope using dc coupling. Describe what happens to the displayed waveform when the scope input is switched to ac coupling.

2.11 How much dc is contained in a perfect square wave that switches between -5 V and $+12$ V?

Sections 2.8–2.9

2.12 List all of the frequency components of the waveform in Figure 2–24 up to the ninth harmonic.

2.13 List all of the frequency components of the square wave of Problem 2.8 up to the ninth harmonic.

2.14 What is the PRF of a pulse waveform whose fourth harmonic is 32 kHz?

2.15 Is it possible for two pulse signals to have exactly the same frequency spectrum and yet produce different responses when applied to the same circuit? Explain.

2.16 A pulse waveform has $t_p = 4\ \mu s$, $T = 12\ \mu s$, and $t_r = t_f = 1\ \mu s$. Determine *all* of the significant harmonic frequency components of this signal.

2.17 The highest significant frequency component of a certain waveform is 25 MHz. What is the signal's rise time?

2.18 *True* or *False:*

(a) A pulse signal's low frequency harmonics have little effect on the signal transition times.

(b) Its high frequency harmonics, if attenuated, can cause tilt on the pulse.

(c) Low frequency harmonics, if attenuated, can cause the dc average value to decrease.

(d) Shifting the phase of a harmonic component will not affect a signal's shape, as long as the harmonic amplitude is unchanged.

(e) A signal's highest significant frequency (f_H) increases as the PRF of the signal increases.

Section 2.10

2.19 If the waveform of Figure 2–24 has $t_r = t_f = 100$ ns, will a low-pass circuit with $f_{hc} = 20$ MHz cause any distortion in the signal?

2.20 Will a high-pass circuit with $f_{lc} = 20$ Hz cause significant tilt on the waveform of Figure 2–24?

2.21 If the pulse of Figure 2–10 is applied to a low-pass circuit that has $f_{hc} = 1$ kHz, what would you expect to see at the low-pass output? (*HINT:* Check the signal's fundamental frequency.)

2.22 If the pulse of Figure 2–10 is applied to a high-pass circuit that has $f_{lc} = 100$ Hz, what would the output look like? (*HINT:* Use the signal's average value.)

2.23 Determine the f_{lc} and f_{hc} limits for a band-pass circuit that has to pass a 20 kpps signal with the following characteristics without producing significant distortion:

duty-cycle = 15%   peak value = +8 V

baseline = 0 V $t_r = 15$ ns; $t_f = 7$ ns

2.24 Figure 2–25 shows a pulse waveform (v_{IN}) and three different output waveforms. For each output, indicate what type(s) of circuit could produce that output. Explain each answer.

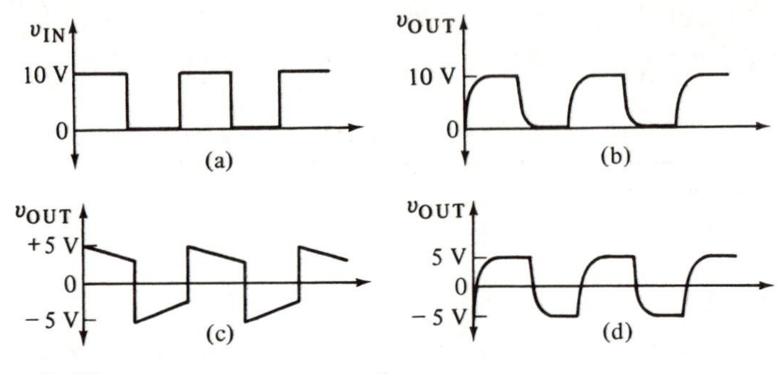

Figure 2–25

Sections 2.11–2.12

2.25 A pulse generator with transitions times of 2 ns is to be displayed on an oscilloscope CRT. What bandwidth should the scope have in order to make accurate measurement of these transition times? Solve this problem using f_H. Then solve it by using t_r (scope).

2.26 A certain oscilloscope has a specified rise time of 10 ns. Can this scope display a 10 MHz sinewave without significantly affecting its amplitude or phase?

2.27 A technician uses an oscilloscope to measure the transition times of a certain pulse waveform and measures $t_r = t_f \approx 150$ ns. He then measures the transition times of another waveform that he knows is much faster than the first one, and he gets the same results.

Explain what has happened. Can you find the approximate bandwidth of the scope from the previous information?

3

RC Circuits with Pulse Inputs

3.1 Introduction

Simple *RC* circuits occur quite often in pulse and digital circuits. Sometimes they are used deliberately to change the shape of a pulse or to couple the pulse from one circuit to another. At other times, they occur naturally because of the presence of resistance and capacitance throughout any circuit. In most circuits, unwanted *parasitic* capacitance is present due to PN junction capacitances and capacitance between adjacent conductors. These parasitic capacitances are usually very small (1–10 pF) and have negligible effect on low frequency operation. However, they can adversely affect the signals in pulse circuits where signal transitions must occur in a matter of nanoseconds.

In this chapter we will review the basic characteristics of capacitors and simple *RC* circuits, including the exponential equations. We will then examine the response of simple *RC* circuits to pulse waveforms. In particular, we will see how changing the circuit time constant affects the shape of the circuit output for a given input pulse waveform.

3.2 Basic Capacitor Characteristics

The behavior of a capacitor in a circuit is defined by the relationship between the capacitor current and the rate of change of its voltage. That is,

$$\frac{dv}{dt} = \frac{i}{C} \tag{3–1}$$

Figure 3–1 shows a capacitor with a voltage v_{xy} across its terminals and having a current i flowing toward its upper terminal (plate) and away from its lower terminal. Recalling that i represents conventional current (positive charges), this flow of current will charge the capacitor voltage to the polarity shown. The *rate* at which this voltage, v_{xy}, changes depends on the value of i. Note that lowercase symbols are used for v_{xy} and i since, in general, both will be time-varying quantities.

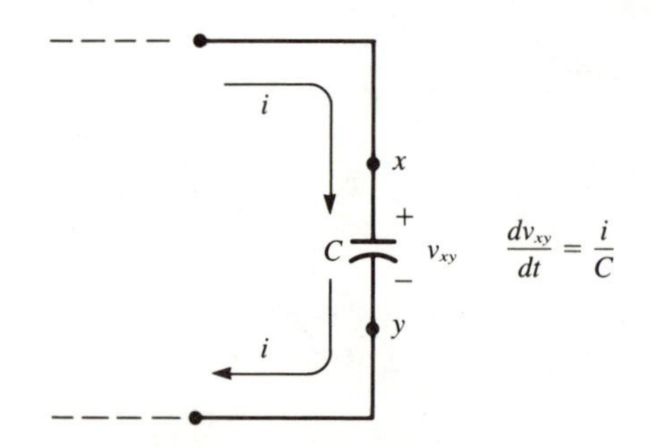

Figure 3–1 **Capacitor current-voltage relationship**

The relationship $dv/dt = i/C$ can be used to establish two well-known capacitor characteristics.

(a) *A capacitor's voltage cannot change instantaneously* but must change at a rate determined by the amount of current available.

(b) *When a capacitor voltage is constant ($dv/dt = 0$), its current must be zero.* This fact is why the capacitor's current eventually becomes zero in a dc circuit.

3.3 Charging a Capacitor

With both switches open in Figure 3–2(a), there will be no flow of current and therefore no voltage across the resistors or the capacitor. When SW1 is closed, the capacitor will begin to charge as current flows from the source through the 10 kΩ resistor as shown in part (b). Eventually, it will charge to the full 20 V. The familiar waveforms for the current, capacitor voltage (v_{xy}), and resistor voltage (v_{wx}) are shown in part (c). Recall the following important points:

(a) At the instant the switch closes ($t = 0$) all of the source voltage appears across the resistor and none across the capacitor, and the current is at its maximum value.

(b) Thereafter, the resistor voltage (and the current) will exponentially decay toward zero, while the capacitor voltage charges to 20 V. Each will reach its steady-state value after approximately *five* time constants (5τ) where $\tau = RC = 10$ ms for the given R and C values.

3.4 Discharging a Capacitor

In Figure 3–2 the capacitor will eventually charge to v_{xy} = +20 V after 5τ. If SW1 is then opened, the capacitor will, theoretically, maintain its charge indefinitely because there is no discharge path. Of course, in practice, any capacitor will gradually discharge through its own leakage resistance, but this usually takes quite a while.

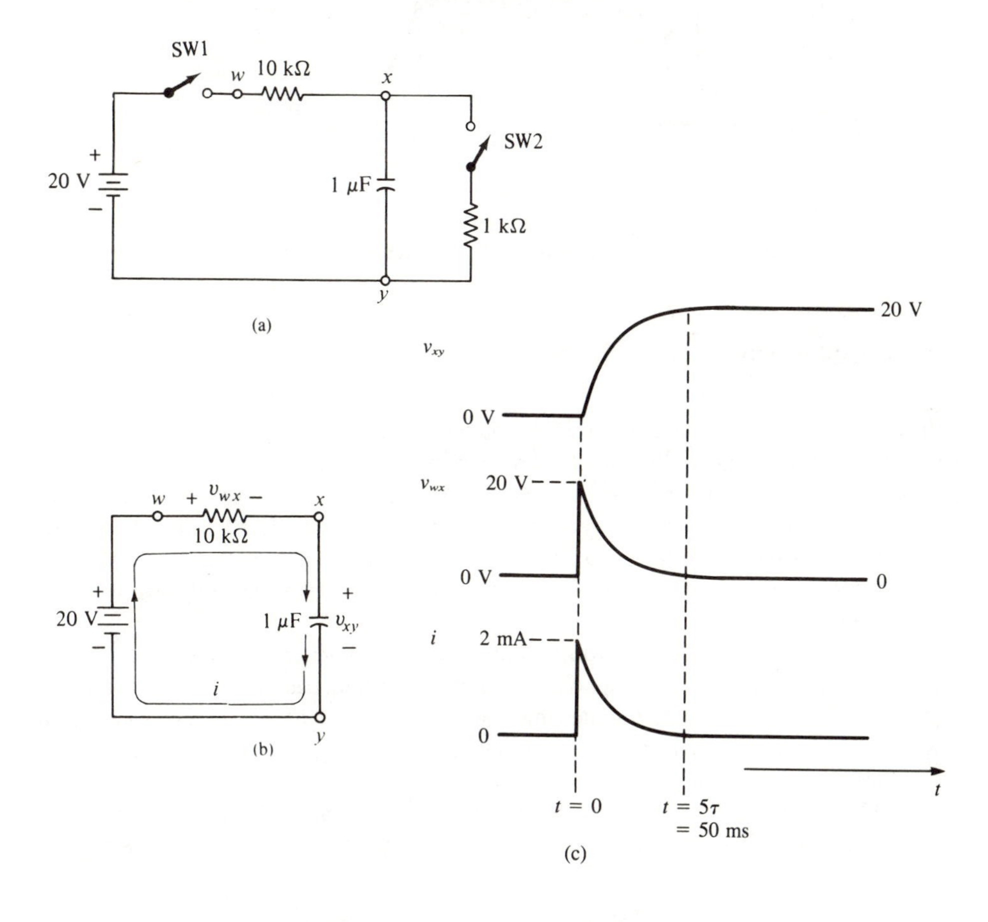

Figure 3–2 **Capacitor charging when SW1 closes**

If SW2 is closed, a complete discharge path exists as shown in Figure 3–3(a). The capacitor's charge will supply current around this path in the direction shown. The waveforms for the capacitor voltage and current are shown in part (b). The voltage begins discharging from 20 V at t = 0 (the instant SW2 is closed) and follows an exponential decay to 0 V after 5τ. Simultaneously, the discharge current jumps to 20 mA at t = 0. It then gradually decays and reaches zero after 5τ. Note that the discharge current in Figure 3–3 flows in the opposite direction (relative to the capacitor) compared to the charging current in Figure 3–2.

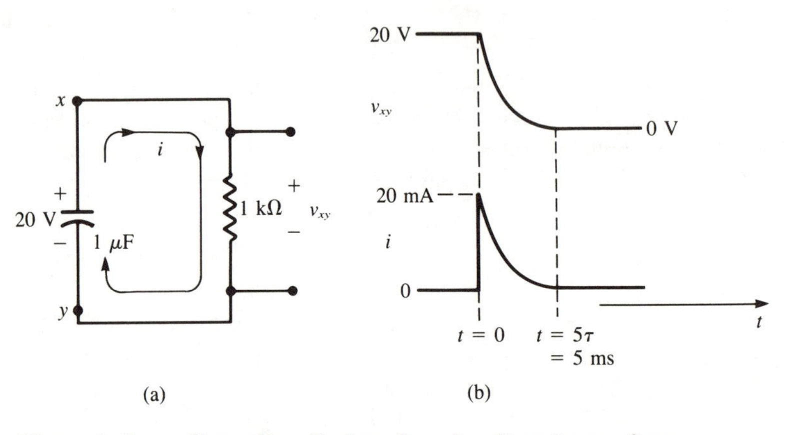

Figure 3–3 **Capacitor discharging circuit and waveforms**

3.5 The Exponential Form

When solving circuits such as those in the previous sections, it is convenient to be able to mathematically express the variation of currents and voltages with time. Although graphical representations are more descriptive, mathematical expressions are easier to manipulate and are usually more accurate and require less space. In addition, the expressions for any current or voltage in a single time constant circuit (one that contains either one capacitor *or* one inductor) have the same general form. This general form is called the *exponential form*.

The key to the exponential form is the *exponential function* $\epsilon^{-t/\tau}$ in which ϵ (Greek letter epsilon) stands for the Naperian (natural) logarithm base and is approximately equal to 2.718. The t stands for the time variable, which can assume any value from 0 to ∞, and the constant τ is the time constant. Both t and τ have to be expressed in the same time units.

The value of $\epsilon^{-t/\tau}$ for a given value of t can be determined by raising 2.718 to the appropriate power. For example, to determine the value of this function when $t = \tau$,

$$\epsilon^{-t/\tau} = \epsilon^{-1} = \frac{1}{\epsilon} = \frac{1}{2.718} = 0.368$$

For $t = 2\tau$,

$$\epsilon^{-t/\tau} = \epsilon^{-2} = \frac{1}{\epsilon^2} = \frac{1}{2.718^2} = 0.135$$

Notice from these examples that as t increases, the value of the exponential function decreases. These examples were relatively simple because the exponents were whole numbers. For fractional values of t/τ, either mathematical tables or a calculator must be used. Here are the results of the tabulation of some values of $\epsilon^{-t/\tau}$ for various integer values of t:

t	$\epsilon^{-t/\tau}$
0	1.000
τ	0.368
2τ	0.135
3τ	0.050
4τ	0.018
5τ	0.007
10τ	0.000045

Again notice that increasing values of t result in rapidly decreasing values of $\epsilon^{-t/\tau}$. Even though the function will actually reach zero only when $t = \infty$, for the purposes of this text the function will be assumed to be zero when $t \geq 5\tau$. A plot of the exponential function versus t is shown in Figure 3–4. The curve in Figure 3–4 has the same shape as the waveform of capacitor discharge in Figure 3–3 and the waveform of resistor voltage in Figure 3–2.

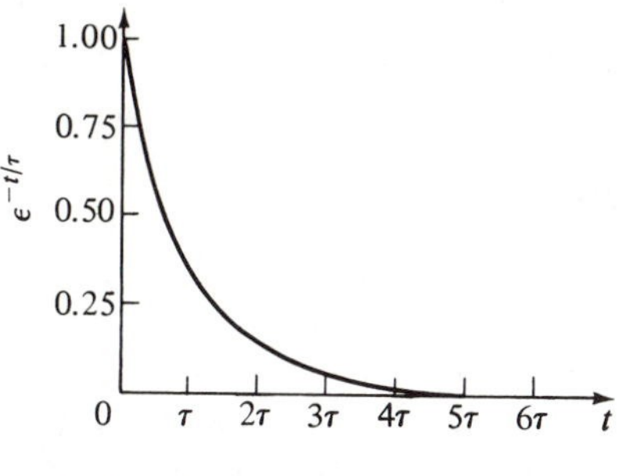

Figure 3–4 **Plot of $\epsilon^{-t/\tau}$**

Exponential form for v The exponential function can now be used to express the behavior of currents and voltages in single time constant circuits. It can be shown using the methods of calculus and differential equations that in a linear circuit containing only one capacitance (or inductance), the voltage v across *any* element in the circuit in response to any *sudden change* can be expressed as follows:

$$v = V_F + (V_0 - V_F)\epsilon^{-t/\tau} \qquad \text{(3–2)}$$

Equation (3–2) shows how the voltage (v) varies with time, where V_0 represents the *initial* value of v at the instant the *change* is made (at $t = 0$), and V_F represents the *final* value of v attained in steady state (at $t \geqslant 5\tau$).

The definitions of V_0 and V_F can be verified by using Equation (3–2) to calculate the initial and steady-state values of v. The initial value of v is found at $t = 0$, using Equation (3–2).

$$\begin{aligned} v(t = 0) &= V_F + (V_0 - V_F)\epsilon^{-0} \\ &= V_F + (V_0 - V_F) \times 1 = V_0 \end{aligned}$$

The final steady-state value of v is found at $t \geqslant 5\tau$, where $\epsilon^{-t/\tau}$ is essentially zero.

$$v(t \geqslant 5\tau) = V_F + (V_0 - V_F) \times 0$$
$$= V_F$$

This general voltage expression can be applied to the capacitor charging circuit discussed earlier and redrawn in Figure 3–5. Consider first the capacitor voltage v_{xy}. Assume the capacitor is initially discharged to zero volts. At $t = 0$ the switch is closed instantaneously (this is the sudden change), and since the capacitor voltage, v_{xy}, cannot change instantaneously, it must be zero at $t = 0$. Thus, $V_0 = 0$ V.

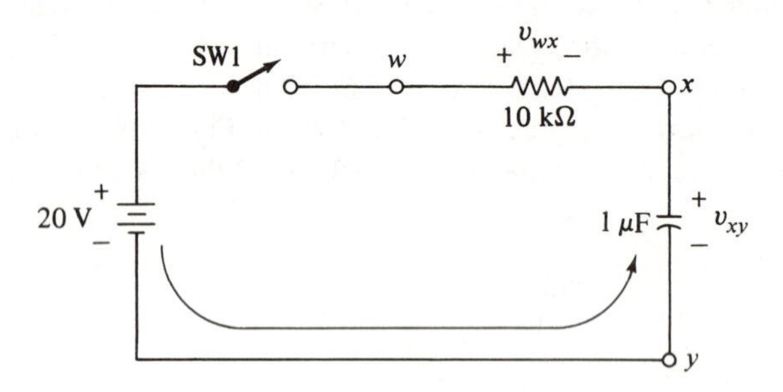

Figure 3–5 **Capacitor charging circuit**

To determine V_F, the steady-state value of v_{xy}, it is necessary to determine the voltage to which the capacitor will charge. This can be done by replacing the capacitor with an open circuit, since in steady state the capacitor draws no current (for a constant source voltage). In this condition, the value of v_{xy} will be 20 V, which is V_F.

The circuit time constant (τ) is determined by the capacitance value and the equivalent resistance seen by the capacitor (which is the *Thevenin* resistance seen between x and y). In this circuit the resistance is 10 kΩ. Thus, τ is (10 kΩ)(1 μF) = 10 ms.

These values of V_0, V_F, and τ are inserted into Equation (3–2) to produce the mathematical expression for the variation of v_{xy} with time:

$$v_{xy} = V_F + (V_0 - V_F)\epsilon^{-t/\tau}$$
$$= 20 + (0 - 20)\epsilon^{-t/10\text{ ms}}$$
$$= 20 - 20\epsilon^{-t/10\text{ ms}}$$
$$v_{xy} = 20(1 - \epsilon^{-t/10\text{ ms}})\text{V} \quad (3\text{–}2)$$

This last expression, if plotted versus t, will result in the same curve as v_{xy} in Figure 3–2. The student should verify this by taking several values of t and evaluating the expression. The expression reflects the fact that v_{xy} *increases* with time; the exponential term is *subtracted* from one so that as the exponential decreases, the total term in parentheses increases.

Example 3.1 Using the preceding expression, determine the voltage across the capacitor 5 ms after SW1 is closed.

Solution: Set $t = 5$ ms in the expression for v_{xy}

$$v_{xy}(t = 5 \text{ ms}) = 20(1 - \epsilon^{-0.5}) \text{ V}$$
$$= 20(1 - 0.606) \text{ V}$$
$$v_{xy}(t = 5 \text{ ms}) = \mathbf{7.88 \text{ V}}$$

Example 3.2 Determine how long it will take for the capacitor in Figure 3–5 to reach 10 V.

Solution: Set $v_{xy} = 10$ V and solve for t.

$$10 = 20(1 - \epsilon^{-t/10 \text{ ms}})$$
$$0.5 = 1 - \epsilon^{-t/10 \text{ ms}}$$

Rearranging the preceding gives

$$\epsilon^{-t/10 \text{ ms}} = 0.5$$

Taking the natural log (ln) of both sides of this equation yields

$$-t/10 \text{ ms} = \ln (0.5) = -.693$$

so that

$$t = \mathbf{6.93 \text{ ms}}$$

is the time at which v_{xy} reaches 10 V.

Example 3.3 Referring again to Figure 3–5, determine the mathematical expression for the resistor voltage v_{wx} in response to the closing of the switch.

Solution: The values of V_0, V_F, and τ must be determined and substituted into the general expression for v_{wx}.

$$v_{wx} = V_F + (V_0 - V_F)\epsilon^{-t/\tau}$$

The initial voltage across the resistor, as determined previously, is 20 V; $V_0 = 20$ V. The steady-state resistor voltage will be zero, since the capacitor acts as an open circuit in steady state allowing no current to flow. Thus, $V_F = 0$ V.

The time constant τ is 10 ms, as determined previously. *For a given circuit the time constant is unique no matter what voltage or current we happen to be looking at*. The complete expression for the resistor voltage is therefore

$$v_{wx} = 0 + (20 - 0)\epsilon^{-t/10 \text{ ms}}$$
$$= \mathbf{20\epsilon^{-t/10 \text{ ms}}}$$

If this expression were plotted versus time, the result would have the shape of v_{wx} shown in Figure 3–2. The student should verify this.

Expression for *i* The technique used in the preceding discussion can also be used to find the mathematical expression for *current* in a circuit. The general expression is

$$i = I_F + (I_0 - I_F)\epsilon^{-t/\tau} \qquad \textbf{(3–3)}$$

I_0 is the initial value of current, and I_F is the steady-state value. The following example will illustrate the use of this expression:

Example 3.4 Referring again to Figure 3–5, determine the equation for the current that flows in response to the switch closure.

Solution: The circuit time constant is known (10 ms), and I_0 and I_F must be determined. The initial current flow is determined by the initial resistor voltage drop divided by the resistance. Thus, $I_0 = 20\text{ V}/10\text{ k}\Omega = 2\text{ mA}$.

The value of I_F is zero because in steady state the capacitor is open. The expression for the circuit current then becomes

$$\begin{aligned} i &= 0 + (2 - 0)\epsilon^{-t/10\text{ ms}} \\ &= \mathbf{2\epsilon^{-t/10\text{ ms}}} \end{aligned}$$

where the units are mA. This result shows that the current decreases exponentially with a time constant of 10 ms. The same result could have been obtained using Ohm's law and the expression for resistor voltage obtained in Example 3.3. That is,

$$i = \frac{v_{wx}}{R} = \frac{20\epsilon^{-t/10\text{ ms}}}{10\text{ k}\Omega} = 2\epsilon^{-t/10\text{ ms}}(\text{mA})$$

3.6 More Applications of the Exponential Form

The procedure followed in the preceding section and to be used in all *single capacitor* circuits to determine the expression for a voltage (current) in an *RC* circuit in response to a change is summarized as follows:

(a) Find the *initial* value of the voltage (current) of interest the instant after the change is made. This is done by solving the circuit with the capacitor voltage held constant at the value it had just *prior* to the change.

(b) Find the *final* value of the voltage (current). This is accomplished by solving the circuit, using the capacitor as an *open circuit*. (This step is always performed, even if the circuit is not allowed to reach steady state because of the occurrence of a second change.)

(c) Determine the equivalent Thevenin resistance seen by the capacitor. Using this resistance value, calculate the time constant $\tau = RC$.

(d) Substitute the values determined in steps 1 through 3 into the appropriate expression [Equation (3–2) or (3–3)].

Example 3.5 Consider the circuit in Figure 3–6(a). Before the switch is closed, the capacitor has been charged to $v_{xy} = -5$ V. At $t = 0$ the switch is closed. Determine the expressions for capacitor voltage and current. Plot v_{xy} versus time.

Solution: When the switch is closed, the capacitor voltage momentarily holds at $v_{xy} = -5$ V. Thus, $V_0 = -5$ V. In steady state after closing the switch, the capacitor will

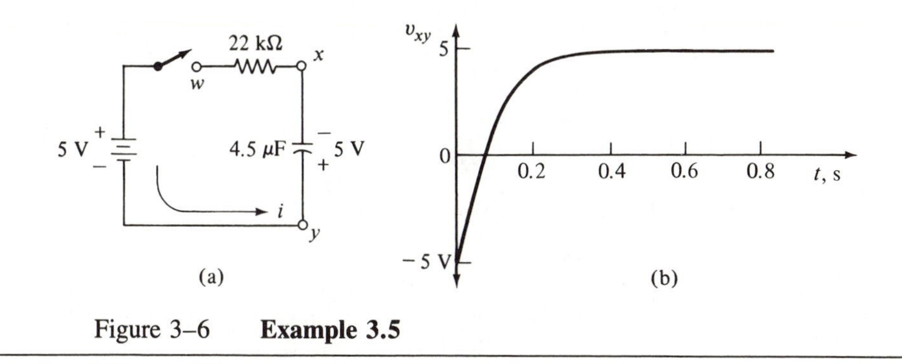

Figure 3–6 **Example 3.5**

---

reach $v_{xy} = +5$ V, since the entire supply voltage will appear across the open circuit between x and y (remember in steady state the capacitor acts as an open circuit to dc). Thus, $V_F = +5$ V. The time constant is

$$\begin{aligned}\tau &= (22\ \text{k}\Omega)(4.5\ \mu\text{F}) \\ &= (22 \times 10^3)(4.5 \times 10^{-6})\ \text{s} \\ \tau &= 0.1\ \text{s}\end{aligned}$$

The complete expression for v_{xy} is therefore

$$\begin{aligned}v_{xy} &= 5\ \text{V} + (-5\ \text{V} - 5\ \text{V})\epsilon^{-t/0.1\ \text{s}} \\ &= \mathbf{5 - 10}\boldsymbol{\epsilon}^{-t/0.1\ \text{s}}\end{aligned}$$

The plot of v_{xy} versus time is shown in Figure 3–6(b). In this case neither the initial nor final values of capacitor voltage are zero. This case occurs quite often in practical applications.

To find the expression for i, we must first find I_0, the current at the instant the switch closes. At that instant the voltage across the resistor will be $v_{wx} = +10$ V because the capacitor voltage aids the source voltage. Thus, $I_0 = 10\ \text{V}/22\ \text{k}\Omega = 0.45$ mA. Of course, $I_F = 0$ since i will be zero in steady state. The expression for current is

$$\begin{aligned}i &= I_F + (I_0 - I_F)\epsilon^{-t/\tau} \\ &= 0 + (0.45\ \text{mA} - 0)\epsilon^{-t/0.1\ \text{s}} \\ \boldsymbol{i} &= \mathbf{0.45}\boldsymbol{\epsilon}^{-t/0.1\ \text{s}}\ \textbf{mA}\end{aligned}$$

Example 3.6 Determine how long it takes the capacitor voltage to reach zero in the circuit of Figure 3–6.

Solution: (a) Using the expression for v_{xy} determined in Example 3.5, set $v_{xy} = 0$ and solve for t.

$$\begin{aligned}0 &= 5 - 10\epsilon^{-t/0.1\ \text{s}} \\ 5 &= 10\epsilon^{-t/0.1\ \text{s}} \\ 0.5 &= \epsilon^{-t/0.1\ \text{s}}\end{aligned}$$

Taking the natural log of both sides

$$-.693 = -t/0.1 \text{ s}$$

or

$$\mathbf{t = 69.3\ ms}$$

as the time required for the capacitor voltage to reach zero.

Example 3.7 The voltage source in Figure 3–7 is a step from 2 V to 10 V. We will assume that its rise time is very short compared to the circuit time constant, so that the circuit will not respond during the step transition. In other words, as far as the circuit is concerned, the source is an ideal step signal.

Assume that the input has been sitting at 2 V for a long time before $t = 0$, and draw the v_{xy} and v_{wx} waveforms.

Solution: Since the input voltage has been at 2 V for some time, it is assumed that the capacitor will have charged to a steady-state value *before* $t = 0$. To determine this value, simply treat the capacitor as an open circuit and calculate v_{xy} for a e_i of 2 V. The result can be obtained by using the simple voltage-divider method.

$$\begin{aligned} v_{xy}(\text{before } t = 0) &= \left(\frac{1\text{ k}\Omega}{1\text{ k}\Omega + 1\text{ k}\Omega}\right) 2\text{ V} \\ &= 1\text{ V} \end{aligned}$$

At $t = 0$ the input suddenly changes to 10 V, but the capacitor voltage must hold at the value it had just prior to the change. Therefore, the value of the capacitor voltage at $t = 0$ is $V_0 = 1$ V. To find the final capacitor voltage after the input has switched to 10 V, replace the capacitor by an open circuit and find v_{xy}.

$$v_{xy}(\text{final}) = V_F = \frac{1\text{ k}\Omega}{2\text{ k}\Omega}(10\text{ V}) = 5\text{ V}$$

To determine τ, it is necessary to find the Thevenin equivalent resistance seen by the capacitor. In this case, the resistance is easily determined as

$$\begin{aligned} R &= 1\text{ k}\Omega \,\|\, 1\text{ k}\Omega \\ &= 500\ \Omega \end{aligned}$$

The time constant is therefore

$$\begin{aligned} \tau &= (500\ \Omega)(15\ \mu\text{F}) \\ &= (5 \times 10^{2})(1.5 \times 10^{-5})\text{ s} \\ \tau &= 7.5\text{ ms} \end{aligned}$$

The capacitor voltage will exponentially charge from 1 V to 5 V starting at $t = 0$, as shown in Figure 3–7(b). It will reach 5 V in approximately $5\tau = 37.5$ ms.

Once the capacitor waveform has been determined, we can use Kirchoff's voltage law (KVL) to determine the v_{wx} waveform. That is,

$$v_{wx} + v_{xy} = e_i$$

or

$$v_{wx} = e_i - v_{xy}$$

The result is drawn in Figure 3–7(b). It is obtained as follows:

Before $t = 0$, $v_{wx} = 2\text{ V} - 1\text{ V} = 1\text{ V}$. At $t = 0$, e_i jumps to 10 V while v_{xy} stays at 1 V, so that v_{wx} jumps to $10\text{ V} - 1\text{ V} = 9\text{ V}$. After 5τ, v_{xy} reaches 5 V, so that $v_{wx} = 10\text{ V} - 5\text{ V} = 5\text{ V}$. The v_{wx} waveform will exponentially decay from its peak of 9 V at $t = 0$ to its steady state value of 5 V.

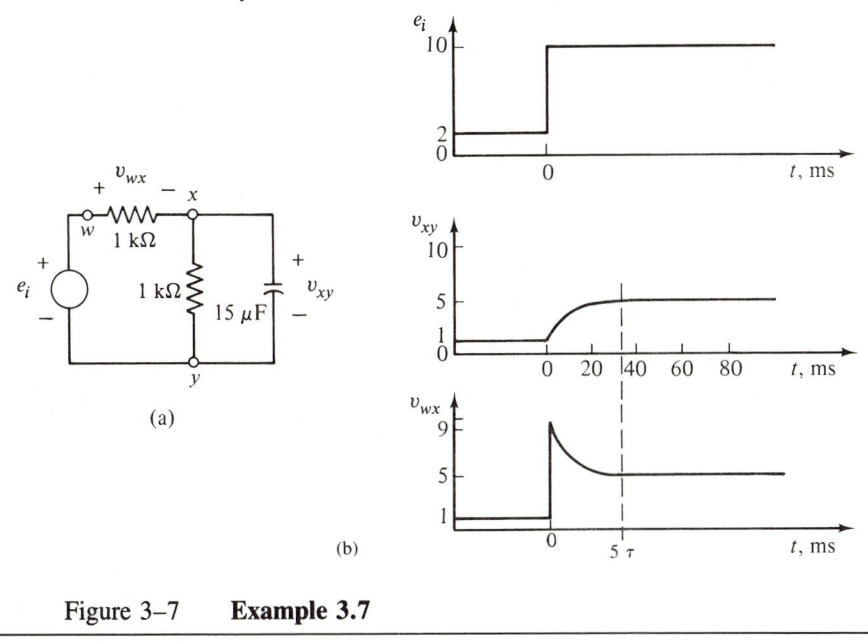

Figure 3–7 **Example 3.7**

3.7 *RC* Low-Pass Circuits

Two simple *RC* low-pass circuits are shown in Figure 3–8. Their low-pass characteristic derives from the fact that the output signal is taken from across a capacitor; at high frequencies this capacitor effectively shorts out the signal. The high-frequency cutoff for both circuits is given by

$$f_{hc} = \frac{1}{2\pi\tau} \tag{3–4}$$

where τ is the circuit time constant.

Both circuits will pass low frequencies to the output because at low frequencies, the capacitive reactance will be very high compared to the resistance values. The circuit in (b), however, will pass only a fraction of these low frequencies because of the voltage divider action between R_1 and R_2.

3.8 *RC* High-Pass Circuits

Figure 3–9 shows two *RC* high-pass circuits. They are similar to the low-pass circuits except that the outputs are taken across a resistor. Their high-pass characteristic occurs

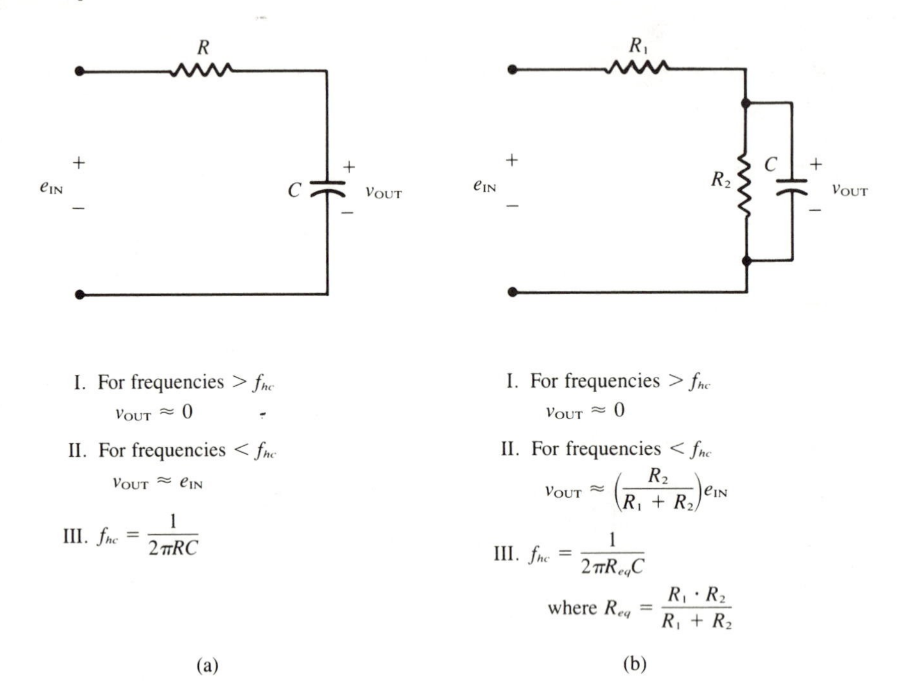

Figure 3–8 **RC low-pass circuits**

because the capacitor acts like a short circuit to high frequencies so that all of the input reaches the output. The low-frequency cutoff for both circuits is given by

$$f_{lc} = \frac{1}{2\pi\tau} \tag{3–5}$$

which is the same as f_{hc} for the low-pass circuits.

Although both are classified as high-pass circuits, the circuit in (b) will not attenuate low frequency signals completely because of the voltage divider action of R_1 and R_2.

It should be apparent that the high-pass circuits of Figure 3–9 are exactly the same as their low-pass counterparts of Figure 3–8; it's just a matter of how the output terminals are defined. Since both types of *RC* circuits occur often in pulse circuitry, it's important to know their effects on pulse input signals.

3.9 Low-Pass Circuit with Pulse Input

Figure 3–10(a) shows an *RC* low-pass being driven by a pulse input. From our discussions in Chapter Two, we know that a low-pass circuit can distort the pulse transitions under certain conditions. For the *RC* low-pass, this slowing-down of the signal transitions is caused by the time required for the capacitor to charge and discharge; that is, the circuit time constant. The degree to which the output pulse is distorted (compared to the input) depends on the value of τ and will increase as τ increases. We will examine the several degrees of pulse distortion that occur as τ ranges from very small to very large.

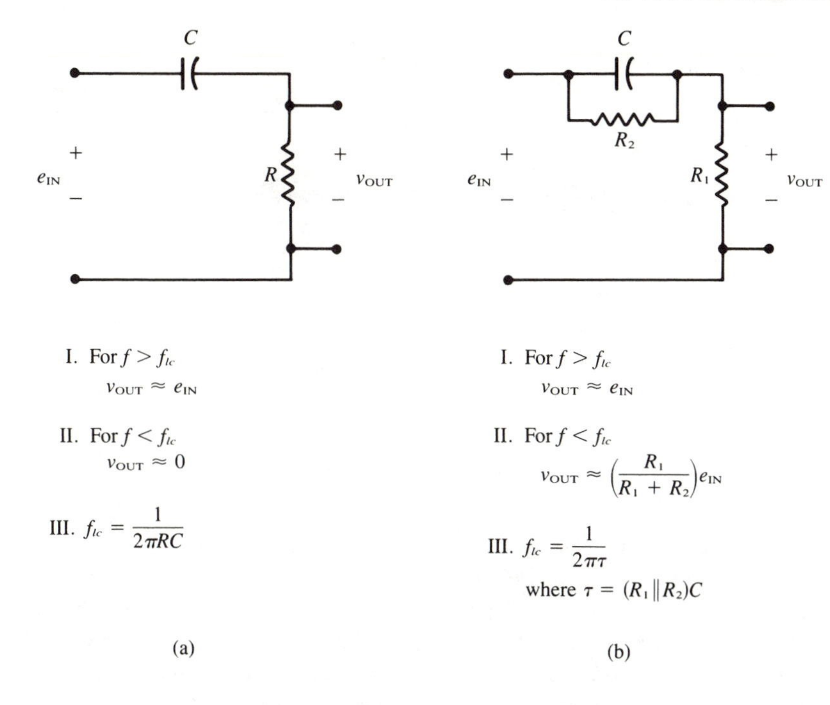

Figure 3–9 ***RC* high-pass circuits**

Case I: $\tau \ll t_r$ (and t_f)

When the circuit time constant is much shorter than the signal transition times, the low-pass will have very little effect on the signal. Figure 3–10 shows this situation. We can reason as follows. The short time constant allows the capacitor voltage to charge rapidly as e_i rises to +10 V and discharge rapidly when e_i drops to 0 V. If τ is much shorter than the signal rise and fall time, the capacitor voltage can almost follow the changes in the input signal.

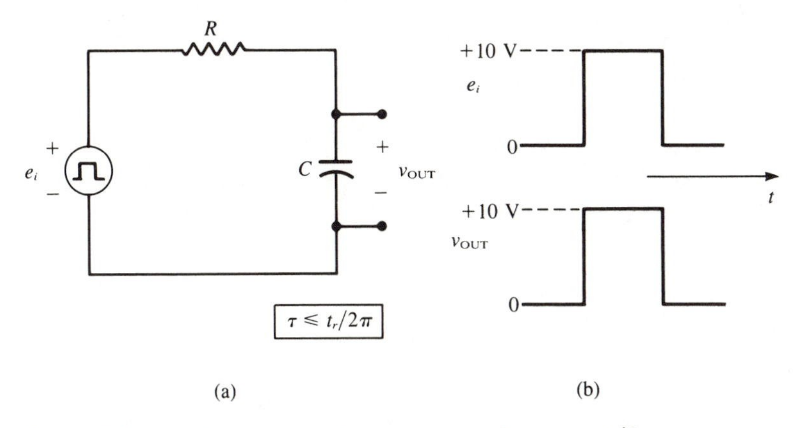

Figure 3–10 **Low-pass pulse response when $\tau \leq t_r/2\pi$**

We can determine the largest value of τ that will produce insignificant output distortion by using the criterion from Chapter Two.

$$f_{hc} \geq 2f_H \tag{2–7}$$

The low-pass cutoff frequency is given by

$$f_{hc} = \frac{1}{2\pi\tau}$$

The signal's highest significant frequency (assuming $t_r = t_f$) is

$$f_H = 1/2t_r$$

Substituting into (2–7)

$$\frac{1}{2\pi\tau} \geqslant 2\left(\frac{1}{2t_r}\right)$$

Rearranging, we have

$$\tau \leqslant \frac{t_r}{2\pi} \quad \textbf{(3–6)}$$

When this criterion is satisfied, the *RC* low-pass will produce very little distortion, and the output pulse will be essentially identical to the input pulse. A very close inspection of the waveforms with an oscilloscope will show that the output transitions are actually slightly longer than the input transitions because the capacitor voltage will always *lag* behind the input. This difference, however, will become smaller and smaller as τ is decreased, until input and output are essentially identical.

Example 3.8 The low-pass has $R = 2.2\ \text{k}\Omega$ and $C = 100$ pF. What is the minimum transition times that an input pulse can have before the low-pass produces high-frequency distortion?

Solution:

$$\tau \leqslant \frac{t_r}{2\pi} \quad \textbf{(3–6)}$$

Rearranging,

$$t_r \geqslant 2\pi\tau = 2\pi RC$$
$$= \mathbf{1.38\ \mu s}$$

Thus, the input transition times should be no less than 1.38 μs.

Example 3.9 What will the output look like for the circuit of Figure 3–11(a) if the input has $t_r = t_f = 40$ ns?

Solution: This is a low-pass circuit with a time constant

$$\tau = (560\ \Omega \| 810\ \Omega) \times 27\ \text{pF}$$
$$= 8.98\ \text{ns}$$

We also have

$$\frac{t_r}{2\pi} = \frac{40\ \text{ns}}{6.283} = 6.37\ \text{ns}$$

Since (3–6) is *not* quite satisfied, there will be some small distortion of the output transitions as shown in Figure 3–11(b).

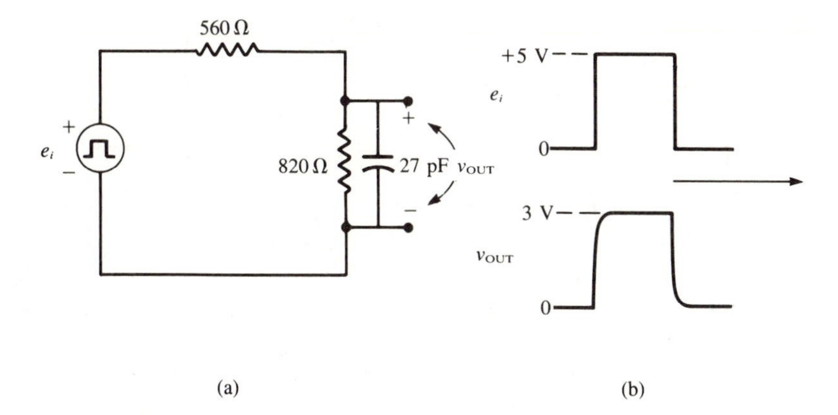

Figure 3–11 **Example 3.9**

In addition to the effect on t_r and t_f, this low-pass circuit will attenuate the pulse amplitude because of the voltage division between the resistors. The capacitor will charge up to only

$$\left(\frac{820}{820 + 560}\right) \times 5 \text{ V} \approx 3 \text{ V}$$

Case II: $\tau \leqslant (t_p/5)$

This last example was a case where τ was large enough to cause some distortion, but small enough so that the capacitor could reach its steady-state voltage during the input pulse duration. The capacitor will always be able to fully charge to its steady state value if the input stays at its peak value longer than 5τ. In other words, whenever

$$t_p \geqslant 5\tau$$

or

$$\tau \leqslant \frac{t_p}{5}$$

the low-pass output will be a complete (though possibly distorted) pulse.

Of course, the output transition times could be considerably lengthened compared to the input. This is illustrated in Figure 3.12 where $t_p = 1$ ms and $\tau = 0.1$ ms. The capacitor will reach steady-state in $5\tau = 0.5$ ms, exactly halfway through the pulse duration.

For cases such as this where τ is much greater than the signal transition times, the output rise and fall times can be determined from the following equation:

$$t_r = t_f = 2.2\tau \qquad \textbf{(3–7)}$$

This formula can be derived from the exponential equation for the capacitor voltage. We will leave this derivation as an end-of-chapter exercise.

Example 3.10 What is the rise time of the output pulse in Figure 3–12? What is the fall time?

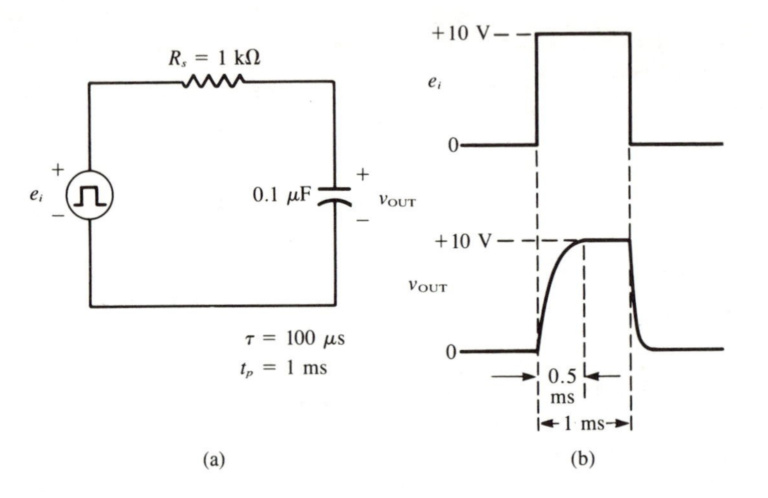

Figure 3–12 **Low-pass with $\tau \leqslant t_p/5$ so that v_{OUT} can reach its peak value**

Solution:

$$t_r = t_f = 2.2 \times 0.1 \text{ ms} = \mathbf{0.22 \text{ ms}}$$

The rise and fall times here are the same because the time constant is the same value regardless of whether the capacitor is charging or discharging. That is, the signal source has the same internal resistance when it is at 0 V as when it is at 10 V. This is not necessarily true for every signal source; in those cases, t_r and t_f will be different.

Case III: $\tau > t_p/5$

When τ is so large that the capacitor cannot fully charge during t_p, the low-pass output will be attenuated in addition to having very slow transition times. In fact, the output shape will be significantly rounded and will bear no resemblance to the input pulse. This situation is illustrated in Figure 3–13.

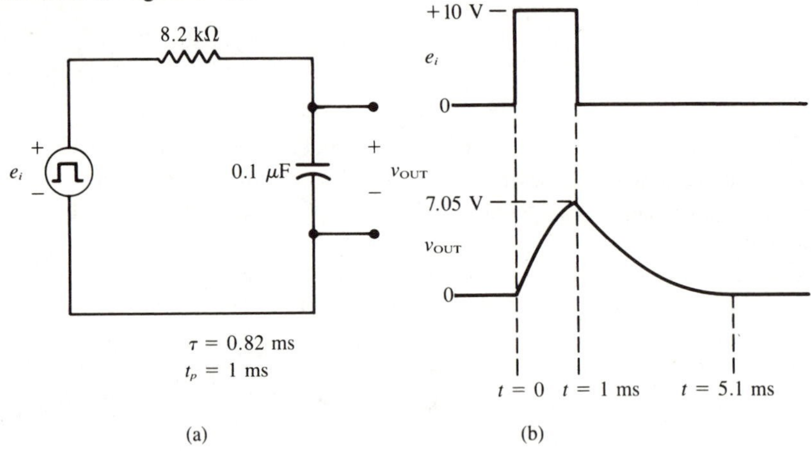

Figure 3–13 **Considerable output distortion and attenuation when $\tau > t_p/5$**

Here the input stays at +10 V for 1 ms while the circuit τ is 0.82 ms. The capacitor voltage will *not* be able to charge to the full 10 V within 1 ms. We can determine the capacitor's peak voltage by setting up the exponential equation for v_{OUT}.

When e_i jumps to +10 V, v_{OUT} will stay at 0 V before the capacitor starts charging. Thus, $V_0 = 0$ V. As the capacitor starts charging, it heads for a final value of +10 V. Thus, $V_F = +10$ V. Note that we still use +10 V as the final steady-state value even though we know that the capacitor will not be allowed to reach that value (the capacitor doesn't know that e_i will return to zero before 5τ). The equation for v_{OUT} is therefore,

$$\begin{aligned} v_{OUT} &= V_F + (V_0 - V_F)\epsilon^{-t/\tau} \\ &= (10 - 10^{-t/.82 \text{ ms}}) \text{ V} \end{aligned}$$

To determine the capacitor voltage at 1 ms, we simply set $t = 1$ ms and solve for v_{OUT}.

$$\begin{aligned} v_{OUT}(\text{peak}) &= 10 - 10\epsilon^{-1/.82} \\ &= 7.05 \text{ V} \end{aligned}$$

Thus, v_{OUT} exponentially rises to 7.05 V while e_i is at +10 V. After 1 ms, e_i rapidly returns to zero and the capacitor starts discharging from its peak of 7.05 V toward 0 V with the same τ.

Example 3.11 Determine the peak value of v_{OUT} if C is increased to 1 μF.

Solution: τ is now 8.2 ms. Thus,

$$\begin{aligned} v_{OUT} &= 10 - 10\epsilon^{-t/8.2 \text{ ms}} \\ v_{OUT}(\text{peak}) &= 10 - 10\epsilon^{-1/8.2} \\ &= \mathbf{1.15 \text{ V}} \end{aligned}$$

Example 3.12 Repeat for $C = 100$ μF.

Solution:

$$\begin{aligned} \tau &= 820 \text{ ms} \\ v_{OUT} &= 10 - 10\epsilon^{-t/820 \text{ ms}} \\ v_{OUT}(\text{peak}) &= 10 - 10\epsilon^{-1/820} \\ &\approx \mathbf{12.19 \text{ mV}} \end{aligned}$$

As these examples show, the output amplitude becomes smaller and smaller as τ increases. It is important to understand, however, that the capacitor will always pick up some amount of charge during the t_p interval regardless of the value of τ. This fact will be of value to us a little bit later.

Summary

Figure 3–14 summarizes the response of the *RC* low-pass circuit to a pulse input. Note that the output becomes more distorted and attenuated as τ increases. When τ is greater than $10t_p$ the capacitor will pick up very little charge during the e_i pulse duration.

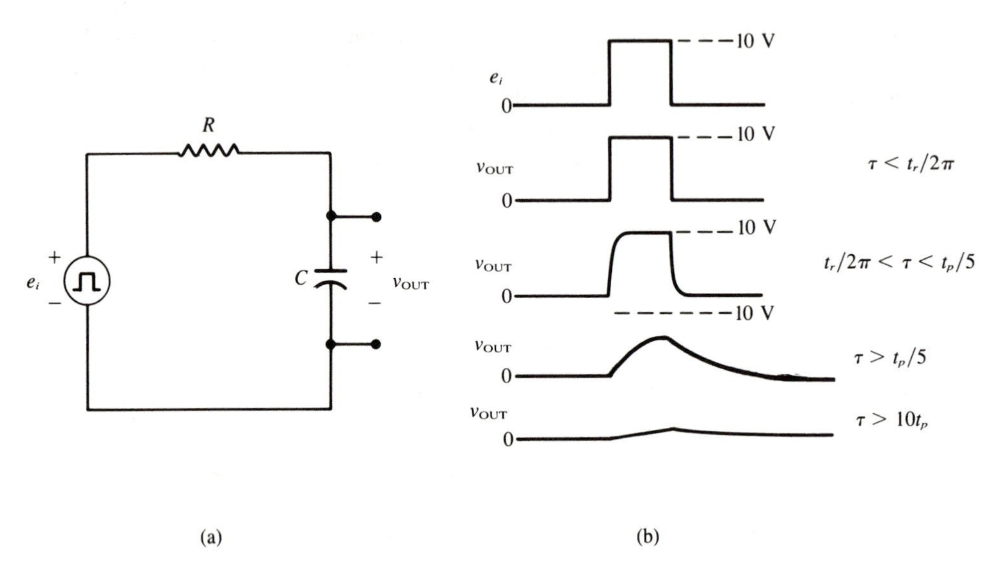

Figure 3–14 **Summary of low-pass (integrator) circuit pulse response as τ increases**

It should be mentioned here that the *RC* low-pass is often referred to as an *integrator* circuit, a term derived from a mathematical operation that is beyond the scope of our work. We will use this term from time to time so that you will become familiar with it.

3.10 High-Pass Circuit with Pulse Input

We have seen that an *RC* low-pass (integrator) will pass an input pulse without significant distortion if τ is small enough, but will cause distortion and attenuation as τ increases. The *RC* high-pass behaves in just the opposite manner. Since the output is taken across the resistor, the resistor voltage must equal the difference between the input signal and the capacitor voltage. That is,

$$v_R + v_{cap} = e_i$$

so that

$$v_R = e_i - v_{cap}$$

We will use this KVL relationship to help us find v_R since we now know how to find v_{cap}.

Case I: $\tau > t_p/5$

Figure 3–15 illustrates the situation where τ is much greater than t_p. In this case, the capacitor will charge up to only a small voltage during the time that e_i is at its peak value. We can determine V_p once we know τ and t_p by using the exponential form as we did in the preceding section. For the values shown, the capacitor will charge up to 1.8 V (reader should verify). When e_i returns to 0 V, the capacitor voltage will slowly discharge from 1.8 V to 0 V as shown in Figure 3–15(b).

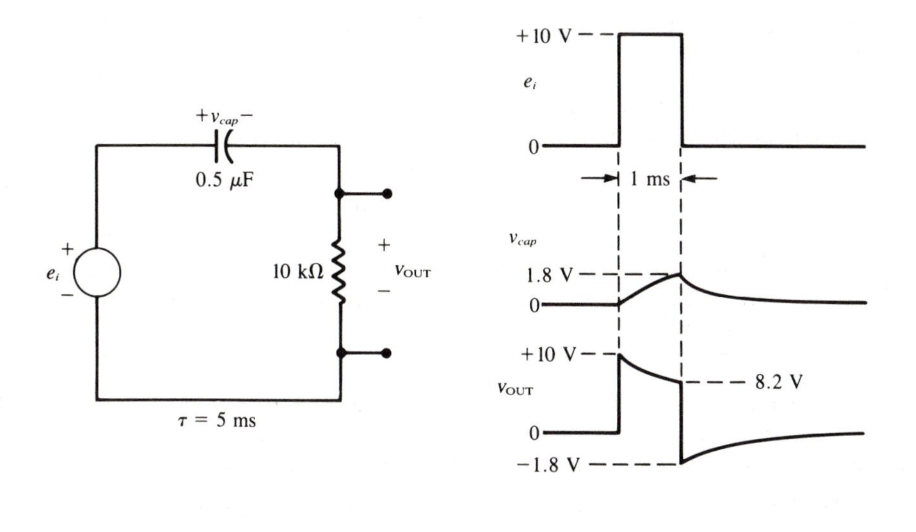

Figure 3–15 **High-pass response when $\tau > t_p/5$ so that capacitor cannot fully charge**

The waveform for v_{OUT} is simply the algebraic difference between e_i and v_{cap} and is obtained as follows:

(a) Prior to $t = 0$, e_i is at zero, so $v_{OUT} = 0$ V.

(b) At $t = 0$, e_i jumps to +10 V, and (since τ is much greater than t_r) v_{cap} will not change during the e_i transition time. Thus, v_{OUT} will jump to +10 V to satisfy KVL.

(c) Between $t = 0$ and $t = t_p$, v_{cap} gradually increases. Since e_i is fixed at +10 V, v_{OUT} must decrease in exactly the same manner that v_{cap} increases. Thus, at $t = t_p$ the value of v_{cap} will have increased to 1.8 V and the value of v_{OUT} will have decreased to $10 - 1.8 = 8.2$ V.

(d) At $t = t_p$, the input rapidly returns to 0 V. Since $v_{OUT} = e_i - v_{cap}$, the output will equal $-v_{cap}$. That is, $v_{OUT} = -v_{cap}$ so that v_{OUT} jumps to −1.8 V at $t = t_p$ and then gradually rises to 0 V as the capacitor discharges.

There are several important points to learn from this example. First, note that the high-pass output has exactly the same transitions as the input. When the input makes a rapid jump in either direction, v_{OUT} makes the same jump. This makes sense since the high-pass should pass all of the fast portions of e_i. Another way to look at it is that the capacitor cannot follow the rapid changes in e_i, and so it passes or *couples* these rapid transitions to the output.

Second, note that whenever e_i is at a constant level (either 0 V or +10 V), the resistor voltage waveform goes through the exact *opposite* change as the capacitor voltage waveform.

Third, notice that v_{OUT} goes *negative* for a portion of time even though e_i is always positive. This fact is an important characteristic of a high-pass circuit.

Example 3.13 What is the percent tilt on the output signal of Figure 3–15? How can this tilt be decreased?

Solution: The v_{OUT} pulse goes from a peak of +10 V down to 8.2 V during the input pulse duration. Thus, the amount of tilt is 1.8 V and the percent tilt is

$$\frac{1.8\text{ V}}{10\text{ V}} \times 100\% = 18\%$$

The tilt can be reduced by increasing R or C or both so that τ is greater. The greater value of τ means that the capacitor voltage will not increase as much during t_p, and therefore v_{OUT} will not decrease as much.

The question now arises as to how large τ should be in order to produce negligible tilt in v_{OUT}. There will always be some tilt no matter how large we make τ. A 10 percent tilt will occur for $\tau = 10t_p$. We can prove this fact by using the exponential equation for v_{cap} and setting $\tau = 10t_p$ to see how far v_{cap} will have charged to at $t = t_p$.

$$v_{cap} = 10 - 10\epsilon^{-t/\tau} = 10 - 10\epsilon^{-t/10t_p}$$

$$v_{cap}(\text{at } t = t_p) = 10 - 10\epsilon^{-t_p/10t_p} = 10 - 10^{\epsilon-0.1}$$

$$\approx 1\text{ V}$$

Thus, v_{OUT} will decrease by 1 V while v_{cap} increases by 1 V. This tilt would be 1 V, or 10 percent.

Thus, we have proved that the tilt will be 10 percent or less whenever $\tau \geqslant 10t_p$. We could have arrived at this same result using the criterion for low percentage tilt:

$$f_{lc} \leqslant 1/63t_p$$

The high-pass frequency is $1/2\pi\tau$. Thus, this criterion becomes

$$\frac{1}{2\pi\tau} \leqslant \frac{1}{63t_p}$$

Using $2\pi \approx 6.3$, we obtain

$$\frac{1}{6.3\tau} \leqslant \frac{1}{63t_p}$$

or

$$\tau \geqslant 10t_p$$

Case II: $\tau < t_p/5$

As τ is made smaller, the high-pass output will exhibit more and more tilt, thereby becoming more distorted compared to e_i. Figure 3–16 illustrates what happens when τ is short enough for the capacitor to fully charge during the input pulse duration. We will still assume, however, that τ is much longer than the input transition times, so that e_i is essentially an ideal pulse.

For the circuit values shown, $\tau = 0.1$ ms so that the capacitor will charge to +10 V in 0.5 ms, halfway through the t_p interval, and remain there until e_i returns to zero, at which time v_{cap} will discharge to zero in 0.5 ms.

The resistor waveform, v_{OUT}, can again be determined using $v_{OUT} = e_i - v_{cap}$. At $t = 0$, when e_i jumps to +10 V, v_{OUT} will jump to +10 V since $v_{cap} = 0$ V. During t_p,

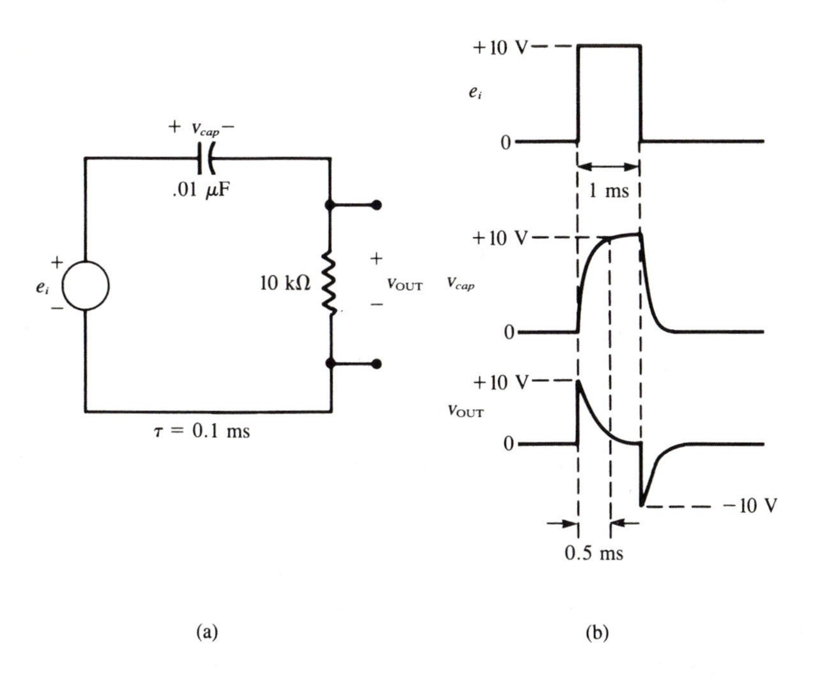

Figure 3–16 **High-pass circuit with $\tau < t_p/5$ produces positive and negative spikes corresponding to input transitions**

while v_{cap} charges from 0 to 10 V, v_{OUT} will decay exponentially from 10 V to 0 V. At $t = t_p$, when e_i rapidly jumps to zero, $v_{OUT} = -v_{cap} = -10$ V so that the output rapidly jumps from 0 V to −10 V. Then, as the capacitor discharges from 10 V to zero, the resistor voltage will exponentially rise from −10 V to zero.

Once again, note that v_{OUT} contains the same rapid transitions as e_i, and that the exponential portions of v_{OUT} are opposite to those of v_{cap}. The v_{OUT} transitions, while they are of the same magnitude as the e_i transitions, start at 0 V and extend in both the positive and negative polarities. These transitions are often called *spikes.*

A high-pass circuit is often used to convert a pulse input into two spikes, one positive and one negative, corresponding to the input pulse transitions. When the circuit is used for this purpose, it is often called a *differentiator* circuit, another term derived from higher mathematics.

Example 3.14 What will happen to the output spikes of the *RC* differentiator in Figure 3–16 if τ is reduced?

Solution: A shorter τ will cause the capacitor voltage to charge and discharge more rapidly. Correspondingly, the resistor waveform, v_{OUT}, will have more rapidly decaying exponential portions. Figure 3–17 shows the result for $\tau = 0.05$ ms. Note that the output spikes are narrower and can be made even more narrow by reducing τ further.

Case III: $\tau < t_r/2\pi$

As the last example illustrated, as τ is reduced the output spikes will maintain the same amplitude but will have a shorter duration. This effect will continue as τ is reduced further, but only up to a point. At some point, illustrated in Figure 3–18, further reductions in τ will cause a *decrease* in the amplitude of the output spikes.

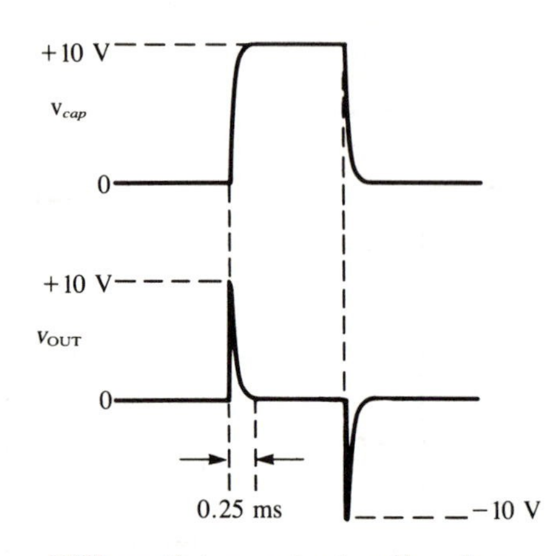

Figure 3–17 **Differentiator output spikes become narrower as τ is decreased (compare with Figure 3–16)**

When τ is made very small, we can no longer assume that the capacitor voltage will not change during the input transitions. As stated in the low-pass discussion, the capacitor can partially charge up during the input rise time when τ is very small. When this happens, the voltage across the resistor will necessarily be smaller than e_i since v_{cap} is greater than zero. A similar effect occurs during the input fall time.

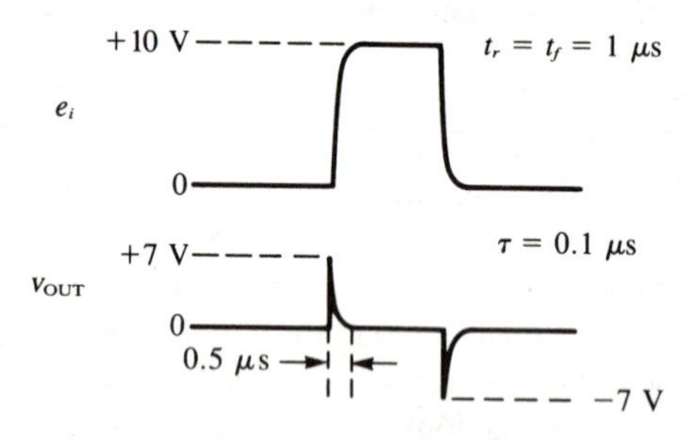

Figure 3–18 **Differentiator output spikes are reduced in amplitude when $\tau < t_r/2\pi$**

The reduction in spike amplitude will occur whenever

$$\tau < t_r/2\pi$$

There is no convenient formula for calculating the spike amplitude for these cases, and there is normally no need for one. The important thing is to know that the amplitude will diminish if τ is made too low. When a differentiator is designed to produce narrow spikes, τ should be kept small but not lower than $t_r/2\pi$ if the spike amplitude is to be preserved.

Example 3.15 An *RC* differentiator is to be designed to convert a 6 V pulse having $t_r = t_f = 4\ \mu s$ to two narrow 6 V spikes. The output of the differentiator is to drive an amplifier having $R_{IN} = 20\ k\Omega$. What is the smallest C that can be used?

Solution: The situation is depicted in Figure 3–19. The amplifier input resistance serves as the R portion of the *RC* circuit. The value of C should be chosen to ensure that

$$\tau > t_r/2\pi = 637\ \text{ns}$$

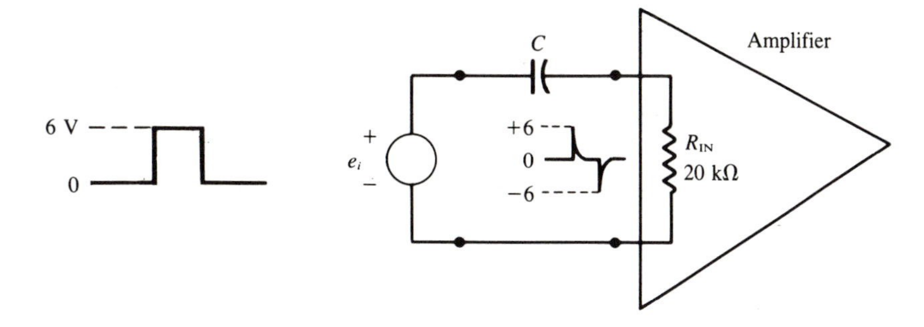

Figure 3–19 **Example 3.16**

so that

$$RC > 637 \text{ ns}$$

$$C > \frac{637 \text{ ns}}{20 \text{ k}\Omega}$$

$$C > \mathbf{31.85 \text{ pF}}$$

Any value of C greater than this value can be used. Of course, a larger C will increase the duration of the exponential portion of the spikes.

Example 3.16 What will happen to the signal at the amplifier input if the pulse source has an internal resistance of 1 kΩ?

Solution; The 1 kΩ source resistance will be in series with the 20 kΩ amplifier resistance, thereby producing a voltage divider effect. The signal across R_{IN} will be reduced by a factor of 20/21. Thus, the spikes will have an amplitude of approximately ±5.7 V.

The added resistance will also increase τ so that the capacitor charge and discharge will take longer, therefore increasing the duration of the spikes.

Summary

Figure 3–20 summarizes the pulse response of an RC high-pass circuit as τ decreases from very large to very small values. Note that the output is almost identical to the input for large values of τ and exhibits more and more tilt as τ decreases. Also note that the output spikes for small values of τ diminish in amplitude as τ decreases below $t_r/2\pi$.

We are now ready to investigate the response of RC circuits to periodic pulse inputs. Before we do that, however, we must understand the difference between a circuit's natural and forced responses.

3.11 Natural and Forced Responses

Every circuit containing at least one energy-storage element (capacitance or inductance) will exhibit a *natural response* to almost any input. This natural (transient) response is determined by the circuit time constants (many circuits have more than one time constant)

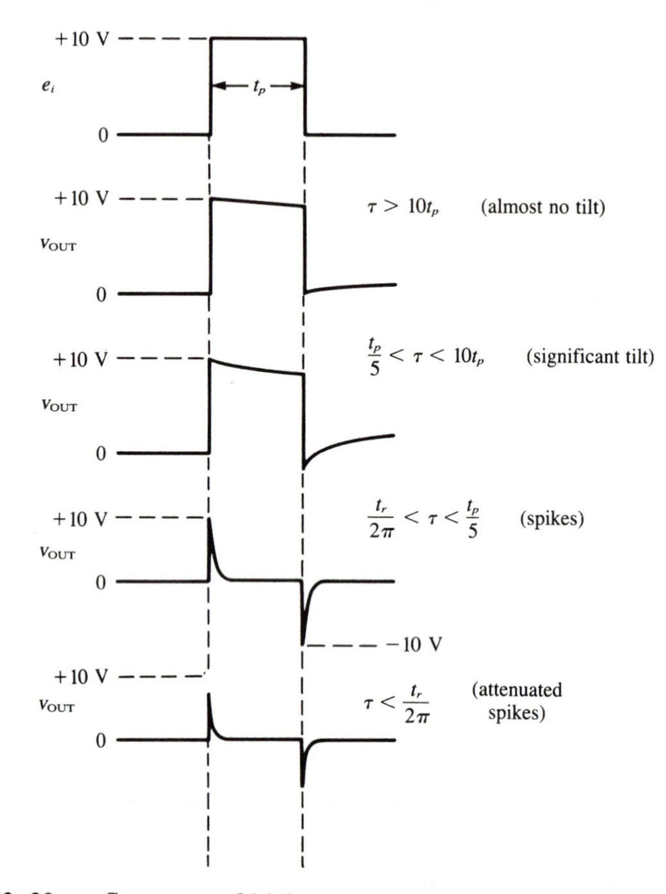

Figure 3–20 **Summary of high-pass pulse response**

and usually consists of one or more exponential components. The nature of the natural response is solely determined by the circuit components and *not* by the input. On the other hand, a *forced* response is caused by the input. In this case the nature of the circuit's *forced* (steady-state) response will depend on the input. In general, a circuit's total response to an input is made up of both a natural and a forced response. For example, in the case of an *RC* circuit being driven by a step voltage, the output across the capacitor will have an exponentially rising portion and then a constant portion. The exponential portion is due to the natural response of the circuit (depends on *RC*), while the constant portion is due to the fact that the input is constant.

Another example occurs when the input is a *sinusoidally* varying voltage. The ac circuit theory states that with a sinusoidal input the *RC* network will always have a sinusoidal output. This is exactly true only in steady state *after* the *natural* response of the circuit has died out. Actually, the instant after a sinusoidal input is applied to an *RC* circuit (or any linear reactive circuit), the output will exhibit an exponential (natural) response as well as a sinusoidal (forced) response. In practice, the natural-response portion is not often observed on the oscilloscope because it dies out within 5τ, while the forced response continues until the input is removed. A typical voltage output from an *RC* network whose input voltage is

$$v_i = A \cos \omega t$$

will be

$$v_o = B\cos(\omega t - \theta) + C_\epsilon^{-t/\tau}$$

The first term in the output expression is the forced response due to the input. The second term is the natural response of the *RC* circuit. A plot of this expression is given in Figure 3–21. Notice that the output does not reach its final sinusoidal shape until after 5τ. After 5τ, then, the circuit has reached its steady state.

The concept of steady state can now be more precisely defined as the time it takes for the natural responses of the circuit to die out. For a single time-constant circuit, steady state will always be obtained in 5τ, no matter what type of input is present. If any periodic input is applied to the circuit, the circuit's output will also become periodic (at the same frequency) after five time constants have passed.

In examining the *RC* circuit's response to periodic pulse inputs, the steady-state response will be the main concern. However, in order to provide a more complete picture, the technique involved in finding the *total* response will be presented in the first case under consideration. After that, only the steady-state responses will be discussed, since they are frequently the only ones of interest.

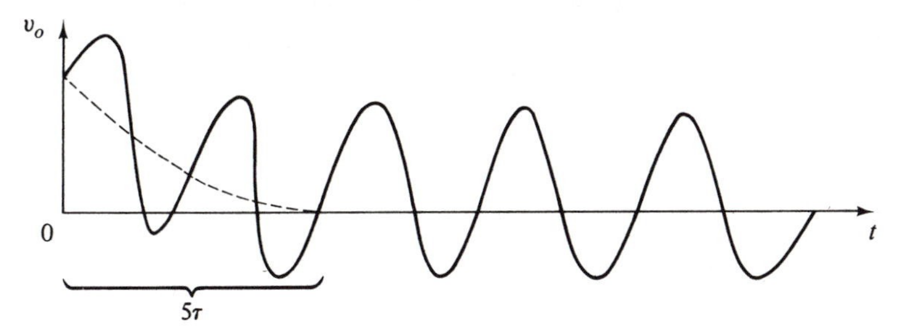

Figure 3–21 ***RC* circuit response to sinusoidal input**

3.12 *RC* Circuit Response to Periodic Inputs

Everything that we have said concerning the *RC* circuit response to single pulses is equally applicable to periodic pulse inputs. The response to periodic pulses, however, has the following additional characteristics:

(a) The output will reach steady-state (become periodic) after 5τ, regardless of the frequency, harmonic content, etc., of the input signal.

(b) The low-pass* output will have a dc component equal to the dc component of the input.

(c) The high-pass* output will not have a dc component. It will therefore alternate between being positive, then negative with respect to ground. We already saw evidence of this with the high-pass response to a single pulse.

The complete response of an *RC* circuit to a periodic pulse input has to be performed step-by-step after each input transition occurs. It can be a very tedious process,

*We are referring to the simple series *RC* circuit.

especially when τ is much longer than the input period (T), so that many cycles of the input have to occur before the outputs reach steady state. We will do one example to illustrate this process and to show how the waveforms reach their steady-state conditions.

Figure 3–22 shows a 1,000 pps square wave driving an *RC* circuit that has $\tau = 0.72$ ms. We will assume that the input transition times are very fast so that the square wave can be considered ideal. We will also assume that the capacitor is initially at 0 V. Thus at $t = 0$, when e_i jumps to +10 V, the capacitor starts charging according to the equation

$$v_{xy} = 10 - 10\epsilon^{-t/0.72 \text{ ms}}$$

After 0.5 ms it will have reached approximately 5 V. Thus, in 0.5 ms it will have charged to 50 percent of the final value it was heading towards.

When e_i suddenly jumps to zero at $t = 0.5$ ms, the capacitor begins to discharge toward zero. This time, however, it will not reach zero because e_i will remain at zero for only 0.5 ms. Thus, the capacitor will discharge for 0.5 ms. We can set up the exponential equation for the capacitor discharge and evaluate it after 0.5 ms. Alternatively, we can use the fact that the capacitor voltage will go 50 percent of the way toward its final value in 0.5 ms as we determined during the charging interval. Thus, v_{xy} will be 2.5 V at $t = 1$ ms.

At $t = 1$ ms, e_i jumps back to +10 V and the capacitor starts charging again, but this time starting at +2.5 V. Since it is heading toward 10 V, it has 7.5 V to go. In 0.5 ms,

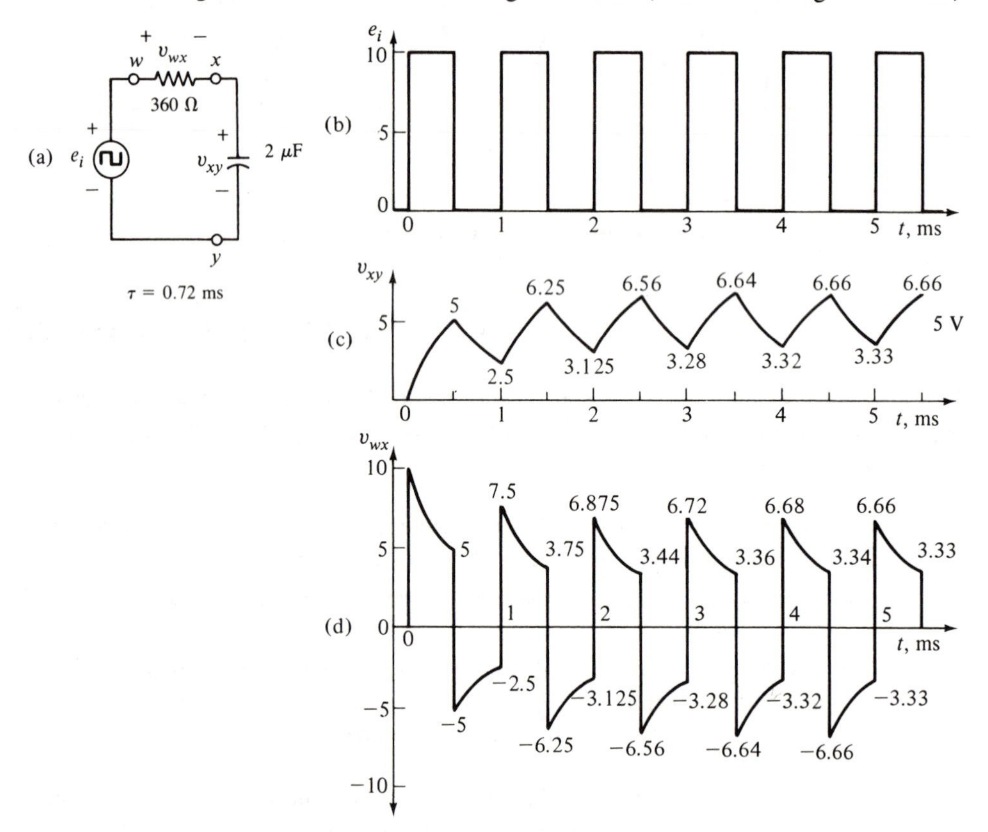

Figure 3–22 **Square wave driving an *RC* network, and resulting waveforms**

it will charge 50 percent of this distance so that at $t = 1.5$ ms it will be at 6.25 V when e_i again returns to zero.

At $t = 1.5$ ms, the capacitor begins discharging from 6.25 V toward zero and will discharge 50 percent of the way in 0.5 ms. Thus, at $t = 2$ ms, $v_{xy} = 3.125$ V.

The pattern should now be clear. Each time e_i jumps to 10 V for 0.5 ms, the capacitor charges 50 percent of the distance from its current value toward +10 V. Each time e_i jumps to 0 V for 0.5 ms, the capacitor will discharge 50 percent of the way toward zero. If we continue this pattern for the next few cycles of e_i, the result for v_{xy} appears as shown in the figure.

Note that after the fourth cycle, the v_{xy} waveform begins to repeat itself; that is, it reaches steady state. We could have predicted this occurrence since $5\tau = 3.6$ ms. In steady-state, v_{xy} will alternately charge and discharge between 6.66 V and 3.33 V.

The waveform of resistor voltage can now be determined using KVL as we did for the single pulse case. At $t = 0$, v_{xy} will follow the rapid transition of e_i to +10 V. Then, as the capacitor charges from 0 to 5 V, v_{wx} decays from 10 V to 5 V. At $t = 0.5$ ms, v_{wx} rapidly jumps from +5 V to −5 V in response to the rapid 10 V decrease of e_i. (Recall that the rapid transitions of e_i appear across the resistor because the capacitor voltage cannot change instantaneously.)

Between $t = 0.5$ ms and $t = 1.0$ ms, v_{wx} gradually rises from −5 V to −2.5 V as the capacitor voltage drops from +5 V to +2.5 V. At $t = 1$ ms, v_{wx} again responds to the positive transition of e_i by jumping from −2.5 V to +7.5 V (a + 10 V change).

If we continue this procedure for the v_{wx} waveform, the result will be as shown in the figure. After approximately four input cycles, the waveform reaches steady state and becomes repetitive.

Now that we have determined the waveforms, we should note several important points.

(a) The capacitor waveform in steady state consists of increasing and decreasing exponential portions alternating between the limits of 6.66 V and 3.33 V. Thus, the average or dc value of the v_{wx} waveform is approximately 5 V. This value is exactly the same as the average or dc value of the input. This situation is to be expected because the capacitor will act as an open circuit to dc and will not allow dc current to pass in steady state. Therefore no dc voltage component will appear across the resistor, which can be seen by an examination of the v_{wx} waveform. Since the v_{wx} waveform in steady state consists of equal positive and negative portions, its average or dc value is zero.

(b) The low-pass output has very long t_r and t_f because τ is so large compared to the input transition times.

(c) The high-pass output has fast transitions like the input but has significant tilt because $\tau < 10t_p$ (or, alternatively, because $f_{lc} > 1/63t_p$).

Example 3.17 What will happen to the *time* required for the waveforms to reach steady state if the input frequency is increased?

Solution: Nothing, since τ is still the same. But more cycles of the input will occur during the transient interval before steady state is reached.

The example of Figure 3–22 produces two distorted output waveforms because τ is too large to produce an undistorted low-pass output and too small to produce an undistorted high-pass output. Recall from our discussion of the single pulse cases that the low-pass output will become less distorted as τ is made smaller and more distorted as τ is increased. Conversely, the high-pass output will become more distorted as τ decreases and less distorted as τ increases. These same ideas hold true for periodic pulse inputs. Figure 3–23 shows examples of the various cases for both low-pass and high-pass outputs in *steady-state*. Study them carefully.

Example 3.18 What should the steady-state output look like in Figure 3–24? How long will it take for v_{OUT} to reach steady-state?

Solution:

$$\tau = 4.7\ \text{k}\Omega \times 20\ \mu\text{F}$$
$$= 94\ \text{ms}$$

Since $\tau \gg 10t_p$, the low-pass (integrator) output will be almost pure dc equal to the input dc component. The input has a 25 percent duty cycle, so its dc average value is

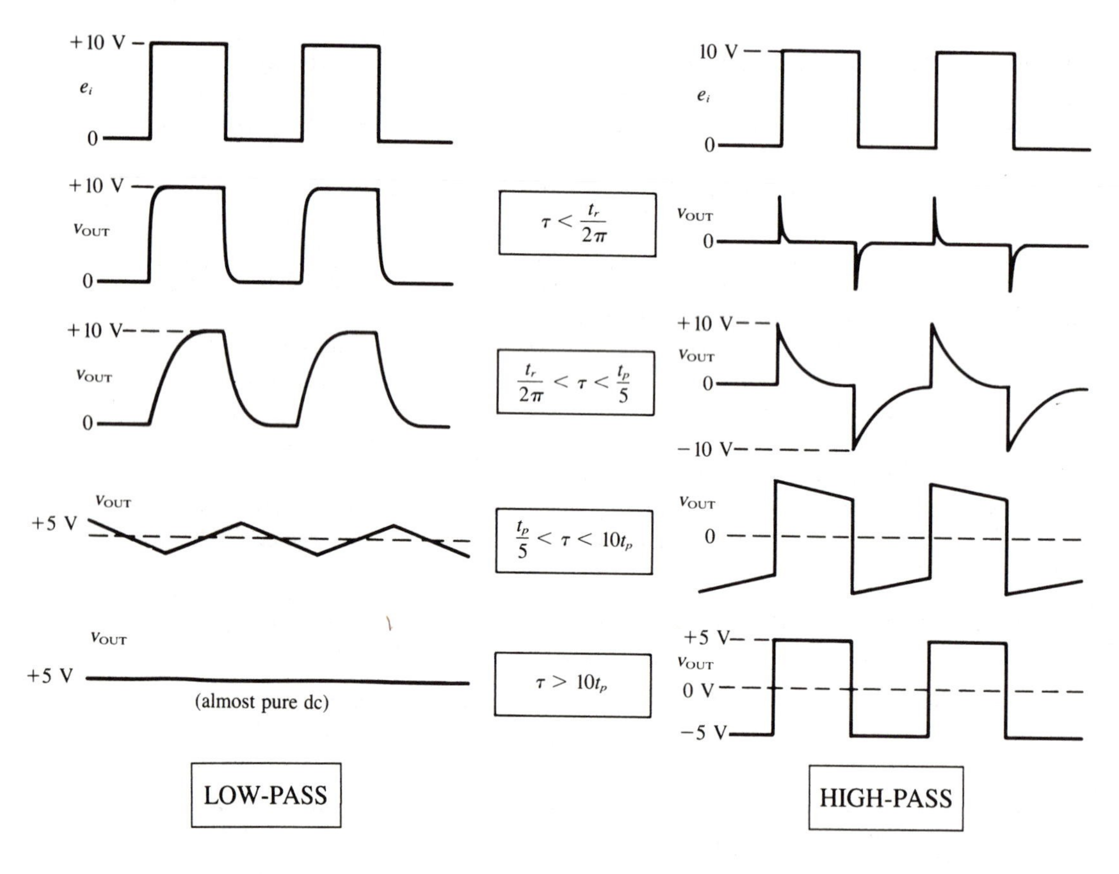

Figure 3–23 **Comparison of low-pass and high-pass responses**

$$25\% \times 10 \text{ V} + 75\% \times 2 \text{ V} = 4 \text{ V}$$

Thus, v_{OUT} will be an almost constant +4 V DC level with a small variation caused by the fact that the capacitor will charge and discharge somewhat no matter how large the value of τ. The output will reach this steady-state condition after an interval of $5\tau =$ **470 ms.**

This integrator circuit with large τ is used whenever it is desired to extract only the dc component of the input while eliminating all of the other frequency components.

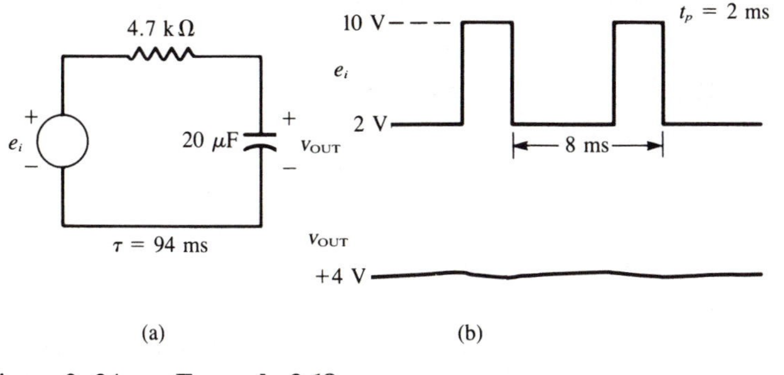

Figure 3–24 **Example 3.18**

Example 3.19 What will the steady-state output look like in Figure 3–25 if the input has $t_r = t_f = 10$ ns?

Solution:

$$\tau = 15 \text{ k}\Omega \times 510 \text{ pF}$$
$$= 7.65 \ \mu\text{s}$$

Since $t_r/2\pi < \tau < t_p/5$, the high-pass output will consist of positive and negative spikes coinciding with the transitions of e_i. The amplitude of these spikes will be 18 V since the e_i transitions are 18 V jumps. Note that v_{OUT} has no dc component even though e_i does. This differentiator is used to couple only the fast transitions of the input to the output.

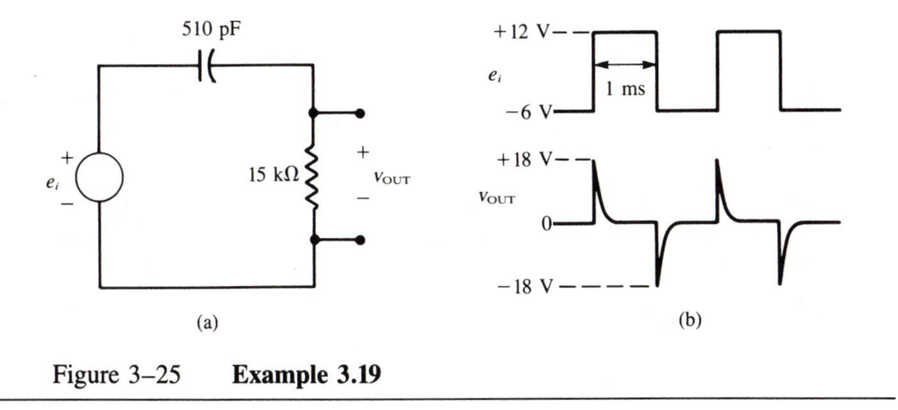

Figure 3–25 **Example 3.19**

Example 3.20 Figure 3–26 shows what happens when an oscilloscope vertical amplifier is set for *AC* input coupling. A series dc blocking capacitor is inserted between the input signal and the vertical amplifier, which has $R_{IN} = 1\ M\Omega$. Clearly this forms a high-pass circuit that, under certain conditions, could produce tilt on a pulse waveform. This tilt will show up on the scope display.

If $C = 1\ \mu F$, determine the maximum pulse duration that can be applied to the scope input without producing significant tilt (>10 percent).

Solution:

$$\tau = 1\ M\Omega \times 1\ \mu F$$
$$= 1\ s$$

For no significant tilt, we want

$$\tau > 10t_p$$

or

$$t_p < \frac{\tau}{10} = 0.1\ s$$

Thus, the maximum t_p that can be used without causing significant tilt is **0.1 s.**

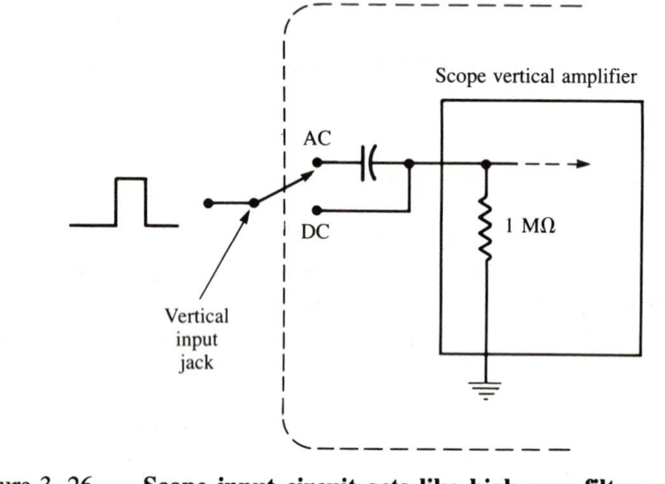

Figure 3–26 **Scope input circuit acts like high-pass filter when AC coupling is used on vertical input**

Example 3.21 Figure 3–27 shows a pulse signal source driving a logic circuit. The input resistance of the logic circuit is very large and can be neglected. The input capacitance is approximately 10 pF. The pulse source has $t_r = t_f = 10$ ns. Determine the largest value of source resistance, R_s, which should be used if the pulse is to reach the logic circuit input without a significant increase in its transition times.

Solution: R_s and C_{IN} form a low-pass circuit. For no high frequency distortion, we want

$$\tau \leqslant \frac{t_r}{2\pi} = 1.59\ ns$$

Thus,

$$R_s \times 10\text{ pF} \leq 1.59\text{ ns}$$

or

$$\mathbf{R_s \leq 159\ \Omega}$$

Example 3.22 With $R_s = 100\ \Omega$ in Figure 3–27, a technician measures a rise time of 23 ns at the logic circuit input instead of the expected 10 ns. What are some possible reasons for such a large discrepancy?

Solution: The two most likely reasons involve the characteristics of the oscilloscope that was used to make the measurement. The first reason could be the effect of scope input capacitance. When the scope is connected across the logic circuit input, the scope capacitance is in parallel with the logic circuit C_{IN}. This connection increases the total circuit capacitance and, therefore, τ. This effect can be reduced by using a low capacitance scope probe.

The second reason could be that the oscilloscope amplifier's bandwidth is too small, attenuating the pulse's high frequency harmonics before the pulse is displayed on the CRT. This situation can be avoided by using a scope with sufficiently large bandwidth.

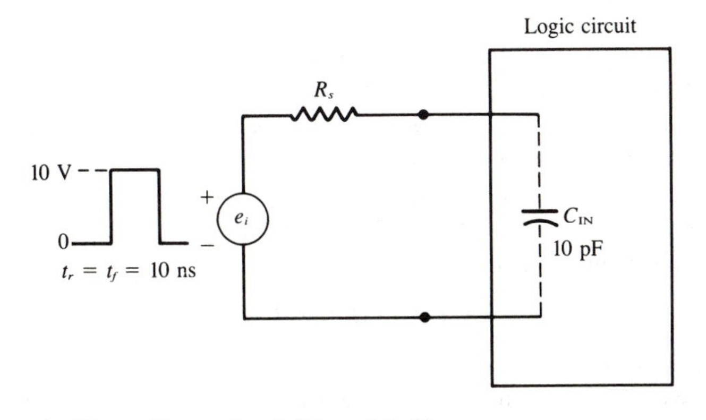

Figure 3–27 **Examples 3.21 and 3.22**

3.13 Loading Effects

We will now examine the effects of various kinds of loading on some of the circuits we have been analyzing. It is important that a technician recognize when loading is taking place in a circuit, and be able to anticipate when loading might be a problem. The situation of Example 3.22 is a very typical kind of loading problem. In general, there are three types of loading that are the most common: resistive, capacitive, and a combination of the two.

A resistive load across a circuit's output will generally cause *attenuation* of the output signal. Of course, attenuation depends on the size of the load resistance as compared to the circuit's internal resistance. The load resistance will also reduce the circuit time constant, and this reduction can affect the shape of the output signal. For our first example consider the circuit of Figure 3–28, which is being driven by a step signal source having an internal resistance of 1 kΩ. The circuit is functioning as a high-pass differentiator, which is producing a spike output corresponding to the input transition. We can assume that the input transition is very fast compared to the circuit time constant.

Ignoring the load resistance for now, the analysis proceeds as follows. At $t = 0$ when the input jumps to 10 V, the capacitor will momentarily hold at 0 V. The 10 V source voltage must divide between the 1 kΩ source resistance and the 10 kΩ output resistance. Thus at $t = 0$ the output voltage v_{xy} will jump to

$$v_{xy}(t = 0) = \left(\frac{10\ \text{k}\Omega}{10\ \text{k}\Omega + 1\ \text{k}\Omega}\right) \times (10\ \text{V})$$
$$= 9.1\ \text{V}$$

This result means that the output voltage (v_{xy}) will immediately jump to 9.1 V in response to the input step. As the input remains at 10 V, the 0.001 μF capacitor will begin to charge toward 10 V with a time constant

$$\tau = (0.001\ \mu\text{F}) \times (11\ \text{k}\Omega) = (1 \times 10^{-9}) \times (11 \times 10^{3})$$
$$= 11 \times 10^{-6}$$
$$\tau = 11\ \mu\text{s}$$

In 55 μs (5τ) the capacitor will have fully charged, and the circuit will be in steady state. The output voltage during the 55 μs transient interval will be exponentially decreasing toward zero as shown in part (c) of Figure 3–28. The v_{xy} output is a 9.1 V spike with a very fast rise time and a falling portion that reaches steady state in 55 μs.

Now consider what happens when the output of this circuit is subjected to a 5 kΩ load [shown dotted in Figure 3–28(a)]. The 5 kΩ load essentially changes the output resistance between points x and y to

$$5\ \text{k}\Omega \,\|\, 10\ \text{k}\Omega = 3.3\ \text{k}\Omega$$

Therefore, the previous analysis can be applied using 3.3 kΩ instead of 10 kΩ. The resulting v_{xy} waveform is shown in Figure 3–28(d). As can be seen, the output spike has an amplitude of only 7.7 V. This could be easily calculated as follows:

$$\left(\frac{3.3\ \text{k}\Omega}{3.3\ \text{k}\Omega + 1\ \text{k}\Omega}\right) \times 10\ \text{V} = 7.7\ \text{V}$$

Note also that the v_{xy} waveform reaches steady state in only 21.5 μs because the load resistor has lowered the time constant of the circuit to

$$\tau = (3.3\ \text{k}\Omega + 1\ \text{k}\Omega) \times (0.001\ \mu\text{F})$$
$$= 4.3\ \mu\text{s}$$

The total effect of the load resistor, then, is to cause a decrease in both the amplitude and duration of the output spike. As the value of R_L is decreased, these effects become more pronounced.

Example 3.23 If R_S were reduced, how would this help eliminate some of the loading effects?

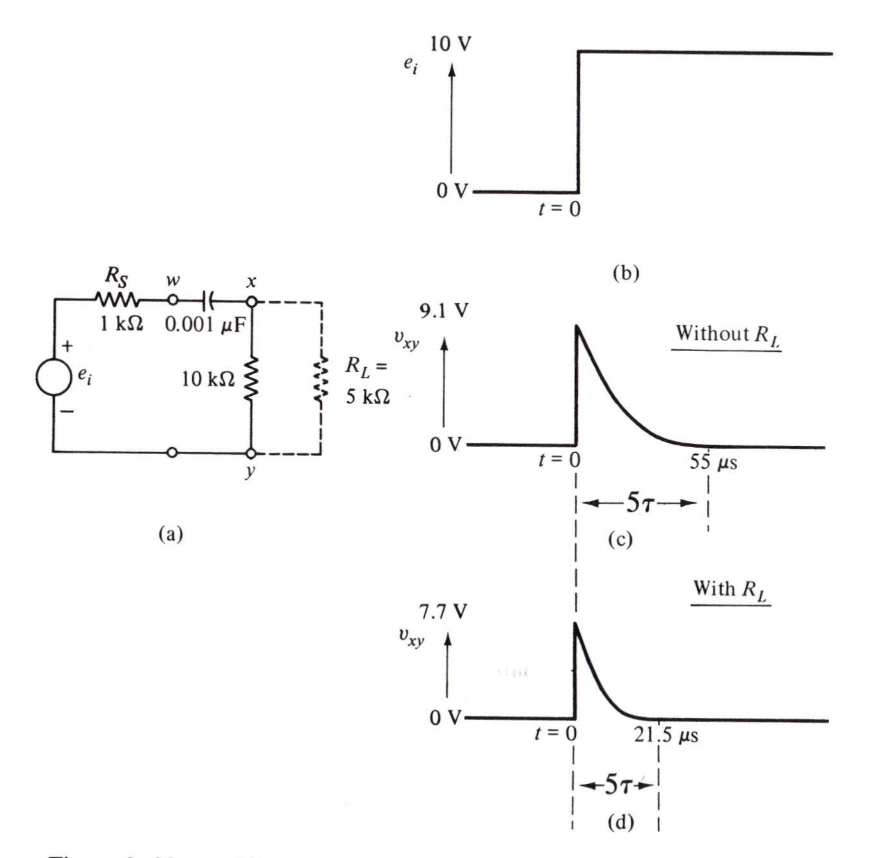

Figure 3–28 **Effect of resistive loading on a high-pass circuit**

Solution: Decreasing the source resistance will help to stabilize the output *amplitude* against changes in output resistance due to loading. This is because a smaller value of R_S reduces the effect on the voltage-divider action caused by the output resistance. For example, if R_S is decreased to 100 Ω, the output pulse amplitudes in the two cases of Figure 3–28 would be 9.9 V and 9.7 V, respectively (these values should be verified by the student). Of course, in the impractical situation of $R_S = 0$, the output amplitude would *always* be 10 V.

The decrease in R_S will not, however, eliminate the effect of a load resistance on the output pulse duration. The load will always cause a decrease in the output resistance and thus a decrease in τ.

Capacitive loading When a capacitive load is placed across a circuit output, it will tend to slow down the output signal transitions. Of course, this effect is more pronounced as C_L becomes larger. In fact, if C_L is small enough, it will have no noticeable effect on the output transition times. A very common situation is shown in Figure 3–29. Here a voltage divider is being used to reduce the amplitude of the step input signal. The input step has a rise time of 0.1 μs and a 20 V amplitude.

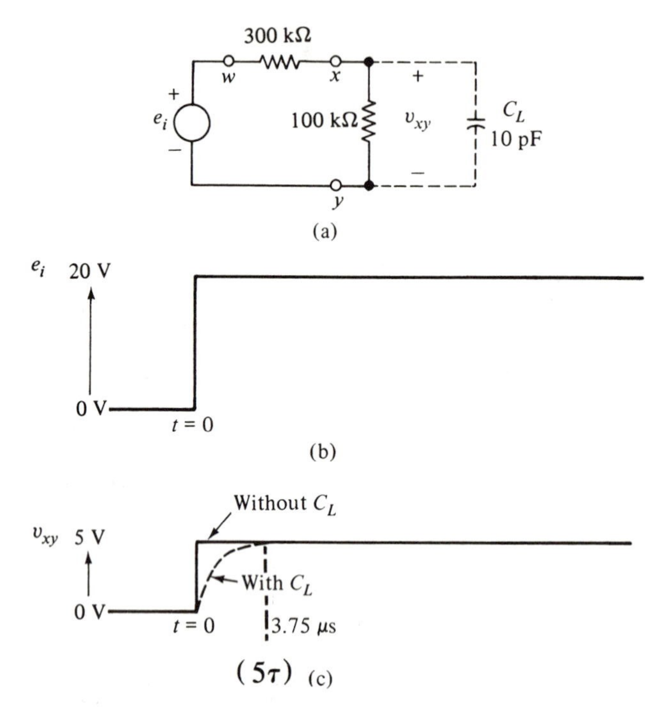

Figure 3–29 **Voltage-divider network with capacitive load**

Without the capacitive load, the output signal at v_{xy} will be an exact duplicate of e_i except that it will be reduced in amplitude by the factor

$$\frac{100\text{ k}\Omega}{100\text{ k}\Omega + 300\text{ k}\Omega} = 0.25$$

Thus, v_{xy} will be a 5 V step as shown in Figure 3–29(c).
With a 10 pF capacitor load across the output terminals, the circuit becomes a low-pass circuit with a time constant given by

$$\begin{aligned}\tau &= (300\text{ k}\Omega \,\|\, 100\text{ k}\Omega) \times (10\text{ pF}) \\ &= 75\text{ k}\Omega \times 10\text{ pF} = 0.75\ \mu\text{s}\end{aligned}$$

Therefore, in response to the 10 V step input, the v_{xy} output will rise toward 5 V exponentially and will reach steady state in $5\tau = 3.75\ \mu$s (see dotted v_{xy} waveform in Figure 3–29). The effect of the capacitor load, then, is to prevent the output (v_{xy}) from accurately following the rapid transitions of the input. The voltage-divider ratio is still 0.25, as far as the low frequencies are concerned; but the high frequencies are attenuated to a greater degree. In most cases such as this, the resultant distortion is unacceptable and must be compensated for.

One effective means of compensating for the effects of a capacitive load is illustrated in Figure 3–30, where the same circuit of Figure 3–29 is shown except for a compensating capacitor (C_c), which has been placed across the 300-kΩ leg of the voltage divider. The compensating capacitor is used to try to restore the voltage-divider ratio to its original value (0.25) *for all frequencies*. This can be done if C_c is adjusted so that its impedance is three times greater than that of C_L. If this is done, the total impedance Z_{wx}

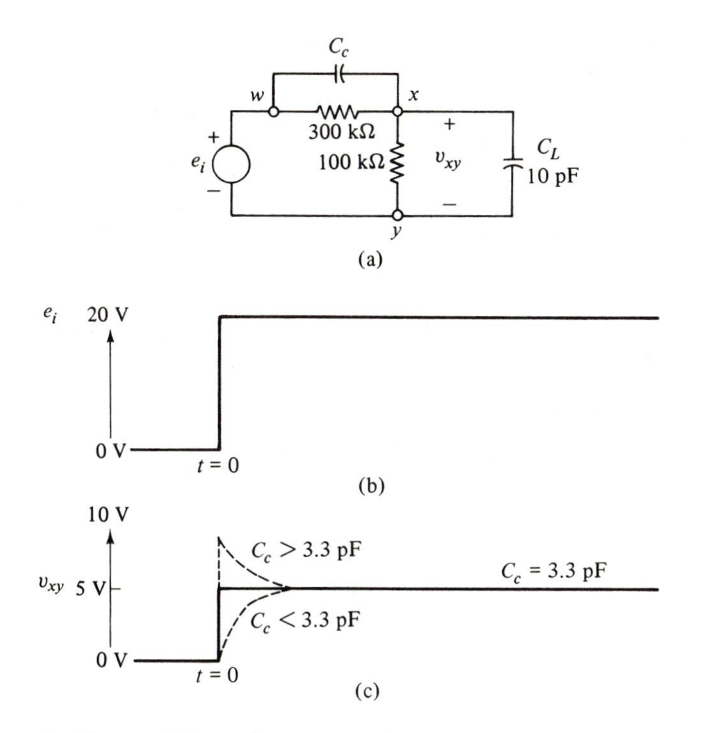

Figure 3–30 **Effect of compensating capacitor on output waveform**

between w and x will be three times greater than the total impedance Z_{xy} between x and y at all frequencies, because both the resistive and reactive components of Z_{wx} will be three times greater than those of Z_{xy}. To obtain this result, then, the capacitance of C_c must be one-third of C_L, since capacitive impedance is *inversely* proportional to capacitance. Therefore, $C_c = C_L/3 = 3.3$ pF for perfect compensation. *The* RC *product of each leg in the voltage divider is the same*. That is,

$$300\ \text{k}\Omega \times 3.3\ \text{pF} = 0.99\ \mu\text{s} \approx 1\ \mu\text{s}$$
$$100\ \text{k}\Omega \times 10\ \text{pF} = 1\ \mu\text{s}$$

This is always the case if perfect compensation is to be obtained.

To summarize, if $C_c = 3.3$ pF, then

$$Z_{wx} = 300\ \text{k}\Omega \| 3.3\ \text{pF}$$
$$Z_{xy} = 100\ \text{k}\Omega \| 10\ \text{pF}$$

which results in

$$Z_{wx} = 3Z_{xy}$$

and

$$v_{xy} = \left(\frac{Z_{xy}}{Z_{xy} + Z_{wx}}\right)e_i$$
$$= 0.25e_i$$

at *all* frequencies. Thus, no matter what the input is, the output will be exactly the same except for the 0.25 attenuation [see solid waveform in Figure 3–30(c)]. The correct value of C_c is quite critical and is usually accomplished using a variable capacitor that is adjusted until the desired output shape is obtained. If C_c is too large, too much of the high frequency content of the input will be passed to the output; the output will then exhibit overshoot as shown in Figure 3–30(c). If C_c is too small, not enough of the input high frequency content will reach the output. The result will be an output with slower transitions than the input.

Example 3.24 If C_L is increased to 30 pF, and the 100 kΩ resistor is decreased to 50 kΩ, determine the value of C_c needed for compensation.

Solution: Using the fact that the *RC* products of each leg must be equal,

$$300 \text{ k}\Omega \times C_c = 50 \text{ k}\Omega \times 30 \text{ pF}$$

or

$$C_c = \frac{50 \text{ k}\Omega \times 30 \text{ pF}}{300 \text{ k}\Omega} = 5 \text{ pF}$$

Compensated oscilloscope probe One of the major applications of the compensated voltage divider is the 10X oscilloscope probe, illustrated in Figure 3–31. The 10X probe increases the effective scope input resistance by placing a 9 MΩ resistor in series with the oscilloscope's 1 MΩ input resistance. The total of 10 MΩ at the probe tip will produce less of a resistive loading effect on the circuit being measured than the 1 MΩ. Inclusion of the 9 MΩ in the probe, however, produces a voltage divider that attenuates the signal by a factor of 0.1 before it reaches the scope amplifier. This attenuation would not be a problem if it were not for the scope's input capacitance (shown as 45 pF).* The presence of this capacitance will cause high frequency distortion of any pulses applied to the probe tip.

To minimize this distortion, a compensating capacitor is placed across the 9 MΩ resistance in the probe. In this case the correct value for C_c will be 5 pF (reader should verify). In practice, the scope input capacitance is not accurately known and will vary from one scope to another. For this reason, C_c is made a variable capacitor, which has to be adjusted by the operator before using the scope for any pulse measurements. The adjustment is made while applying a fast-transition pulse signal to the probe and varying C_c for the best displayed waveform.

The inclusion of C_c has an additional beneficial effect. Since it is effectively in series with C_{IN}, the total capacitance at the probe tip will be reduced. For example, with $C_c = 5$ pF and $C_{IN} = 45$ pF,

$$C_{IN}(\text{probe}) = \frac{5 \text{ pF} \times 45 \text{ pF}}{5 \text{ pF} + 45 \text{ pF}} = 4.5 \text{ pF}$$

Thus, the 10X probe *reduces* C_{IN} by a factor of *ten*. This reduction means the scope input capacitance will have less of a loading effect on the circuit being measured.

---

*We will assume that the cable capacitance is included in C_{IN}.

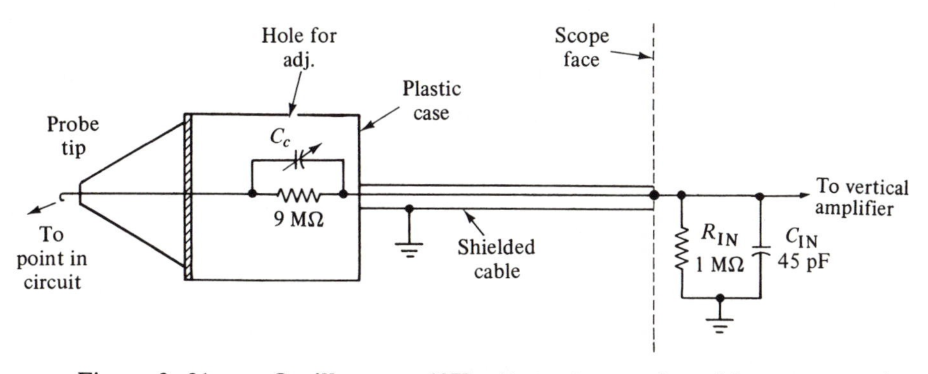

Figure 3–31 **Oscilloscope 10X attenuator probe with compensating capacitor**

Imperfect compensation Before leaving this topic, it should be pointed out that perfect compensation is rarely possible due to the presence of source resistance. The internal resistance of the source is generally not accessible and so cannot be paralleled with a compensating capacitor. Thus, there will be some degree of high frequency distortion in a compensated voltage divider, but it will be much less than in an uncompensated voltage divider.

3.14 Inductive Effects

In addition to resistance and capacitance, every circuit contains some amount of inductance. Most often, circuit inductance in pulse circuitry is of the *parasitic* type; that is, it is not deliberately placed in the circuit but is present in the wiring. In some applications inductors are used if a short time constant is called for. They are seldom used in a long time-constant application because a large value of inductance can be obtained only with an iron-core inductor, which is much larger, heavier, and more expensive than the corresponding capacitor for a similar application. Furthermore, such a large inductor would be shunted with a significant amount of interwinding capacitance, and it would also possess nonlinear characteristics because of the iron core. Both of these effects are usually undesirable.

Small, inexpensive air-core inductors are used in some short time-constant applications. Even when inductance is not purposely used, however, the effects of parasitic inductance can be significant in circuits where current is being switched at a rapid rate. For example, if a current changes from 0 mA to 100 mA in 0.1 μs ($di/dt = 10^6$ A/s), it will induce 1 V in a wire that has only 1 μH inductance ($v = L\, di/dt$). This occurrence is especially important in high-speed logic circuits, where rapid changes in current flowing over the power and ground lines can induce significant voltage spikes in the wire inductance. These voltage spikes act as noise (unwanted signals) that can seriously affect circuit operation.

Ringing Another effect of parasitic inductance is ringing, a phenomenon occurring under certain conditions in a circuit that contains both capacitance and inductance. Consider the circuit of Figure 3–32(a), where a fast-rising step waveform is applied to a circuit that contains a small series wiring inductance and a small load capacitance. The series resistance is the signal source resistance.

The output voltage across the capacitor will be a step waveform that exhibits overshoot and ringing before settling down to its steady-state 10 V level, as shown in Figure 3–32(b). The ringing represents sinusoidal oscillations that gradually diminish in amplitude.

Ringing will occur whenever the circuit R, L, and C values satisfy the criterion

$$\frac{1}{R}\sqrt{\frac{L}{C}} > \frac{1}{2}$$

The left-hand expression may be recognized as the series RLC circuit's frequency selectivity factor, Q. Thus, ringing will be present whenever

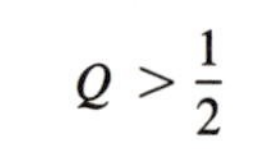

$$Q > \frac{1}{2}$$

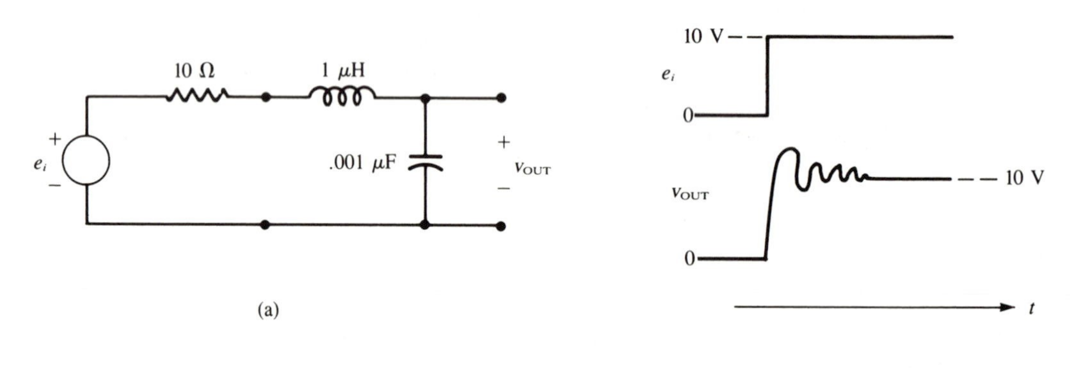

Figure 3–32 **Ringing appears on v_{OUT} waveform when series *RLC* has Q > 0.5**

This criterion is easily satisfied by the values in Figure 3–32.

$$Q = \frac{1}{R}\sqrt{\frac{L}{C}} = \frac{1}{10}\sqrt{\frac{10^{-6}}{10^{-9}}} = 3.16$$

If Q is lower than 0.5, the output will have smooth transitions with no overshoot or ringing. In many cases, overshoot and ringing are undesirable because of the effect they may have on the circuit being driven. One way to eliminate overshoot and ringing is to reduce Q to below 0.5 by adding series resistance. This strategy is called *damping out* the ringing oscillations.

Example 3.25 How much resistance should be added to the circuit of Figure 3–32 to eliminate overshoot and ringing?

Solution: We want

$$\frac{1}{R}\sqrt{\frac{L}{C}} \leqslant \frac{1}{2}$$

for no ringing. Thus,

$$\frac{1}{R}\sqrt{\frac{10^{-6}}{10^{-9}}} \leqslant \frac{1}{2}$$

or

$$R \leqslant 63.2\ \Omega$$

This means that we have to add **53.2 Ω** to the 10 Ω already in the circuit.

Addition of too much series resistance to damp out the ringing will have a detrimental effect on the output rise time, since the addition slows down the charging of the capacitance. For this reason the total circuit resistance should not be made too much larger than the value needed to reduce the overshoot to an acceptable level. A good rule of thumb is that $Q = 0.8$ will produce a 10 percent overshoot without a significant increase in rise time.

Figure 3–33 shows the pulse response of a series *RLC* circuit with $Q = 0.8$. Note that there is an overshoot and some ringing on each output transition. The overshoot is approximately 10 percent on each transition, which is usually acceptable. If it is not, then Q must be reduced further, at the price of longer transition times.

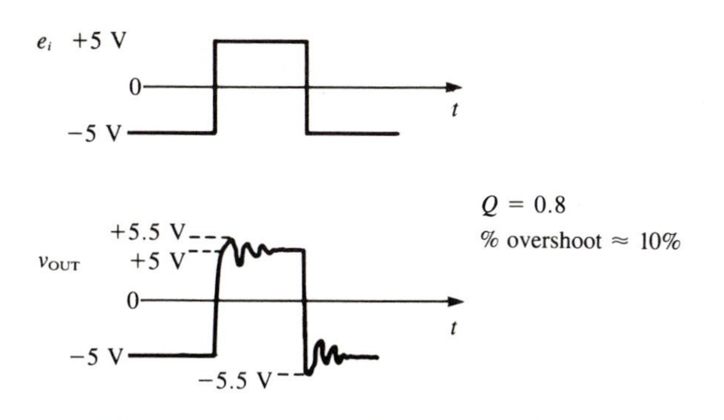

Figure 3–33 ***RLC* response to an input pulse when Q = 0.8**

Questions

Sections 3.1–3.4

3.1 Determine the time constant of the circuit of Figure 3–34, and sketch the v_{xy} and v_{wx} waveforms in response to the switch closure.

3.2 What will be the approximate values of v_{wx} and v_{xy} at a time equal to one time constant after the switch is closed? After *two* time constants?

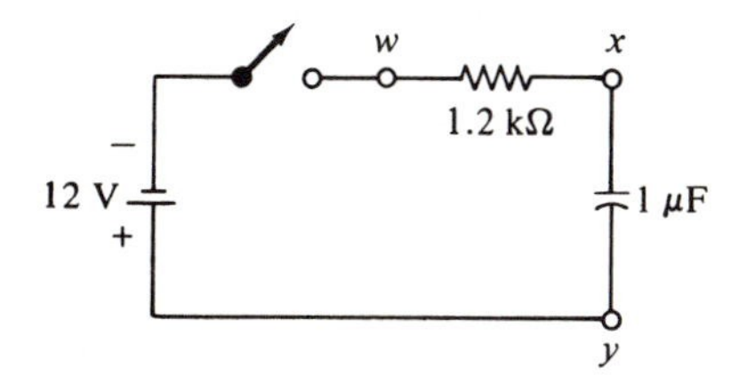

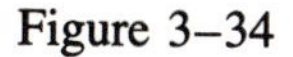
Figure 3–34

3.3 Sketch the waveforms of v_{xy} and i in response to the switch opening in Figure 3–35.

3.4 As the capacitor in Figure 3–35 discharges through the 150 Ω resistor, power is being lost in the resistor. If the 6 V source is now out of the circuit, where is the power coming from and how did it get there?

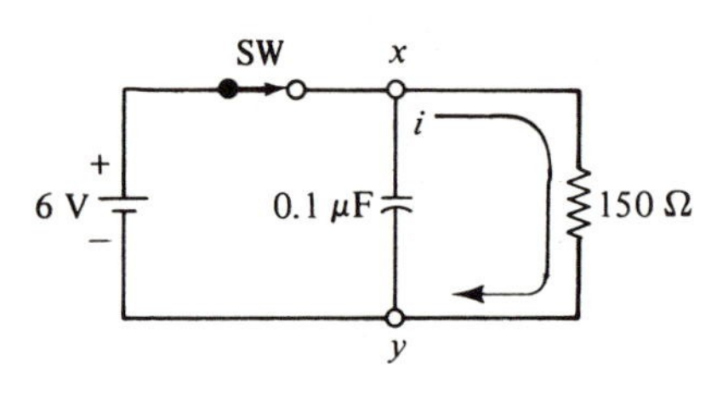

Figure 3–35

3.5 Determine the following values: $\epsilon^{-1.3}$, $\epsilon^{-0.25}$, $\epsilon^{-6.7}$, $\epsilon^{-0.02}$.

3.6 Determine the value of x where $\epsilon^{-x} = 0.25$.

3.7 Determine the mathematical expression for v_{xy} in Figure 3–34.

3.8 Use the result of Problem 3.7 to determine the value of v_{xy} at $t = 1$ ms and at $t = 10$ ms.

3.9 Use the expression from Problem 3.7 to determine how long it will take the capacitor to reach −7.7 V.

3.10 Determine the mathematical expression for the current in the circuit of Figure 3–35.

3.11 Consider the circuit in Figure 3–36. Before the switch is closed, the capacitor has been charged to 10 V. At $t = 0$ the switch is closed. Determine the expression for the resistor voltage (v_{xy}), and plot it versus time.

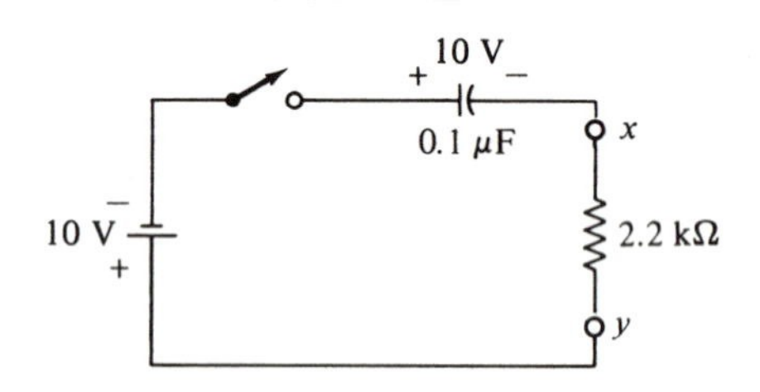

Figure 3–36

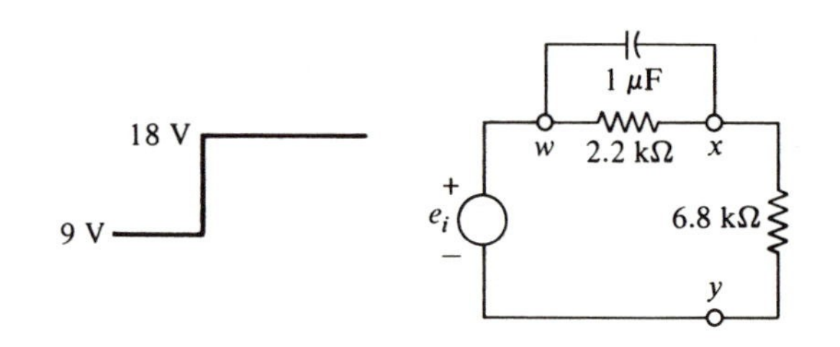

Figure 3–37

3.12 In the circuit of Figure 3–37, the input is an ideal step that is initially at 9 V and jumps to 18 V at $t = 0$. Sketch the waveforms of v_{xy} and v_{wx} in response to this step. Assume that e_i has been at 9 V for a long time.

3.13 Determine the exponential expressions for v_{xy} and v_{wx} in response to the step input in Figure 3–37.

Sections 3.7–3.9

3.14 If the output is taken at v_{xy} in Figure 3–37, will the circuit be a low-pass or a high-pass? Determine its 3-dB cutoff frequency.

3.15 A signal source with $R_s = 500\ \Omega$ produces pulses with $t_r = t_f = 100$ ns. Determine the maximum load capacitance that this signal source can drive before the pulse transitions are significantly distorted.

3.16 Determine the v_{xy} waveform in Figure 3–38 for each of the following cases. Assume e_i has been at 2 V for a long time prior to the positive transition.

(a) $R = 1\ \text{k}\Omega, \quad C = 1\ \mu\text{F}$

(b) $R = 100\ \Omega, \quad C = 1\ \mu\text{F}$

(c) $R = 10\ \text{k}\Omega, \quad C = 2\ \mu\text{F}$

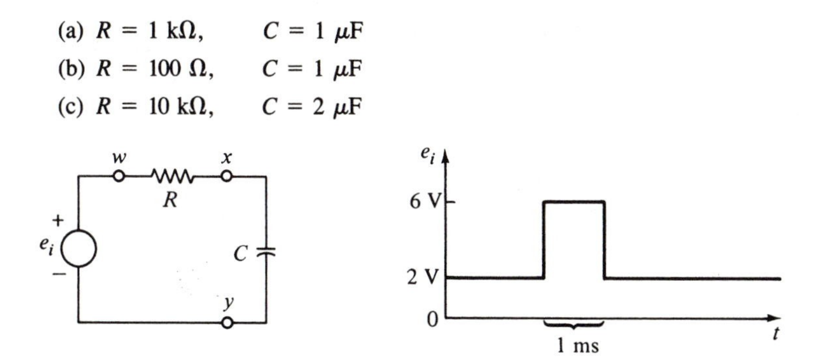

Figure 3–38

3.17 Determine t_r and t_f for the v_{xy} waveform of Problem 3.16(b).

3.18 Figure 3–39 shows a typical situation in digital logic circuits. Each circuit being driven has an extremely high input resistance (1,000 MΩ) that can be neglected, and a small input capacitance of 5 pF. The total load capacitance on the signal output can significantly slow down the signal transitions if too many loads are being driven.

Assume that the circuit driving the loads has an output resistance of 1 kΩ and produces pulses with $t_r = t_f = 2$ ns under a no-load condition. Determine how many loads it can drive before the pulse transition times increase to 100 ns.

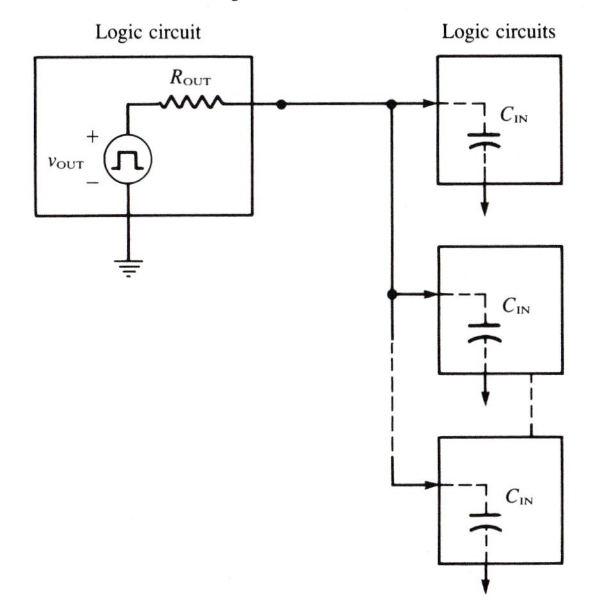

Figure 3–39

3.19 When a *narrow* pulse is applied to an *RC* low-pass circuit, the capacitor output will have a very small peak value compared to that of the input pulse. This output amplitude will be almost directly proportional to the input pulse duration as long as $t_p < \tau/20$. Prove this fact by calculating the peak voltage of the low-pass output response to a 10 V input pulse for each of the following cases:

(a) $t_p = 0.01\tau$
(b) $t_p = 0.02\tau$
(c) $t_p = 0.03\tau$
(d) $t_p = 0.04\tau$
(e) $t_p = 0.05\tau$

Section 3.10

3.20 Using the results of Problem 3.16, determine the v_{wx} waveforms for the three cases of Problem 3.16.

3.21 Determine the percentage tilt in the v_{wx} waveform of Problem 3.20(c).

3.22 The input pulse in Figure 3–40(a) is being converted to spikes by the differentiator circuit.

(a) Draw the output waveform, showing the amplitude and duration of the spikes. Assume e_i has been at 5 V for a long time.
(b) Using the pulse in Figure 3–40(b) as the input to the same circuit, draw v_{OUT}.
(c) Compare the results of (a) and (b). Why are they the same?

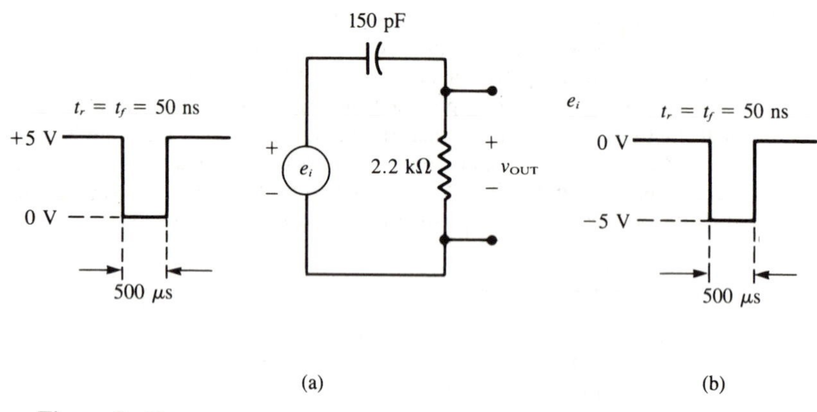

Figure 3–40

3.23 Which of the following will cause a *decrease* in the *amplitude* of the output spikes in Figure 3–40?

(a) increase of C to 330 pF
(b) increase of R to 3.3 kΩ
(c) decrease of C to 2 pF
(d) source resistance of 600 Ω

3.24 What should be done to the value of the capacitor in Figure 3–40(a) if the output is to be a pulse with virtually no tilt? Draw the v_{OUT} waveform for this situation.

3.25 For each of the following statements, indicate whether it pertains to an *RC* low-pass circuit, high-pass circuit, both, or neither, in response to a pulse input.

(a) The output will have less and less distortion as τ is *decreased*.
(b) The output will exhibit considerable tilt if $\tau \gg 10\, t_p$.

(c) τ should be made very large to minimize output distortion.

(d) The output can have both positive and negative portions even if the input is of one polarity.

(e) The output will become smaller in amplitude as t_p is reduced below 5τ.

(f) The output amplitude will diminish if the input t_r increases above a certain value.

(g) The output will always have a 0 V baseline, regardless of the input.

(h) The output becomes more distorted as the circuit's 3-dB frequency is increased.

3.26 Figure 3–41 shows a *biased* differentiator used to convert an input pulse to spikes that are riding on a baseline other than 0 V. The circuit contains two voltage sources: the e_i pulse, and the +5 V DC bias source. It can be analyzed as follows by using the superposition *principle*.

(a) Replace the +5 V source by a short to ground. Then determine v_{OUT} in response to e_i.

(b) Re-insert the +5 V source, but replace e_i by a short to ground. Then determine v_{OUT} produced by just the +5 V source.

(c) Determine total v_{OUT} by superimposing (adding together) the results of (a) and (b). Assume that the e_i pulse occurs long after the +5 V has been applied to the circuit.

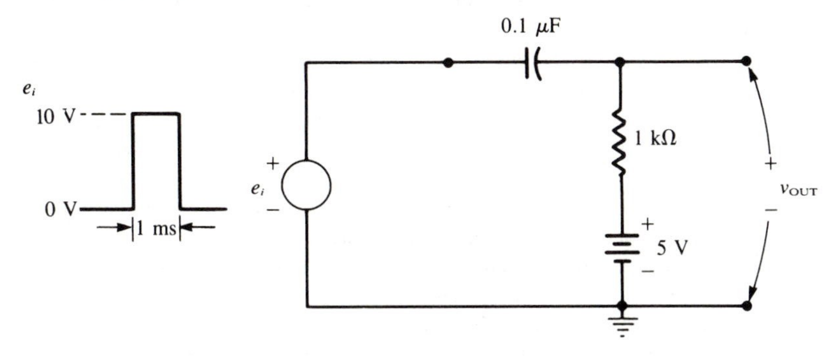

Figure 3–41

3.27 Repeat 3.26 for an input pulse having an amplitude of only 3 V. (This exercise should show that a high-pass output will not necessarily go positive and negative when the output is biased with a DC source.)

Sections 3.11–3.12

3.28 *True or False*: An *RC* low-pass circuit's output will take less time to reach steady state for a 1 kpps square wave than for a 100 kpps square wave.

3.29 The square wave of Figure 3–42 is applied to an *RC* circuit with $R = 4.7\text{ k}\Omega$ and $C = 3.3\ \mu\text{F}$. Draw the waveforms for the voltages across C and R up until steady state is reached.

3.30 Approximately how many periods of the input waveform does it take for the *RC* circuit of Figure 3–42 to reach steady state? How many would it take if the input period were decreased by half? If the input amplitude were decreased to 4 V?

3.31 Referring to Figure 3–42, what is the dc value of the input? What is the dc value of the steady-state capacitor waveform v_{xy}? Why are these two values the same?

3.32 Refer to Figure 3–42. Explain what has to be done to the value of τ if v_{xy} is to contain *only* dc and no appreciable ripple. If this is done, what would happen to the length of the transient interval? Sketch v_{xy} and v_{wx} (steady state only) for this situation.

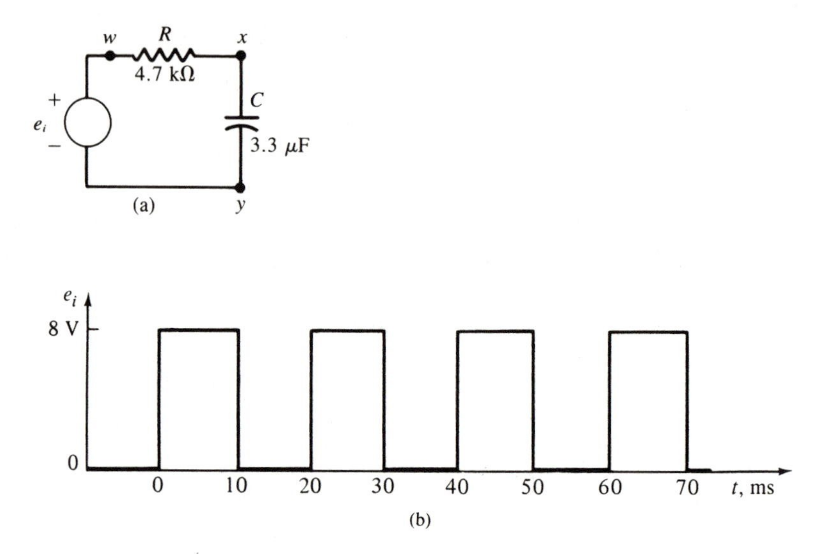

Figure 3–42

3.33 Explain what would have to be done to the τ in Figure 3–42 if it is desired that the v_{xy} output more closely duplicate the input. Sketch v_{xy} and v_{wx} (steady state) for this situation.

3.34 Referring again to Figure 3–42, would it be possible to change τ so that the resistor waveform (v_{wx}) *exactly* duplicates the input? Explain.

3.35 A certain oscilloscope has $R_{IN} = 1\ M\Omega$. What value of blocking capacitor should it use for ac coupling if pulses as wide as 50 ms are to be displayed without significant tilt?

3.36 Figure 3–43 shows a circuit that forms the basis for a simple frequency meter. The unknown frequency signal is applied to a circuit called a *monostable multivibrator* (to be studied later), which produces output pulses at the *same* frequency as the input but has a *fixed* pulse duration, t_p. These pulses are applied to an *RC* integrator having a very large τ that will pass only dc to the capacitor output. A VOM across the capacitor will indicate this dc voltage, which should be proportional to the input frequency.

(a) Assume t_p = 10 ms. What voltage will the meter indicate if the input frequency is 10 Hz?

(b) Repeat for 20 Hz and 50 Hz.

(c) How long will it take the meter reading to become constant once the input frequency is suddenly changed from one value to another?

(d) How can this response time be reduced, and what will happen if it is reduced too far?

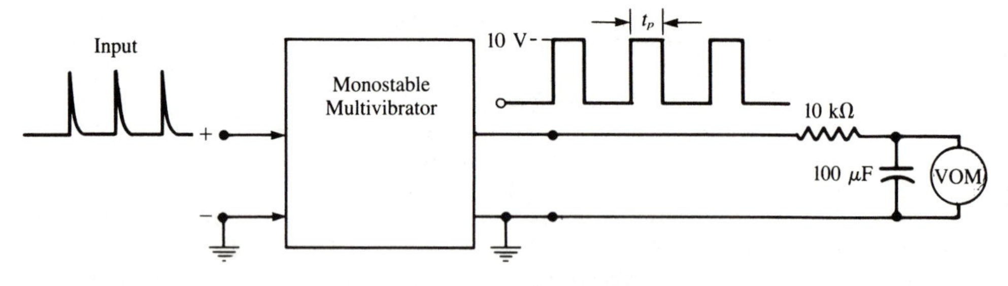

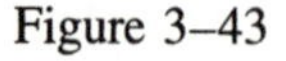
Figure 3–43

3.37 A technician applies a 1 kpps square wave from a signal generator to a scope input and measures the signal rise time. His measured value of t_r, however, is much longer than the expected value. Since the scope bandwidth of 1 MHz is much greater than 1 kHz, he concludes that the signal generator must be faulty. Where is the error in his reasoning?

Section 3.13

3.38 The voltage-divider circuit of Figure 3–44(a) is driven from the square wave input shown in part (b) of the figure. The output of the *unloaded* voltage divider would be 0.75 times the input. For each of the possible output waveforms shown in parts (c) through (e), indicate what type of loading (resistive, capacitive, or both) would produce such an output.

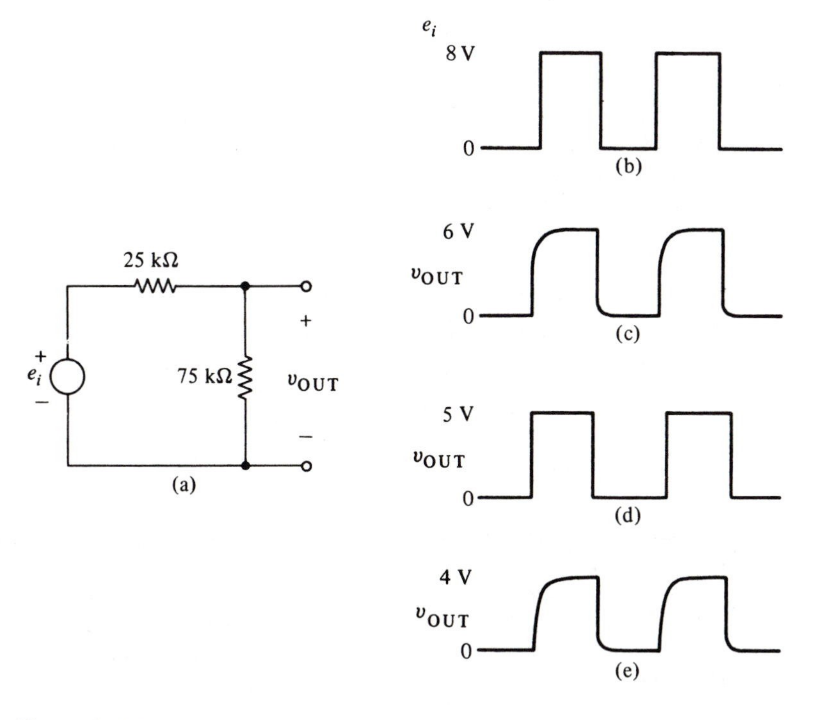

Figure 3–44

3.39 A positive-going 6 V step waveform is applied to the high-pass circuit in Figure 3–45. Sketch the output waveform with no load and with a 1 kΩ load. (Assume input rise time is much smaller than τ.)

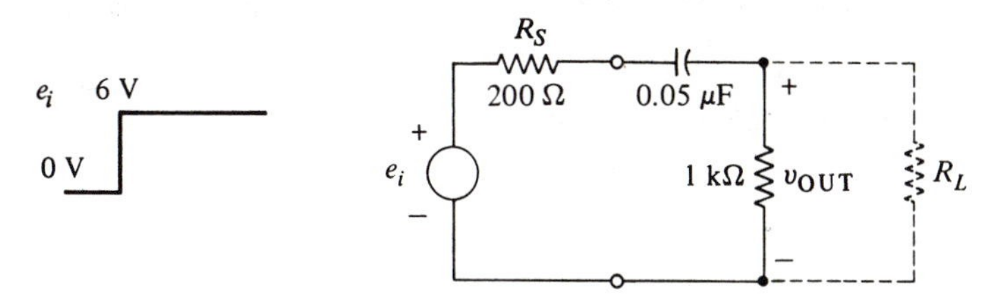

Figure 3–45

3.40 Repeat Problem 3.39 with $R_s = 0$.

3.41 Figure 3–46 shows a voltage divider being loaded by a 15 pF load capacitance. What value of compensation capacitor should be placed across the 6 kΩ resistance to minimize distortion in the v_{OUT} signal?

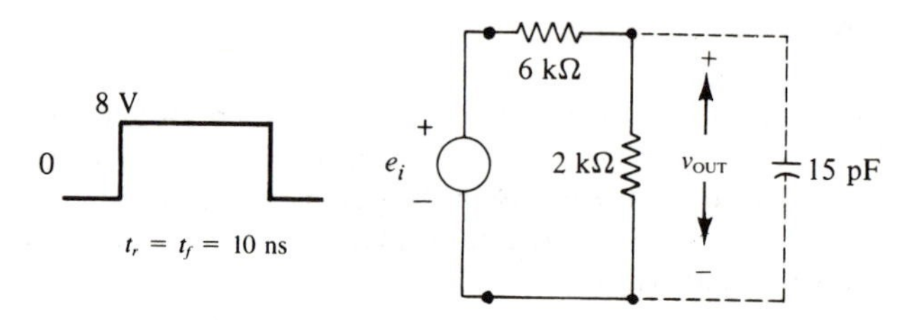

Figure 3–46

3.42 Assume the calculated value of compensation capacitor has been placed across the 6 kΩ resistance. A technician then connects an oscilloscope to observe the v_{OUT} waveform and finds that it has significant high frequency distortion. The following list contains some possible causes of this observation. For each one, explain why it could produce the observed result, and what should be done next to prove or disprove that it is the actual cause.

(a) Internal resistance of input signal source.

(b) Resistor values are not exact.

(c) Technician forgot to compensate scope probe.

(d) Technician remembered to compensate scope probe, but the probe input capacitance is still large enough to load down the output signal.

(e) The oscilloscope has insufficient bandwidth.

3.43 Explain how a compensated 10X oscilloscope probe presents less loading on a circuit being measured.

3.44 A certain oscilloscope has R_{IN} = 1 MΩ and C_{IN} = 33 pF at its vertical input. A 50X probe is connected to the vertical input and its internal capacitance is adjusted for perfect compensation. What is the input resistance and input capacitance at the probe tip?

Section 3.14

3.45 Refer to Figure 3–27. The signal source has R_s = 1 kΩ. The conductors that connect the source to the logic circuit have an inductance of 0.8 μH per inch. What is the maximum distance the source can be from the logic circuit before ringing appears on the signal at the logic circuit input?

3.46 Assume that the total inductance in Problem 3.45 is 10 μH. Determine how much resistance should be added to the circuit so that the overshoot and ringing are eliminated.

3.47 Repeat Problem 3.46, but this time find the extra resistance needed to reduce the overshoot to approximately 10 percent.

4

Switching Devices

4.1 Introduction

In Chapter One we discussed the ideal switching device. Such a device can change between two states, either a perfect open circuit or a perfect short circuit, depending on the status of an input control voltage or current. The ideal switch also has zero switching time, which means that it can change states instantaneously.

In this chapter we will study many of the semiconductor devices that can be operated as switches. None of them will have all of the elements of the ideal switching device, but each of them has characteristics that are well-suited for switching circuit applications. After discussing the individual devices, we will look at approximation techniques that can be used in analysis and troubleshooting of switching circuits. These techniques can be used in most of the situations a technician will encounter.

4.2 The Diode as a Switch

The PN diode is a two-terminal device that conducts current more easily in one direction than the other. Since it has only two terminals, the diode does not have a third control input terminal that can be used to control the state of the switch. Instead, the voltage across the diode terminals determines whether the diode will conduct (closed switch) or not conduct (open switch). Thus, in a sense, the diode is a voltage-controlled switch controlled by the polarity of voltage across the switch terminals.

Figure 4-1(a) shows the electronic symbols for the PN diode. The arrow-type symbol is the one commonly used in circuit diagrams. The general volt-ampere characteristic for a typical silicon* PN diode is shown in part (b) of the figure.

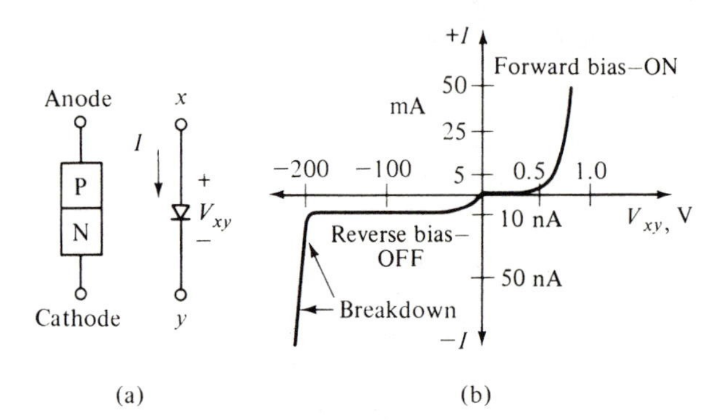

Figure 4-1 **Symbols and volt-ampere curve for a silicon PN diode**

The two regions of operation correspond to the two polarities of anode-cathode voltage. When V_{xy} is positive, the anode is positive relative to the cathode, and this is called *forward bias*; when V_{xy} is negative, the anode is negative relative to the cathode, and the diode is *reverse-biased.*

In the forward-bias region of operation the silicon diode conducts current very readily for voltages greater than 0.5 V. This diode has a *threshold voltage* (V_T) of 0.5 V which is typical for silicon. A *germanium* diode typically has V_T = 0.2 V. For values of forward bias greater than V_T, the diode current increases rapidly for small changes in voltage. When the diode is operated in this region, it is said to be ON (or in the ON state).

In the reverse-bias region of operation the diode current is very low (note the different current scale for the reverse current) except for voltages greater than the reverse breakdown voltage (V_{BR}). In this case V_{BR} is about 200 V. When operated below the breakdown voltage, the silicon diode reverse current (leakage) is approximately constant at 10 nA (0.01 μA). This leakage is typical for silicon switching diodes. Germanium switching diodes typically exhibit reverse currents of 1 μA. For reverse-bias voltages greater than V_{BR}, the diode current increases rapidly. In general, a switching diode is not operated in breakdown. When operated in the low-current reverse-bias region, it is said to be in the OFF state.

The PN diode is not a perfect switch, since its ON-state resistance is not zero, and its OFF-state resistance is not infinite. However, it comes close to being perfect. This fact can be shown by calculating its resistances in both states. For example, for a forward-bias voltage of 0.7 V, a forward current of approximately 50 mA is produced. This current corresponds to a resistance in the ON state of

$$R_{\mathrm{ON}} = \frac{0.7\ \mathrm{V}}{50\ \mathrm{mA}} = 14\ \Omega$$

On the other hand, for a reverse-bias voltage of 100 V, a leakage current of only 0.01 μA is produced. This gives a resistance in the OFF state of

$$R_{\mathrm{OFF}} = \frac{100\ \mathrm{V}}{0.01\ \mu\mathrm{A}} = 10{,}000\ \mathrm{M}\Omega$$

*All devices will be assumed to be silicon unless otherwise noted.

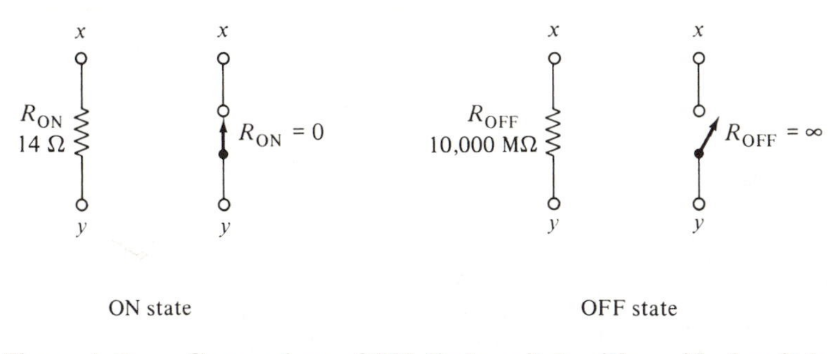

Figure 4–2 **Comparison of PN diode switch with an ideal switch**

Figure 4-2 illustrates the comparisons between the PN diode and the ideal switch.

Approximating the diode characteristics The calculations of R_{ON} and R_{OFF} were done solely to illustrate the great difference between the diode's ability to conduct current in one direction versus the other. The values for R_{ON} and R_{OFF} are not constant and will depend on the operating point on the *I-V* curve. It is not useful to think of the diode as a resistance that can be switched from a very low to a very high value. It is more useful to take advantage of the shape of the *I-V* curve and approximate the diode's operation as follows:

(a) When the diode is forward-biased by more than V_T (0.5 V for silicon), the diode voltage drop will not change appreciably with changes in forward current. Thus, the diode will act like a constant voltage of approximately +0.7 V (for silicon) from anode to cathode.

(b) When the diode is reverse-biased by *less* than its breakdown voltage, the current through the diode will be extremely small. For most cases we can call it zero.

(c) When the diode is reverse-biased by a voltage that is larger than V_{BD}, the diode will be in reverse breakdown and will maintain a nearly constant reverse voltage across its terminals equal to V_{BD}.

What we are saying, then, is that we can approximate the diode's *I-V* characteristic with the one shown in Figure 4–3. For most analysis purposes, this approximate characteristic will be sufficient.

4.3 Analyzing Diode Circuits

We can use the approximate *I-V* characteristic of Figure 4–3 to help analyze many types of diode circuits. First, determine whether the diode is forward biased or reverse biased. Then replace the diode by either an open circuit (if the diode is OFF) or by a constant voltage across its terminals. The flowchart of Figure 4–4 shows the steps in approximate analysis. We will use these steps in the next examples.

Example 4.1 The circuit in Figure 4–5(a) uses a silicon diode with $V_{BD} = 200$ V. Determine the circuit current and voltages.

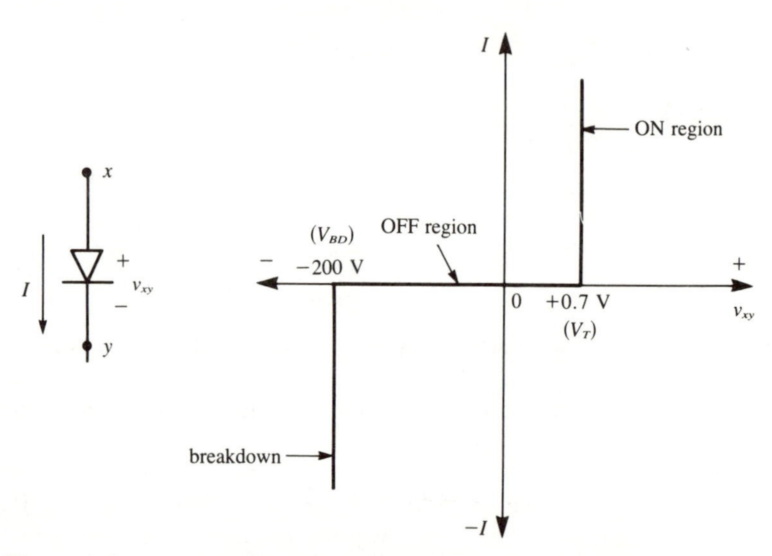

Figure 4–3 **Approximating the diode *I-V* curve (silicon)**

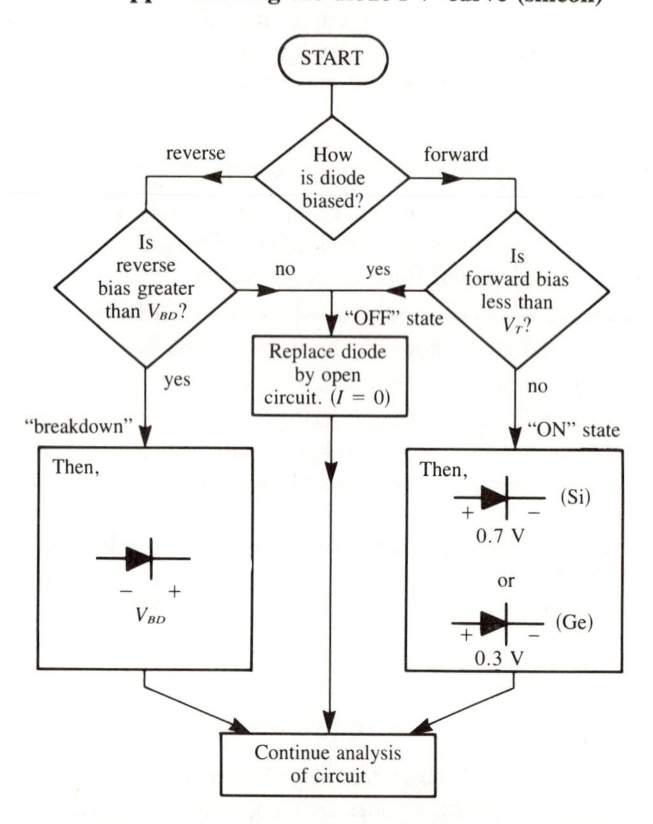

Figure 4–4 **Flowchart of steps in approximate analysis of diode circuits**

Solution: The diode is forward-biased since the negative terminal of the source is connected to the cathode. Since the source voltage is 10 V, the diode is sufficiently forward biased to conduct. We can therefore assume a 0.7 V drop across the diode as

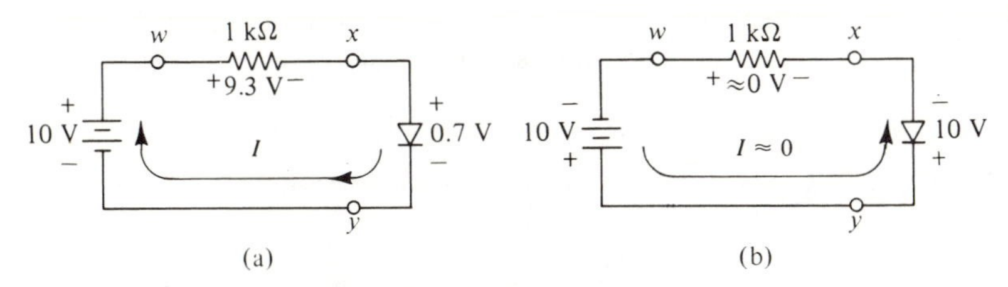

Figure 4–5 **Examples 4.1 and 4.2**

shown in the diagram. The remainder of the applied 10 V has to be across the resistor. Thus, the circuit current is

$$I = \frac{9.3\text{ V}}{1\text{ k}\Omega} = \boldsymbol{9.3\text{ mA}}$$

Example 4.2 The same circuit is shown in Figure 4–5(b) with the source polarity reversed. Find the circuit current and voltages.

Solution: The diode is now reverse biased, and since the source voltage is less than the reverse breakdown voltage, the diode will be OFF and we can mentally replace the diode by an open circuit. Thus, the circuit current is effectively zero, and there can be no voltage across the resistor. The full source voltage will be across the diode. That is, $\boldsymbol{V_{xy} = -10\text{ V}}$.

Example 4.3 Repeat Example 4.2 for a source voltage of 250 V.

Solution: The diode is now reverse-biased by more than its breakdown voltage, so we can assume that the voltage across the diode will be effectively fixed at the value of V_{BD}. That is, $V_{xy} = -200$ V, as shown in Figure 4–6. It follows that the remaining 50 V of the source must be across the resistor so that $V_{wx} = -50$ V. The circuit current therefore becomes

$$I = \frac{50\text{ V}}{1\text{ k}\Omega} = \mathbf{50\text{ mA}}$$

In reverse-breakdown operation the current is no longer limited by the diode but instead depends on the size of the series resistor. For this reason, if a diode is to be operated in

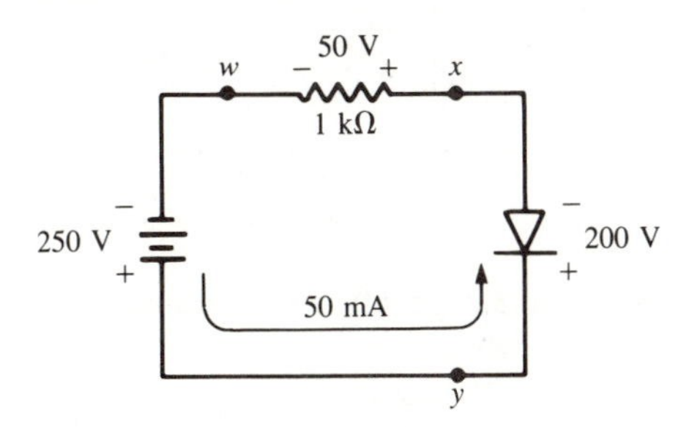

Figure 4–6 **Example 4.3 with diode in reverse breakdown**

breakdown, the series resistor must be large enough to insure against excessive diode power dissipation. For example, in this case the diode power would be

$$P_D = V_{xy} \times I = 200 \text{ V} \times 50 \text{ mA} = 10 \text{ W}$$

Such a large P_D can cause the diode to heat up and possibly burn itself out.

Certain types of diodes are designed specifically to operate in the breakdown region in order to use the constant-voltage property to advantage. *Zener diodes,* as they are called, are manufactured with very sharp breakdown regions and are used almost exclusively in voltage regulators and as voltage references.

4.4 Diode Switching Characteristics

In this section the causes and effects of the diode switching times will be studied. Consider the simple circuit in Figure 4–7(a). The input is a step voltage that is initially at -10 V and jumps to $+10$ V at $t = 0$. Before $t = 0$ the diode is reverse biased, and the input (-10 V) is dropped across the diode, so $v_{wx} = -10$ V, and $v_{xy} = 0$ across R. If an ideal diode is assumed, then when the input jumps to $+10$ V at $t = 0$, the diode will immediately turn ON, so $v_{wx} = +0.7$ V (silicon), and the resistor voltage will become $v_{xy} = 9.3$ V.

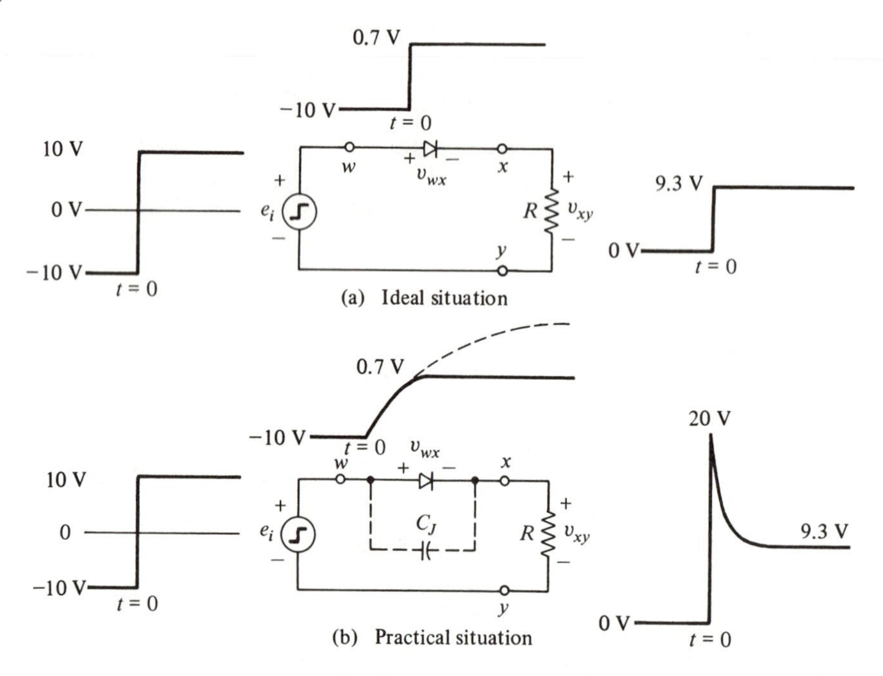

Figure 4–7 **Effect of PN junction capacitance when switching diode from OFF to ON**

A practical PN diode, however, possesses a certain amount of capacitance across its reverse-biased PN junction. You may recall that this capacitance is caused by the majority ions on either side of the junction that have been uncovered by the removal of majority carriers from the junction region due to reverse bias. Figure 4–7(b) shows the same circuit just considered, but now it includes the diode's junction capacitance, C_J. While the input is at -10 V, the diode is OFF, and C_J is charged to $v_{wx} = -10$ V. At $t = 0$ when the input switches to $+10$ V, the capacitor cannot change its voltage instantaneously but instead will begin charging toward $v_{wx} = +10$ V. As C_J charges toward $+10$ V, it will eventually reach $+0.7$ V; at this point, the diode turns ON and holds its forward voltage at $+0.7$ V. The waveform of diode voltage is therefore slowed down by the presence of C_J.

The voltage v_{xy} across R can now be found using KVL. When e_i jumps from -10 V to $+10$ V, the v_{xy} will jump to $+20$ V. This jump occurs because

$$v_{xy} + v_{wx} = e_i$$

or

$$v_{xy} = e_i - v_{wx}$$

With $v_{wx} = -10$ V, v_{xy} will be $+20$ V when $e_i = +10$ V. Then as v_{wx} gradually rises to $+0.7$ V as C_J charges, the resistor voltage will drop to $+9.3$ V as shown. Thus, the presence of C_J has produced an overshoot on the v_{xy} waveform.

The effects of C_J illustrated in Figure 4–7(b) will not occur if the time constant $\tau = RC_J$ is short compared to the input rise time. If $\tau < t_r/2\pi$, the capacitance will be able to charge up as e_i is making its transition so that there will be no overshoot on the v_{xy} waveform. In other words, if τ is short enough, the circuit behaves as if the diode is ideal.

In general, τ will be relatively small because C_J is typically in the low picofarad range (1–40 pF) for small switching diodes. C_J can be much larger for high-current diodes that are not designed for high-speed switching applications.

The following example shows how to calculate how long it takes for the diode to switch from OFF to ON when e_i makes its transition. We will call this value of time t_{ON}.

Example 4.4 If $C_J = 2$ pF, and $R = 10$ kΩ, what is the value of t_{ON} for the circuit of Figure 4–7(b)?

Solution: The procedure is to determine the exponential expression for the capacitor voltage as it charges from -10 V toward $+10$ V, and then calculate how long it takes to reach $+0.7$ V. Using the exponential form gives

$$v_{wx} = +10 - 20\epsilon^{-t/\tau}$$

where τ is given by (10 kΩ) × (2 pF) = 20 ns. Setting $v_{wx} = +0.7$ V yields

$$0.7 \text{ V} = +10 - 20\epsilon^{-t/20 \text{ ns}}$$

or

$$\epsilon^{-t/20 \text{ ns}} = 0.465$$

Taking natural logs of both sides

$$\frac{-t}{20 \text{ ns}} = -0.75$$

or

$$t = \mathbf{15\ ns} = t_{ON}$$

Thus, 15 ns is the value of the diode turn-ON time in this circuit.

Example 4.5 How can the diode turn-ON time be decreased in the circuit of Figure 4–7(b)?

Solution: The problem is to make the capacitor reach +0.7 V faster. This can be done by decreasing R, which will decrease τ. Another method is to increase the amplitude of the positive portion of the input. By doing this, the capacitor will reach +0.7 V faster since it will be charging toward a higher voltage. For example, if the positive portion is increased to 20 V, the turn-ON time will decrease to t_{ON} = 8.8 ns (student should verify). On the other hand, decreasing the negative portion of the input will have the same effect since C_J will be charged to a lower negative voltage during reverse bias and will therefore reach +0.7 V sooner after the input switches.

The preceding calculation is meant only to illustrate the factors determining the diode turn-ON time. The value of C_J is rarely known and, in fact, will not be a constant value. It will vary with the voltage across the diode. Typically, for a fast switching diode, C_J may increase from 1 pF to 5 pF as the diode voltage goes from 10 V reverse bias to 0.7 V forward bias.

Diode turn-off time, t_{OFF} Figure 4–8(a) shows the situation where the diode is being switched from the ON state to the OFF state when e_i jumps from +10 V to −10 V at t = 0. As we would expect, the diode's junction capacitance slows down the switching operation because it has to gradually charge up to the reverse voltage. In addition, there is another effect that slows down the diode turn-OFF. It is the phenomenon of *charge storage,* which occurs in a forward-biased PN junction.

The charge storage is caused by majority carriers that flow across a forward-biased PN junction (electrons from N region to P region, holes from P to N). Many of these carriers recombine with oppositely charged carriers in the new region and remain there while the diode is ON. When a reverse voltage is applied to turn the diode OFF, these stored charges must return to their respective majority regions. This flow of stored charges produces a substantial reverse current for a time duration called the *storage time,* t_s.

The waveforms in Figure 4–8(b) illustrate the effects of both C_J and storage time. Prior to t = 0, the diode is forward-biased and a current I_F will flow depending on the value of R. At t = 0, when e_i switches to −10 V, the diode voltage remains at ≈0.7 V for a time duration equal to t_s. During this interval a reverse current I_R will flow with a value limited by R. This reverse current is removing the diode's stored charge and it will flow for a time, t_s, until all the stored charge has been removed.

At the end of the t_s interval the diode begins to turn OFF, but the junction capacitance causes the diode voltage and current to change gradually with $\tau = RC_J$. Once C_J is fully charged to the source voltage, the diode is OFF and blocks the flow of reverse current. The total time for the diode to switch from ON to OFF is called t_{OFF}, and is equal to the sum of t_s and the capacitor charging time as shown.

The value of t_{OFF} is difficult to predict because the storage time is dependent on several factors involved in the PN junction fabrication process. However, it is known that t_s will increase as the amount of diode current I_F is increased. In other words, more current through the diode in the ON state produces more stored charges that require a longer time (t_s) to be removed when the diode is turned OFF. On the other hand, a larger applied reverse voltage will reduce t_s since it will produce a larger I_R during the storage removal interval.

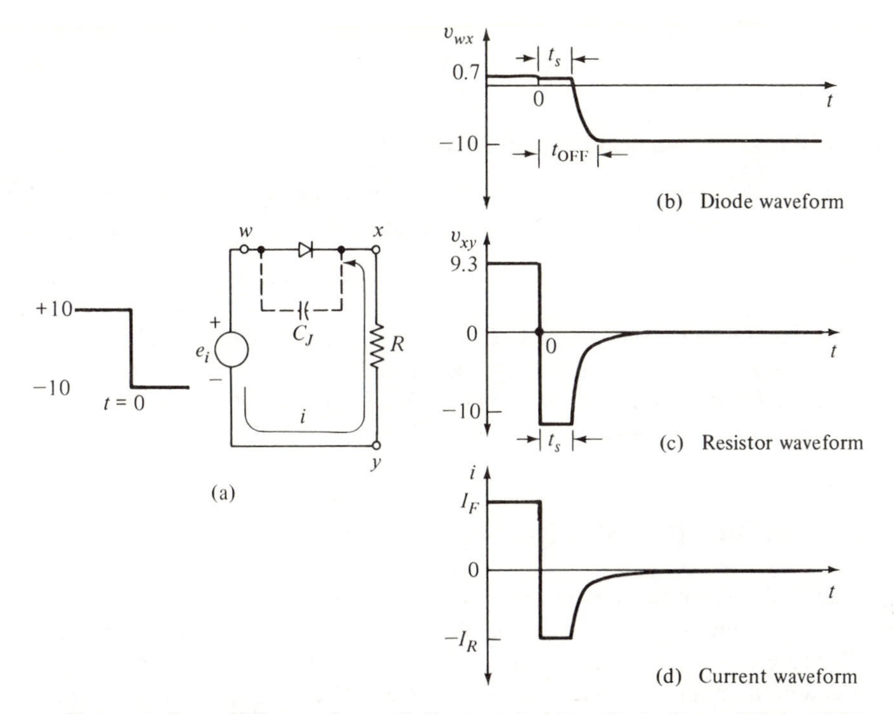

Figure 4–8 **Effects of t_s and C_J on switching diode from ON to OFF**

Typical values for t_{OFF} will be 10–500 ns for switching diodes. High-speed switching diodes can be much faster, while general-purpose diodes and heavy-current diodes will be much slower. Of course, the value of t_{OFF} will also depend on other circuit values.

Example 4.6(a) Assume $C_J = 4$ pF and $R = 10$ kΩ in Figure 4–8. Also assume that e_i has a near-zero transition time. Determine:

(a) the current flowing before $t = 0$

(b) the amplitude of the reverse current pulse during t_s

(c) the total turn-OFF time, t_{OFF}, if $t_s = 50$ ns

Solution:

(a) $I_F = \dfrac{10\text{ V} - 0.7\text{ V}}{10\text{ k}\Omega} = \mathbf{0.93\text{ mA}}$

(b) When e_i switches to -10 V, the total voltage across R will be 10.7 V because the diode voltage holds at 0.7 V. Thus, the reverse current becomes

$$I_R = \frac{10.7\text{ V}}{10\text{ k}\Omega} = \mathbf{1.07\text{ mA}}$$

(c) After the t_s interval, C_J will take 5τ to charge up to the e_i voltage

$$5\,\tau = 5 \times 10\text{ k}\Omega \times 4\text{ pF} = 200\text{ ns}$$

Thus,

$$t_{OFF} = t_s + 5\tau = \mathbf{250\ ns}$$

Example 4.6(b) What will happen to t_{OFF} if R is reduced?

Solution: Reducing R will increase both I_F *and* I_R and so will probably have little effect on t_s. However, a smaller R will decrease the charging time for C_J thereby reducing t_{OFF}.

Of course, as we stated for the turn-ON case, the effects illustrated in Figure 4–8 will *not* occur if the e_i transition time is much slower than t_{OFF}, because the diode will be able to turn OFF as e_i is making its transition. For such cases, the diode current will go from I_F to 0 as e_i changes levels and there will be no pulse of reverse current.

4.5 The Bipolar Junction Transistor (BJT) as a Switch

NPN and PNP bipolar junction transistors (BJTs) can operate as switching devices when the proper actuating signals are used. The BJT is a more useful switching device than the diode because it is a controlled switch; that is, the switch terminals are controlled by a signal at a third terminal. In addition, the BJT current gain allows a small input current to control a larger switch current.

Review of transistor operation Figure 4–9(a) shows an NPN transistor connected in the common-emitter configuration. Here, the universal schematic symbol for the NPN transistor is used, as it will be throughout the text. In this configuration the emitter is shown connected to ground, or common. The base is connected to positive V_{BB} volts through resistor R_B, and the collector is connected to positive V_{CC} volts through R_C.

Because the base is positive relative to the emitter, the E–B junction is forward biased. In this situation the E–B junction will act as a diode and conduct current readily as long as V_{BB} is greater than the junction turn-ON voltage, V_T. If the transistor is silicon, V_T will be 0.5 V. Usually V_{BB} is much larger than V_T; as such, the E–B junction will conduct and maintain a forward voltage drop of about 0.7 V (for silicon). This will produce a flow of base current I_B whose magnitude can be calculated as

$$I_B = \frac{V_{BB} - 0.7\text{ V}}{R_B} \tag{4–1}$$

The flow of base current will produce a collector current, I_C, whose magnitude can be determined by drawing a load line on the common-emitter output curves. Figure 4–9(b) illustrates this by using $V_{CC} = 12$ V and $R_C = 1$ kΩ. The transistor operating point is determined by the intersection of the load line and the curve corresponding to the value of I_B that is flowing in the base. For example, if I_B is 50 μA, the operating point would be at x. At point x the value of I_C is 5 mA, and the value of collector-to-emitter voltage (V_{CE}) is 12 V. This operating point lies in the ACTIVE region because the collector is positive relative to the base ($V_{CB} = V_{CE} - V_{BE} = 11.3$ V), which constitutes reverse bias for the C–B junction.

If the base current is reduced to zero, the operating point will move to y. As can be seen in the figure, this reduces I_C to zero and increases V_{CE} to 12 V. This is the cut-off,

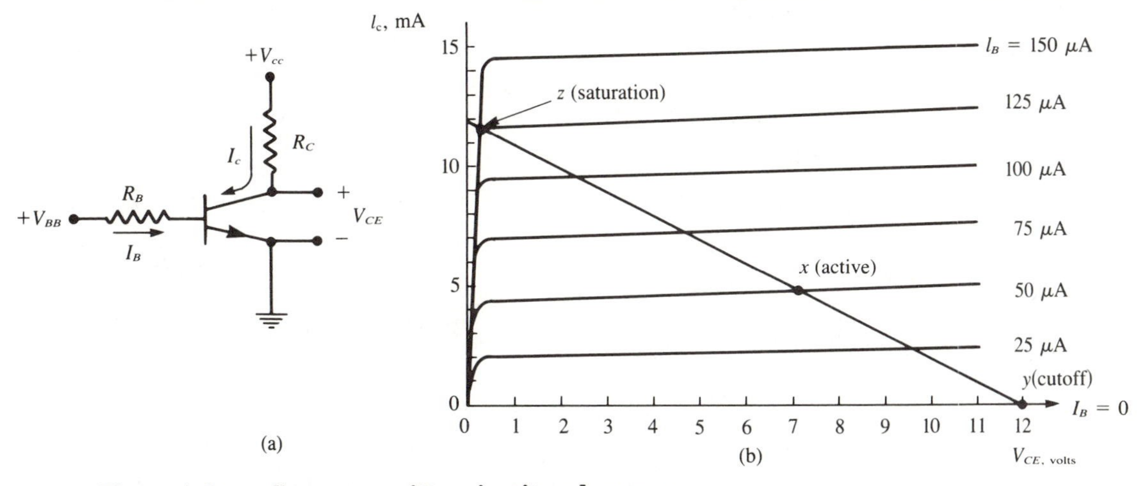

Figure 4–9 **Common-emitter circuit and curves**

or OFF, condition of the transistor. The OFF condition is produced when V_{BB} is less than V_T, or when the $E-B$ junction is reverse-biased so that $I_B = 0$. In either case, I_C will be a very small leakage current, so no voltage drop occurs across R_C; instead, almost the entire V_{CC} supply is dropped across the collector-emitter terminals.

As the base current is increased, the operating point moves up the load line, with I_C increasing and V_{CE} decreasing. Eventually, point z is reached; at this point, I_C is approximately 12 mA, and V_{CE} is 0.2 V. The collector is now negative relative to the base since $V_{CB} = V_{CE} - V_{BE} = -0.5$ V; this means that the C–B junction is *forward biased.* In this condition the collector's ability to collect carriers is diminished to the point where I_C will increase no further with increases in I_B. As can be seen, for values of $I_B \geqslant 125$ μA the operating point stays at z which is in the saturation, or ON, region. In the ON region the collector current is at a maximum since almost all of the V_{CC} supply is dropped across R_C, and very little is dropped across the transistor. The value of V_{CE} in the ON condition is given the symbol V_{CE}(sat) and is typically 0.2 V for silicon and 0.1 V or less for germanium, although this value increases slightly as I_C increases.

The value of I_C in the ON region is symbolized as I_C(SAT). The value of I_B that just produces saturation is called I_B(SAT). In the case shown in the figure, I_B(SAT) = 125 μA, and $I_C(\text{SAT}) \approx 12$ mA.

The transistor switch When used as a *saturated* switch, the transistor operates in either the ON (saturated) region or in the OFF region and switches back and forth between the two. In the ON region the transistor acts as a very low resistance between collector and emitter (typically 1 Ω–50 Ω); in the OFF region the transistor acts as a very high resistance between collector and emitter (typically 10 MΩ for silicon). It is the base current that controls when the transistor switch is ON or OFF. It is helpful to visualize the transistor as a perfect switch between collector and emitter, as illustrated in Figure 4–10.

4.6 Approximate Analysis of the Transistor Switch

In most instances one does not use the transistor's curves when analyzing switching circuits because they are rather cumbersome, even in simple circuits, and often are not

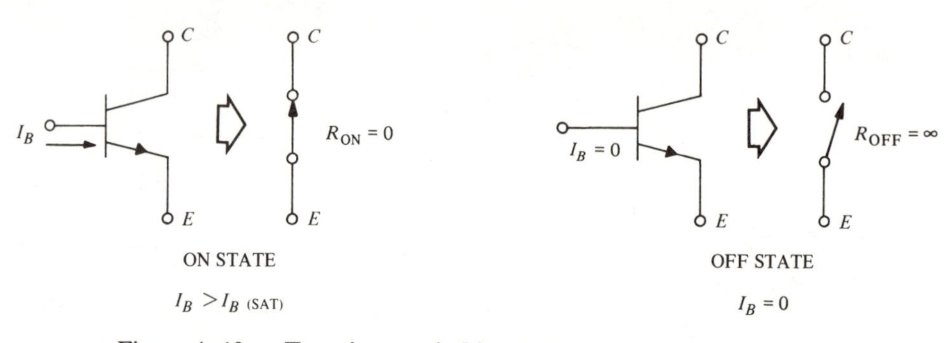

Figure 4–10 **Transistor switching states**

available. Besides, simple approximate methods of analysis have been developed which can be easily applied to all switching circuits, as will now be illustrated.

The ON condition Figure 4–11 shows a transistor connected in the common-emitter configuration. The transistor will operate in the ON state when a positive e_{IN} is applied to the circuit to produce sufficient base current to saturate the transistor. For a given value of e_{IN}, the amount of base current will be

$$I_B = \frac{e_{IN} - V_{BE}}{R_B}$$

This I_B will be amplified by the transistor's forward current gain, h_{FE}, to produce a collector current of

$$I_C = h_{FE} I_B \qquad \text{(In ACTIVE region)} \qquad \textbf{(4–2)}$$

This is the amount of I_C that will flow *if* the transistor is in the ACTIVE region ($V_{CE} \gg 0$). The amount of I_C that can actually flow is limited by the collector resistor (R_C). When the transistor is fully saturated, all of the collector supply voltage (V_{CC}), except for the small V_{CE}(sat), appears across R_C. This gives a collector current at saturation equal to

$$I_C(\text{SAT}) \approx \frac{V_{CC}}{R_C} \qquad \textbf{(4–3)}$$

As I_B is increased from zero, the collector current increases proportionately according to Equation (4–2) until it reaches the limit specified by equation (4–3). Further increases in I_B will not produce an increase in I_C; instead I_C will be saturated at the value of I_C(SAT).

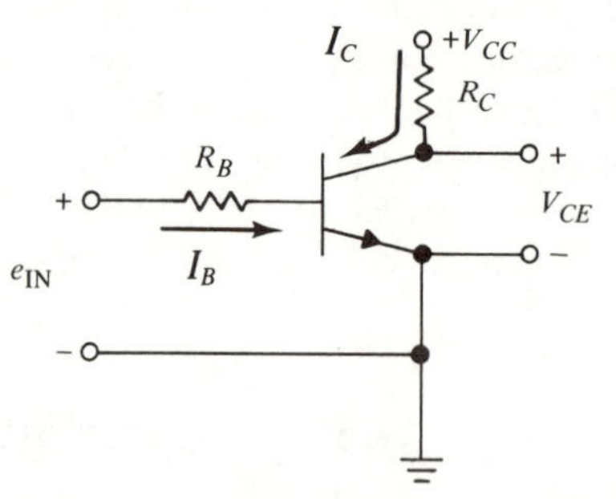

Figure 4–11 **Common-emitter circuit**

The value of base current that will just cause saturation can be determined by using Equation (4–2) and setting $I_C = I_C(\text{SAT})$. Thus,

$$I_C(\text{SAT}) = h_{FE}I_B(\text{SAT})$$

or

$$I_B(\text{SAT}) = \frac{I_C(\text{SAT})}{h_{FE}} \quad \textbf{(4–4)}$$

Whenever the value of I_B produced by e_{IN} equals or exceeds $I_B(\text{SAT})$, the transistor will be saturated. Remember, in this ON state $V_{CE} \approx 0$, and the transistor has a very low resistance between collector and emitter.

The OFF condition The transistor will be OFF whenever $I_B = 0$. In this state only a small leakage current will flow in the collector and emitter. This leakage can usually be neglected and we can assume that $I_C = I_E = 0$ and the transistor is open between collector and emitter. For example, with $I_B = 0$ in Figure 4–11, we can assume that $I_C = 0$ so that $V_{CE} = V_{CC}$. In other words, the transistor acts like an open switch between collector and emitter so that all of V_{CC} appears across these terminals.

Example 4.7 Determine I_C and V_{CE} for the circuit of Figure 4–11 for the following values: $e_{IN} = 6$ V, $V_{CC} = 12$ V, $R_C = 470\ \Omega$ and $R_B = 10$ kΩ. Assume the transistor has $h_{FE} = 60$.

Solution: First, calculate the actual base current

$$I_B = \frac{6\text{ V} - 0.7\text{ V}}{10\text{ k}\Omega} = 0.53\text{ mA}$$

Now calculate the value of $I_B(\text{SAT})$ required to produce saturation.

$$I_C(\text{SAT}) = \frac{12\text{ V}}{470\ \Omega} = 25.5\text{ mA}$$

$$I_B(\text{SAT}) = \frac{25.5\text{ mA}}{60} = 0.425\text{ mA}$$

Since $I_B \geqslant I_B(\text{SAT})$, the transistor is ON. Thus,

$$I_C = I_C(\text{SAT}) = \mathbf{25.5\text{ mA}}$$

$$V_{CE} \approx \mathbf{0\text{ V}}$$

Example 4.8 Assume that the transistor of the preceding example is specified to have an h_{FE} anywhere in the 30–120 range. What is the smallest value of e_{IN} that should be used to ensure that the transistor is ON?

Solution: Since h_{FE} can be any value from 30 to 120, we have to ensure that the base current supplied by e_{IN} is enough to saturate the transistor at the lowest value of h_{FE}. Then, for larger values of h_{FE} the transistor will definitely be ON. Thus, we will use $h_{FE} = 30$ to calculate the required $I_B(\text{SAT})$.

$$I_B(\text{SAT}) = \frac{I_C(\text{SAT})}{h_{FE}} = \frac{25.5\text{ mA}}{30} = 0.85\text{ mA}$$

The value of e_{IN} that will produce this amount of base current can be calculated using KVL.

$$\begin{aligned} e_{IN} &= I_B R_B + V_{BE} \\ &= 0.85 \text{ mA} \times 10 \text{ k}\Omega + 0.7 \text{ V} \\ &= \mathbf{9.2 \text{ V}} \end{aligned}$$

Thus 9.2 V is the smallest e_{IN} that will guarantee the transistor will saturate for h_{FE} values in the given range.

Example 4.9 What is the value of V_{OUT} in Figure 4–12(a) if the transistor is silicon with $h_{FE} = 300$?

Solution: The voltage divider in the base circuit can be Thevenized and replaced by a single source and resistance as shown in Figure 4–12(b).

$$E_{TH} = \frac{1 \text{ k}\Omega}{11 \text{ k}\Omega} \times 5 \text{ V} = 0.455 \text{ V}$$

$$R_{TH} = 1 \text{ k}\Omega \| 10 \text{ k}\Omega = 909 \ \Omega$$

This result shows that there is not enough voltage to adequately forward bias the *E–B* junction. Thus, $I_B = 0$ and the transistor is OFF with $I_C = 0$ and $V_{CE} = V_{OUT} = \mathbf{5 \text{ V}}$.

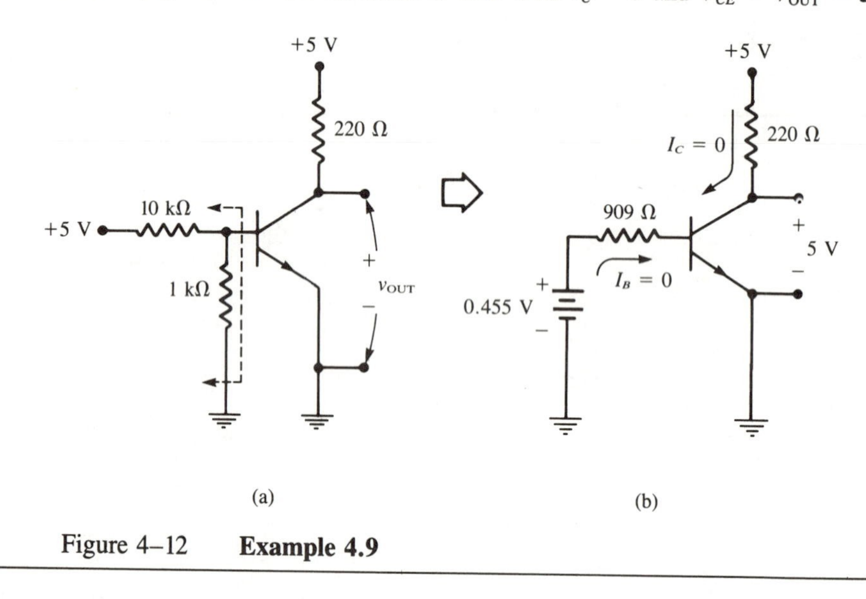

Figure 4–12 **Example 4.9**

Example 4.10 Figure 4–13(a) shows a PNP transistor in an "upside-down" configuration that is commonly used. Determine V_{OUT} for $e_{IN} = +5$ V. Repeat for $e_{IN} = 0$ V.

Solution: When $e_{IN} = +5$ V [Figure 4–13(b)] the transistor's *E–B* junction will not be forward biased. In fact, since the emitter is at +5 V, $V_{BE} = 0$ V. The transistor is OFF with $I_C = 0$. Thus, there will be no voltage across the 3.3 kΩ collector resistor so that $V_{OUT} = \mathbf{0 \text{ V}}$.

When $e_{IN} = 0$ V [Figure 4–13(b)] the *E–B* junction is forward-biased since the transistor is a PNP. The base current is therefore

$$I_B = \frac{5 \text{ V} - 0.7 \text{ V}}{100 \text{ k}\Omega} = 43 \ \mu\text{A}$$

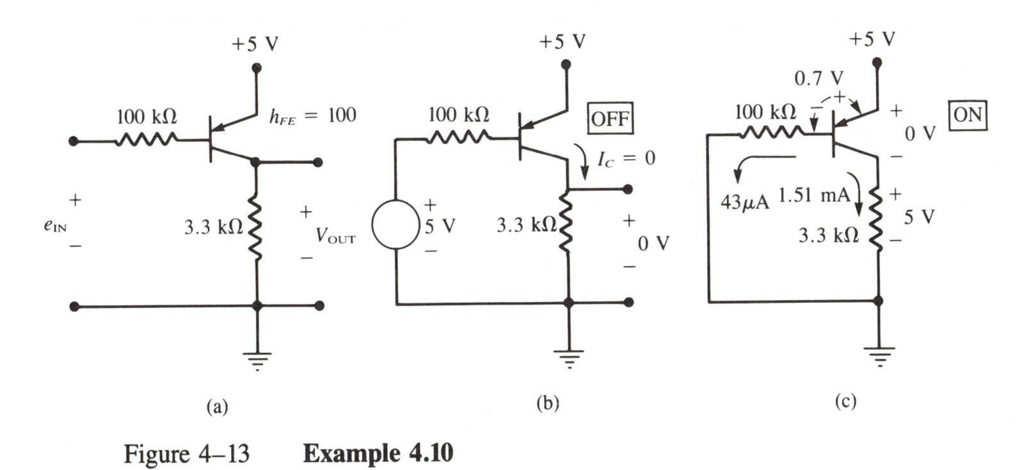

Figure 4–13 **Example 4.10**

We have to now calculate I_B(SAT) to determine whether or not the transistor is ON.

If the transistor is ON, it will essentially be a short between collector and emitter (V_{CE} = 0), so that the full +5 V supply voltage will be across the collector resistor. Thus,

$$I_C(\text{SAT}) = 5\text{ V}/3.3\text{ k}\Omega = 1.51\text{ mA}$$

and

$$I_B(\text{SAT}) = 1.51\text{ mA}/100 = 15.1\ \mu\text{A}$$

Since $I_B > I_B$(SAT) the transistor is ON and $I_C = I_C$(SAT) = **1.51 mA** and V_{OUT} = **+5 V.**

4.7 Transistor Switching Times

The same factors that slow down a diode's switching speed have an identical effect on the transistor. These factors are junction capacitance and charge storage. The analysis is more complex because the transistor has two PN junctions. For this reason, the following discussion will be more descriptive than quantitative. The important thing is to understand how the circuit values affect the transistor switching times.

Turn-on time, t_{ON} In Figure 4–14(a), the transistor is being held OFF by the −10 V input, so no base current or collector current flows, and $v_0 = v_{CE}$ = 10 V. In this situation both transistor junctions are reverse biased. The *E–B* junction will be reverse biased by 10 V because the emitter is at ground while the base is at −10 V with respect to ground. As such, the *E–B* junction capacitance (C_{be}) is charged to −10 V(V_{be} = −10 V), as shown in Figure 4–14(b).

The *C–B* junction is reverse biased by 20 V since the collector is +10 V above ground while the base is at −10 V. Thus, the *C–B* junction capacitance C_{cb} is charged to +20 V(V_{cb} = +20 V).

When the input jumps to +10 V in order to turn the transistor ON, these charged junction capacitances cause a delay. The transistor cannot begin to turn ON until C_{be} can be charged to +0.7 V by the +10 V input [see Figure 4–14(c)]. When C_{be} reaches +0.7 V, the transistor starts turning ON but cannot completely turn ON until C_{cb} has discharged to ≈0 V. Because of these junction capacitances, the output takes time to go

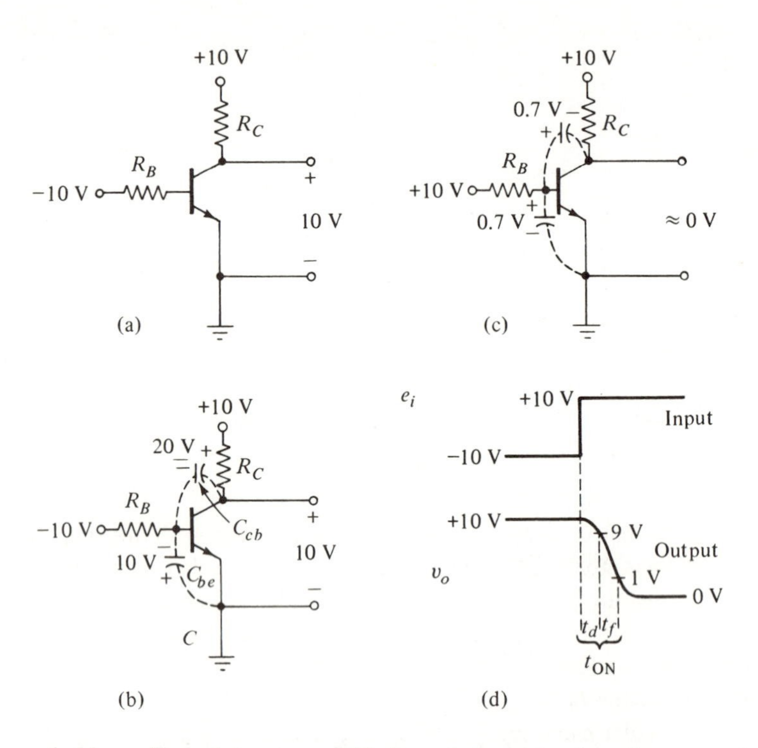

Figure 4–14 **Transistor turn-ON slowed down by junction capacitances**

from +10 V to 0 V. Part (d) of Figure 4–14 illustrates the output response to the input jump from −10 V to +10 V at $t = 0$.

The time that it takes the output to drop by 10 percent is called *delay time* (t_d); the time it takes the output to drop an *additional* 80 percent is called the *fall time* (t_f). The sum of t_d and t_f is considered to be the turn-ON time (t_{ON}).

Example 4.11 Discuss the effects of R_B and R_C on the value of t_{ON}.

Solution: A *smaller* value of R_B will reduce t_{ON} for two reasons. First, it reduces the time required to charge up C_{be}. Second, it increases the amount by which I_B exceeds I_B(SAT), the minimum value needed to saturate the transistor. This effect is called *overdriving* the transistor, and it generally reduces t_{ON}.

A *larger* value of R_C will reduce t_{ON} because it reduces I_C(SAT) and, therefore, I_B(SAT), so that the transistor will be more overdriven for a given input base current.

Turn-off time, t_{OFF} As long as the input remains at +10 V, the transistor stays ON, with the situation as shown in Figure 4–14(c). When the input drops to −10 V in order to turn the transistor OFF, once again the junction capacitances cause a delay in the transition. In addition, another effect is produced due to stored charge in the base region. When the transistor is saturated, any base current above the value I_B(SAT) does not produce an increase in I_C; as a result, there is an excess of charges stored in the base region. When the input tries to turn the transistor OFF, the excess charges in the base are still available to produce collector current. As a result, collector saturation current keeps flowing until the excess base charges are completely removed. This time interval is called the *storage time* (t_s).

At the end of the storage interval the transistor starts to turn OFF but cannot do so completely until C_{cb} charges back up to +20 V through R_C. Eventually, the transistor ends up in the OFF state [Figure 4–14(b)]. The output response to the input jump from +10 V to −10 V is illustrated in Figure 4–15. The storage time (t_s) is the time it takes v_o to rise to 10 percent of 10 V; the rise time (t_r) is the additional time it takes to reach 90 percent of 10 V. The total turn-OFF time (t_{OFF}) is the sum of t_s and t_r.

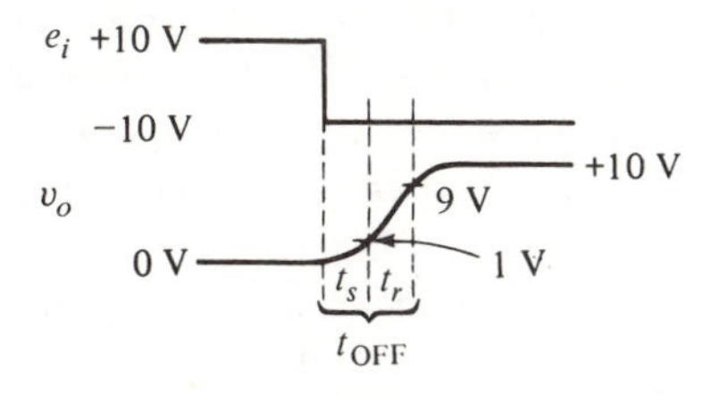

Figure 4–15 **Transistor turn-OFF slowed down by base-stored charges and junction capacitance**

Example 4.12 Discuss how to change R_B and R_C so as to reduce t_{OFF}.

Solution: The storage time, t_S, is usually the largest part of t_{OFF} and can be reduced by overdriving the transistor as little as possible. The actual I_B should not be made too much greater than I_B(SAT) if t_S is to be minimized. An *increase* in R_B will reduce the actual I_B [without affecting I_B(SAT)], thereby reducing t_S.

On the other hand, a *decrease* in R_C will increase I_C(SAT) and I_B(SAT) without affecting the actual I_B. This decrease will reduce t_S because it reduces the amount by which I_B will exceed I_B(SAT). In addition, a smaller R_C will reduce t_r by charging up C_{cb} more rapidly. Of course, the increase in I_C(SAT) means more current drain on the collector power supply.

Speed-up capacitor From the preceding examples, it may be concluded that a faster turn-ON requires a small value of R_B while a faster turn-OFF requires a larger value of R_B. This apparent dilemma can be somewhat resolved by using a *speed-up capacitor* across R_B as shown in Figure 4–16. The effect of C is to provide a surge of base current when the input jumps positively in order to cause a faster turn-ON. When C eventually completes its charging, the base current will be limited by R_B. Thus, a large value of R_B can be used to reduce overdriving and the associated long storage time. During turn-OFF the capacitor couples the rapid drop in the input voltage directly to the base so as to cause a more rapid turn-OFF.

The speed-up capacitor, then, helps decrease both of the transistor switching times. A large enough value of C must be selected to provide a long enough pulse of base current during turn-ON, but small enough so that it can completely charge up before the input

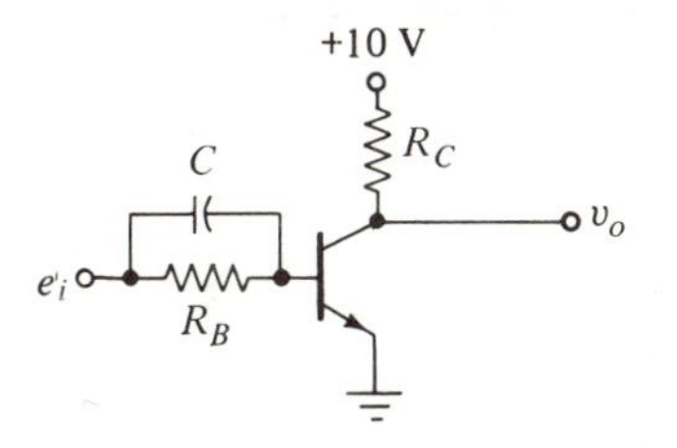

Figure 4–16 **Speed-up capacitor used to decrease switching times**

reverses. The value of C is usually best selected experimentally. A good first estimate can be obtained by selecting C so that the R_BC time constant is around 1 μs.

Modern switching transistors, especially those in integrated circuits, are designed with very fast switching characteristics, so the effects of transistor switching times are usually negligible, except in ultra-high-speed applications.

Typical switching times Modern switching transistors are designed for fast switching and can typically have t_{ON} and t_{OFF} values in the 10–500 ns range. In circuits where the transistor switching times are much shorter than the input transition times, the transistor behaves essentially as an ideal switch as far as switching speed is concerned.

Non-saturated operation We have considered only the transistor as a saturated switch because this is its most common mode of operation. The advantages include simplicity of design, and predictable output voltage levels that are relatively insensitive to transistor parameter variations. As we saw, however, saturation causes storage time delays that slow down the transistor switching speed.

If the transistor can be operated as a non-saturated switch, the storage time delay will be zero and the transistor will have a much shorter t_{OFF}. Figure 4–17 shows one method for preventing saturation. Here a *clamping* diode is connected in parallel with the collector-base PN junction. The diode is chosen to have a smaller V_T than the silicon transistor's C–B junction. Thus, when the C–B junction starts to become forward-biased at the onset of saturation, the diode will conduct and the C–B forward bias will be clamped (limited) to the small diode ON voltage. For example, if the diode is germanium, the C–B forward bias will be limited to about 0.2 V. This limitation prevents the collector region from injecting carriers into the base region, therefore eliminating storage time delay at turn-off.

The inclusion of the clamping diode will slightly increase the value of V_{CE} in the ON state. Since $V_{BE} = +0.7$ V and the diode has 0.2 V in the polarity shown, the V_{CE} value will be 0.5 V. This value is slightly higher than it would be if the clamping diode were not used and the transistor were allowed to saturate [recall, V_{CE}(SAT) is typically 0.2 V for silicon].

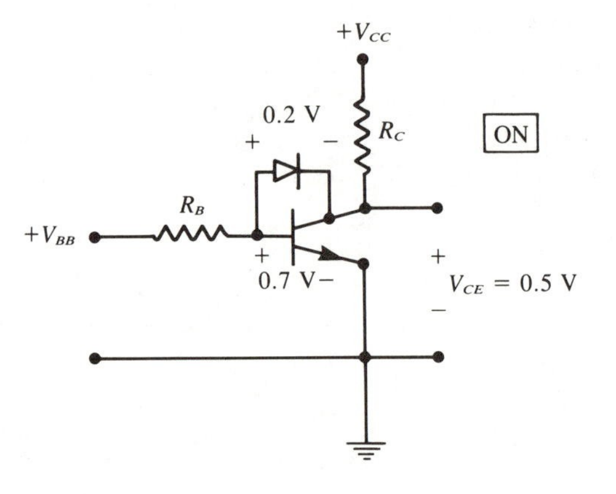

Figure 4–17 ***C–B* clamping diode prevents saturation and reduces t_s to zero**

4.8 Field-Effect Transistor (FET) Switches

The bipolar transistor we just discussed is a *current-controlled* switch because the input base current controls the state of the switch. Junction field-effect transistors (JFETs) and MOS field-effect transistors (MOSFETs) can be operated as *voltage-controlled* switches, and both find widespread use in switching circuit applications. MOSFETs are especially important in digital integrated circuits.

JFET as a switch Figure 4–18 shows an N-channel JFET connected in the common-source configuration. When it is to be operated as a switch, the JFET input gate-source voltage V_{GS} has to switch between 0 V and V_{GS}(OFF). With $V_{GS} = 0$ V*, the JFET will conduct drain current since its N-channel will not be depleted. The drain-source resistance for $V_{GS} = 0$ V is called r_{DS}(ON) and represents the channel resistance when the JFET is ON. Typically r_{DS}(ON) values range from 10–100 Ω.

When V_{GS} is made more negative than V_{GS}(OFF), the channel will be completely pinched OFF and there will be very little drain current. V_{GS}(OFF) will be negative for an N-channel device and typically ranges from −1 V to −10 V. In this OFF state the JFET has a very large drain-source resistance, typically 1000 MΩ or more.

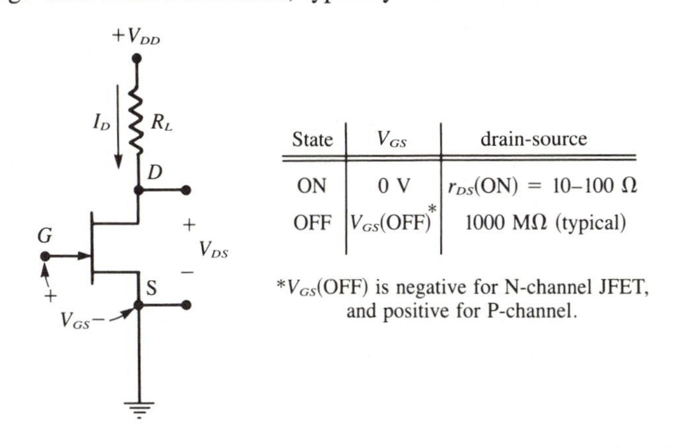

State	V_{GS}	drain-source
ON	0 V	r_{DS}(ON) = 10–100 Ω
OFF	V_{GS}(OFF)*	1000 MΩ (typical)

*V_{GS}(OFF) is negative for N-channel JFET, and positive for P-channel.

Figure 4–18 **N-channel JFET in common-source configuration**

Example 4.13 The JFET of Figure 4–18 has $V_{GS}(\text{OFF}) = -2$ V and $r_{DS}(\text{ON}) = 40\ \Omega$. In the OFF state the drain-source resistance is 2000 MΩ. Determine the V_{DS} output voltage when $V_{DD} = +5$ V, $R_L = 1$ kΩ and $V_{GS} = 0$ V. Repeat for $V_{GS} = -5$ V.

Solution: With $V_{GS} = 0$ V the JFET is ON and we can replace it by $r_{DS}(\text{ON}) = 40\ \Omega$ between drain and source as shown in Figure 4–19(a). The value of V_{DS} can be calculated using the voltage-divider rule.

$$V_{DS}(\text{ON}) = \frac{40\ \Omega}{40\ \Omega + 1\ \text{k}\Omega} \times 5\ \text{V}$$

$$= \mathbf{192\ mV}$$

---

*V_{GS} can be made slightly positive (≤0.5 V). Any V_{GS} higher than this will forward-bias the gate-channel PN junction and cause heavy gate current flow.

With $V_{GS} = -5$ V, the JFET will be OFF since V_{GS} is more negative than $V_{GS}(\text{OFF}) = -2$ V. In the OFF state the JFET acts like a 2000 MΩ resistance between drain and source, as shown in Figure 4–19(b). Again, using the voltage-divider rule

$$V_{DS}(\text{OFF}) = \frac{2000\ \text{M}\Omega}{2000\ \text{M}\Omega + 1\ \text{k}\Omega} \times 5\ \text{V}$$

$$\approx \mathbf{5\ V}$$

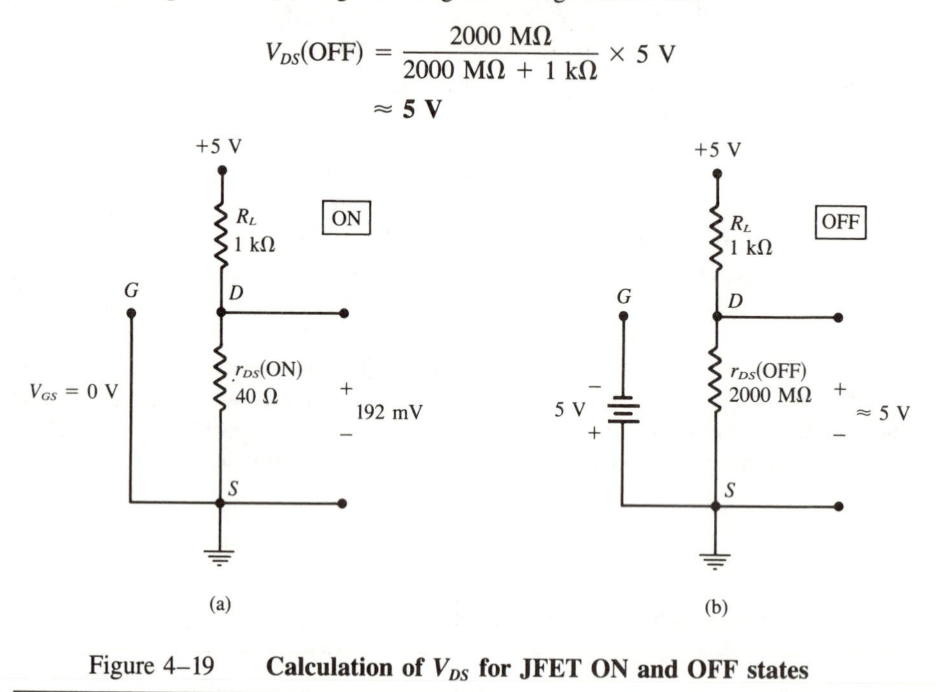

Figure 4–19 **Calculation of V_{DS} for JFET ON and OFF states**

You may have noted in this example that it is the gate-source voltage (not current) that determines whether the resistance between the switch terminals (drain and source) will be low or high. There will be a small current flowing through the gate input terminal, but this current is generally in the nA range and can normally be neglected. One of the major advantages of the JFET switch over the bipolar switch is that it draws negligible current from the input signal source.

A disadvantage of the JFET switch is that the gate input voltage has to have a polarity opposite to that of the drain voltage. Thus, the output from one JFET switching circuit cannot directly drive the input of another. In comparison, the enhancement MOSFET switch does have this feature.

E-MOSFET as a switch The enhancement-type MOSFET (E-MOSFET) is extremely well suited for digital applications, because the gate input voltage polarity is the same as the drain output polarity. Figure 4–20 shows an N-channel E-MOSFET connected in the common-source configuration. The E-MOSFET is a normally OFF device because there is no drain-source conductive channel when $V_{GS} = 0$ V. The device turns ON when V_{GS} is made more positive than $V_{GS}(\text{th})$, the gate-source threshold voltage required for conduction. In practice, V_{GS} is made much greater than $V_{GS}(\text{th})$ in order to reduce the ON resistance between drain and source. $V_{GS}(\text{th})$ is positive for N-channel devices and is typically 0.5–5 V.

Typically, the OFF state resistance is 10,000 MΩ or more, while the ON state resistance is in the 100 Ω–1 kΩ range. The input resistance at the gate terminal is extremely high (10^{12} Ω) so that a high-resistance signal source can drive the input without being loaded down.

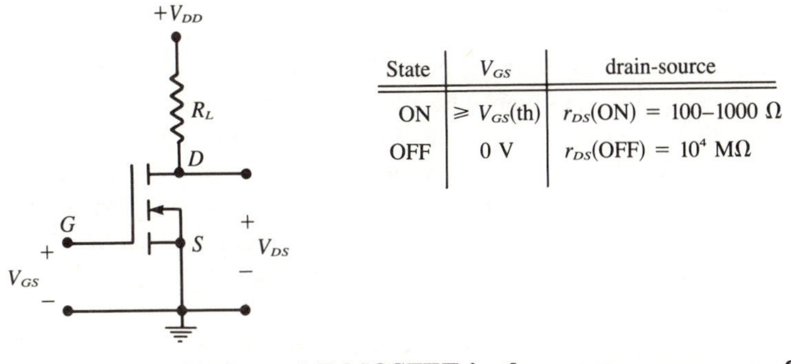

State	V_{GS}	drain-source
ON	$\geq V_{GS}(\text{th})$	$r_{DS}(\text{ON}) = 100\text{–}1000\ \Omega$
OFF	0 V	$r_{DS}(\text{OFF}) = 10^4\ \text{M}\Omega$

Figure 4–20 **N-channel E-MOSFET in the common-source configuration**

Example 4.14 The MOSFET in Figure 4–20 has $V_{GS}(\text{th}) = 1.5$ V and $r_{DS}(\text{ON}) = 100\ \Omega$. Using $V_{DD} = +5$ V and $R_L = 100\ \text{k}\Omega$, determine V_{DS} when $V_{GS} = 0$. Repeat for $V_{GS} = +5$ V.

Solution: With $V_{GS} = 0$ the E-MOSFET is OFF so that essentially all of V_{DD} will be across the high-resistance channel. Thus,

$$\mathbf{V_{DS} \approx +5\ V}$$

With $V_{GS} = +5$ V the E-MOSFET is ON so that the drain-source acts like a 100 Ω resistance. The output voltage will therefore be

$$V_{DS} = \frac{100\ \Omega}{100\ \Omega + 100\ \text{k}\Omega} \times 5\ \text{V}$$
$$\approx \mathbf{5\ mV}$$

These results show that the circuit is acting as a voltage level inverter (discussed in Chapter 1) since a 0 V input produces a +5 V output, and a +5 V input produces approximately a 0 V output. We will cover inverter circuits more thoroughly in a later chapter.

P-channel E-MOSFETs operate in basically the same manner as the N-channel devices, except that the drain and gate voltages are *negative* relative to the source, and the drain current flows in the opposite direction. Figure 4–21 shows a P-channel E-MOSFET circuit. Note that V_{DD} is negative. Also note that V_{GS} has to be made more negative than $V_{GS}(\text{th})$ to turn the device ON.

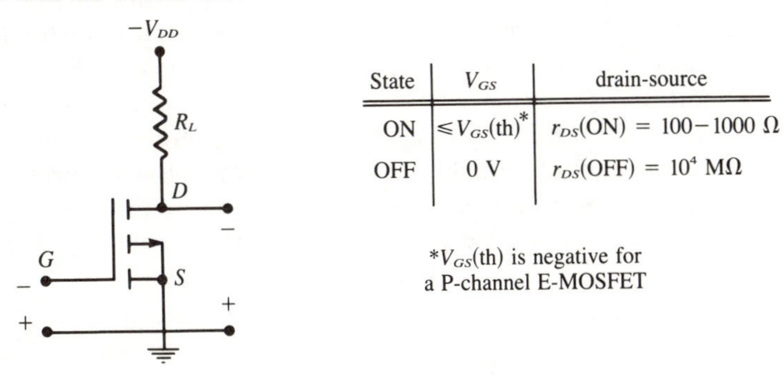

State	V_{GS}	drain-source
ON	$\leq V_{GS}(\text{th})^*$	$r_{DS}(\text{ON}) = 100\text{–}1000\ \Omega$
OFF	0 V	$r_{DS}(\text{OFF}) = 10^4\ \text{M}\Omega$

*V_{GS}(th) is negative for a P-channel E-MOSFET

Figure 4–21 **P-channel E-MOSFET**

4.9 Approximate Analysis of Switching Circuits

When we have to follow a switching circuit schematic diagram while troubleshooting or while trying to determine the circuit operation, it is rarely necessary to perform exact calculations in order to obtain useful results. In many switching circuits, we can often use approximations that simplify the analysis without introducing any significant errors. After all, the switching devices will be in either the fully conducting ON state, or the non-conducting OFF state. Some of the common approximations that can be used are given here. Others will be introduced in the examples to follow.

(a) Assume that the input to a switching device will either keep the device ON or keep the device OFF. Of course this assumption is based on knowing that the devices are actually being used as switches. In most cases the use will be obvious from the application. If it is not obvious, then a more exact analysis is required.

(b) A switching device in the ON state can be replaced by a short circuit between the switch terminals (e.g. C–E terminals for a BJT, D–S terminals of a JFET or MOSFET). In the case of a diode or a transistor *E–B* junction, you may want to include the forward-bias voltage drop if you think it will introduce too much error if neglected. This decision becomes easier to make as you gain more experience.

(c) A switching device in the OFF state can be replaced by an open circuit between the switch terminals.

(d) Assume that the gate inputs of JFETs and MOSFETs draw no current.

(e) Assume that the bipolar transistors have very large values of h_{FE} so that a very small I_B will drive them into saturation.

The following examples will illustrate how these approximations can be used to simplify the analysis of switching circuit operation.

Example 4.15 Refer to the circuit in Figure 4–22(a). Inputs v_1 and v_2 are logic signals that can either be at 0 V or +5 V with respect to ground. Determine the approximate value of v_{OUT} for the following cases: (a) $v_1 = v_2 = 0$ V; (b) $v_1 = 0$ V, $v_2 = 5$ V; (c) $v_1 = 5$ V, $v_2 = 0$ V; (d) $v_1 = 5$ V, $v_2 = 5$ V.

Solution:

(a) With $v_1 = 0$ V, transistor Q_1 will be ON because its *E–B* junction is forward-biased. With $v_2 = 0$ V, transistor Q_2 will be OFF. Replacing Q_1 by a short and Q_2 by an open, we can redraw the circuit as shown in Figure 4–22(b). Resistors R_2 and R_5 form a voltage-divider that provides the base voltage for Q_3, which is connected in the *common-collector* configuration as an *emitter-follower.* We can assume that Q_3's h_{FE} is high enough to allow us to neglect the effect of its input resistance ($\approx h_{FE} \times R_6$) on the base voltage divider. Thus, the voltage at Q_3's base will be

$$V_{B_3} \approx \frac{R_5}{R_5 + R_2} \times 5\text{ V} = 3\text{ V}$$

v_{OUT}, the voltage at Q_3's emitter, will be 0.7 V less than V_{B_3}.

$$v_{OUT} = 3\ \text{V} - 0.7\ \text{V} = \mathbf{2.3\ V}$$

(b) With v_2 changed to 5 V, Q_2 will be ON and this situation will place R_4 in parallel with R_5 as drawn in Figure 4–22(c). The parallel combination of R_2 and R_5 will have a resistance of

$$\frac{120 \times 270}{120 + 270} \approx 83\ \Omega$$

This resistance will give a Q_3 base voltage of

$$V_{B_3} = \frac{83}{83 + 180} \times 5\ \text{V} = 1.57\ \text{V}$$

and

$$v_{OUT} = 1.57 - 0.7 = \mathbf{0.87\ V}$$

(c) and (d) With $v_1 = 5$ V, Q_1 will be OFF because its E–B junction will not be forward-biased. With Q_1 essentially open [see Figure 4–22(d)], none of the +5 V bias supply voltage can reach the base of Q_3, regardless of the state of Q_2. Thus, $V_{B_3} = 0$ V and $v_{OUT} = 0$ V.

This table summarizes the results for all four cases:

v_1	v_2	v_{OUT}
0 V	0 V	2.3 V
0 V	5 V	0.87 V
5 V	0 V	0 V
5 V	5 V	0 V

Example 4.16 A technician makes the following measurements on the circuit of Figure 4–22(a).

v_1	v_2	v_{OUT}
0 V	0 V	2.19 V
0 V	5 V	2.19 V
5 V	0 V	0 V
5 V	5 V	0 V

What are some of the possible circuit faults that could produce these erroneous results?

Solution: If we compare this measurement table with the calculated results of the preceding example, we can see that the only *major* discrepancy is in the second entry. The small discrepancy in the first entry is probably due to the tolerances on component values and the approximations used in our calculations.

Since the output for the second case in the table was measured to be the same as the first case, it is obvious that changing v_2 from 0 V to 5 V had no effect on v_{OUT}. There

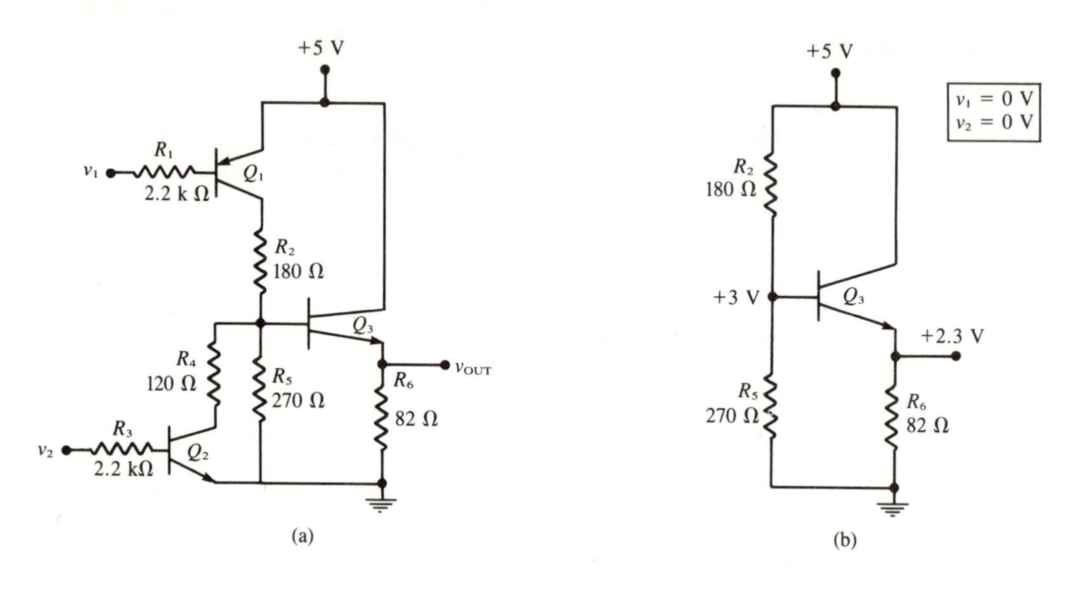

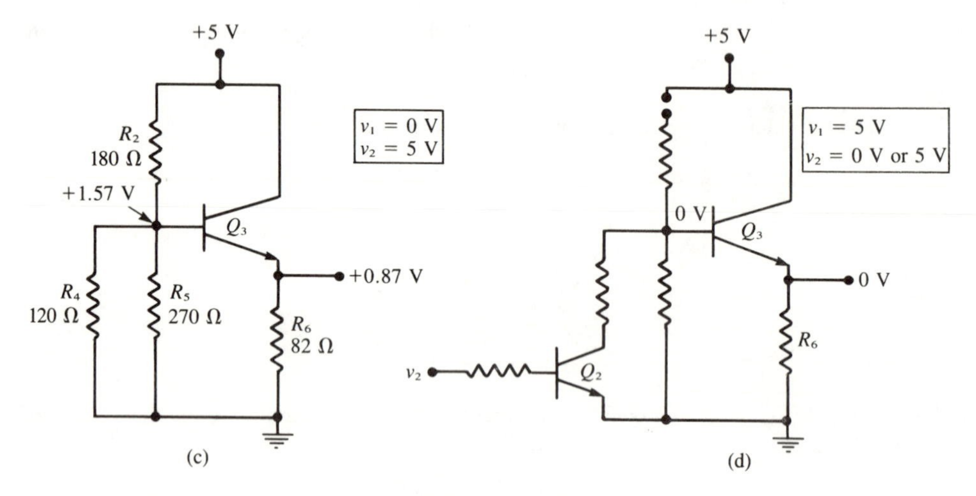

Figure 4–22 **Examples 4.15–4.16**

are several possible circuit faults that could prevent v_2 from having any effect on v_{OUT}. Some of them are

(a) Resistor R_3 is open internally or where it makes external connections to v_2 or to Q_2's base.

(b) R_4 is open internally or where it makes external connections to Q_2's collector or Q_3's base.

(c) Q_2 has an internal open from emitter to base or from collector to base.

(d) There is a break in the connection from Q_2's emitter to ground.

Example 4.17 Refer to the circuit of Figure 4–23. Inputs v_1 and v_2 are logic inputs that can either be 0 V or −5 V. Determine what combination of input levels will activate the motor.

Solution: The motor is activated by the closure of contacts *x-y*. This set of contacts is controlled by relay K_1. When the relay is energized (sufficient current through its coil) it will keep contacts *x-y* closed. The relay is energized when Q_3 turns ON. Q_3 will turn ON when the current I_1 flows from the +5 V supply through R_1 into its base.

JFETs Q_1 and Q_2 can prevent Q_3 from turning ON by shorting the base of Q_3 to ground and shunting I_1 away from the base. This effect will occur when *either* JFET (or both) is turned ON by 0 V at its gate. The only situation where Q_3 will be allowed to turn ON is when Q_1 and Q_2 are each turned OFF by −5 V at their gates. This will open both JFETs between drain and source and will allow base current to flow into Q_3.

Thus, the motor will be energized only when $v_1 = v_2 = -5$ **V.**

Example 4.18 What is the function of the diode across the relay coil in Figure 4–23?

Solution: The diode is used to suppress the large counter EMF ("inductive kick") that would appear across the coil when its current is abruptly interrupted by Q_3 turning OFF. The diode will limit the counter EMF to approximately 0.7 V. Without the diode, the counter EMF can be several hundreds of volts and could permanently damage the transistor.

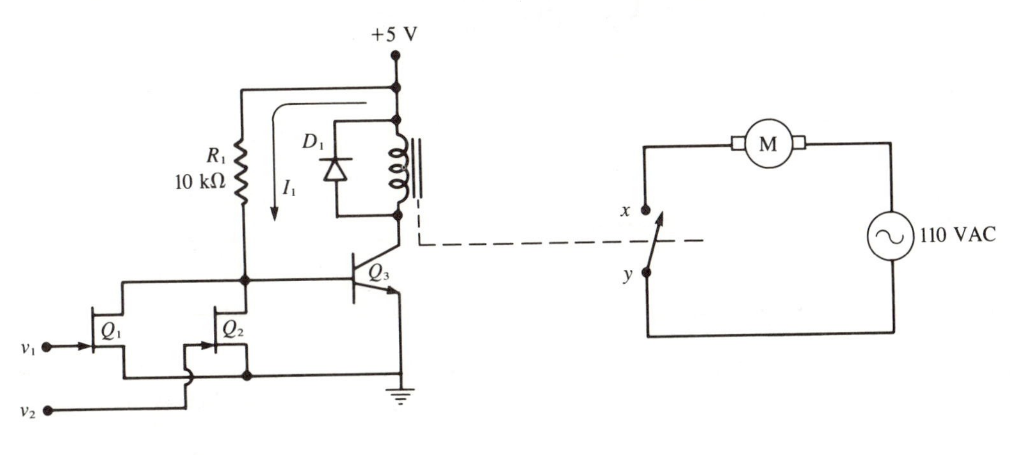

Figure 4–23 **Examples 4.17–4.18**

Example 4.19 Determine v_{OUT} for the circuit of Figure 4–24(a) when $e_{IN} = 0$ V. Repeat for $e_{IN} = +5$ V.

Solution:

(a) With $e_{IN} = 0$ V, Q_1 will be OFF so that there can be no current flow through D_1 and Q_1. Current from the +5 V supply will flow through R_2 and D_2 into the base of Q_2. This current turns Q_2 ON so that $v_{OUT} = 0$ **V.**

(b) With $e_{IN} = +5$ V, Q_1 will turn ON. We can replace Q_1 by a short circuit and redraw the circuit in Figure 4–24(b). With Q_1 ON, current can flow from the +5 V supply through R_2, D_1, and Q_1 to ground. The voltage at point *x* will be approximately +0.7 V (the drop across D_1).

This voltage is not large enough to provide forward bias for both D_2 and the *E–B* junction of Q_2. Thus, Q_2 will be OFF so that no current will flow through R_3, and $v_{OUT} = 12$ **V.**

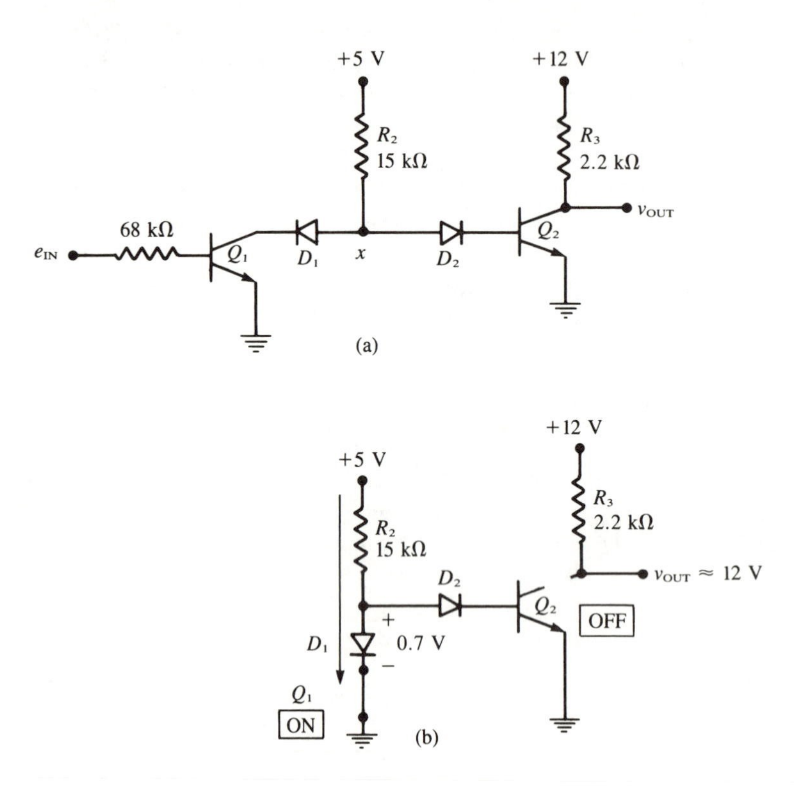

Figure 4–24 **Example 4.19**

Questions

Sections 4.1–4.3

4.1 A certain silicon switching diode with a reverse breakdown rating of 50 V is used in each of the circuits of Figure 4–25. Calculate the diode current and voltage in each case.

4.2 Determine v_{OUT} for each of the e_{IN} values given in Figure 4–26. The diode is silicon.

Section 4.4

4.3 What factors determine a PN junction's switching speed?

4.4 In the circuit of Figure 4–27 the input is alternately turning the diode ON and OFF. Assume that the diode's switching times are *slower* than the input rise and fall times, and sketch the diode and resistor waveforms.

4.5 Indicate the effects of the following changes on the turn-ON time of the diode in Figure 4–27.

(a) Increase R

(b) Increase V_1

(c) Increase V_2

4.6 Repeat the preceding problem for t_{OFF}.

4.7 The circuit of Figure 4–28 is supposed to pass negative input pulses and completely block positive pulses because of the diode. However, the application of fast positive input pulses produces spikes in the output as shown. What is causing these spikes, and what can be done to eliminate them?

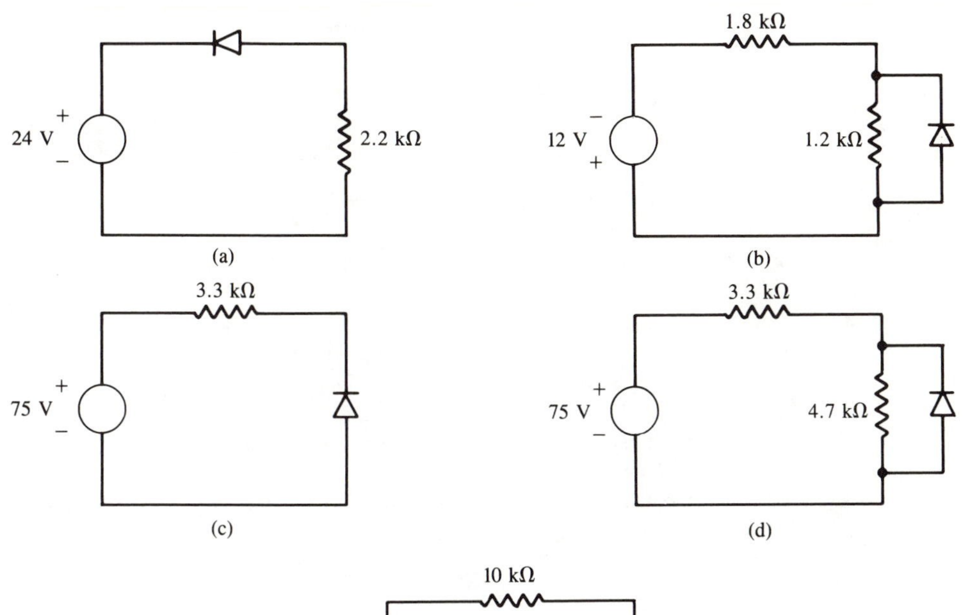

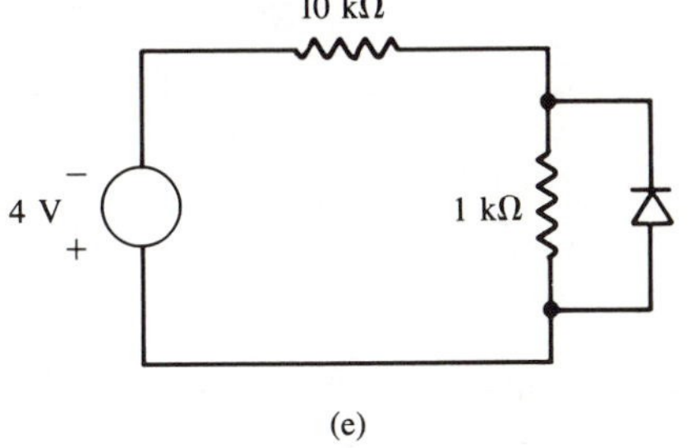

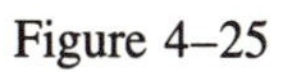
Figure 4–25

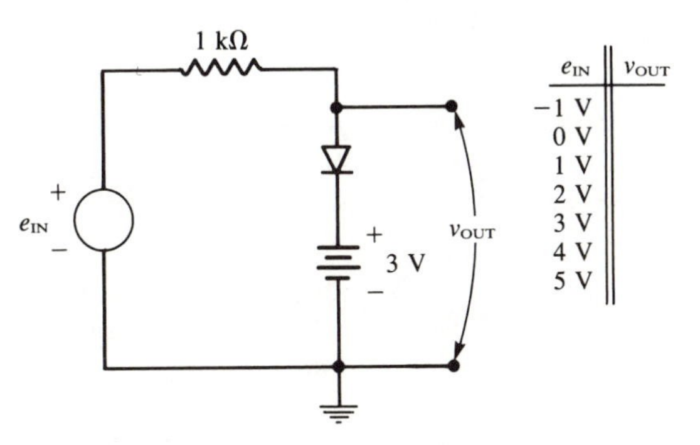

e_{IN}	v_{OUT}
−1 V	
0 V	
1 V	
2 V	
3 V	
4 V	
5 V	

Figure 4–26

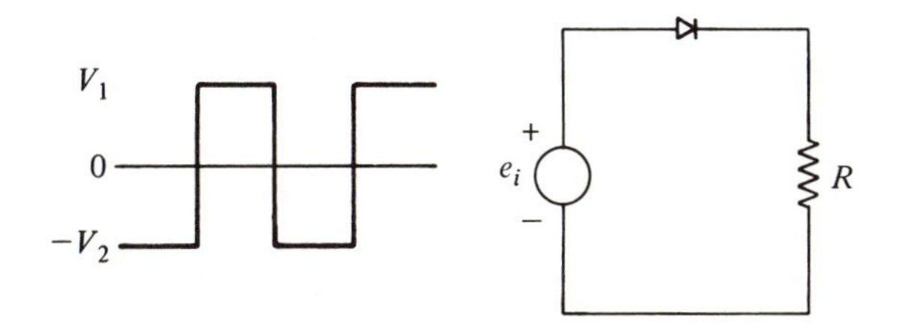

Figure 4–27

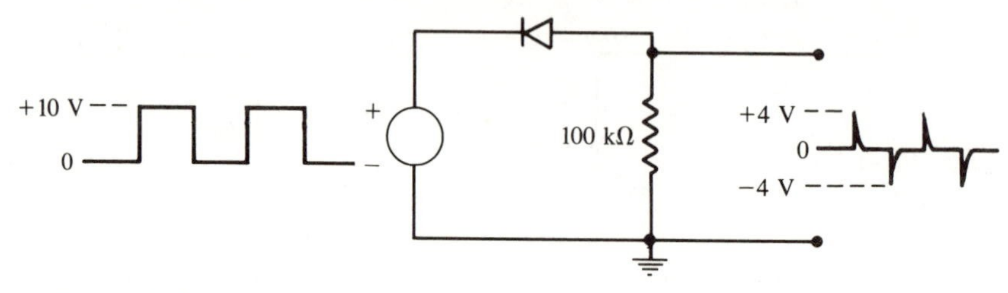

Figure 4–28

Sections 4.5–4.6

4.8 A silicon transistor is connected in the common-emitter configuration (Figure 4–11). Determine I_B, I_C, and V_{CE} when $R_C = 150\ \Omega$, $R_B = 10\ k\Omega$, $V_{CC} = 6$ V, and $e_{IN} = 6$ V. The transistor has a typical h_{FE} of 120.

4.9 One of the main advantages of operating the transistor as a saturated switch is that the ON state will be relatively insensitive to normal changes in transistor characteristics due to temperature.

Assume that due to temperature variations, the transistor of Problem 4.8 has its V_{BE} and h_{FE} values change as shown in the following.

Temp. °C	V_{BE}	h_{FE}
0°	0.75 V	90
25°	0.7 V	120
50°	0.65 V	180

Determine I_C and V_{CE} at each temperature.

4.10 Change the 1 kΩ to 47 kΩ and repeat Example 4.9, Figure 4–12.

4.11 In the circuit of Figure 4–29(a) the transistor has an h_{FE} that can range from 150 to 250. Determine the maximum useable value for R_B if the transistor is to be turned ON whenever e_{IN} is lower than 2 V.

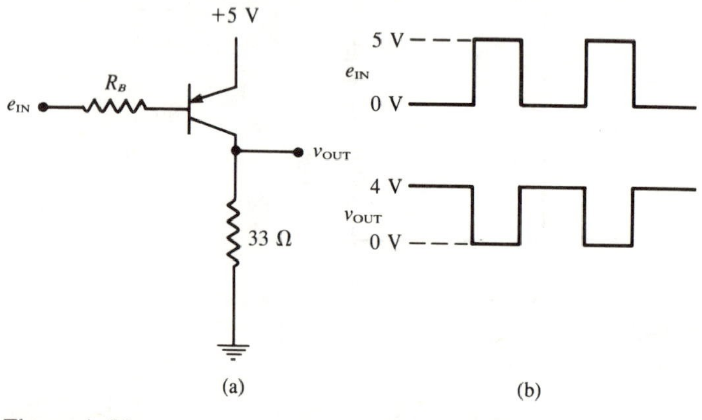

Figure 4–29

4.12 A technician tests the circuit of Figure 4–29(a) by applying the e_{IN} waveform shown in Figure 4–29(b) and by observing the resultant v_{OUT} waveform. What is wrong with the v_{OUT} waveform, and which of the following are possible causes?

(a) The transistor h_{FE} is lower than the specified minimum.

(b) The value of R_B is too large.

(c) The transistor is leaky in the OFF state.

(d) The +5 V supply voltage has decreased due to loading.

Section 4.7

4.13 A 10 V pulse riding on a 0 V baseline is applied to an NPN common-emitter circuit. Sketch the voltage waveform at the collector, taking into account all of the various delays caused by junction capacitance and charge storage.

4.14 For the situation of Problem 4.13, indicate how each of the following changes will affect t_{ON}.

(a) decrease in R_B

(b) increase in amplitude of input pulse

(c) decrease in R_C

(d) increase in h_{FE}

(e) speed-up capacitor across R_B

(f) clamping diode from collector to base

4.15 Repeat Question 4.14 for t_{OFF}. Keep in mind that the storage time, t_s, depends on how overdriven the transistor is. You can use the ratio of I_B(ACTUAL) to I_B(SAT) to determine the degree of oversaturation.

4.16 Explain how a C–B clamping diode decreases a transistor's switching time. Why can't the diode be silicon?

Section 4.8

4.17 The JFET in Figure 4–30 has r_{DS}(ON) = 50 Ω and V_{GS}(OFF) = −3.5 V. Determine v_{OUT} for e_{IN} = 0 V and e_{IN} = −50 V.

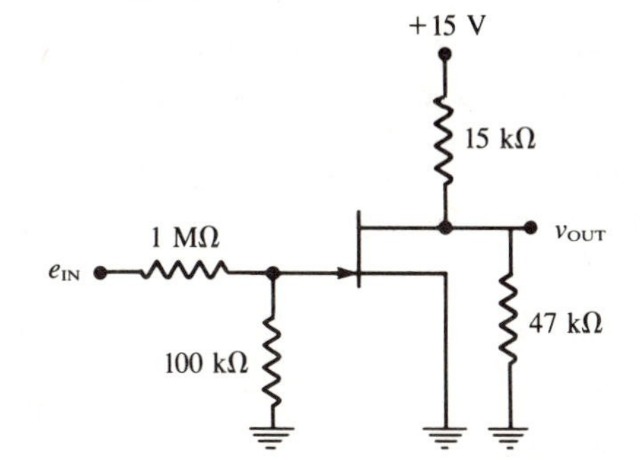

Figure 4–30

4.18 Give one advantage and one disadvantage of the JFET switch as compared to the bipolar transistor.

4.19 How does an E-MOSFET switch differ from a JFET switch?

4.20 The JFET in Figure 4–31 has V_{GS}(OFF) = −2 V, and the E-MOSFET has V_{GS}(th) = +1.5 V. Determine v_{OUT} for e_{IN} = 0 V and e_{IN} = −5 V.

4.21 What is the largest value that R_1 can have in Figure 4–31 if v_{OUT} is to be ≈0 V when e_{IN} = −5 V?

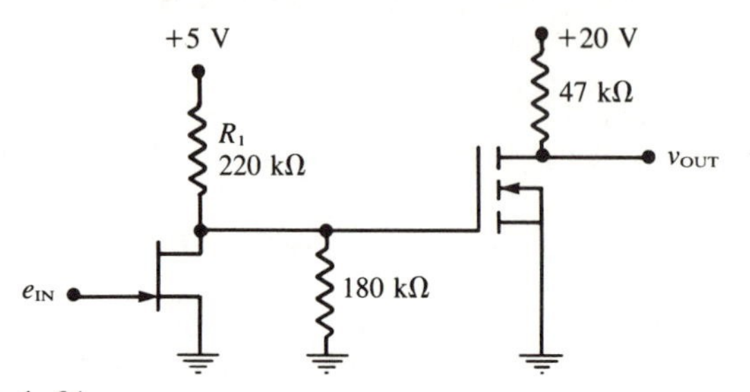

Figure 4–31

Section 4.9

4.22 Refer to the circuit of Figure 4–22(a). A technician makes the following measurements on this circuit.

v_1	v_2	v_{OUT}
0 V	0 V	1.25 V
0 V	5 V	0.9 V
5 V	0 V	0 V
5 V	5 V	0 V

From the items in the following list, find the actual circuit fault that caused these results. For each of the other items, explain why it could not produce the observed results.

(a) Q_1 is shorted between emitter and collector.

(b) Q_3 is shorted between emitter and base.

(c) Q_2 is shorted between emitter and collector.

(d) There is a break in the connection from R_5 to the base of Q_3.

4.23 Determine the approximate value of v_{OUT} in the circuit of 4–32 for each set of input voltages listed in the accompanying table. Assume that all the E-MOSFETs have $V_{GS}(\text{th}) = +2$ V.

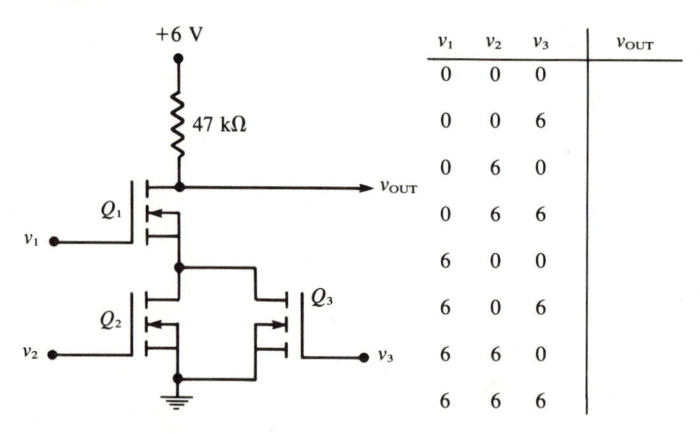

v_1	v_2	v_3	v_{OUT}
0	0	0	
0	0	6	
0	6	0	
0	6	6	
6	0	0	
6	0	6	
6	6	0	
6	6	6	

Figure 4–32

4.24 A technician makes measurements on the circuit of Figure 4–32 and finds that v_{OUT} is correct for all of the cases except $v_1 = 6$ V, $v_2 = 0$ V, and $v_3 = 6$ V, where it reads 6 V instead of 0 V. What are some of the most probable causes for this discrepancy?

4.25 Where would you connect a bipolar transistor switch in the circuit of Figure 4–32 so that it would keep $v_{OUT} \approx 0$ V whenever its input is 6 V, *regardless* of the states of v_1, v_2 and v_3?

4.26 What would probably happen to v_{OUT} in Figure 4–24 if diode D_2 were shorted out?

5

The Operational Amplifier

5.1 Introduction

The operational amplifier (op-amp) is an extremely versatile circuit that has gained wide acceptance as a basic building block in electronic systems. Although it is inherently an amplifier, and so finds many applications in the linear, analog areas, it is also capable of being used in numerous switching circuit applications.

The op-amp is a very complex circuit containing many components. For example, one of the most popular versions, the 741, consists of 20 transistors, 12 resistors, and a capacitor. Fortunately, most op-amps are now packaged as integrated circuits, so it is rarely necessary to be concerned with their internal circuitry. In fact, we can treat the op-amp as a device having certain input and output characteristics. A thorough knowledge of these characteristics is all that is needed to be able to understand and analyze the operation of most op-amp circuits.

The material in this chapter assumes no prior knowledge of op-amps. It is intended to provide a basic understanding of the op-amp characteristics without emphasizing any particular application area. Switching circuit applications of the op-amp will be presented in appropriate chapters later in the text and will use the background developed in this chapter. If you are already familiar with the op-amp, you may wish just to skim the topics and review only those you feel necessary.

5.2 What is an Operational Amplifier?

The term *operational amplifier* normally refers to a high-gain, directly coupled voltage amplifier with a *differential* input (two input terminals, neither of which has to be grounded). Since it is directly coupled (meaning it does not use coupling capacitors to block DC), it can amplify signal frequencies all the way down to 0 Hz. Its upper frequency limit is typically over 1 MHz.

Although the op-amp is a complete amplifier, it is designed so that external components can be connected to its terminals to determine the amplifier operating characteristics. That is, the amplifier voltage gain, input and output impedances, and frequency response depend almost solely on *stable* external components. This dependence makes it relatively easy to tailor the amplifier to fit a particular application, and it is this versatility that has made the use of op-amps so popular in industry.

In many cases an op-amp has its output connected to one of its inputs, either directly or indirectly. This situation is called *feedback;* when feedback is employed we say that the op-amp is operating in the *closed-loop mode*. When no feedback is employed, we say that the op-amp is operating in the *open-loop mode*. These two modes should become clearer as we go along.

Through most of our study of op-amps, we will neglect many of those effects that contribute to non-ideal operation; inclusion of those effects can easily obscure the basic simplicity of op-amp circuits. Furthermore, in many cases, the op-amp does operate very close to ideally; this is especially true of the newer devices on the market. We will discuss some of the effects that cause non-ideal operation just before the end of this chapter.

5.3 Basic Op-amp Characteristics

All op-amps have at least five terminals, as identified on the op-amp symbol in Figure 5–1. There are two input terminals, an output terminal, and two power supply bias terminals. Most op-amps have other terminals used for special purposes that are of no concern at this time.

The op-amp triangle symbol of Figure 5–1 is fairly standard, though there are variations occasionally. Sometimes the negative (−) and positive (+) inputs are reversed, with − on top and + on bottom. In some instances, the + terminal is not shown at all, and it is assumed to be connected to ground. (For the balance of this chapter, the words *positive* and *negative* will be dropped, and the symbols + and − will be used.)

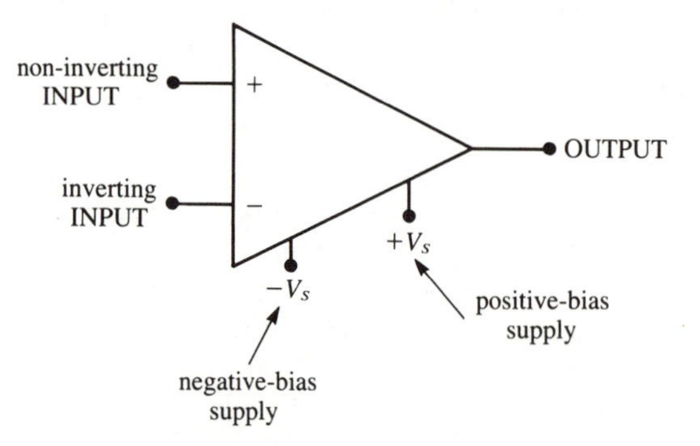

Figure 5–1 **Op-amp symbol showing the most important terminals**

Where is ground? As stated earlier in the definition of the op-amp, neither input terminal has to be connected to ground. Thus, none of the op-amp terminals serve as our ground reference. Where, then, is our ground reference? This question can be answered by looking at Figure 5–2(a), where the two bias supplies have been shown connected to a common-ground reference. This ground serves as the reference for all the terminal voltages of the op-amp as follows:

e_1 —The voltage at the op-amp "+" input with respect to ground.
e_2 —The voltage of the "−" input with respect to ground.
e_{OUT}—The voltage at the output terminal with respect to ground. (The load, if any, is connected between e_{OUT} and ground.)
$+V_S$—The voltage at the positive supply terminal with respect to ground.
$-V_S$—The voltage at the negative supply terminal with respect to ground.

The diagram in Figure 5–2(a) is too cumbersome to use on circuit schematics. It is usually simplified as in Figure 5–2(b), where the ground is not shown but is assumed to be at the common terminal of the positive and negative supplies. All the voltages at the various op-amp terminals are assumed to be referenced to this unseen ground.

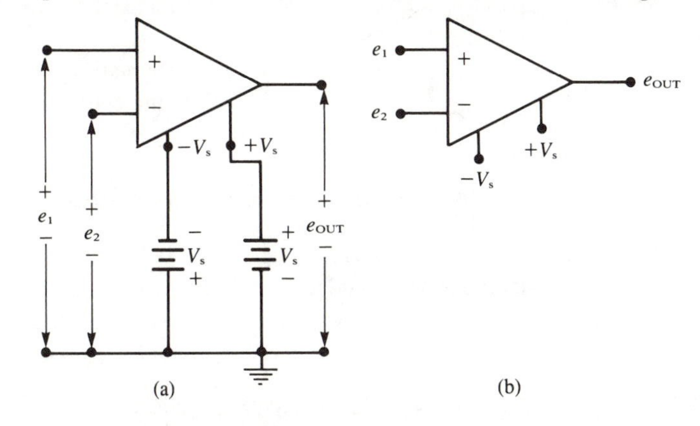

Figure 5–2 **(a) Ground is the common terminal of the positive and negative bias supplies; (b) In most circuit schematics the common ground is not shown, but it is assumed to be at common terminal of the supplies.**

To further simplify the op-amp diagram, many schematics do not show the power supply terminals on the op-amp symbol. We will sometimes show them, however, to remind us of the op-amp's voltage limits.

Bias supply values Most op-amps use *symmetrical* + and − supply voltages; that is, both supplies are the same magnitude. The most common supply voltages are ±15 V. The use of + and − bias supplies allows the op-amp output voltage to swing both positive and negative with respect to ground. Some op-amps use only a single polarity supply; for example, +30 V and 0 V connected to the supply terminals. The polarity of e_{OUT} is then limited to this same polarity.

Input/output characteristics The op-amp is a *differential* amplifier, which means that it amplifies the *difference* in the voltages present at the two input terminals. Specifically, the output voltage is given by

$$e_{OUT} = A_{VOL} \times (e_1 - e_2) \tag{5–1}$$

where e_1 and e_2 are respectively, the voltages at the + and − input terminals with respect to ground. A_{VOL} is the op-amp's *open-loop voltage gain;* that is, the op-amp voltage gain without any external feedback connections. The $(e_1 - e_2)$ term is the *difference* in the voltages present at the input terminals. We will often refer to this difference voltage as the *differential* input, e_d. Thus, we have

$$e_d = e_1 - e_2 \tag{5–2}$$

which is the voltage that we would measure at the + input with respect to the − input (see Figure 5–3). Note that when $e_1 > e_2$, e_d will be positive, and when $e_1 < e_2$, e_d will be negative. It is important to remember that the op-amp *amplifies* this differential and not e_1 or e_2. To emphasize this we can rewrite Equation 5.1 as

$$e_{OUT} = A_{VOL}e_d \tag{5–2}$$

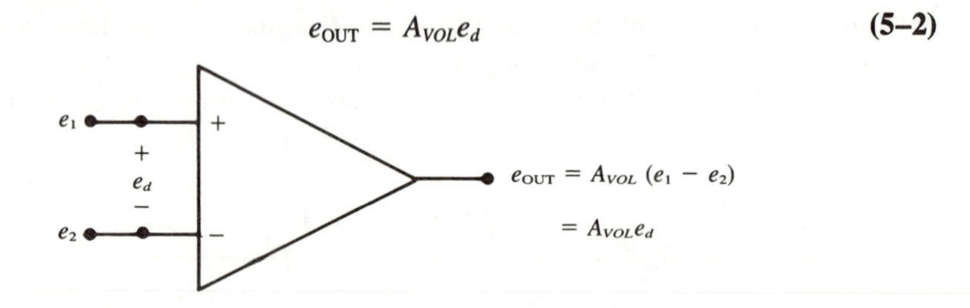

Figure 5–3 **Op-amp amplifies the *difference* between e_1 and e_2**

Modern op-amps have very large values of open-loop voltage gain, A_{VOL}, ranging from 1000 to 1,000,000. It is important to understand that the *differential voltage*, e_d, is amplified by A_{VOL} to produce e_{OUT}. This is *always* true, even when the op-amp is used in the closed-loop mode as we shall see later.

Example 5.1 Determine e_d and e_{OUT} for each of the cases in Figure 5–4, using $A_{VOL} = 1000$.

Solution:

(a) $e_d = e_1 - e_2 = 100\text{ mV} - 90\text{ mV} = \mathbf{+10\ mV}$

$e_{OUT} = A_{VOL}e_d = 1000 \times 10\text{ mV} = \mathbf{+10\ V}$

(b) $e_d = e_1 - e_2 = 12.09\text{ V} - 12.10\text{ V} = \mathbf{-10\ mV}$

$e_{OUT} = 1000 \times (-10\text{ mV}) = \mathbf{-10\ V}$

Note that e_{OUT} is smaller than either input voltage in (b) because the op-amp amplifies the input *differential*.

(c) $e_d = -10\text{ mV} - 0 = \mathbf{-\ 10\ mV}$

$e_{OUT} = 1000 \times (-10\text{ mV}) = \mathbf{-10\ V}$

(d) $e_d = 0 - (-10\text{ mV}) = \mathbf{+10\ mV}$

$e_{OUT} = 1000 \times 10\text{ mV} = \mathbf{+10\ V}$

The results of this example bring out several important points:

(1) When e_d is *positive,* the op-amp's + input is *positive* with respect to its − input, and e_{OUT} is *positive*.

(2) When e_d is *negative,* the op-amp's "+" input is *negative* with respect to its "−" input, and e_{OUT} is *negative*.

(3) With the "+" terminal grounded ($e_1 = 0$), the output has the *opposite* polarity as the voltage applied to the "−" terminal. For this reason, the "−" input is often called the *inverting* input.

(4) With the "−" terminal grounded ($e_2 = 0$), the output has the *same* polarity as the voltage applied to the "+" terminal. For this reason, the "+" input is often called the *non-inverting* input.

(5) The output will be exactly 0 V only when $e_1 = e_2$.

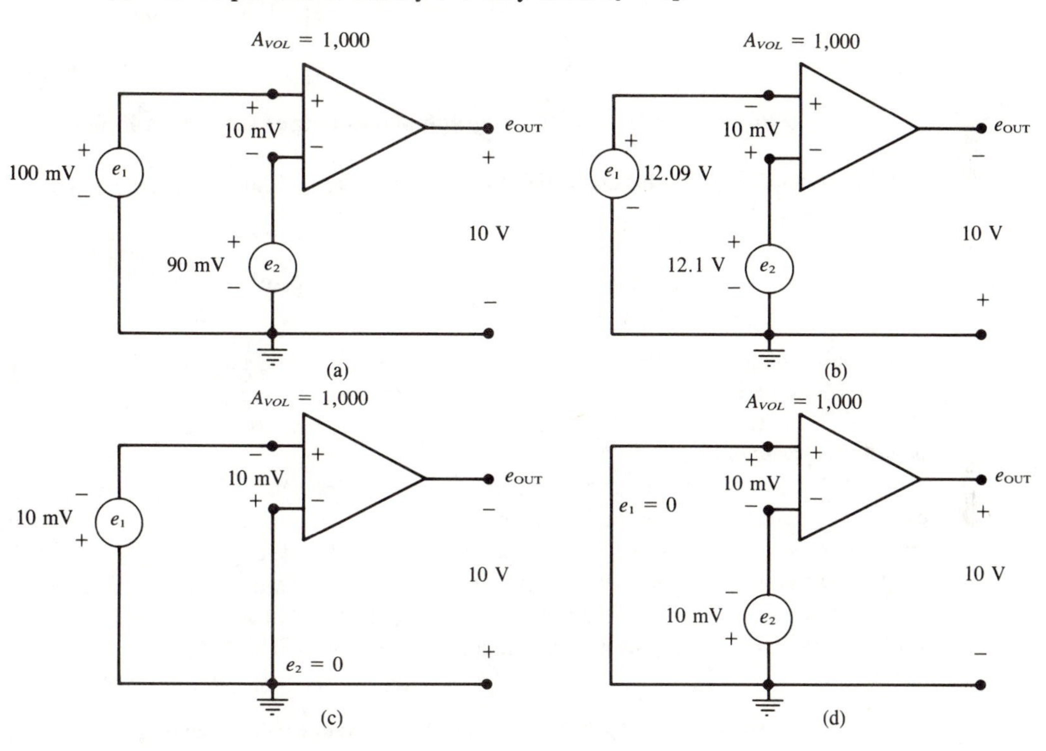

Figure 5–4

Output voltage saturation The maximum amplitude of the op-amp output voltage is limited by the size of the power supply voltages connected to the op-amp bias terminals. Typically, the maximum amplitude is about 1.5 V less than the bias supply. For example, using ±15 V bias supplies, the amplitude of e_{OUT} is limited to ±13.5 V; that is, $-13.5\text{ V} \leqslant e_{OUT} \leqslant +13.5\text{ V}$. This case can be shown graphically by plotting e_{OUT} versus e_d, as shown in Figure 5–5 for an op-amp with $A_{VOL} = 10^4$.

Figure 5–5 shows that e_{OUT} varies linearly with e_d (that is, $e_{OUT} = A_{VOL}e_d$) over a very small range of e_d values. When the magnitude of e_d exceeds a certain limit, e_{OUT} *saturates* at either $+V_{SAT}(+13.5\text{ V})$ or $-V_{SAT}(-13.5\text{ V})$, depending on the polarity of e_d.

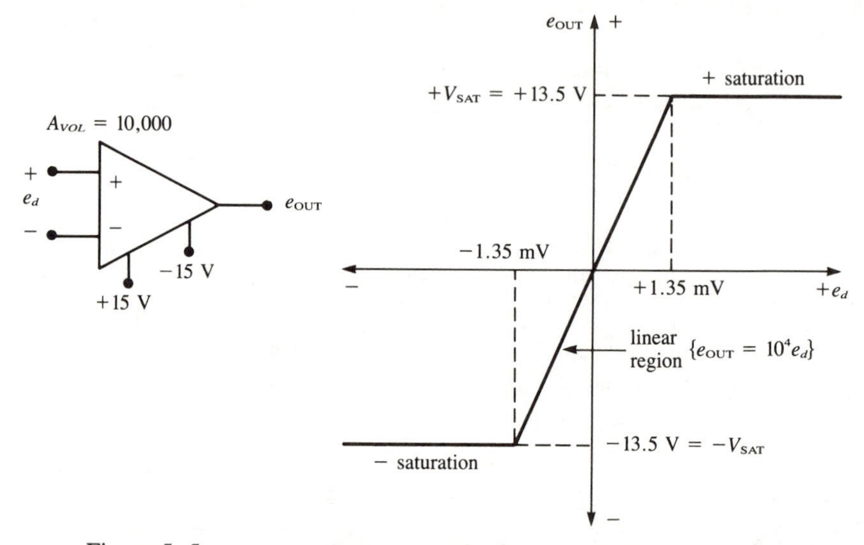

Figure 5–5 **e_{OUT} versus e_d graph shows output saturation limits**

The magnitude of e_d that produces the onset of saturation is found by setting $A_{VOL} \times e_d$ equal to ±13.5 V. Thus,

$$e_{d(\mathrm{SAT})} = \frac{\pm V_{\mathrm{SAT}}}{A_{VOL}} = \frac{\pm 13.5\ \mathrm{V}}{10{,}000} = \mathbf{\pm 1.35\ mV}$$

This result indicates that only 1.35 mV of either polarity across the input terminals will drive the output into saturation. Many op-amps have open-loop gains of greater than 100,000. With A_{VOL} = 100,000, the differential input only has to exceed 135 μV of either polarity to produce saturation.

What this fact means is that the op-amp will operate in its *linear* range only for extremely small differences in the voltages applied to its inputs. This limitation might appear to be serious. However, in many op-amp circuits, the use of *negative feedback* will keep the differential e_d small enough for linear operation (as we shall see later).

In analyzing many op-amp circuits, it is useful to know that e_d is very small when the op-amp output is in its linear range (not saturated). The following rule will be handy later.

> *When an op-amp output is in its linear range, e_d is extremely small; that is, the voltage across the input terminals can be assumed to be virtually zero.*

Example 5.2 An op-amp with $A_{VOL} = 10^5$ uses ±15 V bias supplies and has e_{OUT} = +2 V. If the voltage at the "+" input is 4 V, what is the voltage at the "−" input?

Solution:

$$e_{\mathrm{OUT}} = A_{VOL} e_d$$

Therefore,

$$e_d = \frac{e_{\mathrm{OUT}}}{A_{VOL}} = \frac{+2\ \mathrm{V}}{10^5} = +20\ \mu\mathrm{V}$$

Thus, the voltage at the "−" input must be 4 V − 20 μV, which is virtually the same as the voltage at the "+" input. In practice we can assume the op-amp input terminals are at the same voltage *as long as the output is below its saturation limits.*

5.4 Op-Amp Equivalent Circuit

Like all voltage amplifiers, the op-amp can be represented by a general equivalent circuit (illustrated in Figure 5–6). The input side is shown as an impedance Z_{IN}, which is the op-amp's differential input impedance; that is, the impedance between the input terminals. The output side is shown as a dependent voltage source with an amplitude $A_{VOL} \times e_d$ and an output impedance Z_{OUT}. This equivalent circuit is valid only in the linear region.

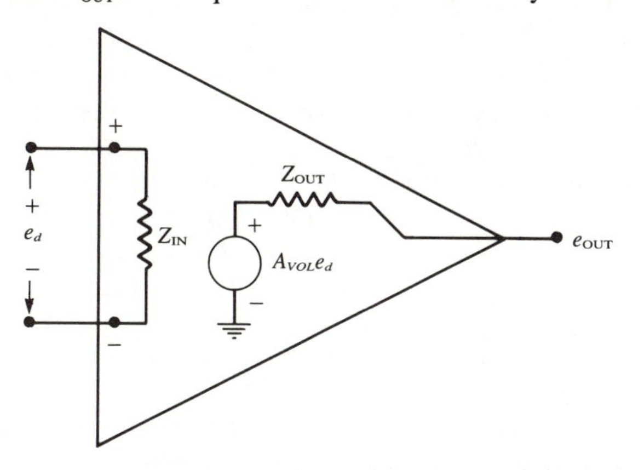

Figure 5–6 **Op-amp equivalent circuit (linear operation)**

Input impedance and input current Z_{IN} is shown as a resistance in Figure 5–6—although at very high frequencies it will have a significant capacitive reactance component. However, for our purposes we can consider it as pure resistance. Values of Z_{IN} are usually rather high for op-amps, generally ranging from 10 kΩ to over 1 MΩ, with most tending toward the higher end of that range.

When an op-amp is operating in its linear region, the combination of high Z_{IN} and small e_d results in an extremely small current flowing through the input terminals. For example, with $Z_{IN} = 1$ MΩ and $e_d = 100$ μV, the input current will be only 0.1 nA. This effect allows us to establish a second rule that is useful in analyzing op-amp circuits.

The current flowing through the op-amp input terminals can be assumed to be virtually zero when the op-amp is in its linear region.

We will make good use of this rule later.

Output impedance and output current The value of Z_{OUT} for a modern op-amp is generally in the range of 50–200 Ω, which is considered relatively low. However, when an op-amp is used with negative feedback, its *effective* output impedance is reduced to less than 1 ohm. This characteristic allows the op-amp to drive most loads with very little loss of output amplitude.

The maximum available output current from an op-amp is typically in the 5–25 mA range. This relatively low output current is not a serious drawback, because in most op-amp applications the output is driving another op-amp circuit with high Z_{IN}. When a high-current load is to be driven, a high-current transistor can be used to amplify the op-amp output current.

5.5 Op-Amp Circuits: the Comparator

The op-amp is normally used in the closed-loop mode (with feedback) because the voltage gain can be accurately controlled with external components. It is seldom used in the open-loop mode because its large open-loop voltage gain drives the output into saturation for any reasonable input signal amplitudes. There is, however, one very important open-loop application called the *voltage comparator.*

A voltage comparator compares one voltage to another and signals which one is greater. In the op-amp comparator, the output normally will be at either saturation limit depending on which input is greater. This situation is shown in Figure 5–7. When e_1 exceeds e_2 by an amount greater than $|e_d(\text{SAT})|$, the output will be in positive saturation at +13.5 V. Conversely, when e_2 exceeds e_1 by more than $|e_d(\text{SAT})|$, e_{OUT} will be saturated at −13.5 V. $|e_d(\text{SAT})|$ is typically very small (135 μV in Figure 5–7) so that normally one of the inputs will exceed the other by more than this value, and e_{OUT} will be at saturation. For this reason, the comparator can be approximately described as shown in the figure. Remember, however, that there is a small linear region (Figure 5–5) where $|e_d| < |e_d(\text{SAT})|$ and $e_{OUT} \neq \pm V_{SAT}$. We will not be concerned with this until we study comparators in more detail in the next chapter.

The voltage comparator has many uses. Figure 5–8 shows two of them. The circuit in (a) is called a *zero crossing* detector since e_{OUT} switches from one saturation level to the other when the e_s waveform crosses 0 V. While e_s is slightly positive, $e_{OUT} = +13.5$ V. When e_s decreases below 0 V, e_{OUT} switches to −13.5 V. When e_s increases to above 0 V, e_{OUT} switches back to +13.5 V. The e_{OUT} waveform is a square wave of the same frequency as e_s.

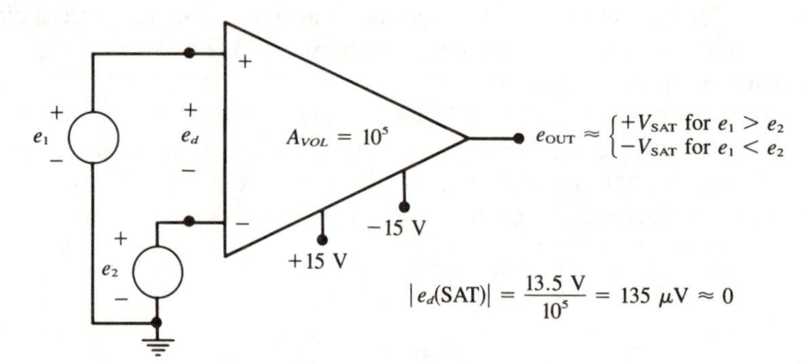

Figure 5–7 **Basic op-amp comparator**

The circuit in Figure 5–8(b), called a *level detector,* detects when a signal, e_{IN}, drops below a DC reference level. The DC reference E_{REF} is connected to the "+" input and e_{IN} is applied to the "−" input. As long as e_{IN} is slightly greater than E_{REF}, the op-amp output will be saturated at −13.5 V. Whenever e_{IN} drops below E_{REF}, the output will switch to +13.5 V.

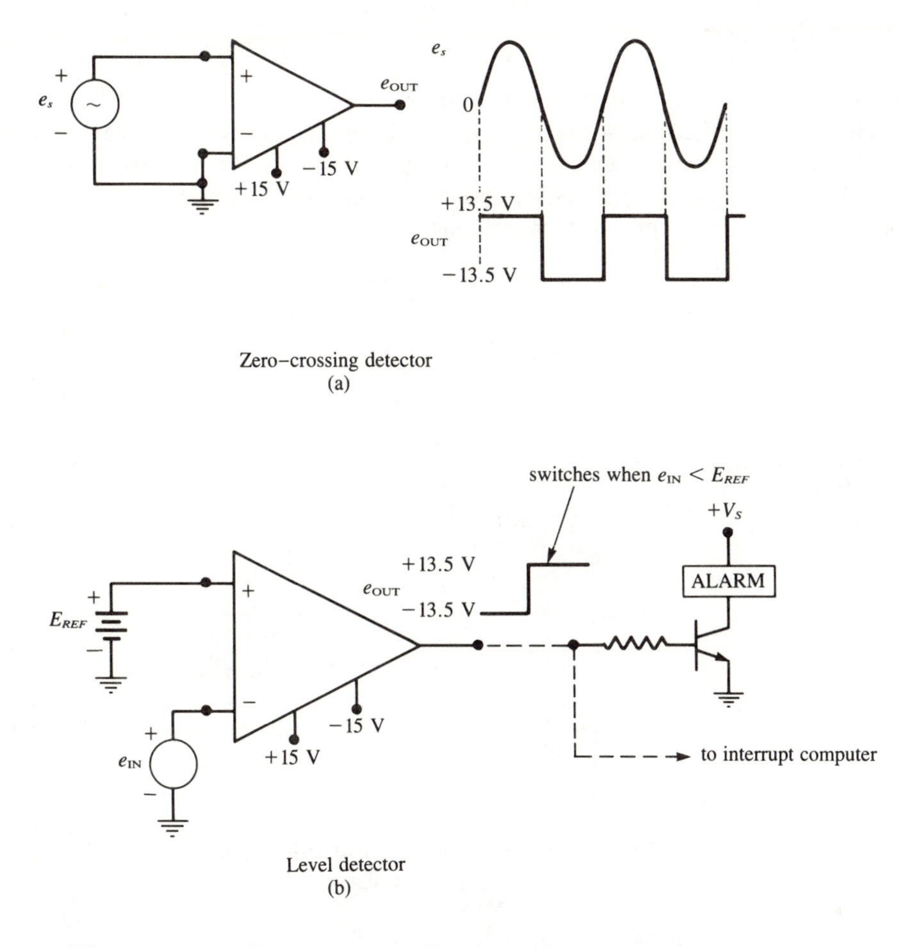

Figure 5–8 **Comparator applications**

The reference voltage E_{REF} can be of either polarity. It can be a separate DC source, or it can be derived from the ± supplies using a potentiometer or zener diode. The input e_{IN} might be a voltage that is proportional to some industrial process variable (e.g. temperature, liquid level, pressure) being monitored. It could also be the voltage from a critical DC power supply in an electronic system, such as for the memory in a digital computer. If this memory supply drops too low, for instance in a power failure, the entire contents of memory could be lost.

The level detector output might be used to activate an alarm, or it might be transmitted to a computer to interrupt its operation to allow the transfer of important information from its main memory to a battery-powered back-up memory.

Comparator switching speed The output waveforms in Figure 5–8 are idealized since they show instantaneous switching from one saturation level to another. Most op-amps are not designed for extremely rapid switching; typically total switching time would be around 50 μs. Some of the newer op-amps are specifically designed for comparator applications and have switching times as low as 25 ns.

5.6 The Voltage-Follower

We will now look at the simplest closed-loop op-amp circuit (Figure 5–9). The input e_{IN} is applied to the "+" input and the output e_{OUT} is connected directly back to the "−" input; this feedback is negative since the output voltage is fed back to the "−" input. Negative feedback may seem strange at first because the output helps to determine the differential input, which is then amplified to produce that output. That is

$$e_d = e_{IN} - e_{OUT}$$

But,

$$\begin{aligned} e_{OUT} &= A_{VOL}e_d \\ &= A_{VOL}(e_{IN} - e_{OUT}) \end{aligned} \quad \textbf{(5–3)}$$

Solving for e_{OUT}, we obtain

$$e_{OUT} = \left(\frac{A_{VOL}}{1 + A_{VOL}}\right) \times e_{IN} \quad \textbf{(5–4)}$$

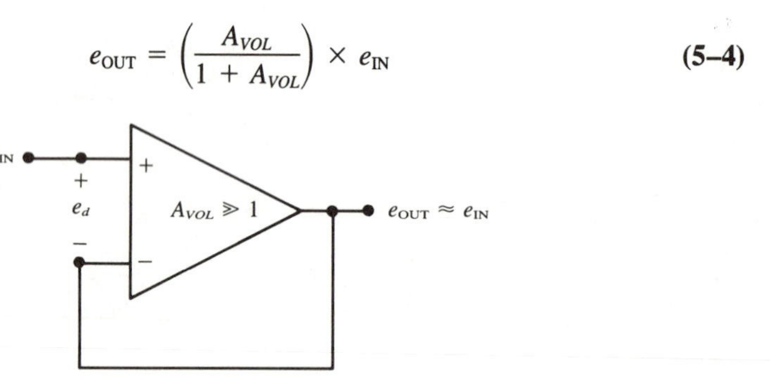

Figure 5–9 **Op-amp voltage-follower uses full negative feedback to produce $e_{OUT} \approx e_{IN}$ for large values of A_{VOL}**

The voltage gain of this circuit from e_{IN} to e_{OUT} is obtained from Equation 5.5 as

$$A_{CL} = \frac{e_{OUT}}{e_{IN}} = \frac{A_{VOL}}{1 + A_{VOL}} \quad \textbf{(5–5)}$$

A_{CL} represents the *closed-loop voltage gain* of the circuit and should not be confused with the A_{VOL} of the op-amp.

Clearly, for large values of A_{VOL}, the closed-loop gain given by Equation 5.5 reduces to

$$A_{CL} = \frac{e_{OUT}}{e_{IN}} \approx 1 \quad \textbf{(5–6)}$$

which indicates that $e_{OUT} \approx e_{IN}$; that is, e_{OUT} follows e_{IN} very closely. The circuit is therefore called a *voltage-follower.*

The nearly unity gain of the voltage follower is applicable to DC or signal inputs of any amplitude up to the ± saturation limits. Of course, when e_{IN} exceeds either of these limits, e_{OUT} will saturate and will not be able to follow e_{IN}.

The exact equation (5.4) for e_{OUT} indicates that e_{OUT} is actually slightly smaller than e_{IN}. This difference is because a small differential input is necessary at the op-amp inputs in order to be amplified by A_{VOL} to produce e_{OUT}. Since A_{VOL} is typically very large, the required e_d is so small that e_{OUT} is equal to e_{IN} for almost all practical purposes.

The use of negative feedback causes e_d to become very small at the value needed to produce $e_{OUT} \approx e_{IN}$. This small e_d will occur whenever negative feedback is used (as long as e_{OUT} is not saturated). We will use this fact later.

Example 5.5 Assume $A_{VOL} = 100{,}000$. Determine the *exact* values of e_d and e_{OUT} for $e_{IN} = +2$ V.

Solution:

$$e_{OUT} = e_{IN}\left(\frac{A_{VOL}}{1 + A_{VOL}}\right)$$
$$= 2\text{ V}\left(\frac{10^5}{1 + 10^5}\right)$$
$$= \mathbf{1.99998\ V} \approx e_{IN}$$
$$e_d = \frac{e_{OUT}}{A_{VOL}} = \mathbf{19.9998\ \mu V}$$

Example 5.6 Repeat for $A_{VOL} = 2{,}000$.

Solution:

$$e_{OUT} = 2\text{ V}\left(\frac{2{,}000}{1 + 2{,}000}\right) = \mathbf{1.999\ V} \approx e_{IN}$$
$$e_d = \frac{1.999\text{ V}}{2{,}000} = \mathbf{999.5\ \mu V}$$

Note that the big decrease in A_{VOL} had very little effect on e_{OUT}, but it did substantially increase the differential input needed to produce e_{OUT}.

Example 5.7 Determine e_{OUT} for a voltage follower if e_{IN} is a 6 V *p-p* sinewave.

Solution: Assume $A_{VOL} \gg 1$. Thus, $e_{OUT} \approx e_{IN}$, which means e_{OUT} is a 6 V *p-p* sinewave *in phase* with e_{IN}.

Voltage-follower input and output impedances The voltage-follower is obviously not used for its voltage gain. Like the bipolar common-collector amplifier (emitter-follower), its voltage gain is approximately 1, but it has a very high input impedance and very low output impedance.

In Figure 5–10 the input signal source e_{IN} sees an *effective* input impedance, $Z_{IN(eff)}$, which can be determined by calculating the signal current i_{IN}. Since i_{IN} also flows through the *op-amp* input impedance, Z_{IN}, we have

$$i_{IN} = \frac{e_d}{Z_{IN}}$$

Substituting $e_d = e_{OUT}/A_{VOL}$, we have

$$i_{IN} = \frac{e_{OUT}}{A_{VOL}Z_{IN}}$$

For the voltage-follower, however, $e_{OUT} \approx e_{IN}$, so this becomes

$$i_{IN} \approx \frac{e_{IN}}{A_{VOL}Z_{IN}}$$

Rearranging,

$$Z_{IN(eff)} = \frac{e_{IN}}{i_{IN}} = A_{VOL}Z_{IN} \tag{5–7}$$

This indicates that the effective input impedance seen by the source connected to the + terminal is A_{VOL} times greater than the op-amp input impedance. Clearly, this impedance can be very large. For example, for an op-amp with Z_{IN} = 1 MΩ and $A_{VOL} = 10^5$, the effective input impedance is 10^5 MΩ! In practice, $Z_{IN(eff)}$ is limited to around 1000 MΩ because of a small leakage current path between the op-amp + input and ground.

The negative feedback essentially multiplies the op-amp Z_{IN} by a factor of A_{VOL}. It also reduces the op-amp Z_{OUT} by the factor A_{VOL}. The effective output impedance of the voltage follower circuit is

$$Z_{OUT(eff)} = \frac{Z_{OUT}}{A_{VOL}} \tag{5–8}$$

which can be extremely low. For example, an op-amp with Z_{OUT} = 75 Ω and $A_{VOL} = 10^5$ will have $Z_{OUT(eff)}$ = 0.75 mΩ.

The high $Z_{IN(eff)}$ and low $Z_{OUT(eff)}$ of the voltage-follower makes it suitable as a buffer amplifier, much like the emitter-follower. The op-amp voltage-follower is superior because of its much higher Z_{IN}, and much lower Z_{OUT}. In addition, it has a constant gain of almost 1 all the way down to DC (0 Hz).

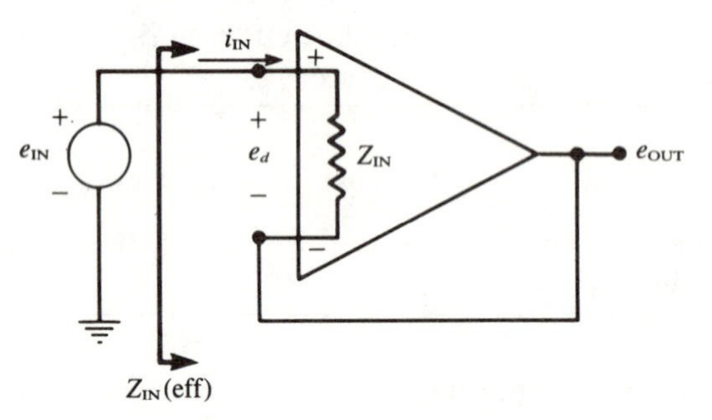

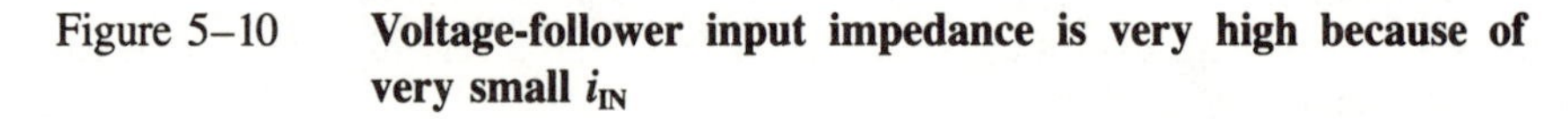
Figure 5–10 **Voltage-follower input impedance is very high because of very small i_{IN}**

5.7 Non-Inverting Amplifier

In the voltage-follower, the output is fed back *directly* to the "−" input. If the output is attenuated before it is fed back, the resultant circuit will have a gain greater than unity. In Figure 5–11, the op-amp output is connected to a voltage-divider R_1-R_2 to produce a voltage e_x. We can determine e_x from

$$e_x = \frac{R_1}{R_1 + R_2} \times e_{OUT}$$

if we recall the rule that the "−" input draws negligible current and will therefore not load down the voltage-divider.

Once again, the negative feedback will cause the differential between the op-amp inputs to be reduced to almost zero. That is,

$$e_d = e_{IN} - e_x \approx 0$$

so that $e_x \approx e_{IN}$. This situation means that e_{OUT} will become that value which causes e_x to equal e_{IN}. Thus,

$$e_x = \frac{R_1}{R_1 + R_2} \times e_{OUT} = e_{IN}$$

Solving for e_{OUT}, we obtain

$$e_{OUT} = \left(1 + \frac{R_2}{R_1}\right) \times e_{IN} \quad \textbf{(5.9)}$$

This shows that e_{IN} is amplified by the factor $(1 + R_2/R_1)$, which is always greater than 1. This factor is the closed-loop gain.

$$A_{CL} = \frac{e_{OUT}}{e_{IN}} = 1 + \frac{R_2}{R_1} \quad \textbf{(5.10)}$$

This formula for closed-loop gain is based on the assumption that $e_d \approx 0$, which implies that A_{VOL} is very large. The greater A_{VOL} is, the more exact this formula becomes. The formula will be accurate to within 99 percent or better as long as the following criterion is satisfied.

$$\frac{A_{VOL}}{A_{CL}} \geqslant 100 \quad \textbf{(5.11)}$$

The ratio A_{VOL}/A_{CL} is the ratio of the op-amp open-loop gain to the circuit's closed-loop gain. It is often referred to as the *loop gain, A_L*. The greater the loop gain, the more exact Equation 5.10 becomes. Closed-loop gains of 100 or less are normally used so that the loop gain is large. This strategy ensures that A_{CL} is independent of the op-amp open-loop gain and is determined by R_1 and R_2.

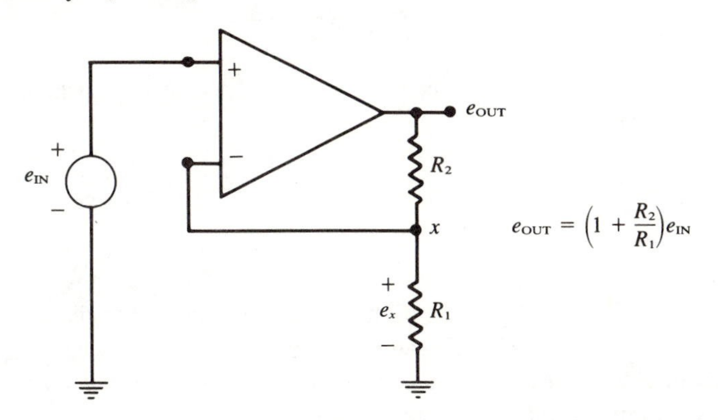

Figure 5–11 **Non-inverting amplifier has a closed-loop gain that depends on external resistors**

Example 5.8 Determine e_{OUT} in Figure 5–12(a).

Solution: The closed-loop gain is

$$A_{CL} = 1 + \frac{390\ \text{k}\Omega}{10\ \text{k}\Omega} = 40$$

Thus,

$$e_{OUT} = 40e_{IN} = \mathbf{8\ V}\ \boldsymbol{p\text{-}p}$$

Note that e_{OUT} is *in phase* with e_{IN}. This is because e_{IN} is applied to the + (non-inverting) input. For this reason, this particular op-amp configuration is called a *non-inverting amplifier.*

Example 5.9 Change e_{IN} to 1 V *p-p* and repeat.

Solution: With $A_{CL} = 40$, it appears that $e_{OUT} = 40$ V *p-p*. However, the output saturates at ±13.5 V as shown in Figure 5–12(b). Thus e_{OUT} is limited to 27 V *p-p* and exhibits clipping on both half-cycles. It is important to note that e_d will *not* be zero when the output saturates. To illustrate, when e_{IN} is at its peak of +0.5 V, e_{OUT} will be at +13.5 V, so that $e_x = +13.5\ \text{V}/40 = +0.35$ V. Thus e_d will be 0.5 V − 0.35 V = 0.15 V. This emphasizes that e_d will be maintained at ≈0 V *only* when the op-amp is in its linear range.

Input and output impedances Like the voltage-follower, the non-inverting amplifier has a very high input impedance and a very low output impedance. It can be shown that the input signal, e_{IN}, sees an effective input impedance given by

$$Z_{IN(eff)} = \frac{A_{VOL}}{A_{CL}} Z_{IN} = A_L Z_{IN} \qquad \textbf{(5–12)}$$

where A_L is the loop gain and Z_{IN} is the *op-amp* input impedance. Clearly, $Z_{IN(eff)}$ depends on loop gain, and it will be greater when A_{CL} is smaller.

The effective output impedance is given by

$$Z_{OUT(eff)} = \frac{Z_{OUT}}{A_L} \qquad \textbf{(5–13)}$$

where Z_{OUT} is the *op-amp* output impedance. The value of $Z_{OUT(eff)}$ is kept low by keeping the loop gain large.

5.8 The Inverting Amplifier

Another very useful closed-loop configuration has e_{IN} applied to the op-amp's − input, while the + input is grounded as shown in Figure 5–13(a). With e_{IN} applied to the − input, the polarity or phase of e_{OUT} will be opposite to that of e_{IN}. For this reason, this configuration is called an *inverting amplifier.*

Let's assume for the moment that e_{IN} is positive with respect to ground, so that e_{OUT} is negative [Figure 5–13(b)]. e_{IN} is connected to the − terminal (point x) through resistor R_1 and tends to make point x positive. However, e_{OUT} is fed back to the − terminal through

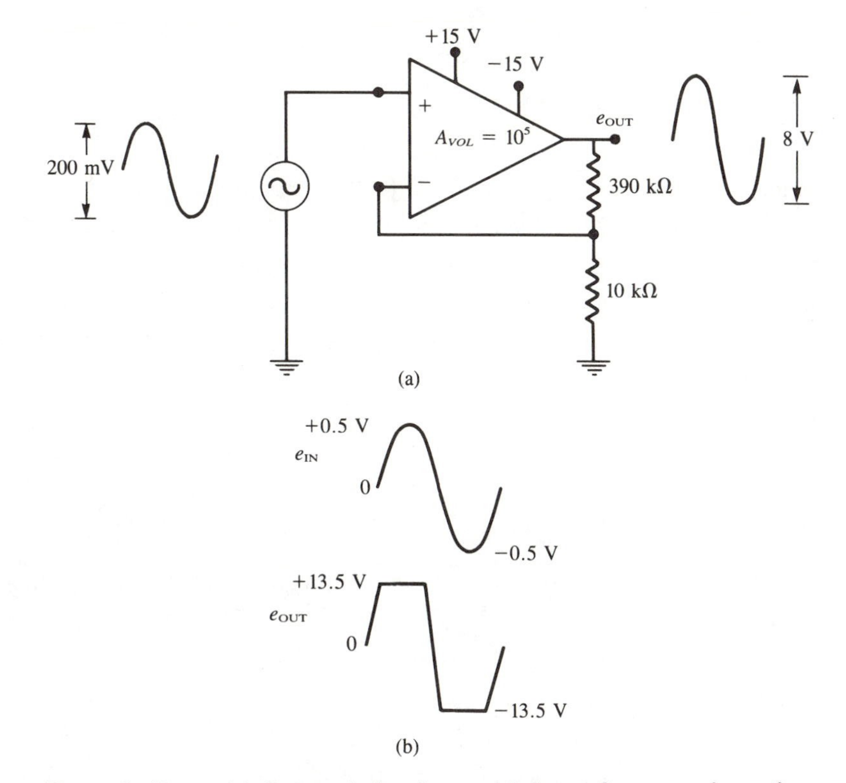

Figure 5–12 **(a) Output is in-phase with input for a non-inverting amplifier; (b) Output saturates if e_{IN} is too large**

R_2 and tries to make point x more negative. The result of this negative feedback is that the − terminal will be only slightly more positive than the + terminal, which is grounded. In other words, e_{OUT} will take on a value that makes e_x just large enough so that the differential across the op-amp inputs, when amplified by A_{VOL}, produces that value of e_{OUT}. For all practical purposes, then, e_x will be at *ground potential*. The − input is sometimes said to be a *virtual* ground in this situation.

We can use the fact that $e_x \approx 0$ to help determine the value of e_{OUT}. We will also use the fact that the current through the op-amp input terminals will be essentially zero because of the extremely small differential input voltage and the large Z_{IN}. With point x at ground, the voltage across R_1 will be equal to e_{IN} and will have the polarity as shown. Thus the current through R_1 is

$$i_i = \frac{e_{IN}}{R_1}$$

and flows in the direction shown.

This same current has to flow through feedback resistor R_2 and produces a voltage across R_2 with the polarity as shown. Since point x is *virtually* at ground, e_{OUT} must equal the voltage across R_2. That is,

$$e_{OUT} = -i_2 R_2 = -i_1 R_2$$

$$= \left(\frac{-e_{IN}}{R_1}\right) R_2 \qquad \textbf{(5–14)}$$

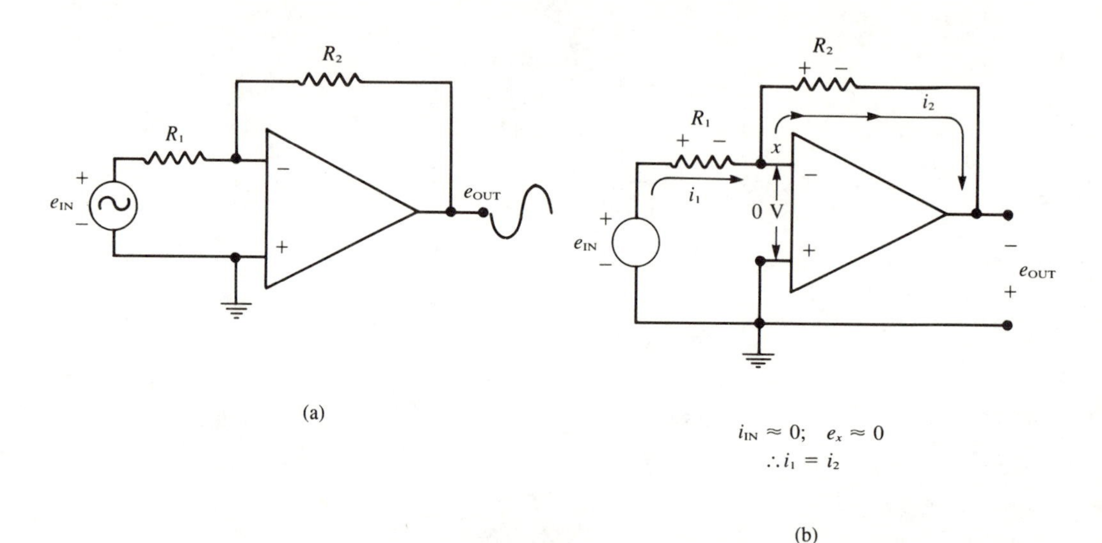

Figure 5–13 **(a) Op-amp inverting amplifier; (b) Currents through R_1 and R_2 are equal, and − terminal is virtually at ground potential ($e_x \approx 0$)**

The ratio of e_{OUT}/e_{IN} is the closed-loop gain A_{CL}.

$$A_{CL} = \frac{e_{OUT}}{e_{IN}} = -\left(\frac{R_2}{R_1}\right) \tag{5–15}$$

The inverting amplifier has a gain determined solely by the ratio of the two resistances, and the negative sign indicates the polarity or phase reversal. This formula for A_{CL} is reasonably exact as long as loop-gain A_{VOL}/A_{CL} is kept very large (≥ 100) and e_{OUT} is *not* saturated. These conditions are necessary to ensure $e_x \approx 0$.

Example 5.10 If $e_{IN} = 120$ mV DC, determine e_{OUT} for $R_1 = 10$ kΩ and $R_2 = 200$ kΩ.

Solution: Assume $A_{VOL} = 10^5$.

$$A_{CL} = -\left(\frac{R_2}{R_1}\right) = -20$$

Thus, $e_{OUT} = -20\ e_{IN} =$ **−2.4 V DC.** This result is almost exact since the loop gain is $10^5/20 = 5000$.

Example 5.11 Consider the circuit of Figure 5–14. The e_{IN} signal source has $R_s = 50\ \Omega$. Determine values for R_1 and R_2 for a gain of −10.

Solution: Since x is virtually at ground potential, the current delivered by the input signal is limited by R_1. To reduce loading effects on the source, we want $R_1 \gg R_s$. Let's use $R_1 = 100R_s = 5$ kΩ. Then since,

$$A_{CL} = -\left(\frac{R_2}{R_1}\right) = -10$$

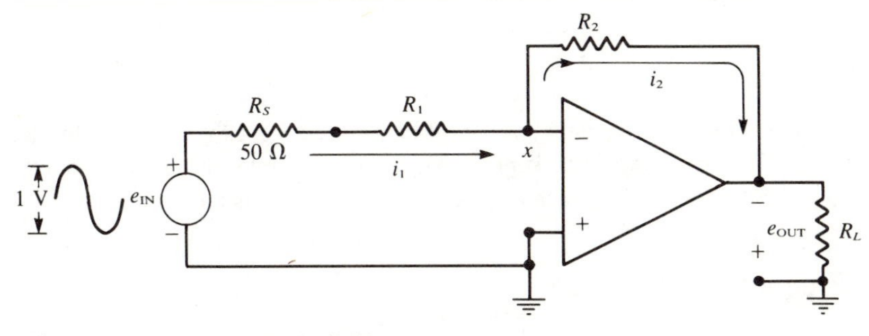

Figure 5–14 **Example 5.11**

we have

$$R_2 = 10R_1 = \mathbf{50\ k\Omega}$$

Other values of R_1 and R_2 will also work; but, as we stated for the non-inverting amplifier, the resistor values are kept at 1 MΩ or less whenever possible.

Input and output impedances The effective input impedance seen by e_{IN} in the inverting amplifier is essentially R_1 because point x is at virtual ground. Thus

$$Z_{IN(eff)} = R_1 \tag{5–16}$$

Clearly, the inverting amplifier does not have as high an input impedance as the non-inverting amplifier or the voltage-follower.

The output impedance is the same as that of the non-inverting amplifier. That is,

$$Z_{OUT(eff)} = \frac{Z_{OUT}}{A_L} \tag{5–13}$$

and is normally very low.

Unity-gain inverter In some applications, the inverting amplifier is used with $A_{CL} = 1$; that is, with $R_2 = R_1$. When used with unity gain, it is called a *phase inverter* or simply an *inverter,* since its sole purpose is to reverse the phase or polarity of the input signal.

5.9 Variations in Open-Loop Gain, A_{VOL}

In closed-loop circuits such as the voltage-follower, the non-inverting and inverting amplifiers, the circuit voltage gain (A_{CL}) is relatively insensitive to the op-amp's open-loop gain, A_{VOL}, provided that the loop gain is very large. This means that a circuit designed to have a specific A_{CL} will operate the same for a wide range of values of A_{VOL}. This is very important because A_{VOL} will vary widely from one op-amp to another and will change with signal frequency, temperature, and aging.

If, for some reason, A_{VOL} becomes low enough to reduce the loop gain (A_{VOL}/A_{CL}) significantly, the closed-loop operation will be affected; in particular, A_{CL} will decrease. As we stated earlier, a loop gain of 100 will produce an error of 1 percent between the actual A_{CL} and the ideal A_{CL}.

In fact, it can be shown that for any closed-loop circuit,

$$A_{CL(\text{actual})} = A_{CL(\text{ideal})}\left[1 - \frac{1}{A_L}\right] \qquad \textbf{(5–17)}$$

This equation shows that the actual A_{CL} will be reduced from the ideal A_{CL} by the factor $(1 - 1/A_L)$. Clearly, A_L should be kept as large as possible by using an op-amp with a very large A_{VOL}, and by designing for an A_{CL} that is no greater than 100 or so.

5.10 Op-amp Frequency Response

We have said that most op-amps have very large values of open-loop voltage gain. This situation is only true at low signal frequencies where the value of A_{VOL} is relatively constant. As frequency is increased, A_{VOL} will eventually decrease and can become very low at very high frequencies. In other words, the op-amp has a low-pass frequency response characteristic.

Figure 5–15 shows the frequency-response curve for a typical op-amp with A_{VOL} plotted versus frequency, f. Note that A_{VOL} is shown both in absolute gain and decibels. Note also the logarithmic scale for frequency. The curve shows that A_{VOL} is constant at 100,000 (100 dB)* for frequencies from 0 Hz up to around 10 Hz. Above 10 Hz, A_{VOL} decreases rapidly with increasing frequency at a rate of −20 dB per decade (i.e., −20 dB for each factor of ten increase in frequency).

The −3 dB point occurs at 10 Hz. This point is called the op-amp's *high-frequency cutoff*, f_{hc}. It is also referred to as the op-amp's open-loop bandwidth, BW_{OL}. The bandwidth refers to the range of frequencies over which the gain is relatively constant. For the op-amp of Figure 5–15, the open-loop bandwidth is $BW_{OL} = 10$ Hz.

This low open-loop bandwidth is typical of many general-purpose op-amps, though it might appear to severely limit the frequency range over which an op-amp will be useful. We must remember, however, that when negative feedback is used, the closed-loop gain is relatively insensitive to A_{VOL} as long as the *loop gain* is *large*. What this means is that even though A_{VOL} decreases drastically with frequency, A_{CL} will not be significantly affected until A_{VOL} causes the loop gain to decrease below 100 or so.

To illustrate, let's suppose we had a *non-inverting* amplifier with $R_1 = 1\ \text{k}\Omega$ and $R_2 = 9\ \text{k}\Omega$, to give an ideal A_{CL} of 10. At frequencies below 10 Hz, the op-amp has $A_{VOL} \approx 10^5$ so that loop gain is $10^5/10 = 10^4$. At a frequency of 1,000 Hz, A_{VOL} has dropped to 1,000 (point x on curve), which reduces the loop gain to $1{,}000/10 = 100$, still large enough to have very little effect on A_{CL}.

Closed-loop bandwidth, BW_{CL} In essence, the closed-loop gain will remain relatively constant over a wider range of frequencies than the open-loop gain. The *closed-loop bandwidth, BW_{CL},* is the range of frequencies over which A_{CL} remains relatively constant. There is a direct relationship between BW_{CL} and BW_{OL}, which is given here:

$$|A_{CL}| \times BW_{CL} = A_{VOL} \times BW_{OL} = \textit{gain-bandwidth product} \qquad \textbf{(5–18)}$$

This equation states that the product of gain and bandwidth is constant regardless of whether the op-amp is operated open-loop or closed-loop. Since A_{VOL} and BW_{OL} are normally given on an op-amp data sheet, the value of BW_{CL} can be determined for any

*The gain, in decibels, is $20\log_{10}(A_{VOL})$. For $A_{VOL} = 100{,}000$, this becomes $20\log_{10}(10^5) = 20 \times 5 = 100$ dB.

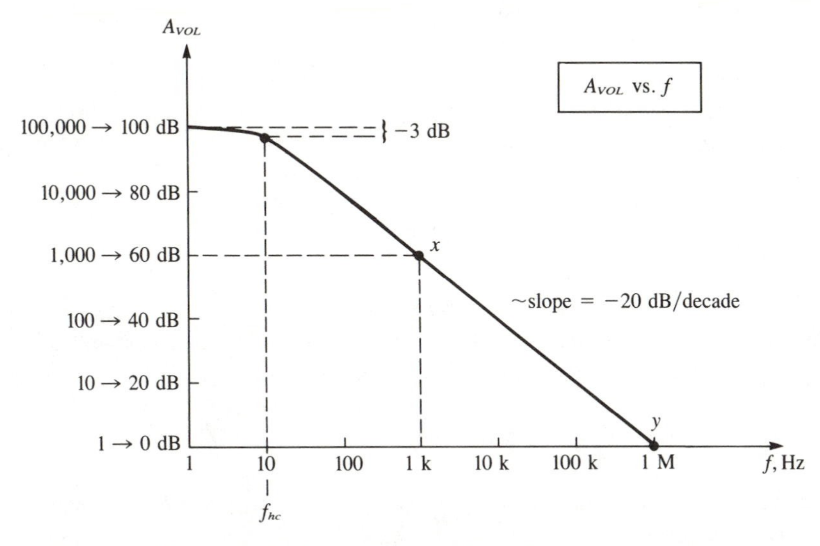

Figure 5–15 **Typical op-amp open-loop frequency response curve**

desired value of A_{CL}. When using this relationship, the *absolute* value of A_{CL} and the low-frequency value of A_{VOL} are used.

Example 5.12 The op-amp with the frequency response curve of Figure 5–15 is used in an inverting amplifier with a closed-loop gain of −10. What will be the bandwidth of this inverting amplifier? Repeat for a gain of −100.

Solution: From the curve, BW_{OL} = 10 Hz and A_{VOL} = 100,000 (the low frequency value for A_{VOL}). Using Equation 5.18 with A_{CL} = 10, we have,

$$10 \times BW_{CL} = 100{,}000 \times 10 \text{ Hz} = 1 \text{ MHz}$$

or

$$BW_{CL} = \mathbf{100\ kHz} \text{ (when } A_{CL} = -10)$$

With $A_{CL} = -100$, we have

$$100 \times BW_{CL} = 1 \text{ MHz}$$

or

$$BW_{CL} = \mathbf{10\ kHz} \text{ (when } A_{CL} = -100)$$

Note that as A_{CL} was *increased,* the bandwidth *decreased.*

In this example, the gain-bandwith product was 1 MHz. This product is a characteristic of that particular op-amp. Every op-amp has a constant gain-bandwidth product. A higher gain-bandwidth product indicates an op-amp that can be used over a higher range of frequencies.

The gain-bandwidth product is sometimes specified on an op-amp data sheet as the *unity-gain bandwidth,* defined as the frequency at which A_{VOL} = 1. Its value always equals the gain-bandwidth product. For example, in Figure 5–15, point *y* shows A_{VOL} = 1 at $f = 1$ MHz. This is the unity-gain bandwidth. Note that it is the same as the gain-bandwidth product of 1 MHz obtained earlier.

To summarize, the use of negative feedback will increase the circuit bandwidth. A greater bandwidth can be obtained by reducing A_{CL}. As larger values of A_{CL} are used, the circuit bandwidth will decrease.

5.11 Slew Rate

Internal circuit capacitances serve to limit how rapidly an op-amp output voltage can change. This limitation is expressed as a *maximum* rate of change of output voltage, dv/dt, and is called the *slew rate*. For example, an op-amp might have

$$\text{slew rate} = \frac{dv}{dt}\,(\text{max}) = 1\ \text{V}/\mu\text{s}$$

This rate indicates that the op-amp output voltage cannot change by more than one volt per microsecond.

Figure 5–16 illustrates the effects of slew-rate limiting when an input pulse is applied to a voltage-follower. The input pulse has very fast rise and fall times, which we can assume to be zero. The op-amp output, however, has much slower transition times. We can determine the rise time by using the slew rate. Let's use a slew rate of 1 V/μs. The rise time will be the time required for e_{OUT} to go from 1 V to 9 V when it is changing at a rate of 1 V/μs. Thus,

$$t_r = \frac{\Delta e_{OUT}}{\text{slew rate}} = \frac{8\ \text{V}}{1\ \text{V}/\mu\text{s}} = 8\ \mu\text{s}$$

In a similar manner, the fall time will be 8 μs.

Commercial op-amps have slew rates ranging from 0.25 V/μs to over 100 V/μs. Clearly, a higher slew rate means that the output pulse will have shorter transition times. This characteristic is important in many pulse circuit applications.

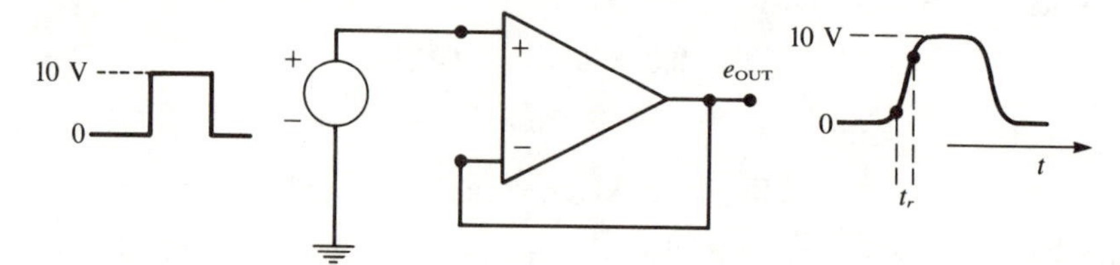

Figure 5–16 **Op-amp pulse response is limited by slew rate**

5.12 Offset Errors

The op-amp amplifies the difference in the voltages present at its input terminals. Ideally, the output voltage will be 0 V when this input differential is 0 V. In practice, however, small inbalances in the op-amp circuitry produce a non-zero output voltage called an *output offset voltage*. For example, Figure 5–17 shows both op-amp inputs tied to ground, but the output is +7.2 V. The value of +7.2 V was chosen at random; it could actually be any voltage between −13.5 V and +13.5 V and varies from one op-amp to another.

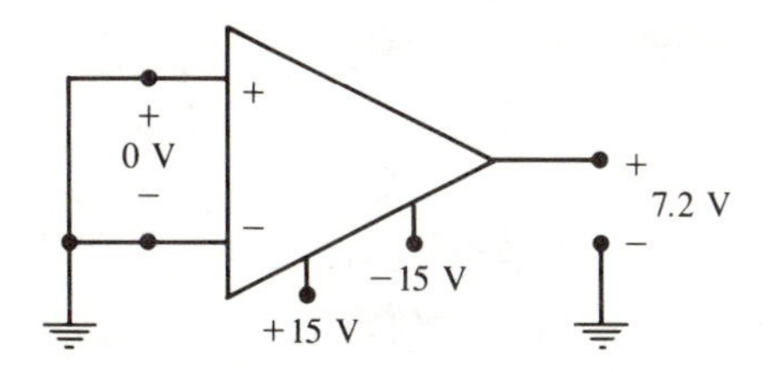

Figure 5–17 **Op-amp output is not zero when e_d = 0 because of voltage offset**

In other words, the op-amp output acts as if there is a voltage difference across the input terminals even when there isn't. It is clearly represented in Figure 5–18, where a voltage source is shown connected to the "+" input. It is not a real source, but we can think of it as the amount of input voltage that (when amplified by A_{VOL}) produces the output offset voltage. This ficticious input source is given the symbol E_{OS} to represent *input offset voltage,* and it can be of either polarity. The size of this input offset voltage can range from a few *microvolts* to a few *millivolts*. In many op-amps, E_{OS} is large enough to drive the output into saturation. Values for E_{OS} are normally given on an op-amp data sheet.

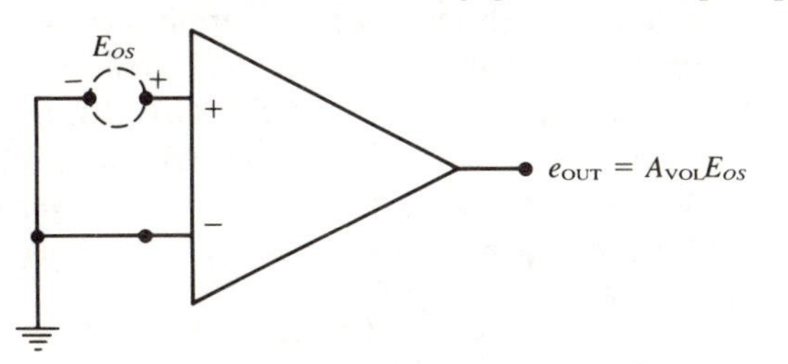

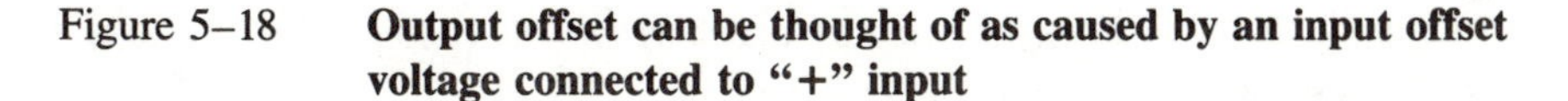

Figure 5–18 **Output offset can be thought of as caused by an input offset voltage connected to "+" input**

The input offset voltage E_{OS} acts just like any other input voltage applied to the op-amp, and its effect on e_{OUT} is added to that of the actual inputs. This is just as true for closed-loop operation as it is for open-loop operation. Even though E_{OS} is very small, it can produce a significant error in e_{OUT}, especially when the op-amp is used with actual input voltages that are in the millivolt range.

In some applications (e.g. input voltages $\gg E_{OS}$), the effect of E_{OS} is not significant and can be neglected. In many applications, however, it has significant effect and must be eliminated or at least minimized. This minimization can be done by injecting a small voltage into the op-amp circuit to cancel out the effects of E_{OS}. This strategy is called *offset compensation,* or *offset nulling*. Figure 5–19 shows this method for an op-amp that has special *offset nulling terminals*. The offset nulling procedure is carried out by setting e_{IN} to exactly 0 V and adjusting the multi-turn potentiometer until e_{OUT} = 0 V. The same procedure is followed for other op-amp circuits. For those circuits with more than one input source, all inputs should be set to zero before adjusting the null potentiometer for e_{OUT} = 0 V.

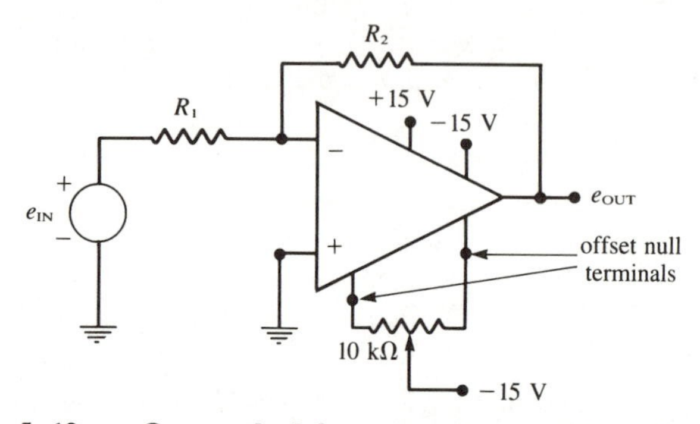

Figure 5–19 **One method for correcting or nulling the effects of offset**

Bias currents Ideally, the op-amp input terminals draw no current. In practice, however, there are very small currents that flow through these terminals. These *bias currents* are generally less than 1 μA. In op-amps using bipolar transistors as the input stage, the bias currents are typically 100 nA. Those using JFET or MOSFET input stages have bias currents in the picoampere (pA) range.

Figure 5–20(a) shows an inverting amplifier with the bias currents, $I_B(-)$ and $I_B(+)$, at the respective input terminals. These currents must flow through any resistance connected to the op-amp terminals. The $I_B(-)$ current flows through a Thevenin equivalent resistance of $R_1 \| R_2$. This current produces at the "−" terminal an unwanted *extra* voltage that causes an offset error in e_{OUT}. This offset error can be reduced by placing a resistor R_3 between the "+" terminal and ground [see Figure 5–20(b)]. The $I_B(+)$ current produces a voltage across R_3 which appears at the "+" terminal and cancels out the extra voltage at the "−" terminal. Since $I_B(-)$ and $I_B(+)$ are typically close in value, this cancellation will be improved by making R_3 equal to $R_1 \| R_2$. That is,

$$R_3 = \frac{R_1 R_2}{R_1 + R_2}$$

to minimize the error in e_{OUT} caused by bias currents. R_3 can be called the *bias-current compensation resistor.*

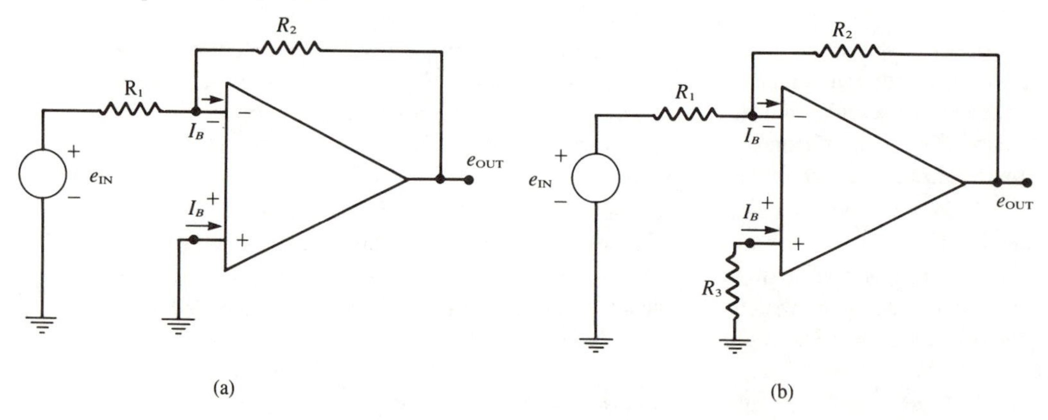

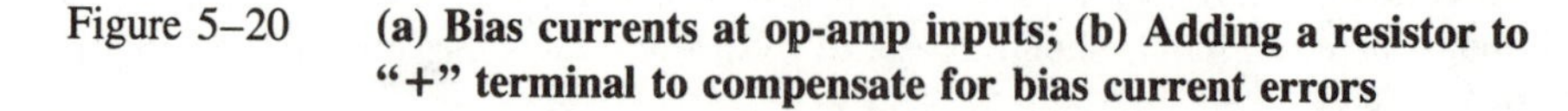
Figure 5–20 **(a) Bias currents at op-amp inputs; (b) Adding a resistor to "+" terminal to compensate for bias current errors**

Offset current The inclusion of R_3 reduces the error due to bias currents, but it does not eliminate it entirely. This is because $I_B(-)$ and $I_B(+)$ are never exactly equal. In fact, their difference is often listed on op-amp data sheets as *input offset current, I_{OS}*. Clearly, a small I_{OS} is desirable since it minimizes the differential voltage produced at the op-amp inputs by the bias currents flowing through the external circuit resistances. Values of I_{OS} typically range from a few picoamperes for op-amps with FET input stages to a maximum of a few hundred nanoamperes for bipolar inputs. Because I_{OS} is not zero, the resistances connected to the op-amp should not be made too large or a sizable differential voltage will be developed. For a bipolar op-amp, the external resistors are generally kept at 1 MΩ or below. Of course, an FET op-amp with its much lower I_{OS} can use larger resistors, even up to 1000 MΩ.

In summary, the output error caused by input bias currents can be minimized by following three steps:

(a) Using an op-amp with a low value of I_{OS}.

(b) Using a bias-current compensating resistor.

(c) Keeping all external resistances at reasonable values.

If these steps do not bring the output offset error to an acceptably low value, the rest of the error can be eliminated by following the offset nulling procedure described earlier for eliminating the effects of E_{OS}. This offset nulling process may have to be performed periodically because of a phenomenon called *offset drift* caused by changes in E_{OS} and I_{OS} due to temperature, variations in bias supplies, and component aging.

5.13 Positive Feedback

We have seen how negative feedback can be used to construct op-amp circuits having predictable, controllable gains that are relatively insensitive to changes in open-loop gain. This strategy is used at the expense of reduced gain because the negative feedback reduces the effect that the input signal has on the output. When negative feedback is employed, the op-amp generally operates in its linear range and saturation is avoided.

Op-amps can also be used with positive feedback; that is, with part of the output fed back to the "+" input, usually through some sort of network. The effects of positive feedback are quite different from those of negative feedback. Positive feedback tends to drive the output *away* from its linear region and *toward* saturation.

We will use the circuit of Figure 5–21(a) to illustrate the effect of positive feedback. At first glance, this circuit looks like a non-inverting amplifier configuration. Upon closer inspection, however, we see that it is not the same as the non-inverting amplifier, because the "+" and "−" inputs are reversed. The input signal is applied to the "−" input and a portion of the output is fed back to the "+" input.

Let's start by setting the $e_{IN} = 0$ V. If we assume that the op-amp has been adjusted for zero offset errors, it would be reasonable to expect that $e_{OUT} = 0$ V, which would make $e_x = 0$ V and $e_d = 0$ V, thereby producing the expected $e_{OUT} = 0$ V. While this expectation sounds reasonable, it is not possible in practice because any noise at the op-amp inputs will drive the output toward saturation.

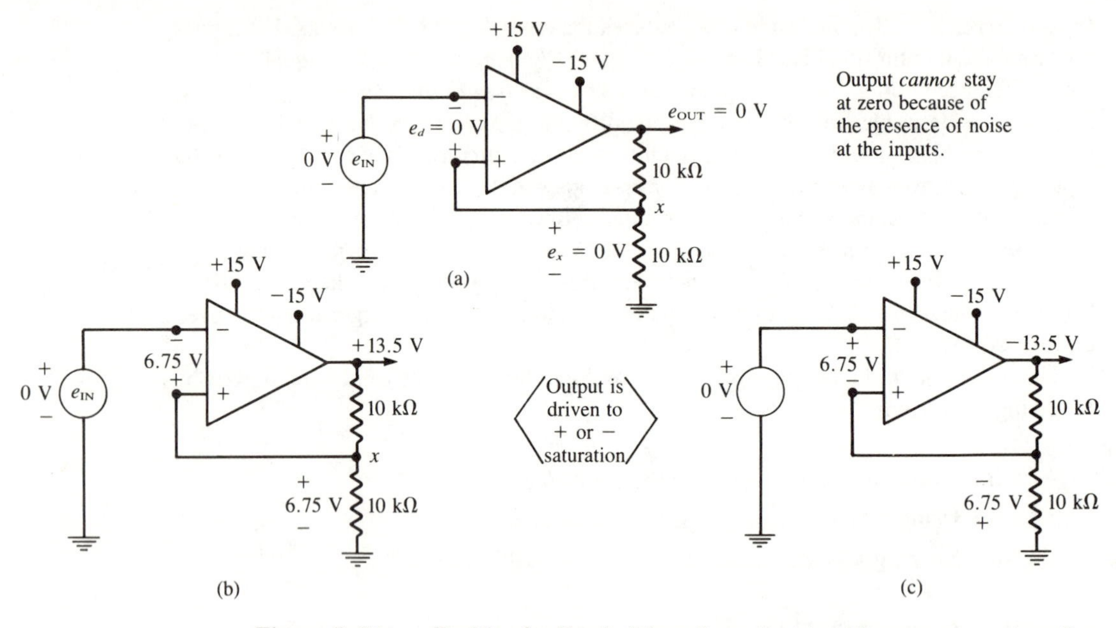

Figure 5–21 **Positive feedback drives the output to either + or − saturation**

For example, if the noise makes the differential input e_d slightly positive, the op-amp will amplify this e_d and cause e_{OUT} to increase positively. As e_{OUT} goes positive, e_x will increase positively, thereby increasing e_d even further, which, in turn, will cause e_{OUT} to increase even further. This process continues until e_{OUT} reaches its positive saturation limit of +13.5 V. If the noise had originally made e_d go negative, the output would have been driven to −13.5 V. This process, called *regenerative action,* takes place very rapidly at a speed determined by the op-amp internal circuitry.

Figure 5–21, parts (b) and (c), shows the two possible end results of the regenerative process. Both results satisfy the op-amp characteristics and neither is more likely than the other. Thus, this circuit with positive feedback has two possible *stable* operating states for the same input voltage. In other words, it is a *bistable* circuit.

Example 5.13 Assume that $e_{IN} = 0$ and $e_{OUT} = +13.5$ V, as in Figure 5–21(b). What will happen to e_{OUT} as e_{IN} is gradually increased positively?

Solution: With $e_{OUT} = +13.5$ V, there will be +6.75 V at the "+" input so that $e_d = +6.75$ V. As e_{IN} is increased, e_d will get smaller and smaller. This situation will have no effect on e_{OUT}, however, until e_d becomes lower than e_d(SAT), the value needed to keep e_{OUT} at + saturation. As e_{IN} increases to almost 6.75 V, e_d will drop below e_d(SAT). Then the output will start to come out of saturation and will decrease into its linear region. As soon as e_{OUT} begins to decrease, regenerative action takes place and drives the output to negative saturation (−13.5 V).

To summarize, as e_{IN} is increased to approximately +6.75 V, the op-amp output will rapidly switch from +13.5 V to −13.5 V through the regenerative action of positive feedback.

What we have seen is that positive feedback prevents the op-amp from having a stable operating point in its linear region. Whenever an input tries to bring the output into the linear region, regenerative action takes place and drives the output to saturation. Some of the op-amp switching circuits that we examine later in the text will use this action to produce desirable results.

Questions

Sections 5.1–5.3

5.1 A certain op-amp has $A_{VOL} = 50{,}000$ and uses ± 12 V bias supplies. Determine e_{OUT} for each of the following cases.

(a) $e_1 = 375\ \mu V$ and $e_2 = 325\ \mu V$

(b) $e_1 = -2.843$ mV and $e_2 = -2.793$ mV

(c) $e_1 = 8.174$ V and $e_2 = 8.177$ V

(d) $e_1 = 0$ V, $e_2 = -1$ V

5.2 The op-amp of the preceding problem produces an output of -3.72 V when the voltage at its "+" input is 1.95 V. What is the voltage at its "−" input?

5.3 What input differential voltage will produce output saturation in an op-amp that uses ± 12 V supplies and has $A_{VOL} = 50{,}000$?

5.4 What can be said about the voltages at the op-amp input terminals whenever the output is in its linear region?

Section 5.4

5.5 A certain op-amp has $Z_{IN} = 50\ k\Omega$ and $A_{VOL} = 100{,}000$. If $e_{OUT} = -3$ V, determine e_d and i_{IN}.

5.6 What can be said about the current flowing through an op-amp input terminal when e_{OUT} is not saturated?

Section 5.5

5.7 Sketch e_{OUT} for the comparator circuit in Figure 5–22(a) if e_{IN} is a 20 V *p-p* sinewave.

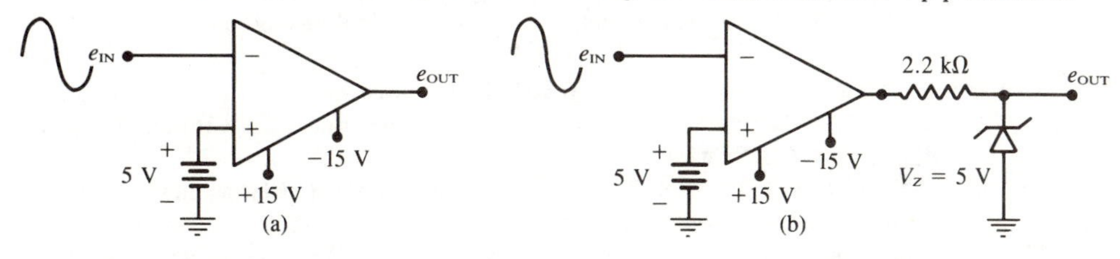

Figure 5–22

5.8 Reverse the input terminals and repeat Problem 5.7.

5.9 Determine the e_{OUT} waveform if the circuit of Figure 5–22(a) is modified as indicated in Figure 5–22(b).

Section 5.6

5.10 What are the main advantages of a voltage-follower?

5.11 Why can't e_{OUT} ever be exactly equal to e_{IN} in a voltage-follower?

5.12 Consider the circuit in Figure 5–23 where a voltage-follower is used as a buffer between the high impedance signal source and the relatively low input impedance of the Com.-E amplifier. The op-amp has $A_{VOL} = 10^4$, $Z_{IN} = 20\ \text{k}\Omega$ and $Z_{OUT} = 120\ \Omega$. The Com.-E amplifier has $Z_{IN} = 1\ \text{k}\Omega$.

(a) Calculate Z_{IN}(eff) and Z_{OUT}(eff) for the voltage-follower.

(b) Calculate the *exact* signal voltage that reaches the op-amp's "+" input.

(c) Calculate the *exact* voltage e_x, at the input of the Com.-E amplifier. (Assume $X_c = 0$).

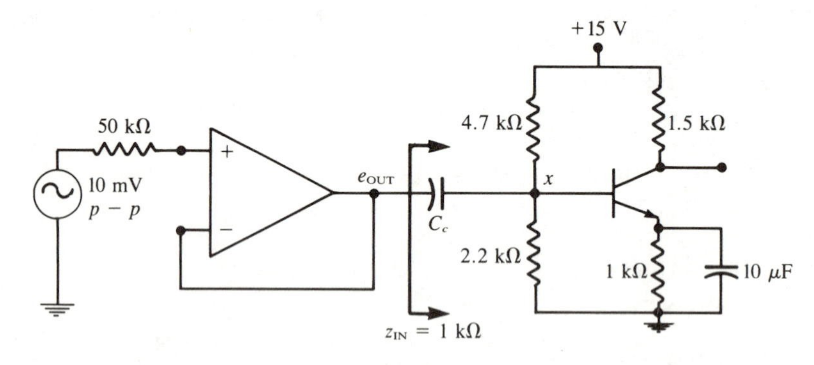

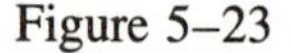
Figure 5–23

5.13 Replace the op-amp in Figure 5–23 with one that has $A_{VOL} = 2{,}000$; $Z_{IN} = 10\ \text{k}\Omega$; and $Z_{OUT} = 200\ \Omega$. Recalculate e_x and note that it changed very little with the different op-amp characteristics. This is one of the benefits of negative feedback.

Section 5.7

5.14 A non-inverting amplifier uses $R_1 = 2.2\ \text{k}\Omega$ and $R_2 = 15\ \text{k}\Omega$. The op-amp uses ± 15 V supplies.

(a) Determine e_{OUT} when e_{IN} is a 500 mV *p-p* signal.

(b) What is the maximum e_{IN} that can be used before saturation occurs in e_{OUT}?

5.15 A certain op-amp has $A_{VOL} = 20{,}000$. If it is used in a non-inverting amplifier, what is the largest closed-loop gain that can be used before the relationship $A_{CL} = (1 + R_2/R_1)$ starts becoming inaccurate?

5.16 What happens to the input and output impedances of a non-inverting amplifier as the R_2/R_1 ratio is increased?

5.17 What approximate voltage should appear across the 10 kΩ resistor in Figure 5–12(a)?

Section 5.8

5.18 A 2 V DC signal is applied to an inverting amplifier that has $R_1 = 10\ \text{k}\Omega$ and $R_2 = 50\ \text{k}\Omega$. Determine e_{OUT}. Assume ± 15 V supplies for the op-amp.

5.19 Change the input in the preceding problem to a 4 V *p-p* sinewave. Sketch e_{IN} and e_{OUT}.

5.20 Change e_{IN} to 8 V *p-p* and repeat.

5.21 A 20 mV signal source with $R_s = 150\ \Omega$ is fed to an inverting amplifier, which is to have a gain of -50. Determine appropriate values for R_1 and R_2 such that the signal source is not significantly loaded down.

5.22 If R_s is changed to 1.5 kΩ in the preceding problem, the required value for R_2 will be very large. This value is undesirable for several reasons. Show how to add a voltage-follower to eliminate the large value for R_2.

Sections 5.9–5.10

5.23 If an op-amp generally has a very small open-loop bandwidth (typically 10 Hz), why is it useful over a much greater frequency range when it is used closed-loop?

5.24 A certain op-amp has a low-frequency A_{VOL} of 50,000, and an open-loop bandwidth of 50 Hz. Determine the bandwidth of each of the following amplifiers using this op-amp.

(a) non-inverting amplifier with $R_1 = 20\ \text{k}\Omega$ and $R_2 = 680\ \text{k}\Omega$

(b) inverting amplifier with $R_1 = 1\ \text{k}\Omega$ and $R_2 = 1\ \text{M}\Omega$

(c) voltage-follower

Section 5.11

5.25 A perfect 4 V pulse is applied to a non-inverting amplifier which has $R_2 = R_1 = 10\ \text{k}\Omega$. Determine the rise time of e_{OUT}, if the op-amp slew rate is 0.5 V/μs.

5.26 The comparator output in Figure 5–22(a) goes from " + " saturation to " − " saturation in 0.5 μs. What is the op-amp's slew rate?

Section 5.12

5.27 What steps should be followed to eliminate all output offset errors in an op-amp circuit?

5.28 Determine the value of the bias-current compensation resistor required for the circuit of Problem 5.14.

5.29 Repeat for Problem 5.18.

Section 5.13

5.30 Assume that $e_{IN} = 0$ and $e_{OUT} = +13.5$ V as in Figure 5–21(b). Describe what happens as e_{IN} is increased negatively with respect to ground.

5.31 Assume that $e_{IN} = 0$ and $e_{OUT} = -13.5$ V as in Figure 5–21(c). Describe what happens as e_{IN} is increased positively with respect to ground. Repeat for e_{IN} increased negatively with respect to ground.

6

Signal Conditioning Circuits

6.1 Introduction

In this chapter we will study some of the wide variety of circuits whose function is to take a signal and change one or more of its characteristics. This general function is called *signal conditioning,* and some of the more common signal conditioning circuits include clippers, limiters, inverters, comparators, Schmitt triggers, and buffers. The analysis of circuits in this chapter illustrates how one can determine the operation of an unfamiliar circuit. You can then apply these techniques in analyzing circuits not discussed herein.

6.2 Diode Clipper (Limiter)

A clipper circuit is used to select for transmission that part of a waveform that lies above or below some reference level. Clippers are also referred to as *limiters* when they are used to limit the positive or negative (or both) excursions of a signal.

Series Clipper Figure 6–1(a) is a series clipper using a silicon* diode in series with the output. Since the diode will conduct only when it is forward-biased, the negative

*Again, all devices will be silicon unless otherwise indicated

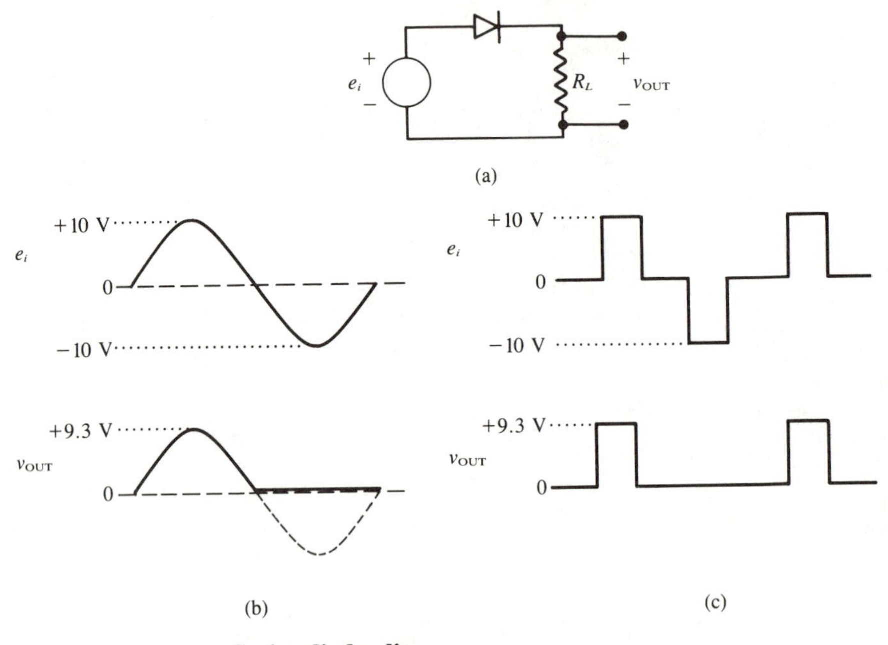

Figure 6–1 **Series diode clipper**

portions of the input signal will not reach the output (assuming no reverse breakdown). Almost all of the positive portion of the input signal will reach the output (except for the V_F drop). Figure 6–1(b) and (c) shows the clipped output signals for two kinds of input signal. Essentially, this circuit clips OFF those portions of e_i that are below 0 V.

Example 6.1 The diode in Figure 6–1 has I_F (max) = 50 mA and a reverse voltage rating, PRV, of 50 V. If R_L = 300 Ω, what is the maximum amplitude sinewave that can be applied to this circuit?

Solution: The maximum positive input amplitude is determined by I_F (max) since the diode will be conducting. Using a silicon diode, we have

$$\begin{aligned} e_i\,(\text{max}+) &= V_F + I_F(\text{max}) \times R_L \\ &= 0.7\text{ V} + 50\text{ mA} \times 300\ \Omega \\ &= 15.7\text{ V} \end{aligned}$$

The maximum negative input amplitude is determined by the PRV rating. When e_i is negative, the diode is reverse-biased and the entire input voltage is across the diode. Thus,

$$e_i(\text{max}-) = \text{PRV} = -50\text{ V}$$

Obviously, in this case, the positive limit on e_i is more stringent. Therefore, the sinewave has to have an amplitude less than **15.7 V** peak or **31.4 V p-p**.

The clipper in Figure 6–1 clips off the *negative* portion of the input signal. By turning the diode around, the circuit will clip OFF the positive portion of e_i. Figure 6–2 shows this arrangement. Note that v_{OUT} contains the negative portion of e_i minus the 0.7 V diode drop.

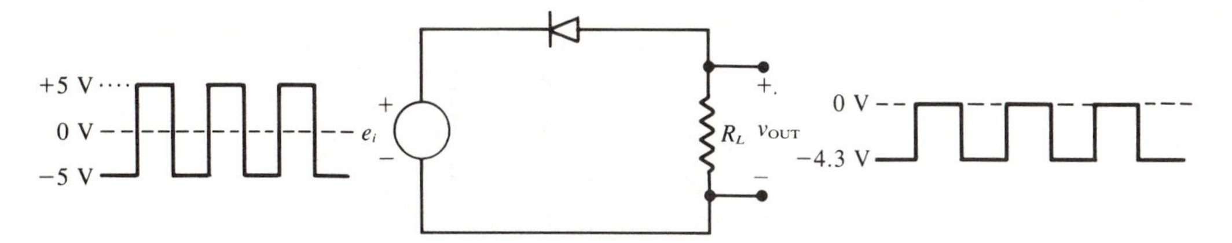

Figure 6–2 **Series clipper with diode turned around**

The series clipper has two drawbacks. One is the small loss in amplitude because of V_F. This loss may not be important, depending on the particular application. Of course, this clipper will not work for e_i signals that are not large enough to supply the forward bias required by the diode.

A second disadvantage is that the diode's junction capacitance, C_J, might couple part of any fast input transitions to the output. For example, Figure 6–3 shows an input pulse with very fast transitions applied to a series clipper. Since the pulse is positive, the clipper output should ideally be zero. Under certain conditions, however, the output will exhibit narrow spikes corresponding to the input transitions.

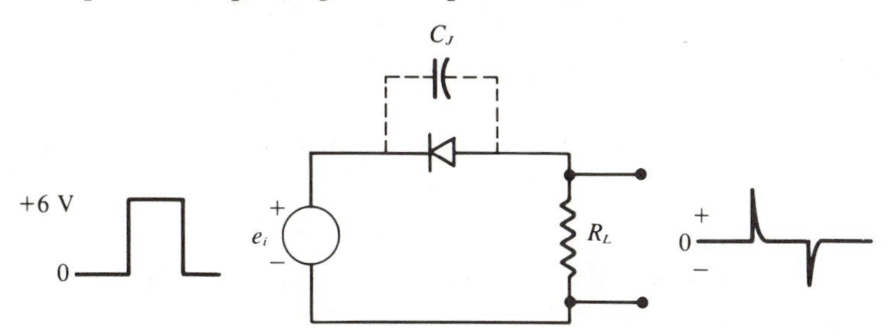

Figure 6–3 **Diode capacitance can couple high-speed transitions to output**

Example 6.2 What conditions are necessary for spikes to occur in the series clipper? How can these spikes be eliminated?

Solution: C_J and R form a high-pass circuit with a time constant $\tau = R_L C_J$. Recall that the output will exhibit spikes if τ is greater than $t_r/2\pi$. Thus, if t_r is small enough, τ can exceed $t_r/2\pi$ even though C_J is usually only a few picofarads.

The spikes can be eliminated by reducing τ. This is most easily done by reducing R_L, if possible. C_J can also be reduced by using a diode specifically designed for high-speed switching applications and so has a small C_J.

Shunt clipper (limiter) In this circuit, the diode is placed across the output as shown in Figure 6–4(a). Its position means that the output can never be more positive than about 0.7 V with respect to ground. The diode will have no effect on any negative output voltage since it would be reverse-biased and will act like an open circuit.

The example waveforms shown in Figure 6–4 are similar to those of the series clipper, but there are notable differences. In the shunt clipper, the signal is not clipped to zero but rather at $\approx$ 0.7 V. (The clipped portion is not perfectly flat at 0.7 V because the diode forward voltage will increase from about 0.5 V to 0.7 V as it goes from barely ON to fully ON.) On the other hand, the shunt clipper does not affect the amplitude of the

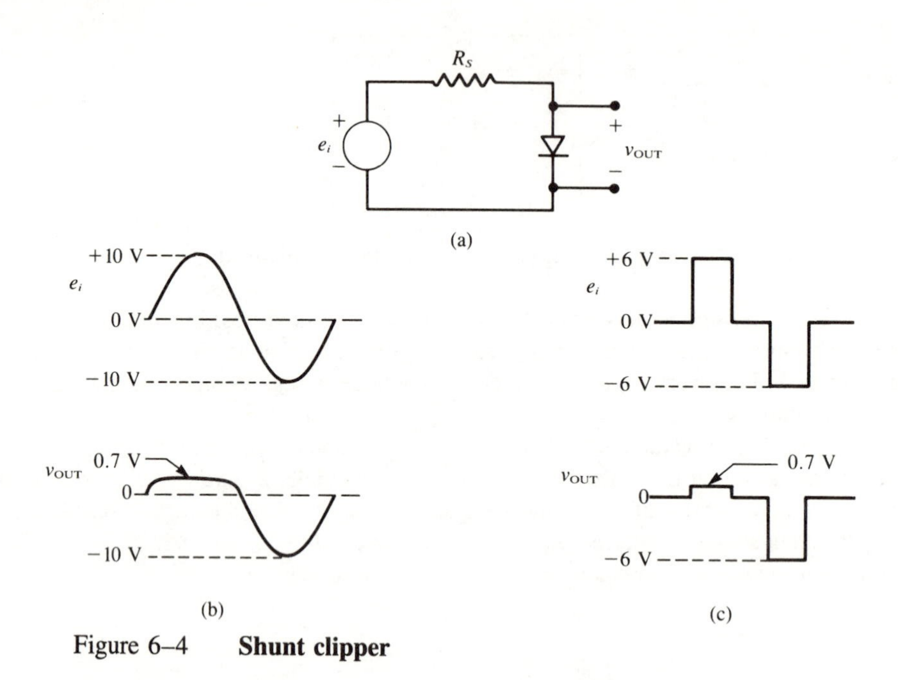

Figure 6–4 **Shunt clipper**

portion of the input signal that it passes to the output. The full e_i signal reaches the output under no-load conditions.

Example 6.3 Determine the waveform across the load in Figure 6–5.

Solution: The diode will limit the negative amplitude of the output to −0.7 V but will have no effect on the positive amplitude. Thus, when e_i is at +30 V, the diode is an open circuit so that v_{OUT} is determined by the voltage divider ratio between R_s and R_L.

$$v_{OUT}(+) = \frac{6.8\ \text{k}\Omega}{2.2\ \text{k}\Omega + 6.8\ \text{k}\Omega} \times 30\ \text{V} \approx +22.67\ \text{V}$$

Without the diode, the output would jump to −22.67 V when e_i jumped to −30 V. The shunt diode, however, will limit v_{OUT} to −0.7 V, as shown in Figure 6–5.

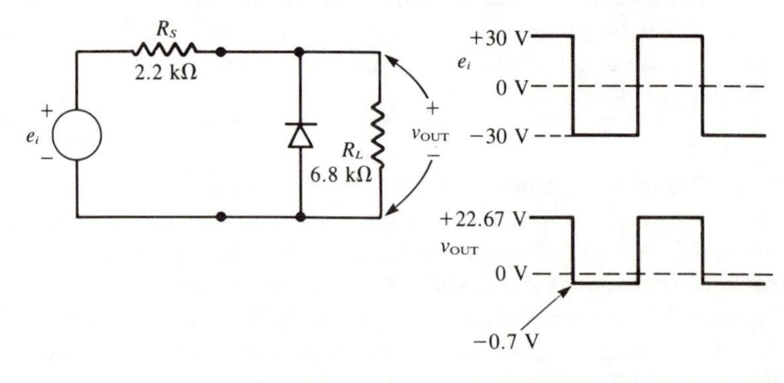

Figure 6–5 **Example 6.3**

The shunt clipper also has some drawbacks. One of them, as already pointed out, is that the output is not clipped at zero, but at 0.7 V. Another is caused by the diode capacitance. This capacitance across the output produces a low-pass action that can slow down the transitions of a very fast input signal. Reducing the circuit time constant by reducing either the circuit resistance or C_J will minimize this effect.

Biased shunt clipper (limiter) All the circuits we have presented so far have clipped or limited the input signal at approximately 0 V, essentially eliminating one polarity of the input signal. This 0 V clipping is used when the load being driven cannot tolerate one of the signal polarities. A good example is Figure 6–6, where a shunt clipper is used to limit the negative portion of e_i to -0.7 V to prevent reverse breakdown of the transistor *E-B* junction. Many transistors have very low values of BV_{EB}, typically around 6 V.

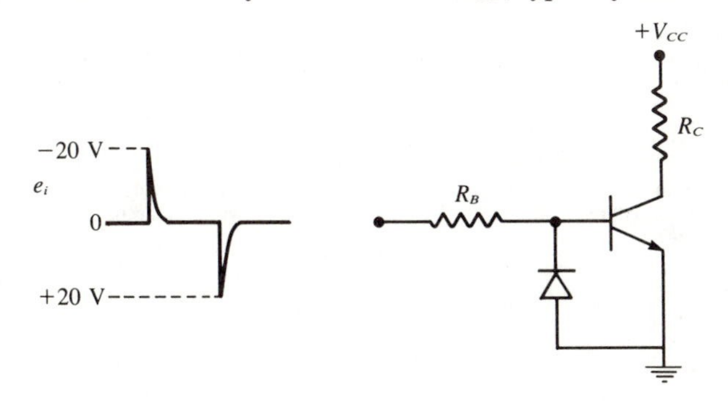

Figure 6–6 **Diode shunt clipper, used to protect transistor from *E-B* reverse breakdown**

In many applications it is necessary to clip or limit an output signal at some voltage other than 0 V. One such circuit is the *biased shunt clipper*, which uses a DC reference or bias source to establish the voltage limit. For example, Figure 6–7(a) shows how a series combination of a diode and a DC reference source, E_R, can be placed across the output terminals. Although there are other possibilities for the direction of the diode and polarity of E_R, we will analyze this one case in detail to illustrate the process by which any of the other cases can be analyzed.

Our analysis will consider the two operating states of the diode. When the diode is conducting, the voltage across the output terminals will be equal to the sum of E_R plus the diode forward voltage, V_F. Since V_F cannot be much more than around 0.7 V, this means that the largest *positive* output voltage will be

$$V_{LIM+} \approx E_R + 0.7 \text{ V}$$

Thus, when the diode is ON the circuit can be replaced by the equivalent circuit of Figure 6–7(b), where $v_{\text{OUT}} \approx E_R + 0.7$ V. The output voltage is independent of the value of e_i *provided that the diode is ON*.

When the diode is not conducting, it acts as an open circuit so that the path containing the diode and E_R is essentially disconnected from the circuit. Thus, when the diode is OFF, the circuit can be replaced by the equivalent circuit drawn in Figure 6–7(c). Clearly, then, the output will follow the input; that is, $v_{\text{OUT}} = e_i$ (with no load across the output).

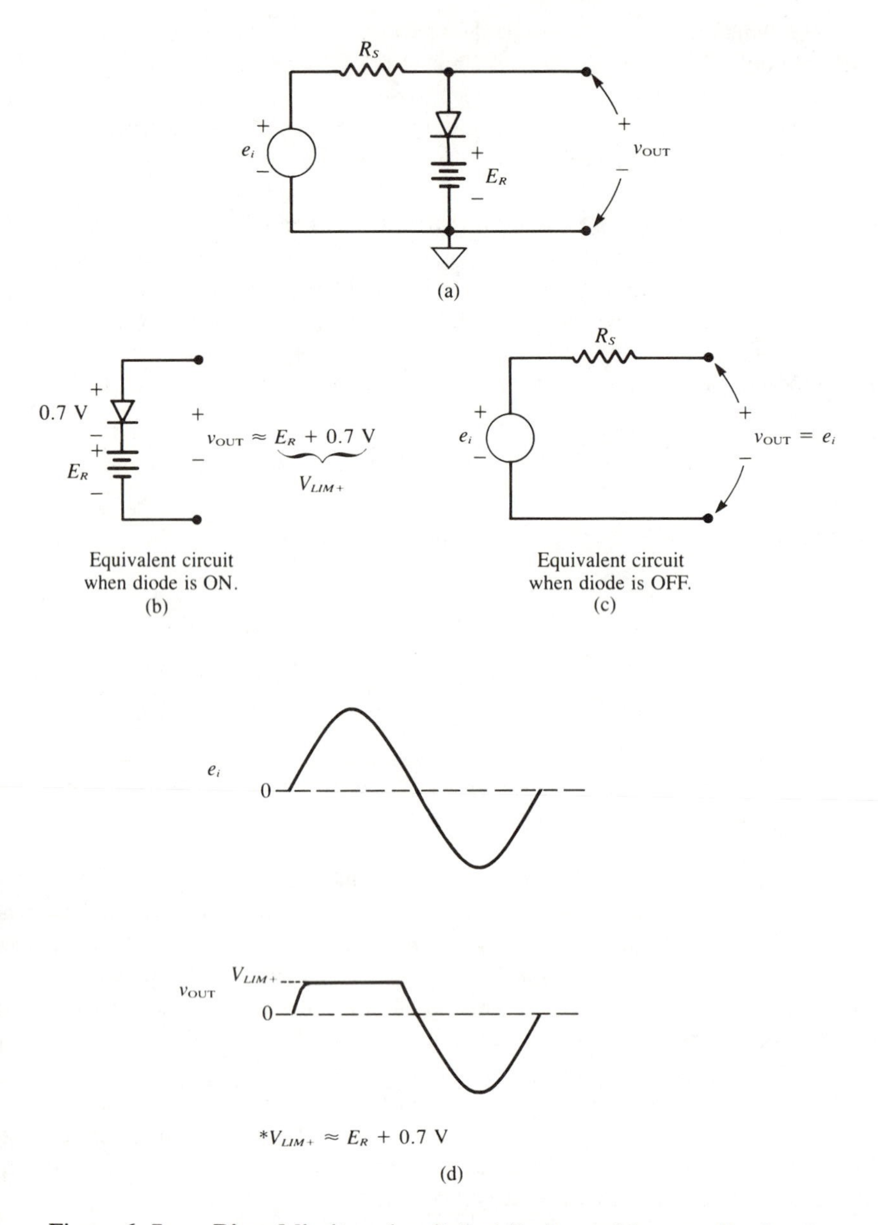

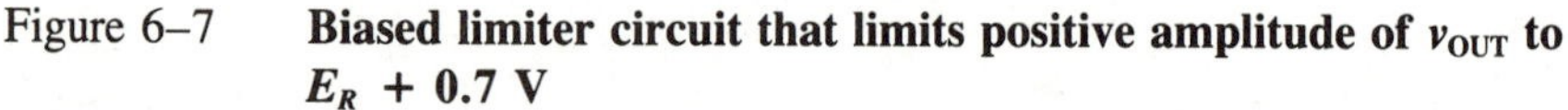

Figure 6–7 **Biased limiter circuit that limits positive amplitude of v_{OUT} to E_R + 0.7 V**

Now that we have determined the circuit's two operating states, we must find the range of input voltages that will produce each state. The E_R source is reverse-biasing the diode. Thus, e_i has to exceed E_R by 0.7 V in order to turn ON the diode. In other words, when $e_i \geq V_{LIM+}$, the diode will be ON and v_{OUT} will be limited to V_{LIM+}, as in Figure 6–7(b). Conversely, when $e_{in} < V_{LIM+}$, the diode will be OFF and v_{OUT} will follow e_i as in Figure 6–7(c).

Figure 6–7(d) shows the output waveform when the input is a sinewave with an amplitude that exceeds V_{LIM+}. Note that v_{OUT} is exactly the same as e_i, except when e_i exceeds V_{LIM+}. For that portion of e_i the output is clipped at V_{LIM+}. Once again the clipped portion is not perfectly flat because the diode actually starts conducting at a forward bias of around 0.5 V.

Example 6.4 Analyze the biased shunt clipper of Figure 6–8(a) and sketch the v_{OUT} signal when e_i is a 24 V *p-p* bipolar square wave and $E_R = 6$ V.

Solution: If we use reasoning similar to that used in the analysis of Figure 6–7, we come up with the following results:

(a) The circuit limits the *negative* portion of the signal such that the output can go no more negative than

$$V_{LIM-} \approx -E_R - 0.7 \text{ V}$$

(b) The output will be limited to V_{LIM-} whenever e_i tries to go more negative than V_{LIM-}.

(c) The output will follow e_i for all other values of e_i.

With $E_R = 6$ V, this circuit will clip at −6.7 V. The v_{OUT} signal for a 24 V *p-p* square wave input is drawn in Figure 6–8(b).

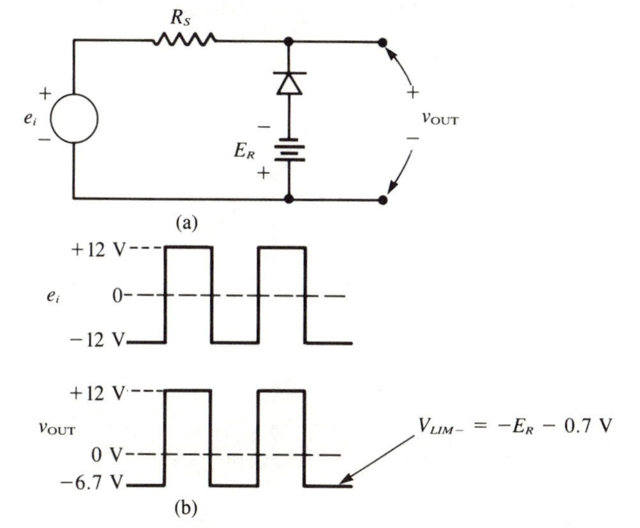

Figure 6–8 **Biased shunt clipper that limits the negative output amplitude**

Example 6.5 Determine the output waveform for the circuit of Figure 6–9.

Solution: This circuit is a combination of the two biased clippers just analyzed. It is a *double-ended clipper* that will limit both the positive and negative excursions of the output. The positive limit is determined by D_1 and the 16 V supply. Thus

$$V_{LIM+} = 16.7 \text{ V}$$

The negative limit is determined by D_2 and the 8 V supply. Thus,

$$V_{LIM-} = -8.7 \text{ V}$$

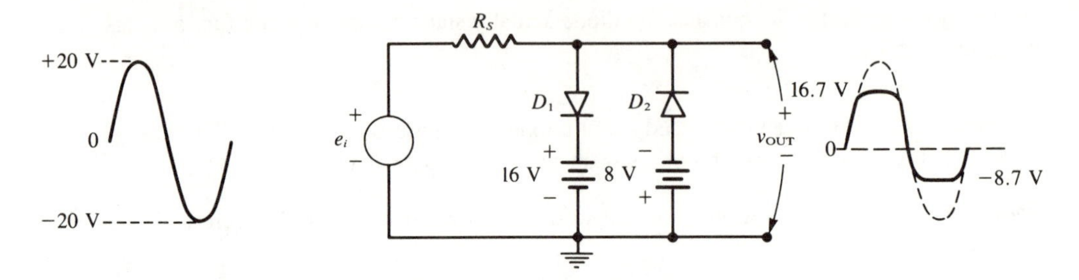

Figure 6–9 **Double-ended shunt clipper**

The output waveform is therefore able to follow e_i unless e_i goes more positive than 16.7 V or more negative than −8.7 V.

Zener diode clippers The biased clipper circuits require a bias source, and this may not always be feasible. Zener diodes can be used to provide the same output limiting action without the need for a bias source. Figure 6–10(a) shows a 5.1 V zener used to limit the positive output swing to 5.1 V. When e_i exceeds +5.1 V, the zener goes into reverse breakdown and keeps $v_{OUT} \approx V_Z = 5.1$ V. The zener will also limit the negative excursion of v_{OUT} to −0.7 V because it will be forward-biased when e_i goes negative.

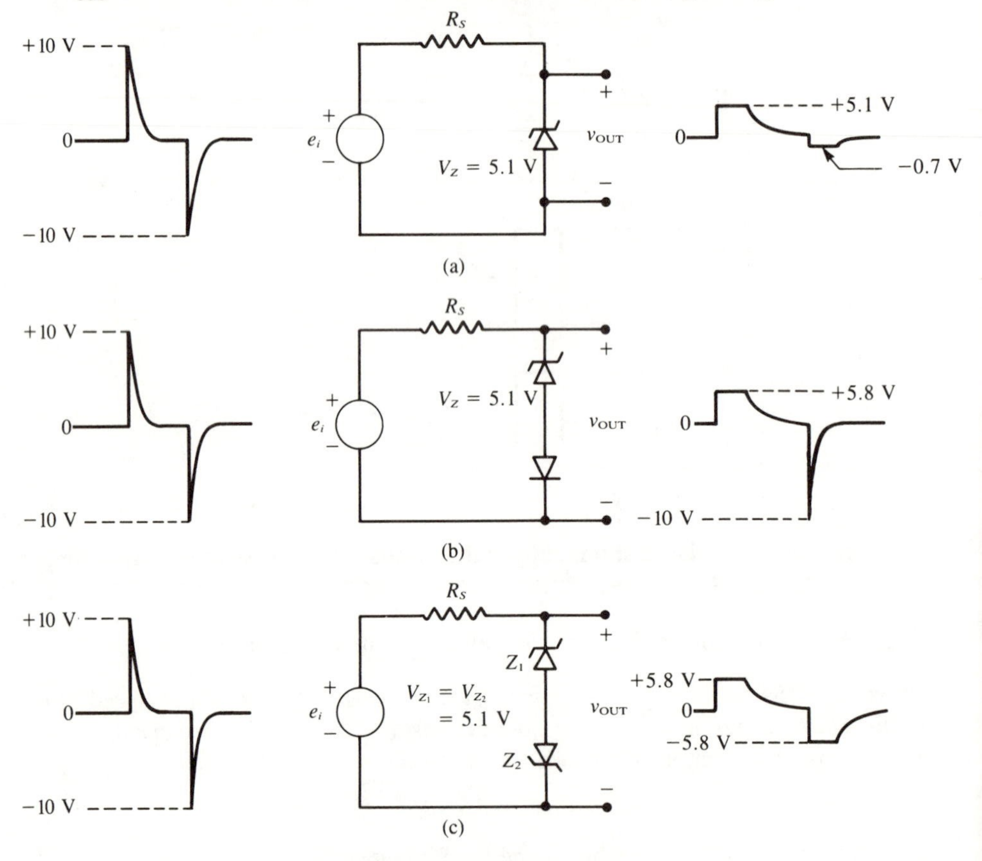

Figure 6–10 **Zener diode clippers**

By adding a diode in series with the zener [Figure 6–10(b)], the circuit will not limit the negative excursion. The diode will be OFF for negative values of e_i, so that $v_{OUT} = e_i$.

If this diode is replaced by another zener diode [Figure 6–10(c)], the circuit functions as a *double-ended limiter*. When e_i is sufficiently positive, zener diode Z_1 will be in breakdown and Z_2 will be forward-biased, so that v_{OUT} is limited to 5.1 V + 0.7 V = +5.8 V. Likewise, when e_i is sufficiently negative, Z_1 will be forward-biased and Z_2 will be in reverse breakdown, so that $v_{OUT} \approx -5.8$ V.

The zener diode clipper circuit has several disadvantages compared to the biased shunt clipper. It is difficult to get precise clipping levels (V_{LIM+} and V_{LIM-}) because of the tolerances on the zener breakdown voltages (typically ±5 percent). Even if the zener voltages were more accurate, there are only certain values available; one may not always be able to find the exact value required.

Since zener diodes are not generally designed for high-speed switching circuit applications, they will have a larger C_J than switching diodes. This situation could produce a significant increase in signal transition times when the input is a fast-changing pulse waveform. In some applications, the increased transition times would be unacceptable, and a biased clipper using a high-speed switching diode should be used.

6.3 Operational Amplifier Clipper Circuit

The high-gain characteristic of an operational amplifier can be used to improve the diode clipper circuit. Figure 6–11(a) shows an arrangement that combines a diode and an op-amp. The op-amp is connected in the inverting amplifier configuration with e_i applied to the op-amp "−" terminal and with the "+" terminal at ground. Note that the circuit output v_{OUT} is not taken directly from the op-amp output, v_x. Also note that the diode is connected in the negative feedback loop from the op-amp output to the "−" input.

To analyze this circuit, we will again consider the two possible diode states. When the diode is ON, the negative feedback path is complete. Since the op-amp is operating with negative feedback, its differential input voltage, e_d, must be approximately zero. Thus, the "−" input is approximately at ground and $v_{OUT} \approx 0$ V [see Figure 6–11(b)].

With $v_{OUT} \approx 0$ V, the op-amp output must be at +0.7 V since the diode is ON; that is, $v_x \approx +0.7$ V. The input signal e_i has to be *negative* to produce a positive op-amp output. We can determine the value of e_i needed to produce $v_x = +0.7$ V by using the following reasoning. As e_i is increased negatively from zero, it produces a negative differential input that is amplified by the op-amp's large open-loop gain, A_{VOL}. When e_i is slightly negative, the op-amp output forward biases the diode that closes the negative feedback loop. Once the negative feedback takes effect, v_{OUT} will be maintained at ≈ 0 V. The value of e_i that produces $v_x = +0.7$ V can be obtained if we know A_{VOL}. For example, with $A_{VOL} = 10^5$, we have

$$A_{VOL} \times (-e_i) = +0.7 \text{ V}$$

Therefore,

$$e_i = \frac{-0.7 \text{ V}}{A_{VOL}} = \frac{-0.7 \text{ V}}{10^5}$$
$$= -7.0\ \mu\text{V}$$

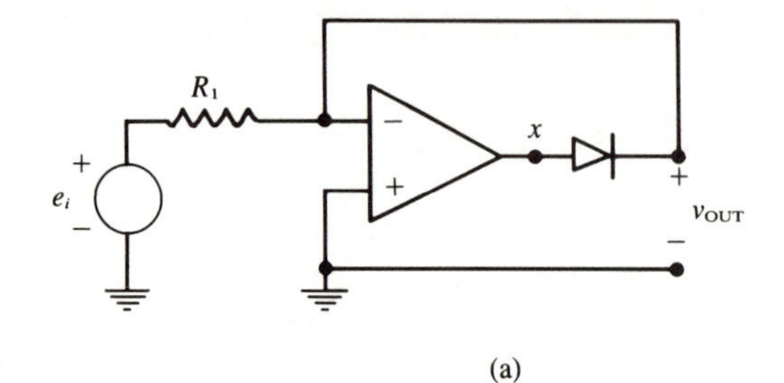

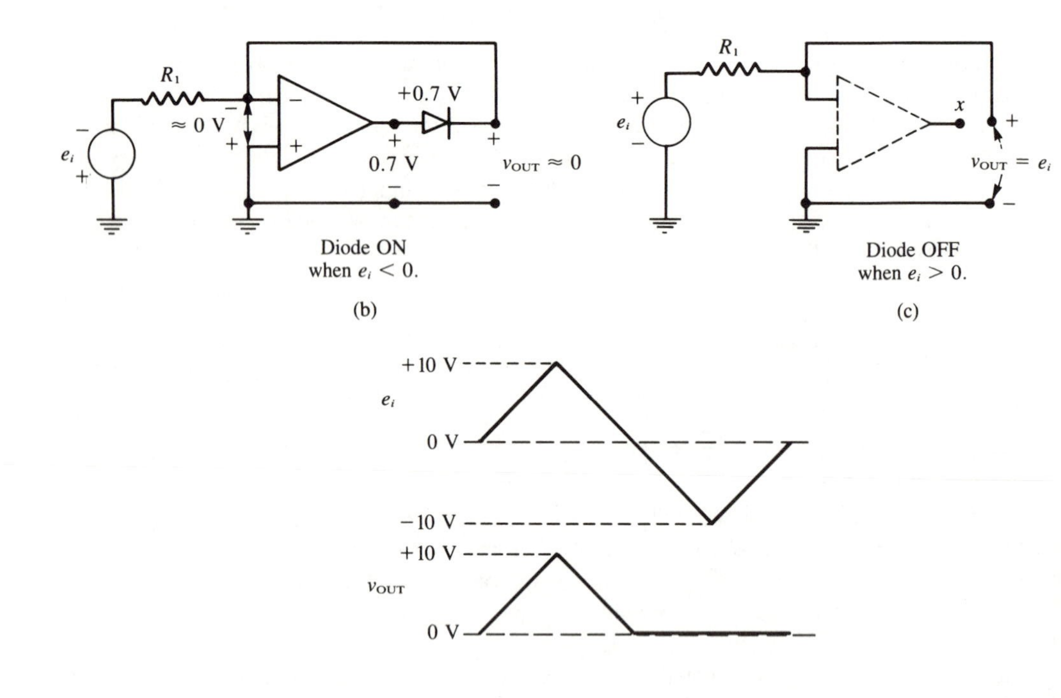

Figure 6–11 **Op-amp/diode clipper circuit clips the input signal at precisely 0 V and eliminates the effect of diode forward voltage**

Thus, for all practical purposes, whenever e_i becomes slightly negative, the op-amp output will become $\approx +0.7$ V, the diode will be ON, and v_{OUT} will be very close to 0 V.

Now let's consider what happens when the diode is OFF. In that case the diode is not conducting and there will be no connection between the op-amp output and its "−" input and therefore no negative feedback. This situation is drawn in Figure 6–11(c) where the diode has been replaced by an open circuit. Without negative feedback, the op-amp just acts like a very large resistance (Z_{IN}), which we can usually ignore. Thus, v_{OUT} is connected to e_i through R_1, so that $v_{OUT} = e_i$ (with no load).

The diode will be OFF whenever the op-amp output is less than +0.7 V. This output will occur when e_i is close to 0 V or when it is positive. For our purposes, then, we can say that v_{OUT} will follow e_i when e_i is ≥ 0 V.

Figure 6–11(d) summarizes the complete circuit operation. Here we can see that $v_{OUT} \approx 0$ for negative e_i, and that $v_{OUT} \approx e_i$ for positive e_i. The clipping point is very close to 0 V, because the large A_{VOL} requires only a small e_i to produce $v_x = +0.7$ V to turn ON the diode.

Note that the diode forward voltage drop does not affect the v_{OUT} waveform as it does in the diode clipper circuits. This is a distinct advantage when the input is a low-amplitude signal. For instance, the op-amp clipper will work even for signals in the microvolt range.

Example 6.6 Determine the v_{OUT} waveform for the circuit of Figure 6–12.

Solution: The diode has been turned around compared to Figure 6–11. Using the same reasoning, we can conclude that the circuit will clip e_i at 0 V and will pass only its *negative* portion to the output.

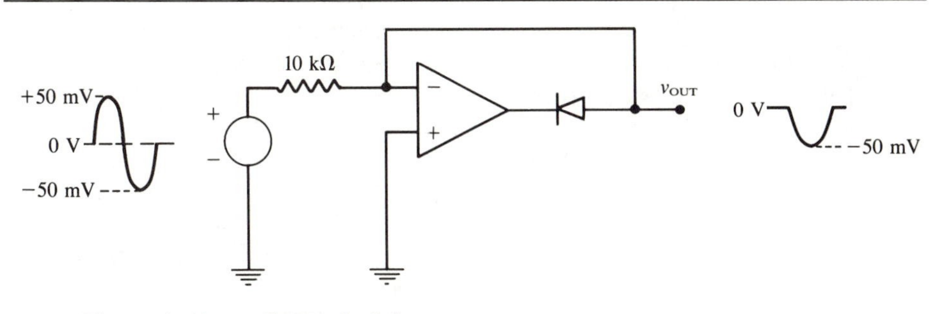

Figure 6–12 **Example 6.6**

Example 6.7 The op-amp in Figure 6–12 uses ±15 V supplies. What is the maximum reverse voltage that the diode has to withstand in this circuit?

Solution: When e_i is negative, the op-amp output will be positive, thereby reverse-biasing the diode. Since the diode is OFF, there is no negative feedback; the op-amp is operating open-loop. Thus, even a very small e_i will cause the op-amp output to saturate at around +13.5 V. The diode, then, will have +13.5 V at its cathode, while its anode is connected to e_i. In this case e_i is very small, so the diode must be able to withstand a reverse voltage of about **13.5 V.**

6.4 The Bipolar Transistor Inverter

An inverter circuit changes the direction of a pulse from positive-going to negative-going, and vice versa. As we shall see later, it is also used in digital logic circuits to convert a low input voltage level to a high voltage level and vice versa. The most widely used inverter circuit uses a bipolar transistor in the common-emitter configuration (Figure 6–13), where the input signal is applied to the base and the output taken at the collector.

The input signal shown is a 10 V pulse riding on a baseline of 0 V. The circuit operates so that when e_i is at 0 V, the transistor is OFF. Therefore, no i_c flows, and no drop is produced across the collector resistor, so the collector is at +10 V above ground,

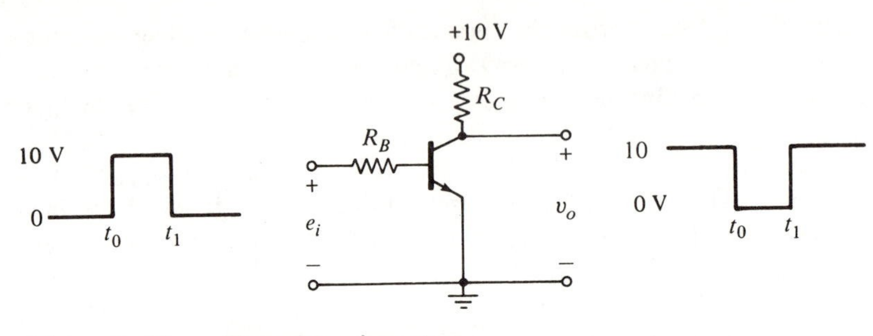

Figure 6–13 **Transistor inverter**

producing v_o=10 V. When the input signal is at 10 V, base current flows, producing a flow of collector current. If i_b is large enough, it will saturate the transistor, so the collector-emitter voltage will be V_{CE}(SAT); this produces $v_o = V_{CE}(\text{SAT}) \approx 0$. Thus, the transistor operates in either the OFF state or the ON state depending on the state of the input. As can be seen from the figure, the positive-going input pulse produces a negative-going output pulse; the circuit essentially *inverts* the input pulse (when the input is low, the output is high, and vice versa).

In this circuit there is no reverse bias of the *E-B* junction to produce cutoff. This situation is satisfactory for modern silicon transistors, since even for a slight *E-B* forward bias (<0.5 V) the base and collector currents would drop to a negligible value. For germanium units a reverse bias would normally be necessary to effectively turn-OFF the transistor and keep the collector leakage low, especially at elevated temperatures. If reverse bias is needed, it is usually obtained by adding another branch to the base circuit, consisting of a negative supply voltage (for NPN) connected to the base through a resistor.

Referring again to the basic circuit of Figure 6–13, it can be seen that for proper operation, the values of R_B and R_C must be chosen so that the transistor will saturate when e_i goes positive. The value of R_B determines the base current, and R_C determines the value of I_C(SAT). The transistor current gain (h_{FE}) determines the value of i_b needed to produce $i_c = I_C(\text{SAT})$.

For example, if $h_{FE} = 50$, then

$$I_B(\text{SAT}) = \frac{I_C(\text{SAT})}{50} \approx \frac{10\text{ V}/R_C}{50} = \frac{0.2\text{ V}}{R_C}$$

This is the value of i_b needed to cause saturation. The actual i_b that flows when e_i is 10 V is

$$I_B(\text{ACT.}) = \frac{10 - 0.7\text{ V}}{R_B} = \frac{9.3\text{ V}}{R_B}$$

For saturation to occur it is necessary that i_b be greater than or equal to I_B(SAT). That is,

$$\frac{9.3\text{ V}}{R_B} \geq \frac{0.2\text{ V}}{R_C}$$

which can be rewritten

$$\frac{R_B}{R_C} \leq 46.5$$

This last expression indicates that for this case the maximum *ratio* of R_B/R_C needed to ensure saturation is 46.5. Obviously, there are an infinite number of values that can satisfy this requirement. In practice, R_C is often chosen to produce a desired I_C(SAT), and then R_B is chosen to produce the required base current. For example, an I_C(SAT) of 10 mA requires that R_C be

$$R_C = \frac{10\text{ V}}{10\text{ mA}} = 1\text{ k}\Omega$$

Thus, to produce saturation, R_B must be chosen so that

$$R_B \leq 46.5\text{ k}\Omega$$

Example 6.7 Assume that the transistor in the preceding discussion has h_{FE} values that can range from 25 to 100. Calculate the value of R_B needed to ensure saturation.

Solution: Since this transistor can have an h_{FE} as low as 25, it is necessary to use a value of R_B that will produce $i_c = I_C(\text{SAT}) = 10$ mA even at the lowest h_{FE}. Thus, at $h_{FE} = 25$, the value of I_B(SAT) becomes

$$I_B(\text{SAT}) = \frac{10\text{ mA}}{25} = 0.4\text{ mA}$$

To produce $i_b = 0.4$ mA, the following is needed:

$$\frac{9.3\text{ V}}{R_B} = 0.4\text{ mA}$$

or

$$R_B = \frac{9.3\text{ V}}{0.4\text{ mA}} = 23.3\text{ k}\Omega$$

This value of R_B will produce saturation even at the lowest value of h_{FE}.

Example 6.8 Consider the inverter circuit shown in Figure 6–14 that uses a PNP transistor. The negative-going input pulse (0 V to −10 V) is supposed to produce a positive-going output pulse (−10 V to 0 V). Suppose, however, that when actually observed, the output pulse is the one shown in Figure 6–14 (−10 V to −2 V). How can the circuit values be modified to bring the circuit into specifications?

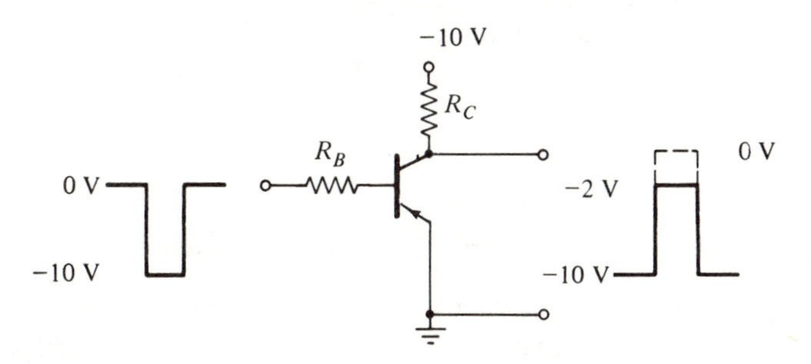

Figure 6–14

Solution: Since the output pulse does not reach 0 V when e_i is at -10 V, it is apparent that the transistor is not saturating. This means that i_b is not large enough to produce the required I_C(SAT). There are *three* possibilities for correcting this situation. First, the value of R_B can be decreased to produce more base current so that the collector current can increase to I_C(SAT). Second, the value of R_C can be increased so that I_C(SAT) is decreased, therefore requiring less base current. Third, a transistor with higher h_{FE} can be used so that the base current requirement is decreased.

Loading We will now examine the effects of resistive and capacitive loading on the inverter output, since in most applications the inverter is subjected to some form of loading. Figure 6–15(a) has an inverter circuit with its output driving a load R_L.

During the positive portion of the input pulse, the transistor conducts and collector current flows. If designed correctly, the output voltage will be approximately zero (transistor saturated), the voltage across R_L will be zero, and no current will flow through the load. In other words, when the transistor is saturated, the load has no effect on the output. This case is illustrated in part (b) of the figure. Here, all the current supplied by the 12 V collector supply (i_1) flows through the transistor collector and none through the load, because the ON transistor will have a very low resistance R_{ON}. Of course, if R_L is made small enough that it is comparable in size to R_{ON} of the transistor, it will draw current away from the transistor, pulling it out of saturation. In most practical cases R_L will not be small enough to have this effect.

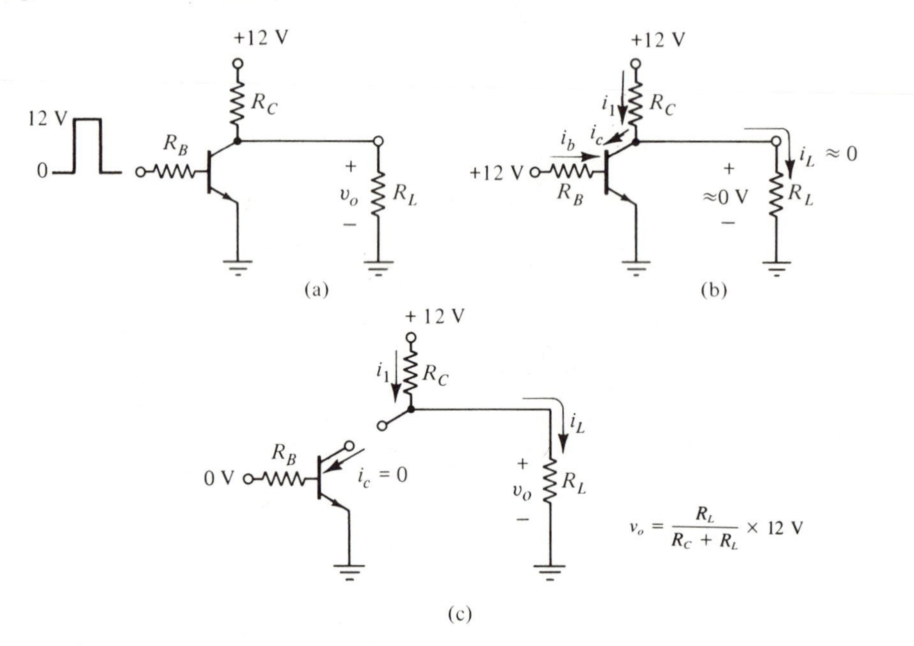

Figure 6–15 **Effect of load R_L on inverter output**

During the 0 V portion of the input, the transistor is OFF, so essentially no i_c flows. This situation is illustrated in Figure 6–15(c), where the collector lead has been opened to indicate that the transistor is acting as an open switch. It can be seen that R_C and R_L are now in series, so v_o will be determined by their voltage divider ratio.

$$v_o = \left(\frac{R_L}{R_L + R_C}\right) 12 \text{ V}$$

Obviously, any R_L will cause v_o to be less than 12 V. However, if R_L is at least twenty times greater than R_C, the value of v_o will be within 5 percent of the 12 V collector-supply voltage.

Example 6.9 A 12 V pulse is applied to the inverter circuit in Figure 6–15, which uses a transistor with $h_{FE}(\text{min}) = 50$. The inverter output must drive a 1 kΩ load without its output dropping below 11 V in the transistor OFF state. Determine appropriate values for R_C and R_B.

Solution: To determine R_C, at cutoff there is

$$v_o = \frac{1\ \text{k}\Omega \times 12\ \text{V}}{1\ \text{k}\Omega + R_C} \geq 11\ \text{V}$$

so

$$R_C \leq 91\ \Omega$$

A value of $R_C = 91\ \Omega$ will give $v_o = 11$ V; if R_C is any larger, v_o will drop below 11 V when the transistor is OFF.

To be safe, let us use $R_C = 75\ \Omega$. Thus, when the transistor is ON we have

$$I_C(\text{SAT}) = \frac{12\ \text{V}}{75\ \Omega} = 160\ \text{mA}$$

Since $h_{FE}(\text{min}) = 50$, a base current of at least

$$I_B(\text{SAT}) = \frac{160\ \text{mA}}{50} = 3.2\ \text{mA}$$

must be supplied. Thus,

$$I_B(\text{ACT.}) = \frac{12\ \text{V} - 0.7\ \text{V}}{R_B} \geq 3.2\ \text{mA}$$

giving a value for R_B of

$$R_B \leq 3.53\ \text{k}\Omega$$

A value of 3.3 kΩ would be sufficient.

Capacitive loading Now consider the effects of a capacitive load on the inverter, as illustrated in Figure 6–16. Assume that the input is initially at 12 V, so that the transistor is ON. Thus, $v_o \approx 0$, and C_L is discharged as in Figure 6–16(a). When the input drops to zero [part (b) of the figure], the transistor turns OFF. At that time, collector current stops flowing; however, C_L will prevent the output from rapidly rising to 12 V since it has to charge up through R_C. The charging time constant is $\tau = R_C C_L$, so the rise time of v_o is $t_r = 2.2 R_C C_L$.

When e_i returns to +12 V [part (c) of the figure], the transistor is turned ON. The capacitor C_L will rapidly discharge through the low resistance of the transistor, so the fall time of the v_o waveform will be relatively fast. We will talk in more detail in Chapter 11 about how to calculate t_f in this situation. For now, we will simply assume that t_f is very fast compared with t_r.

The complete e_i and v_o waveforms appear in Figure 6–16(d). The capacitor C_L has seriously deteriorated the output rise time. If t_r is to be kept below a certain maximum value, R_C has to be chosen small enough to charge C_L within the required time.

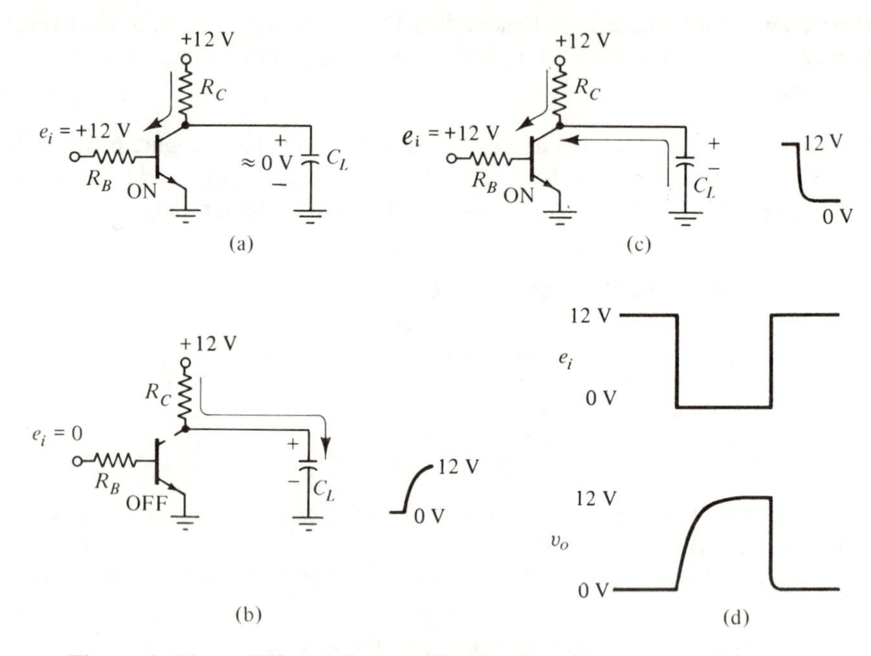

Figure 6–16 **Effects of capacitive load on inverter output**

Example 6.10 If $C_L = 50$ pF in Figure 6–16, determine the values for R_C and R_B needed to provide a rise time of at most 0.1 μs in the output. Use $h_{FE}(\text{min}) = 50$.

Solution: The charging time constant (τ) must be chosen so that $t_r \leq 0.1\ \mu$s. Thus, since $t_r = 2.2\tau$,

$$\tau \leq 0.0455\ \mu\text{s}$$

This means that

$$R_C \times C_L \leq 0.0455\ \mu\text{s}$$

which gives (for $C_L = 50$ pF)

$$R_C \leq 910\ \Omega$$

Using $R_C = 820\ \Omega$ gives an $I_C(\text{SAT})$ of

$$\frac{12\text{ V}}{820\ \Omega} = 14.6\text{ mA}$$

which requires a base current of at least

$$\frac{14.6\text{ mA}}{50} = 0.292\text{ mA}$$

Thus,

$$R_B \leq \frac{11.3\text{ V}}{0.292\text{ mA}} = 38.7\text{ k}\Omega$$

A value of 33 kΩ would be sufficient.

Example 6.11 Assume that values for R_B and R_C have been chosen to produce an output rise time of 0.1 μs, as in the preceding example. A technician connects an oscilloscope to observe the output waveform and measures $t_r \approx 0.2$ μs. Assuming that none of the circuit components is faulty, what are some of the possible reasons for the discrepancy?

Solution: The possible reasons include

(a) The effect of the transistor's junction capacitances, which will increase the total capacitance from collector to ground.

(b) The technician is using an oscilloscope whose bandwidth is not large enough to pass the signal without causing an increase in transition times.

(c) The technician is not using a low-capacitance, 10X probe on the oscilloscope. Thus, the scope input capacitance (typically 47 pF) is placed in parallel with C_L, thereby increasing the total load capacitance on the circuit.

Example 6.12 Figure 6–17 shows another, less common, transistor pulse inverter circuit. It uses two transistors, a PNP (Q_1) and an NPN (Q_2), to provide a low output resistance in both output states. These transistors allow the circuit to drive relatively heavy loads (small R_L, large C_L) without significant loss of amplitude or deterioration (increase) of transition times. Analyze the circuit's operation for the two input voltage levels (0 V and +6 V).

Solution: When $e_i = 0$ V, the *E-B* junction of Q_1 will be forward-biased, allowing base current to flow and turn ON Q_1. On the other hand, $e_i = 0$ V will not provide forward bias for the *E-B* junction of Q_2, so Q_2 will be OFF. With Q_1 ON, its low resistance (from collector to emitter) will essentially connect the 6 V source to the load providing a rapid charging of C_L. v_{OUT} will be slightly lower than 6 V due to the small V_{CE}(SAT) across Q_1.

When $e_i = 6$ V, the *E-B* junction of Q_1 will not be forward-biased, so Q_1 will be OFF. On the other hand, $e_i = 6$ V will forward bias the Q_2 *E-B* junction to turn ON Q_2. With Q_2 ON, its collector-emitter resistance drops to a low value and provides a rapid discharging of the load voltage to ≈ 0 V.

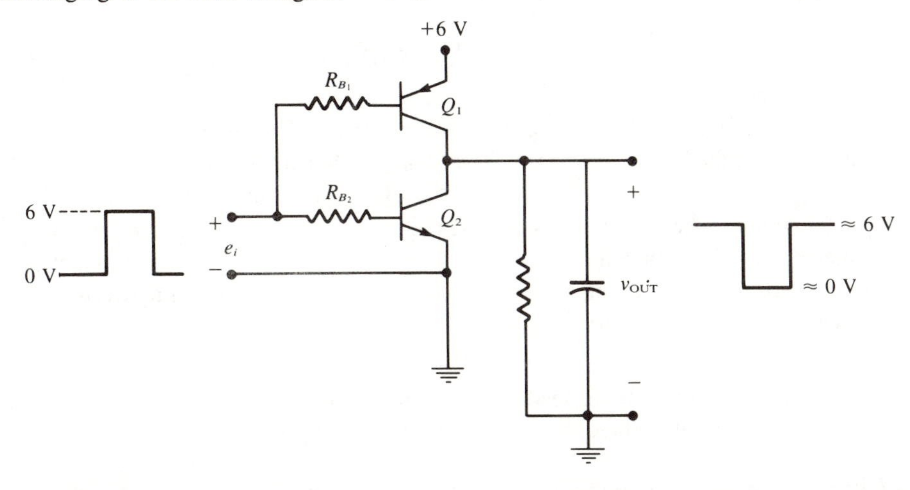

Figure 6–17 **Transistor inverter that uses two transistors to reduce effects of R_L and C_L**

6.5 Buffer Circuits

When a signal source with a relatively high source resistance has to drive a low-resistance load, the signal that reaches the load will be significantly attenuated. In addition, the high source resistance will slow down the charging and discharging of any load capacitance, thereby deteriorating the signal transition times.

Very often, it is necessary to use a *buffer* circuit between a signal source and a load so as to minimize signal attenuation and deterioration of transition times. Ideally, the buffer will have a high input impedance that will not load down the signal source, and a low output impedance for driving low impedance loads. A buffer does not have to have any voltage gain; but it does have to have current gain, because it should draw very little current from the signal source while supplying the current required by the load.

The emitter-follower Figure 6–18(a) shows a 12 V pulse source with an internal resistance of 1 kΩ driving a 100Ω load. The pulse actually reaching the load is only slightly greater than 1 V, because R_L is so much smaller than R_s. Another way to look at this problem is that the amount of current the source can supply to the load is limited by R_s.

One common method for reducing this loading effect is to connect an *emitter-follower* circuit as a buffer between the source and the load, as shown in Figure 6–18(b). The emitter-follower (E-F) is simply a transistor connected in the common-collector configuration, where the load is placed between the emitter and ground.

Recall that in the common-collector configuration, any impedance connected to the emitter appears to be $(h_{FE} + 1)$ times larger when viewed from the base. In other words, the signal source connected to the base sees a *reflected* resistance of $(h_{FE} + 1)R_L$. Thus, the E-F can present a very high resistance to the signal source, especially if h_{FE} is large.

If we assume $h_{FE} = 100$, the total resistance seen by the 12 V input pulse will be $R_s + (h_{FE} + 1)R_L$. That is,

$$1 \text{ k}\Omega + (101)100 \ \Omega = 11.1 \text{ k}\Omega$$

When the pulse is at 12 V, then, the base current will be

$$I_B = \frac{11.3 \text{ V}}{11.1 \text{ k}\Omega} = 1.02 \text{ mA}$$

This base current will produce a drop of approximately 1 V across R_s so that the voltage actually reaching the base will be 11 V. The voltage at the emitter will therefore be about 10.3 V because of the 0.7 V base-emitter drop. Thus, the voltage across R_L is 10.3 V.

Of course, when the input is at 0 V, the transistor will be OFF and there will be no voltage at the base or emitter. The result, then, is that a 12 V positive input pulse produces a positive 10.3 V pulse across the load. This result is a vast improvement over the situation in Figure 6–18(a).

Example 6.13 Calculate the load current in the circuit of Figure 6–18(b) and compare it to the source current. Where is this load current coming from?

Solution: When the output pulse is at its 10.3 V peak, the load current will be

$$I_L = \frac{10.3 \text{ V}}{100 \ \Omega} = \mathbf{103 \text{ mA}}$$

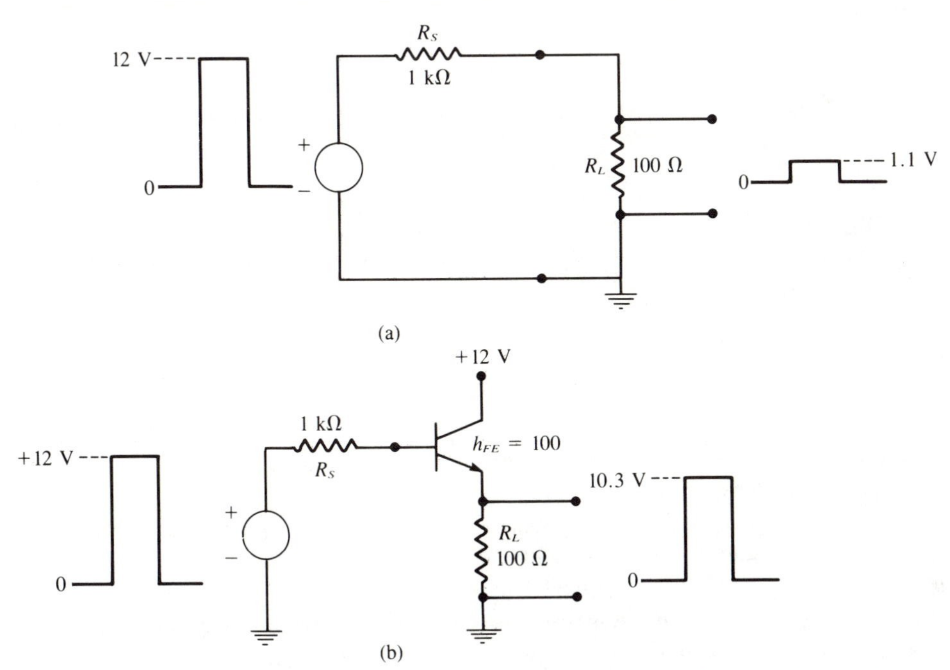

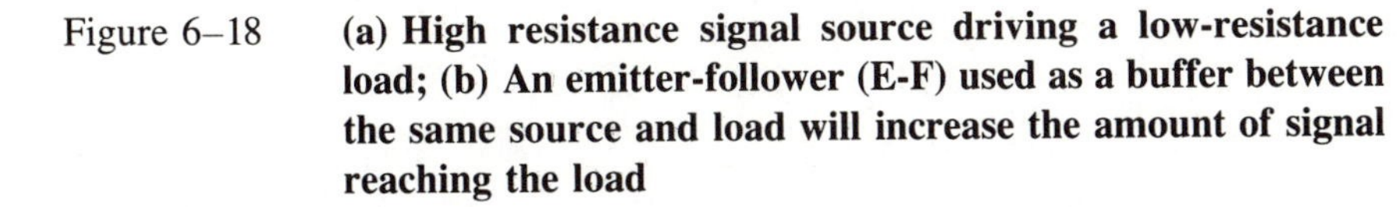
Figure 6–18 **(a) High resistance signal source driving a low-resistance load; (b) An emitter-follower (E-F) used as a buffer between the same source and load will increase the amount of signal reaching the load**

This current is the transistor emitter current. The signal source supplies only the transistor base current of 1.02 mA.

Since $I_E = I_B + I_C$, it is clear that most of the emitter (load) current is supplied by the collector voltage source.

Emitter-follower driving a capacitive load Because it can supply a relatively large current to a load, the E-F can rapidly charge a capacitive load without significantly slowing down the signal rise time. The large emitter current will charge up the load capacitance at a faster rate than the high-resistance signal source. Figure 6–19(a) shows the E-F of Figure 6–18, with a capacitive load in parallel with R_L.

When e_i jumps from 0 V to 12 V, the capacitor will momentarily keep the emitter grounded so that the initial base current will be limited solely by the source resistance. This momentary situation will produce a large initial surge of emitter current to begin charging C_L. This large current will exponentially decrease as C_L rapidly charges to its final value, with a time constant determined by the E-F's very low output impedance.

Example 6.14 What will be the initial surge of emitter current that charges C_L in Figure 6–19(a)? Assume $h_{FE} = 100$.

Solution: When e_i jumps to 12 V, the emitter is held at ground by C_L. Thus, the initial base current will be

$$i_B(@\ t = 0) = \frac{11.3\ \text{V}}{1\ \text{k}\Omega} = 11.3\ \text{mA}$$

which produces

$$i_E = 101 \times 11.3 \text{ mA} = \mathbf{1.14 \text{ A}}$$

This large i_E is only momentary since the capacitor charges rapidly to its final value.

When e_i drops back to 0 V, the transistor will turn OFF and essentially disconnect itself from the loads. The load capacitance then has to discharge through R_L. This discharge will generally produce an output fall time that is slower than the rise time, since R_L usually is not as low as the E-F output impedance through which the capacitor is charged.

Example 6.15 Can R_L be reduced to produce a shorter output fall time?

Solution: Yes, but only up to a point. If R_L is made too small, the output amplitude will drop below its acceptable limit. Of course, a transistor with a higher h_{FE} can be used to counteract the decrease in R_L.

A better way to reduce the output fall time is to use a second transistor to discharge C_L, as shown in Figure 6–19(b). Q_2, a PNP transistor, will be OFF when e_i is at +12 V and will have no effect on the charging of C_L through Q_1. When e_i goes to 0 V, however, Q_2 will be biased ON since its emitter will be positive by virtue of the capacitor's charge. C_L will rapidly discharge through Q_2's low resistance to produce a fast output fall time.

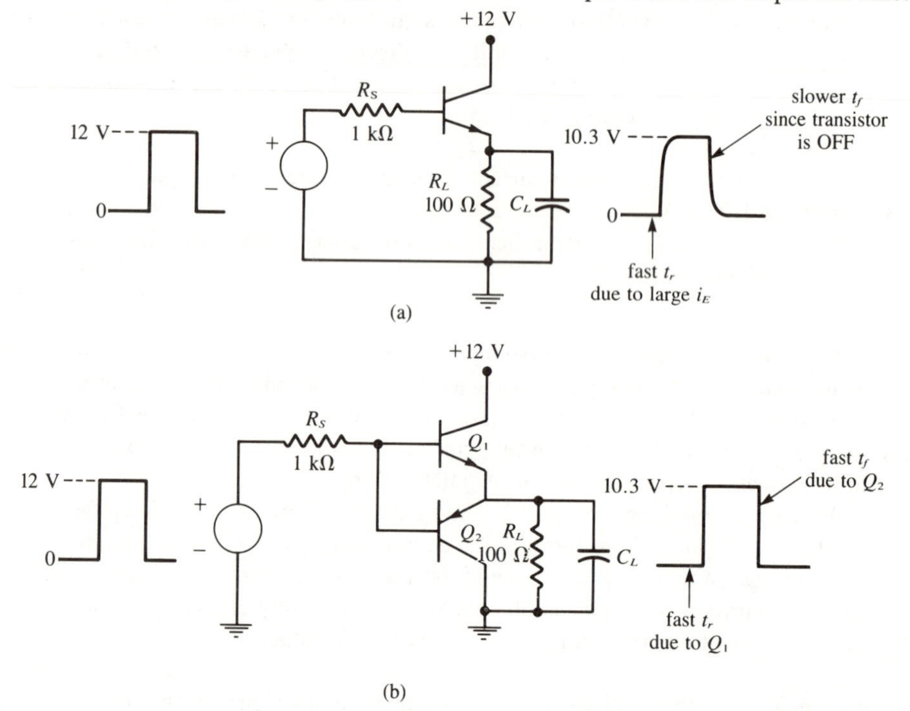

Figure 6–19 **(a) E-F provides rapid charging of capacitive load, but discharge is relatively slow because transistor is OFF; (b) Q_2 serves to rapidly discharge C_L when e_i goes back to 0 V**

Op-amp buffer (voltage-follower) As we saw in Chapter Five, an op-amp connected as a unity-gain voltage follower has extremely high Z_{IN}, extremely low Z_{OUT}, and a voltage gain of approximately one. These factors make it ideally suited for use as a buffer. Figure 6–20 shows it being used as a buffer between a high-resistance signal source and a load.

Since the op-amp Z_{IN} is so large, there will be virtually no current drawn from the signal source. Thus, the full e_i will reach the "+" terminal to produce $v_{OUT} = e_i$. The 1 kΩ load will not load down the voltage-follower whose output impedance is typically less than 1 Ω.

Note the presence of the 22 kΩ resistance in the feedback connection. It is placed there to reduce the effect of the op-amp input bias currents, as was discussed in Chapter Five.

Despite its almost ideal buffer characteristics, the op-amp voltage-follower is limited to low-speed pulse applications (signal transition times of ⩾100 ns) because of its output slew rate limitation. Although there are high-speed op-amps available, they are expensive and cannot match the speed of bipolar switching transistors.

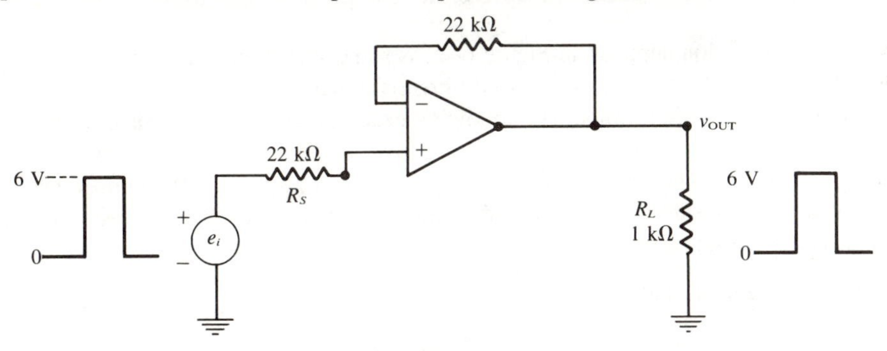

Figure 6–20 **Op-amp voltage-follower used as a buffer**

6.6 The Differential Comparator

A comparator is a circuit that compares an input signal, e_i, to a reference voltage, E_{REF}, and produces an output signal that indicates when e_i exceeds E_{REF}. Typically, a comparator output switches between two levels depending on whether e_i is larger or smaller than E_{REF}.

We will study two basic comparator circuits, the *differential comparator* and the *Schmitt trigger.* Both of these types of comparators can be constructed in several different ways, but we will concentrate on the versions of each which use the op-amp as the basic circuit element.

The differential comparator In Chapter Five, we saw how a high-gain differential amplifier (op-amp) could be used as a differential comparator. The op-amp is operated in the open-loop mode where it has maximum voltage gain. Figure 6–21(a) shows a *non-inverting* differential comparator, where the reference voltage is connected to the "−" input. The output signal will make a positive transition from $-V_{SAT}$ to $+V_{SAT}$ as e_i increases above E_{REF}, and a negative transition when e_i decreases below E_{REF}. Figure 6–21(b) shows an *inverting* comparator where E_{REF} is connected to the "+" terminal. Here the v_{OUT} transitions will be in the opposite direction to the changes of e_i.

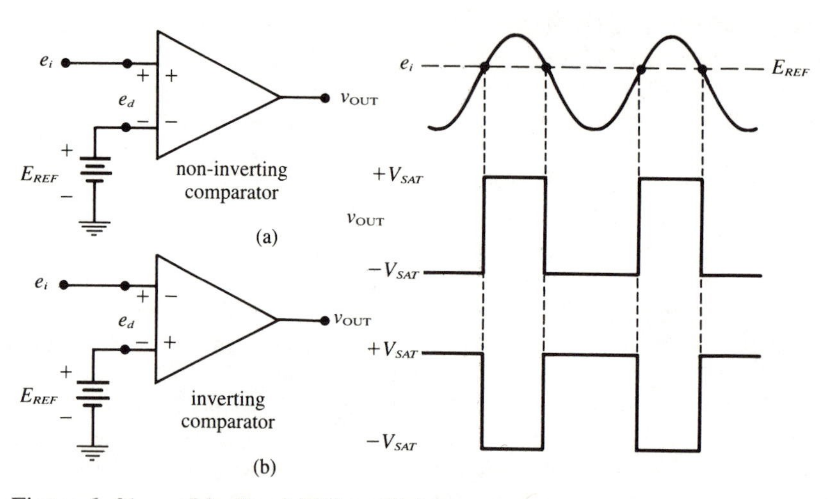

Figure 6–21 **Idealized differential comparator operation**

The operation depicted in Figure 6–21 is somewhat idealized for several reasons. First, there has to be a small, non-zero differential input (e_d) in order to drive the output levels to their saturation limits. The following example will illustrate this case.

Example 6.16 The op-amp of Figure 6–21(a) has $A_{VOL} = 15{,}000$ and operates from ± 15 V supplies. Using $E_{REF} = +5$ V, determine the exact value of e_i needed to produce $+V_{SAT}$ at the output. Repeat for $-V_{SAT}$.

Solution: Recall that

$$v_{OUT} = A_{VOL} \times e_d$$

and that $+V_{SAT}$ is typically 1.5 V below $+V_s$. Thus, $+V_{SAT} = +13.5$ V, and we can determine the necessary differential voltage to produce $+V_{SAT}$ as follows:

$$+V_{SAT} = A_{VOL} \times e_d(\text{SAT})$$
$$+13.5 \text{ V} = 15{,}000 \times e_d(\text{SAT})$$

or

$$e_d(\text{SAT}) = 900\ \mu\text{V}$$

Since, e_d is the difference between e_i and E_{REF}, then e_i has to exceed E_{REF} by 900 μV in order to saturate the output at +13.5 V. Thus, whenever e_i is *greater* than 5.0009 V, v_{OUT} will be at +13.5 V.

In a similar manner we can determine that e_i has to drop 900 μV *below* E_{REF} in order to provide a large enough negative differential voltage to produce an output of −13.5 V. Thus, whenever e_i is *lower* than 4.9991 V, v_{OUT} will be at −13.5 V.

As this example shows, the comparator output will be at one saturation limit when e_i *exceeds* some specific value, which we will call the *upper threshold voltage* (V_{UT}), and at the other saturation limit, when e_i drops *below* some specific value, which we will call the *lower threshold voltage* (V_{LT}). For the preceding example, we have

$$V_{LT} = 4.9991 \text{ V}$$
$$V_{UT} = 5.0009 \text{ V}$$

Note that there is a difference between the upper and lower threshold voltages. This difference is often referred to as the comparator's *hysteresis*, V_H, and it represents the amount by which e_i has to change in order for the output to switch from one saturation level to the other. In general,

$$V_H = V_{UT} - V_{LT} \tag{6–1}$$

For the values above, V_H is 1,800 μV or 1.8 mV.

Figure 6–22 illustrates the effect of V_H on the comparator output by using an expanded time scale to show the variation of e_i. Here it should be clear that the transition times of the v_{OUT} waveform will depend on the rate at which e_i is changing as it goes through the hysteresis region between V_{LT} and V_{UT}. This dependence could be a problem in applications where e_i is changing relatively slowly.

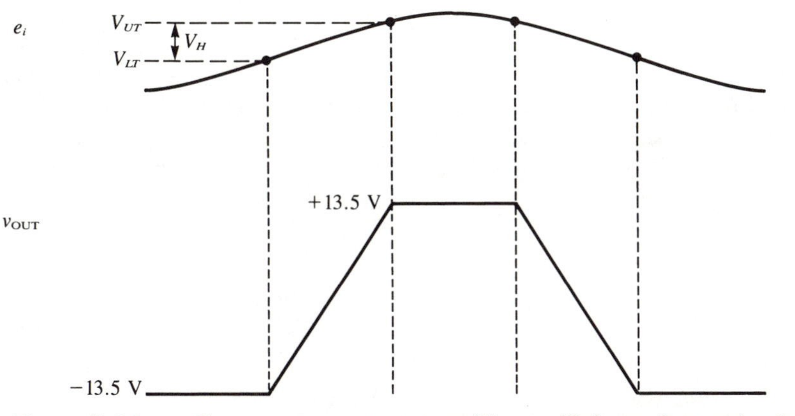

Figure 6–22 **Comparator output transitions will depend on rate at which e_i moves through the hysteresis region between V_{LT} and V_{UT}**

Example 6.17 What can be done to decrease the output transition times for cases where e_i is changing very slowly? What determines the ultimate limit on these transition times?

Solution: An increase in A_{VOL} will reduce the required differential input to produce saturation, and therefore reduce the difference between V_{LT} and V_{UT}. For example, if we use an op-amp with $A_{VOL} = 100{,}000$, the hysteresis will be reduced to only 270 μV (reader should verify). This value means that e_i has to change by only 270 μV in order to cause v_{OUT} to switch from one saturation level to the other.

No matter how large we make A_{VOL}, or how fast e_i is changing, the comparator transition times can be no faster than the op-amp's *slew rate* limitation will allow. When the comparator output transitions are to be as rapid as possible, then an op-amp with a sufficiently high slew rate should be used.

Example 6.18 The LM–311 is a high-speed differential comparator that is designed to have the following typical characteristics:

$$A_{VOL} = 200{,}000$$
$$\text{slew rate} = 25\ \text{V}/\mu\text{s}$$
$$V_{\text{SAT}}(+) = 5\ \text{V}$$
$$V_{\text{SAT}}(-) = 0\ \text{V}$$

Note that its output levels are limited to 0 V and +5 V to be compatible with the levels used by the digital ICs we will study later.

(a) What would be the fastest transition times that this comparator output could have?

(b) What could cause the output transition times to be slower than (a)?

Solution:

(a) The output transitions will have a total voltage change of 5 V, and the fastest that this voltage change can occur is at the rate of 25 V/μs. Thus, the transition times can be no shorter than

$$\frac{5\text{ V}}{25\text{ V}/\mu\text{s}} = \mathbf{200\ ns}$$

(b) Of course, the output transition times can be longer if the input is changing very slowly. For example, to produce an output change of 5 V, there would have to be an input change of

$$\Delta e_i = \frac{\Delta v_{\text{OUT}}}{A_{VOL}} = \frac{5\text{ V}}{200{,}000} = 25\ \mu\text{V}$$

If e_i is changing very slowly as it approaches the value of E_{REF}, it is possible that it may take longer than 200 ns to make this 25 μV change. If so, the v_{OUT} signal transition times will be slower than those predicted by using the slew rate.

Adjustable trigger generator circuit Figure 6–23 uses a differential comparator in a circuit that produces an output consisting of positive spikes corresponding to the points where e_i crosses E_{REF} on its positive-going excursion. The output spikes can be used to trigger some other circuit. For instance, we will use this circuit later as part of an oscilloscope's triggered sweep circuitry.

The op-amp output switches between −10.5 V and +10.5 V as e_i goes above and below E_{REF}. These pulses are then applied to an *RC* differentiator, which converts them to positive and negative spikes. The series diode clipper removes the negative spike to produce the desired output.

E_{REF} is adjustable over the range from −12 V to +12 V. When E_{REF} is set to exactly *zero,* the circuit is called a *zero crossing detector* since the output spikes correspond to points where e_i crosses 0 V on its positive excursion.

Example 6.19 A technician is testing the circuit of Figure 6–23 and observes that the v_x pulses are of the proper amplitude, but the v_y and v_{OUT} spikes are much smaller than expected. What are some of the possible causes for this?

Solution: The spikes at v_y should be 21 V in amplitude, corresponding to the 21 V transitions on the v_x waveform. This value assumes that the *RC* circuit satisfies

$$\tau > \frac{t_r}{2\pi}$$

where t_r is the rise time of the v_x waveform. The spikes at v_y and v_{OUT} would be smaller

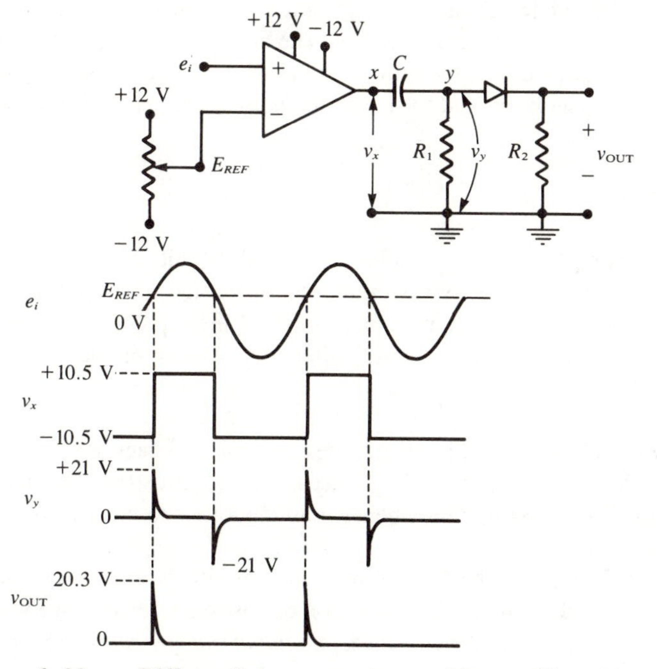

Figure 6–23 **Differential comparator used in an adjustable trigger generator circuit**

than 21 V if τ were too small, or if the v_x signal transition times were too long because of the op-amp slew rate limitation.

Effect of noise on the differential comparator A differential comparator has a very large open-loop voltage gain so that only a small change in e_i will cause the output to switch states. In other words, there is only a very small difference between the upper and lower trigger voltages, V_{UT} and V_{LT}. This characteristic makes the differential comparator susceptible to small amounts of noise on the input signal. If the noise is large enough to cause e_i to go above and below the V_{UT} and V_{LT} points several times as it approaches E_{REF}, then the comparator output will change states several times instead of once.

This phenomenom is illustrated in Figure 6–24, where the e_i signal contains enough noise to cause it to cross the V_{LT} and V_{UT} thresholds more than once, thereby

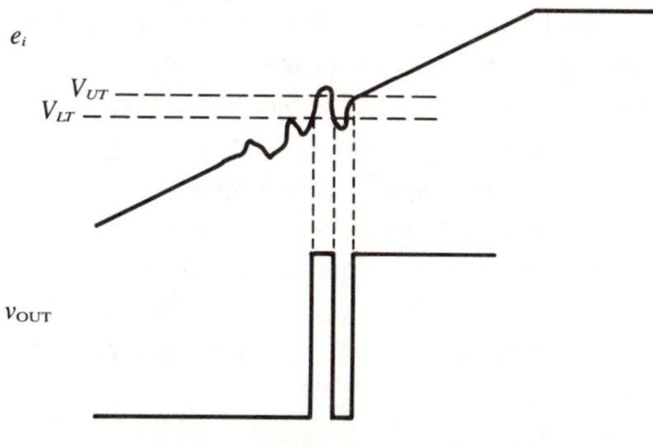

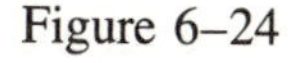
Figure 6–24 **Noise on the input signal can cause multiple transitions on the comparator output**

producing several output transitions. This situation can cause erratic operation of the circuits being driven by the comparator output. The noise will be a problem whenever its peak-peak amplitude exceeds V_H, the difference between the two threshold points. The e_i signal should be adequately filtered to reduce the noise level below this value.

6.7 The Schmitt Trigger

The differential comparator is a simple and useful circuit, but it does have some drawbacks, as we have seen. It is susceptible to noise, and its output transition times will increase for slow-changing input signals. Both of these problems can be overcome by using a Schmitt trigger circuit, named after the inventor of the original vacuum-tube version.

Op-amp Schmitt trigger This circuit uses positive feedback and the regenerative action that was described in Section 5.13. Recall that positive feedback tends to drive the op-amp output toward saturation whenever an input tries to bring the output into the linear region.

Figure 6–25(a) shows the basic Schmitt trigger circuit. It is essentially the same circuit that we analyzed in our earlier discussion of positive feedback. We will analyze it in more detail here, and we will develop the *input/output transfer characteristic* that shows how v_{OUT} changes as e_i changes.

In this circuit, the op-amp output will always be at one saturation level or the other. When it is at $+V_{SAT}$, the voltage at point x will be a positive value determined by R_1 and R_2. This value will be the Schmitt trigger's upper threshold voltage, V_{UT}, as we shall see. Thus

$$V_{UT} = +V_{SAT}\left[\frac{R_2}{R_1 + R_2}\right] \tag{6–2}$$

Similarly, when the op-amp output is at $-V_{SAT}$, v_x will be a negative value, and this value will be the Schmitt trigger's lower threshold voltage, V_{LT}. Thus

$$V_{LT} = -V_{SAT}\left[\frac{R_2}{R_1 + R_2}\right] \tag{6–3}$$

Let's begin by assuming that e_i is at a *negative* value that is sufficient to drive the op-amp to $+V_{SAT}$. Thus $v_x = V_{UT}$ is applied to the op-amp "+" input. This condition will be point a on the v_{OUT} versus e_i transfer curve in Figure 6–25(b). At this point, the differential input e_d is positive. As e_i is increased in the positive direction (points b and c), the op-amp remains at $+V_{SAT}$ until e_i is increased to a value slightly below V_{UT} (point d). At this point the differential input has decreased below the value needed to produce $+V_{SAT}$, and so v_{OUT} starts to decrease into the linear region. As it does so, however, regenerative action takes place and rapidly drives v_{OUT} to $-V_{SAT}$ (point e).

Thus, v_{OUT} rapidly switches from $+V_{SAT}$ to $-V_{SAT}$ with a switching action that is *not* dependent on the rate at which e_i is changing. e_i *initiates* the switching action as it gets sufficiently close to V_{UT} and causes v_{OUT} to start decreasing; then the regenerative action takes over, with the result that v_{OUT} will change at a rate solely dependent on the op-amp's slew-rate limitation.

With v_{OUT} now at $-V_{SAT}$, v_x will change to a negative value equal to V_{LT}, so that the op-amp "+" terminal is now negative. If e_i is now increased above the value V_{UT}

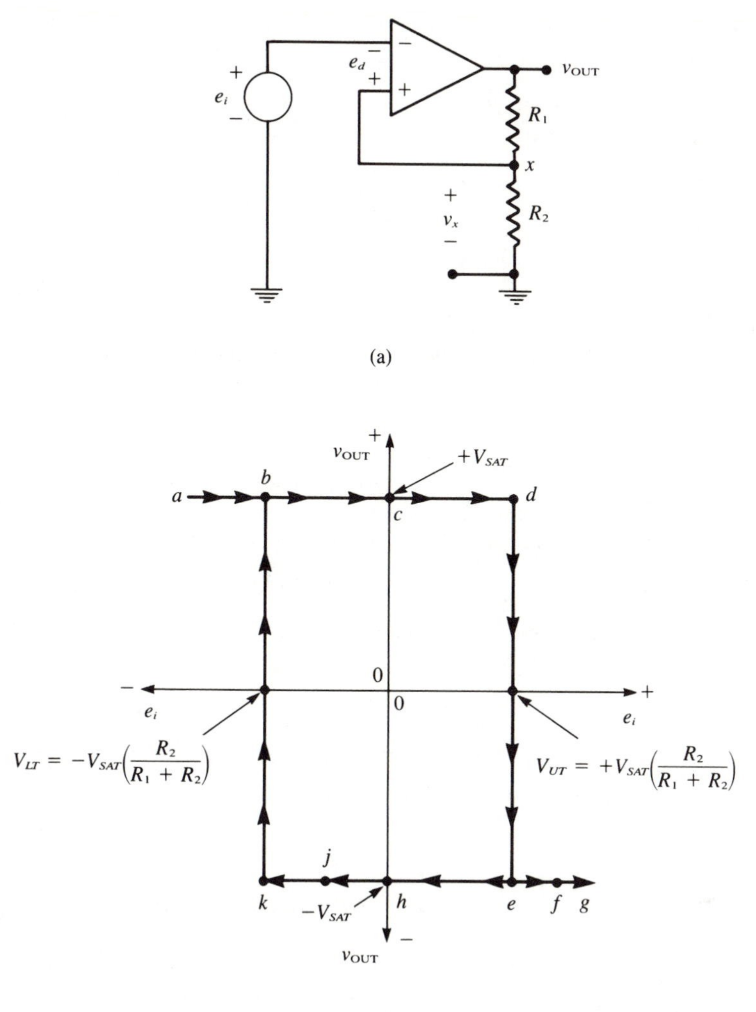

Figure 6–25 **(a) Op-amp Schmitt trigger circuit; (b) v_{OUT} versus e_i relationship**

(points f and g), v_{OUT} will remain at $-V_{SAT}$. If e_i is decreased below V_{UT} (points h and j), v_{OUT} remains at $-V_{SAT}$ until e_i decreases to approximately V_{LT} (point k). At that point, the op-amp starts going into the linear region, and regenerative action rapidly drives it to $+V_{SAT}$ (point b).

The overall operation can thus be summarized as follows:

(a) The output will rapidly switch to $-V_{SAT}$ whenever e_i increases to the value of V_{UT}.

(b) The output will rapidly switch to $+V_{SAT}$ whenever e_i decreases to the value of V_{LT}.

(c) The switching action occurs at a rate determined *solely* by the op-amp's slew-rate limitation.

Example 6.20 The waveform of Figure 6–26 is applied to an op-amp Schmitt trigger that uses ±12 V supplies and has $R_1 = 10\ \text{k}\Omega$ and $R_2 = 1\ \text{k}\Omega$. Sketch the v_{OUT} waveform.

Solution: Determine the values of V_{LT} and V_{UT} when $\pm V_{\text{SAT}} = \pm 10.5$ V. Using equation 6–2 and 6–3:

$$V_{UT} = +0.95 \text{ V}$$
$$V_{LT} = -0.95 \text{ V}$$

Initially, e_i is below -0.95 V so that v_{OUT} will be at $+V_{\text{SAT}} = +10.5$ V. As e_i increases above $+0.95$ V, v_{OUT} rapidly switches to $-V_{\text{SAT}} = -10.5$ V (point x on waveforms). v_{OUT} will remain at -10.5 V until e_i again decreases below -0.95 V (point y), at which time v_{OUT} rapidly switches to $+10.5$ V. This same reasoning is used to determine the rest of the v_{OUT} waveform.

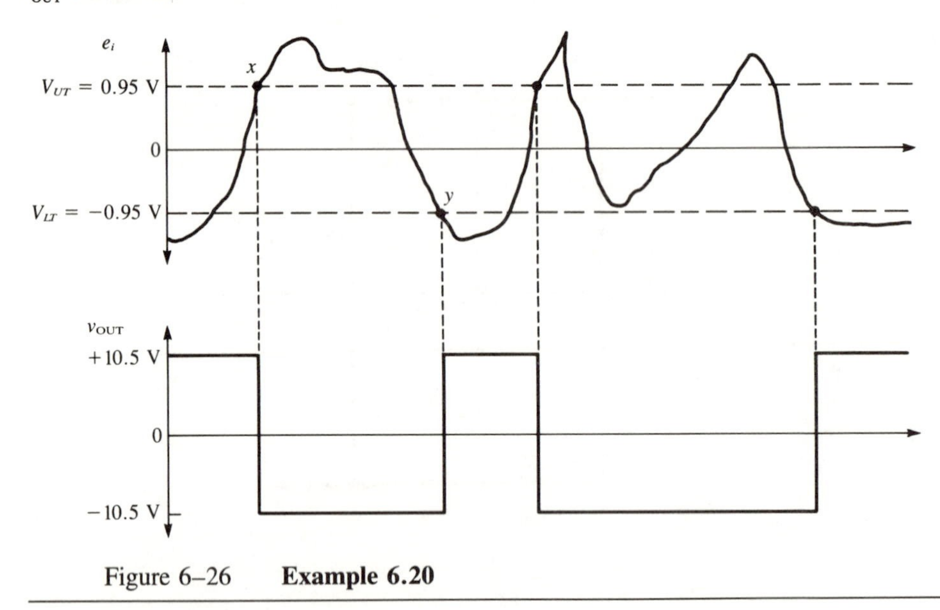

Figure 6–26 **Example 6.20**

Varying V_{LT} and V_{UT} The values of V_{LT} and V_{UT} are clearly dependent on the op-amp saturation voltages and the R_1-R_2 voltage divider ratio. They can be easily varied by changing either R_1 or R_2 or both. This method, however, always produces $V_{LT} = -V_{UT}$, as in the preceding example. The following examples will show two ways to obtain threshold levels that are not subject to this constraint.

Example 6.21 Determine V_{LT} and V_{UT} in Figure 6–27(a) for $E_{REF} = 4$ V.

Solution: The upper threshold voltage, V_{UT}, is determined by solving for the voltage at x when the op-amp output is at $+V_{\text{SAT}}$. We can neglect the effect of the op-amp's small input bias current at the "+" input since the resistor values are so small. Thus, we are left with the circuit of Figure 6–27(b), where we want to calculate V_x. It is a two-source circuit because we can consider the op-amp output to be a source voltage equal to $+V_{\text{SAT}}$. Using the *superposition method,* then, we obtain

$$V_{UT} = V_x = +V_{\text{SAT}}\left(\frac{100\ \Omega}{100\ \Omega + 10\ \text{k}\Omega}\right) + E_{REF}\left(\frac{10\ \text{k}\Omega}{10\ \text{k}\Omega + 100\ \Omega}\right)$$
$$= +13.5\text{ V}(.0099) + 4\text{ V}(0.99)$$
$$\approx \mathbf{+4.09\ V}$$

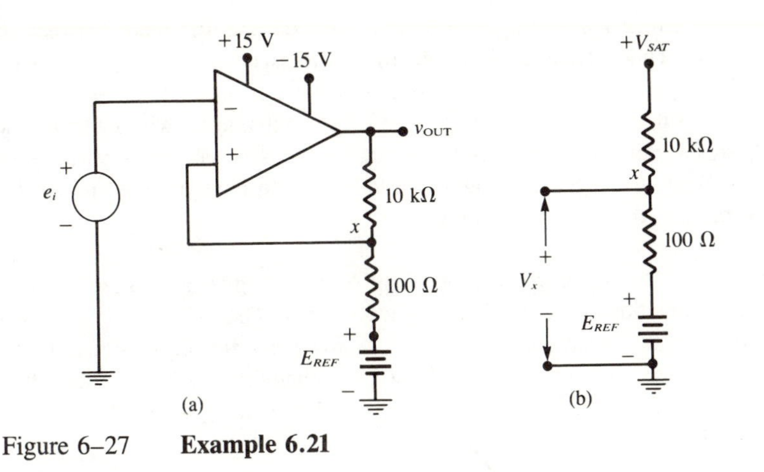

Figure 6–27 **Example 6.21**

In a similar manner, we can calculate V_{LT} as the voltage at point x when v_{OUT} is at $-V_{SAT}$. Thus,

$$V_{LT} = V_x = -V_{SAT}[.0099] + E_{REF}[0.99]$$
$$\approx \mathbf{+3.83\ V}$$

These results show that V_{UT} is slightly greater than E_{REF} and V_{LT} is slightly lower than E_{REF}.

Example 6.22 Determine V_{UT} and V_{LT} in Figure 6–28.

Solution: Once again V_{UT} is determined by finding the voltage at x when v_{OUT} is at $+V_{SAT}$. The diode will be forward-biased and will maintain 0.7 V across its terminals, so that V_x will be 0.7 V below $+V_{SAT}$. Thus

$$V_{UT} = +13.5\text{ V} - 0.7\text{ V} = \mathbf{+12.8\ V}$$

V_{LT} is determined by finding V_x when $v_{OUT} = -13.5$ V. Clearly, the diode will be reverse-biased and we can consider it open. V_{LT} is therefore determined by the resistor voltage divider. Thus,

$$V_{LT} = -13.5\text{ V}\left[\frac{2.2\text{ k}\Omega}{9\text{ k}\Omega}\right]$$
$$= \mathbf{-3.3\ V}$$

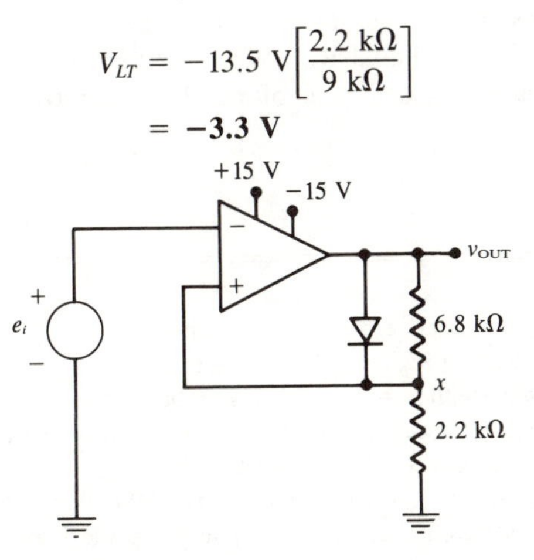

Figure 6–28 **Example 6.22**

Hysteresis Since V_{UT} and V_{LT} can be easily varied, the difference between them can also be varied. This difference is called the hysteresis voltage, V_H, as it was for the differential comparator. As we found in our discussion of noise effects on the differential comparator, it is desirable to have V_H large enough so that noise will not cause the output to switch states more than once as the input increases or decreases past the threshold points. The Schmitt trigger can usually be designed to have a large enough hysteresis voltage to prevent this from happening.

Application The major application of the Schmitt trigger is as a *squaring circuit* to speed up the transitions of slow-changing waveforms. The Schmitt trigger will produce rapid output signal transitions regardless of how slow the input signal is changing. Figure 6–29 shows two examples of how slow-changing waveforms are shaped into fast-changing pulses. These examples use $V_{UT} = 1$ V and $V_{LT} = -1$ V.

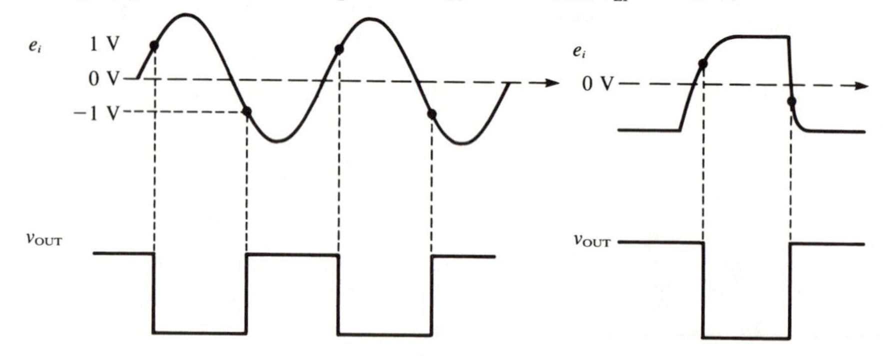

Figure 6–29 **The principal application of the Schmitt trigger is to "square up" slow-changing signals**

There are some other applications of the Schmitt trigger that we will defer until later in the text.

Other Schmitt trigger implementations Schmitt trigger circuits can be constructed using bipolar transistors. This method, however, has become obsolete since op-amp ICs have become inexpensive and easy to use. In addition, as we shall see, Schmitt triggers are readily available as digital ICs. Whatever their form, Schmitt trigger circuits maintain the same basic characteristic of rapid regenerative switching produced by positive feedback.

Questions

Sections 6.1–6.2

6.1 Change R_L to 1.5 kΩ and repeat Example 6.1.

6.2 Determine v_{OUT} in Figure 6–1 when e_i is a 200 mV *p-p* sinewave.

6.3 A square wave with $t_r = t_f = 20$ ns is applied to a series clipper that uses a diode with an average junction capacitance of 4 pF. What is the largest R_L that can be used if there are to be no spikes coupled to the output?

6.4 Determine v_{OUT} in Figure 6–4 when e_i is a 200 mV *p-p* bipolar square wave.

6.5 Turn the diode around and change R_L to 3.3 kΩ in Figure 6–5. Determine v_{OUT}.

6.6 Draw the waveform for the transistor base voltage in Figure 6–6.

6.7 (a) Sketch the output waveform for each of the circuits of Figure 6–30. Use a 40 V *p-p* bipolar square wave for e_i.

(b) Repeat when a 1 kΩ load is connected across the output. (Hint: Use Thevenin equivalent.)

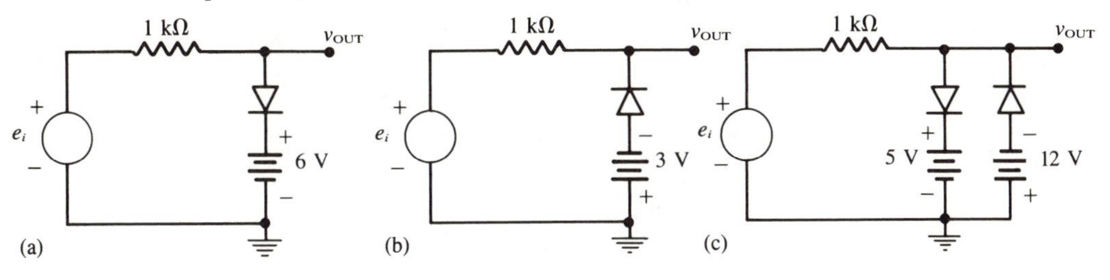

Figure 6–30

6.8 Refer to the circuit of Figure 6–8. The diode has a PRV rating of 25 V and a maximum forward-current rating of 50 mA. Determine the smallest value of R_S that can be used.

6.9 Change E_R in Figure 6–8 to 15 V, and determine v_{OUT} for the given e_i signal. Then find the maximum *positive* amplitude that e_i can have.

6.10 Determine v_{OUT} for the circuit of Figure 6–31.

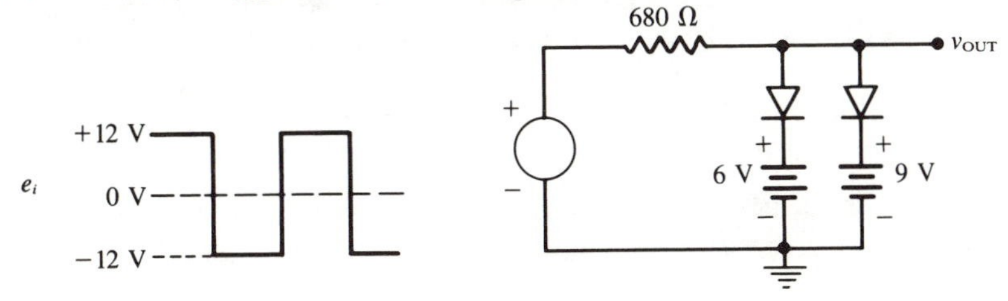

Figure 6–31

6.11 Discuss the relative advantages and disadvantages of the zener diode clipper versus the biased shunt clipper.

6.12 Determine the v_{OUT} waveform for each of the circuits of Figure 6–32. Z_1 has $V_z = 6.8$ V, and Z_2 has $V_z = 3.3$ V. Use a 24 V *p-p* bipolar square wave for e_i.

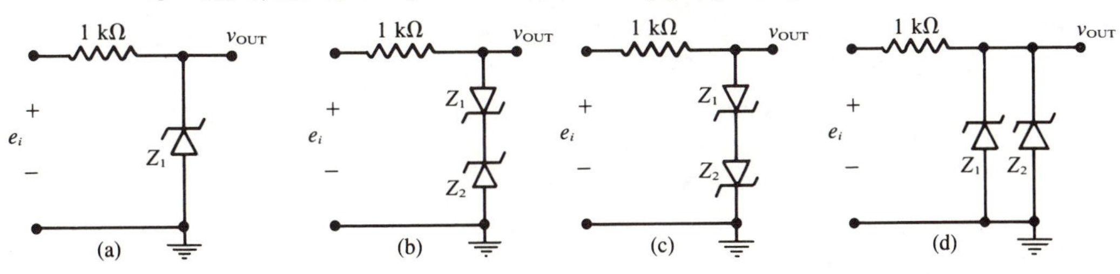

Figure 6–32

Section 6.3

6.13 Determine the exact value of v_{OUT} in the circuit of Figure 6–11(d) when e_i is negative. The op-amp has $A_{VOL} = 20{,}000$.

6.14 Determine the approximate v_{OUT} waveform in Figure 6–11 if $R_1 = 2.2$ kΩ and if a 10 kΩ load is connected across the output.

6.15 Assume $R_1 = 1.5\ k\Omega$ in Figure 6–11. What is the maximum current drawn from the e_i source?

6.16 Determine v_{OUT} for the circuit of Figure 6–33. Assume both op-amps have extremely high open-loop gains.

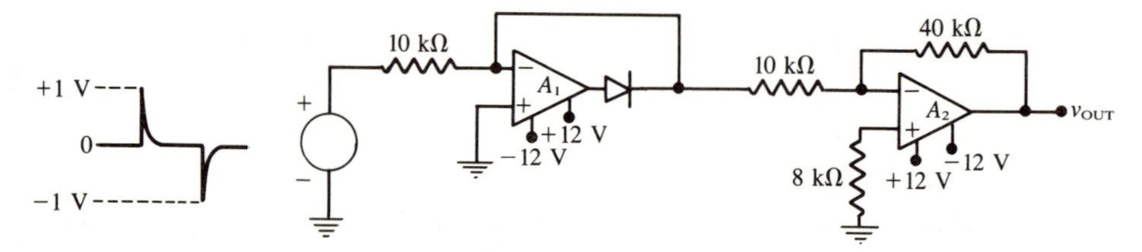

Figure 6–33

6.17 Show how the resistor connected to A_2's positive terminal in Figure 6–33 was determined as 8 kΩ.

6.18 The op-amp in Figure 6–12 uses ±15 V bias supplies. If the input is a 20 V *p-p* sinewave, determine the maximum reverse voltage that the diode has to withstand.

6.19 What is the main advantage of the op-amp clipper as compared to a simple diode clipper circuit?

6.20 A technician is testing the circuit of Figure 6–33 and she measures v_{OUT} as an approximate 0 V level. Which of the following circuit faults could produce this result?

(a) The diode is open.

(b) The diode is internally shorted.

(c) The output terminal of A_1 is shorted to ground.

(d) The 40 kΩ feedback resistor is open.

Section 6.4

6.21 Consider the saturated inverter circuit in Figure 6–34. Determine suitable values for R_B and R_C if $h_{FE}(\text{min}) = 20$, and the transistor has a maximum collector current rating of 50 mA.

6.22 If the inverter in Figure 6–34 is used to drive some type of load, distortion or attenuation of the output could result. Figure 6–35 shows four possible output waveforms. Indicate the type of loading (resistive, capacitive, or both) that could produce each of these waveforms.

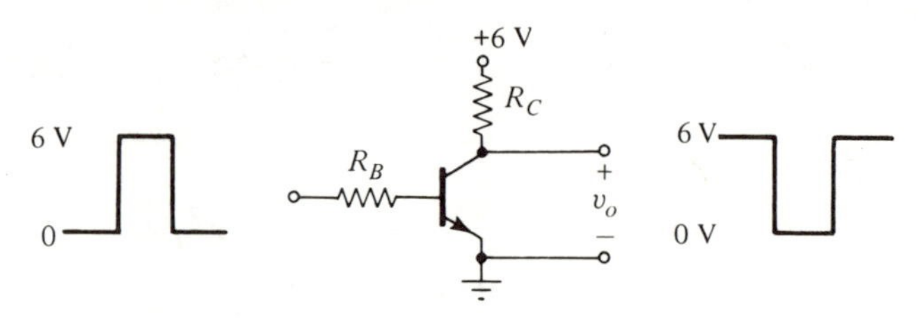

Figure 6–34

6.23 A common application of the inverter is shown in Figure 6–36 where a sinusoidal input is converted to a square wave. Explain the circuit operation.

6.24 Indicate the effect each of the following changes would have on the output t_r and t_f in Figure 6–36.

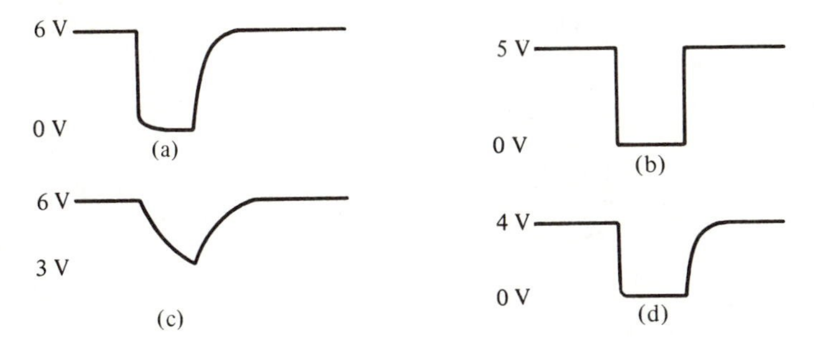

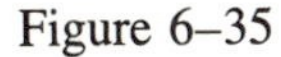
Figure 6–35

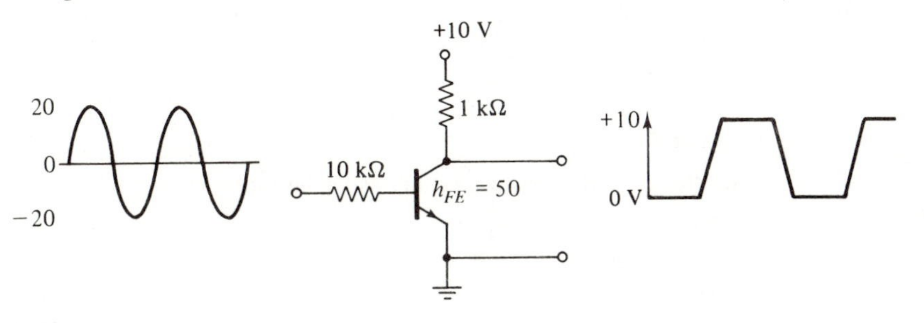

Figure 6–36

(a) Decrease h_{FE}.

(b) Increase input amplitude.

(c) Reduce 1 kΩ.

(d) Increase 10 kΩ.

6.25 The inverter of Figure 6–34 is required to drive a 200 Ω load without v_o dropping below 5.5 V in the OFF state. Determine appropriate values for R_C and R_B. Assume $h_{FE}(\text{min}) = 30$.

6.26 The inverter of Figure 6–34 uses $R_C = 2.2$ kΩ and $R_B = 150$ kΩ. Determine the minimum h_{FE} that can be used if the inverter output levels are to be maintained for input pulse amplitudes as low as 4 V.

6.27 What is the largest capacitive load that the inverter of Problem 6.26 can drive if the output t_r is to be no greater than 100 ns?

6.28 Consider the circuit of Figure 6–17. Whenever e_i makes a transition, one of the transistors goes from saturation to cutoff, and the other goes from cutoff to saturation. In general, a transistor's turn-off time, t_{OFF}, will be longer than its turn-on time, t_{ON}, especially if it is overdriven while ON.

What problem will this difference cause in this circuit? Where would you put a small resistor to prevent a large current surge during the transition states?

Section 6.5

6.29 Figure 6–37(a) shows an inverter driving a load.

(a) Determine the waveform of load voltage.

(b) Figure 6–37(b) shows the same inverter driving the load through an emitter-follower. Determine the load voltage waveform.

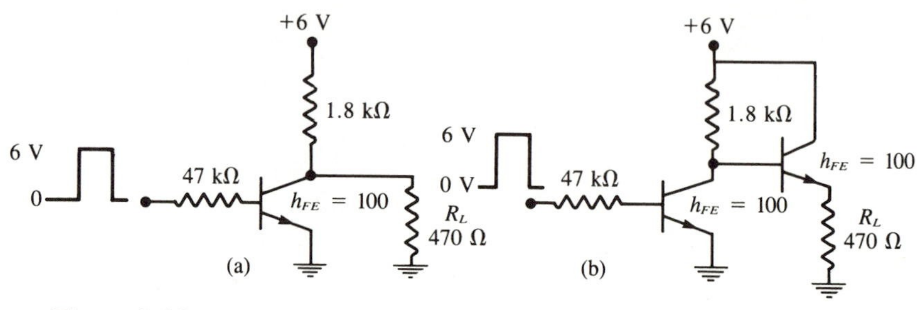

Figure 6–37

6.30 A technician is testing the circuit of Figure 6–19(a) and observes that v_{OUT} is a constant level of approximately 11.3 V. Which of the following is the most probable cause of this result?

(a) R_L is open?

(b) Transistor is shorted between emitter and collector?

(c) Transistor collector is open?

(d) Transistor is shorted between base and emitter?

6.31 A technician tests the circuit of Figure 6–19(b) and observes that v_{OUT} only has an amplitude of about 1 V, and its fall time is much shorter than its rise time. Which of the following is the most probable cause of this result?

(a) Q_1 is shorted between emitter and base.

(b) Q_1's collector is open.

(c) Q_2 is shorted between emitter and base.

(d) Q_2's collector is open.

6.32 Determine the maximum *E-B* reverse voltage that Q_2 has to withstand in the circuit of Figure 6–19(b).

6.33 (a) Calculate the effective output impedance of the op-amp voltage follower in Figure 6–20 if the op-amp has A_{VOL} = 10,000 and Z_{OUT} = 75 Ω.

(b) It may appear then, that this voltage-follower can drive very small values of R_L without loss of output signal. The op-amp output, however, has an output current limitation that should not be exceeded.

Determine the minimum R_L that can be driven if the op-amp has I_{OUT}(max) = 25 mA, and the input signal amplitude can range from 4 V to 11 V.

Section 6.6

6.34 An *inverting* differential comparator uses an op-amp with A_{VOL} = 10,000. If E_{REF} = 3 V, determine V_{UT}, V_{LT} and V_H.

6.35 Repeat Problem 6.34 using A_{VOL} = 50,000.

6.36 Compare the differential comparators of the last two problems.

(a) Which one is more susceptible to noise on the e_i signal? Explain.

(b) Which one is more apt to have its output transitions slowed down because of a slow-changing input? Explain.

6.37 *True* or *False:* A differential comparator's output transitions are determined *solely* by how long it takes the input signal to pass through the hysteresis region.

6.38 The op-amp of Figure 6–38 has A_{VOL} = 20,000 and a slew rate of 5 V/μs. Sketch the waveform at the op-amp output. Determine its rise time and fall time. Then determine appropriate values for R and C to produce the maximum amplitude spikes at v_{OUT}. The op-amp has a maximum output current limit of 10 mA.

6.39 Determine the interval of time between the positive and negative spikes in Problem 6.38. What is the longest the interval can be if we vary the reference voltage?

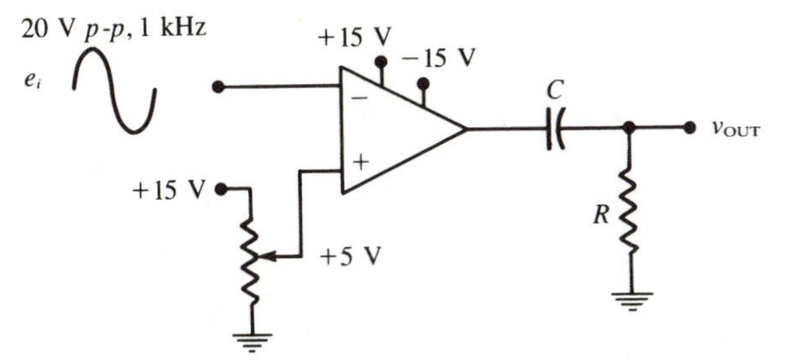

Figure 6–38

6.40 Consider the circuit of Figure 6–23. How would you change it so that the output will be *positive* spikes that occur when e_i decreases below E_{REF}?

6.41 What will happen to the output of the circuit of Figure 6–23 if e_i has an amplitude of 20 V *p-p* and E_{REF} is set to +12 V? Repeat for $E_{REF} = -12$ V.

6.42 The op-amp in Figure 6–23 has a slew rate of 2 V/μs. The circuit uses C = 470 pF, R_1 = 5.1 kΩ and R_2 = 1.5 kΩ. A technician tests the circuit and observes the following:

(a) The v_{OUT} spikes are lower than 20 V.

(b) The positive spikes at v_y are lower than 21 V, but the negative spikes are 21 V.

(c) The v_x waveform is correct.

Can you explain these observations? (*Hint:* Calculate t_r and t_f for the op-amp output. Then compare them to the circuit time constant for each polarity of v_x.)

Section 6.7

6.43 Which of the following are advantages of the Schmitt trigger over the differential comparator?

(a) Its hysteresis can readily be adjusted to provide less sensitivity to noise.

(b) Its output transition times are not limited by op-amp slew rate.

(c) Its output transition times are not affected by the rate at which the input is changing.

(d) Its trigger points can be set more accurately.

6.44 A 10 V *p-p*, 5 kHz sinewave is applied to an op-amp Schmitt trigger that uses ±15 V supplies and has R_1 = 3.3 kΩ and R_2 = 1 kΩ. Sketch the output waveform.

6.45 Change R_2 to 2.2 kΩ and repeat Problem 6.44.

6.46 The Schmitt trigger circuit of Figure 6–27 was determined to have V_{LT} = 3.83 V and V_{UT} = 4.09 V for E_{REF} = 4 V in Example 6.21. This circuit provides a hysteresis, V_H, of 0.26 V.

(a) Prove that this hysteresis voltage is independent of the value of E_{REF}.

(b) Determine what resistance value should be used in place of the 100 Ω in order to obtain $V_H = 1$ V.

6.47 Determine V_{LT} and V_{UT} for the circuit of Figure 6–28 if the diode is reversed.

6.48 Figure 6–39 shows a circuit using a Schmitt trigger in a time-delay application. Determine the delay between opening the switch and current flowing through the load. Neglect the op-amp input impedance.

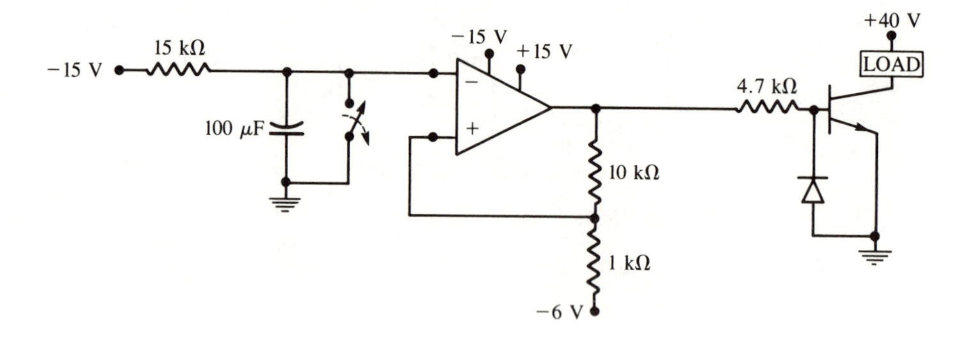

Figure 6–39

7

Logic Gates and Boolean Algebra

7.1 Introduction

A *logic gate* is a switching circuit with two or more inputs and whose output will be either a high voltage or a low voltage, depending on the voltages on the various inputs. Logic gates are widely used in digital computers and in all types of digital circuits and systems. Digital circuits and systems are characterized by the fact that they contain voltages that exist at either of two levels; for example, 0 V and 5 V. In other words, at any instant of time each circuit input and output voltage will either be at some LOW voltage (V_{LO}) or some HIGH voltage (V_{HI}). In practice, the LOW level is actually a range of voltages, as is the HIGH level. For example, between 0 V and 0.8 V might be the LOW level, and between 2 V and 5 V might be the HIGH level. This concept is illustrated in Figure 7–1. The range of voltages between 0.8 V and 2 V is not allowed except during transitions between V_{HI} and V_{LO}.

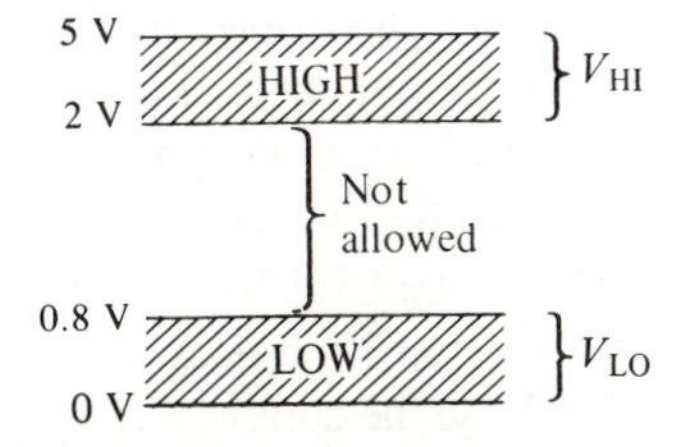

Figure 7–1 **Typical voltage level assignments in a digital system**

There are several types of logic gates, and many different ways to construct each type using discrete components. The wide availability of digital ICs, however, has made it rarely necessary to construct a logic gate from discrete components. As such, we will present only one circuit implementation for each type of gate. Thereafter we will represent each gate by its standard logic symbol, which indicates the gate's input/output logic relationship. These logic symbols are independent of the internal circuitry (e.g. the AND gate has the same symbol whether it is constructed from diodes, or from transistors). In Chapter 9, we will take a look at the internal circuitry of the various IC logic families.

A good portion of this chapter discusses Boolean algebra as a tool for representing and analyzing the operation of logic circuits. As we shall see, Boolean algebra has rules and theorems that are quite different from those of ordinary algebra.

7.2 Diode OR Gate

There are two basic diode logic gates. The first, called an *OR* gate, is shown in Figure 7–2(a). It is actually a two-input OR gate, since v_1 and v_2 are both inputs. The value of the output (v_x) depends on both inputs. Each input can be at either 0 V or 6 V, so there are four possible input combinations. These combinations are listed in Figure 7–2(b), along with the output voltage value for each case. This table of values is called a *truth table;* this term comes from the study of logic where often one voltage level is called the *true* level, and the other level is called the *false* level.

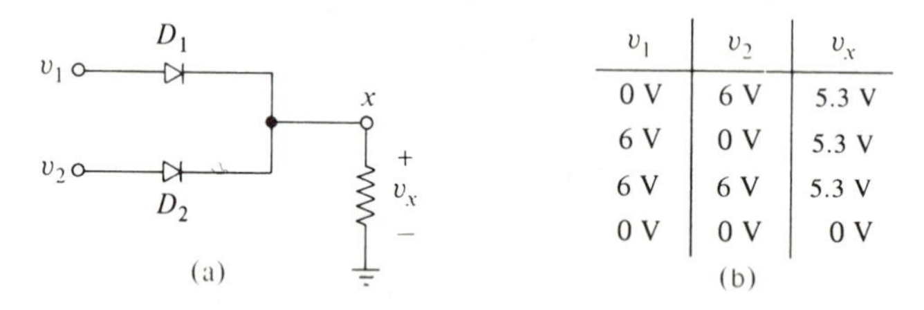

v_1	v_2	v_x
0 V	6 V	5.3 V
6 V	0 V	5.3 V
6 V	6 V	5.3 V
0 V	0 V	0 V

Figure 7–2 **Two-input OR gate and corresponding truth table**

Examination of this truth table shows that the output will be at a HIGH level when either v_1 *or* v_2 *or* both are at a HIGH level. The value of v_x is LOW only when both inputs are at a LOW level.

Consider first the case, where $v_1 = 0$ V and $v_2 = 6$ V, which is shown in Figure 7–3(a). In this case it is obvious that diode D_2 is forward-biased because its anode is made positive relative to its cathode. As such, current will flow through D_2 and R as shown. If the diodes are assumed to be silicon, the forward-voltage drop across D_2 will be 0.7 V, so v_x must equal 6 V − 0.7 V = 5.3 V. The diode D_1 is reverse-biased because its cathode is at +5.3 V relative to ground, and its anode is at 0 V; therefore, it conducts only a small leakage current, which will be neglected. The current through D_2 is easily calculated as $i_2 = i = 5.3\ \text{V}/R$. If the output of this OR gate drives a load (R_L), the parallel combination of $R \| R_L$ must be used to determine i_2.

The next case, where $v_1 = 6$ V and $v_2 = 0$, will obviously be the same as the first except that D_1 will be ON, and D_2 will be OFF. The third case, where both v_1 and v_2 are 6 V, is illustrated in Figure 7–3(b). Here both diodes are ON, so each will have a 0.7 V drop. Again, the output will be 5.3 V, so the current through R is still $i = 5.3\ \text{V}/R$. This time, however, the current is supplied by both inputs through their respective diodes. If

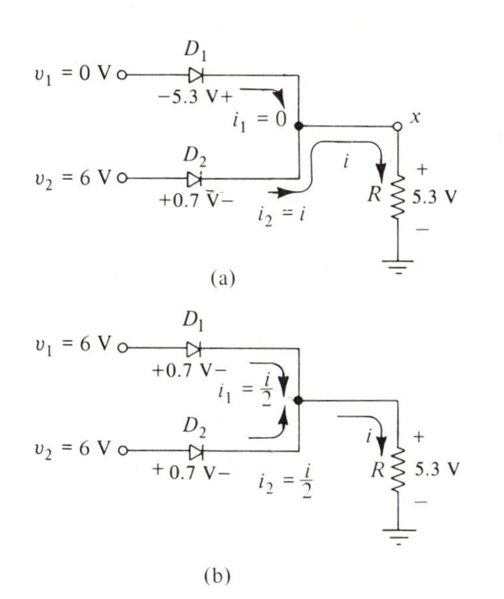

Figure 7–3 **(a) Only D_2 conducting (b) D_1 and D_2 conducting**

we assume that D_1 and D_2 are identical diodes, then the current through each will be the same, so $i_1 = i_2 = i/2$.

In the final case, where $v_1 = v_2 = 0$ V, it is obvious that neither diode will turn ON; thus, no current flows in the circuit, and the output voltage is zero. In summary, *the OR circuit output will be HIGH whenever one or more of its inputs is HIGH.*

Example 7.1 Determine the output of the two-input OR gate for the input waveforms shown in Figure 7–4(a).

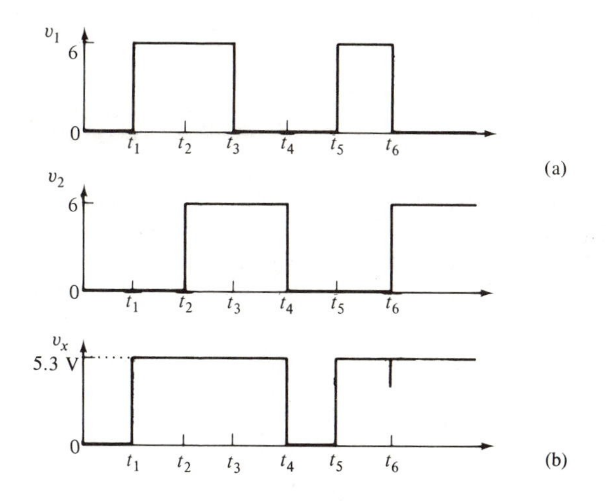

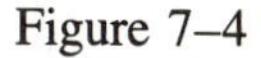
Figure 7–4

Solution: In determining the output (v_x), it is only necessary to look at the two inputs, and whenever either or both is at 6 V, the output will be HIGH (5.3 V). The result is shown in Figure 7–4(b). Before $t = t_1$, both inputs are at 0 V, so $v_x = 0$. At $t = t_1$, the v_1 input goes to 6 V, so $v_x = 5.3$ V. Sometime later, at $t = t_2$, the v_2 input also goes to 6 V; however, since v_1 was already at 6 V, the output simply remains at 5.3 V. At t_3, the v_1 input drops to zero, but v_2 remains at 6 V, so v_x remains HIGH. At t_4, the v_2 input also drops to zero, causing the output to drop to zero. The output remains at zero until t_5, when v_1 jumps back to 6 V.

An interesting and common situation occurs at $t = t_6$, where the inputs are simultaneously changing: v_1 from 6 V to 0 V, and v_2 from 0 V to 6 V. Because both transitions take a certain amount of time, there is a short instant of time when neither input is at 0 V or 6 V but rather somewhere in between. As a result, the output momentarily drops to some value between 0 and 6 V. When the inputs finally complete their transitions, v_x becomes 5.3 V. This *glitch* on the output waveform at t_6 is typical of digital circuits and is a potential source of trouble, depending on the type of circuit the output is driving.

Diode OR gates can be constructed with more than two inputs simply by adding a diode for each input. The same logic holds true for any number of inputs; namely, the output is HIGH if any input is HIGH. Figure 7–5 shows a 3-input diode OR gate.

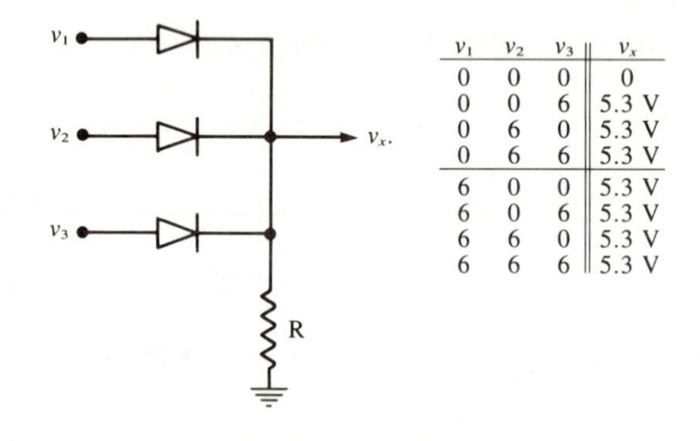

v_1	v_2	v_3	v_x
0	0	0	0
0	0	6	5.3 V
0	6	0	5.3 V
0	6	6	5.3 V
6	0	0	5.3 V
6	0	6	5.3 V
6	6	0	5.3 V
6	6	6	5.3 V

Figure 7–5 **Three-input diode OR gate**

7.3 Diode AND Gate

The second type of diode logic gate, called the *AND* gate, is shown in Figure 7–6 along with its associated truth table. Examination of this truth table shows that the output will be at a HIGH level only when v_1 *and* v_2 are both at a HIGH level. When either input is at zero, the output is LOW.

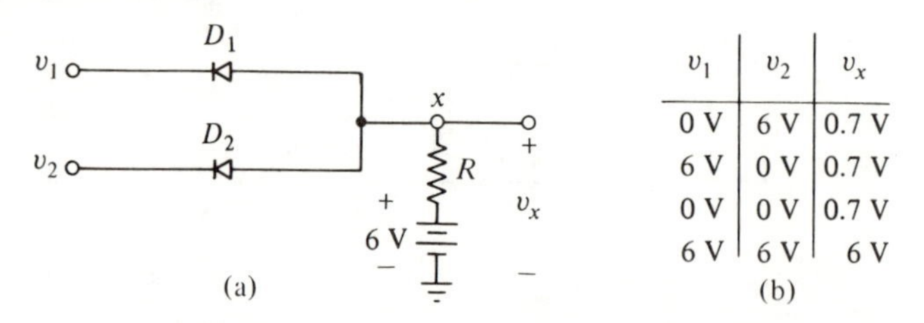

v_1	v_2	v_x
0 V	6 V	0.7 V
6 V	0 V	0.7 V
0 V	0 V	0.7 V
6 V	6 V	6 V

Figure 7–6 **Two-input diode AND gate and corresponding truth table**

Case I: $v_1 = 0$ V, $v_2 = 6$ V

Consider first the case where $v_1 = 0$ V and $v_2 = 6$ V, which is shown in Figure 7–7(a). Diode D_1 has its cathode at 0 V, and its anode is tied to point x, which is connected to +6 V through resistor R. Thus, D_1 is forward-biased, and it will turn ON. As such, current will flow from the 6 V supply through R and D_1 as shown. Diode D_2 is in the OFF state, since its cathode is at +6 V, and it will conduct no current. The current through D_1 and R can be calculated by noting that there will be 5.3 V across R, assuming a 0.7 V ON voltage for D_1 (silicon). Thus, $i_1 = 5.3 \text{ V}/R$.

With 5.3 V across R, it is clear that the output voltage (v_x) will be 6 V − 5.3 V = 0.7 V for this case. This is a LOW level.

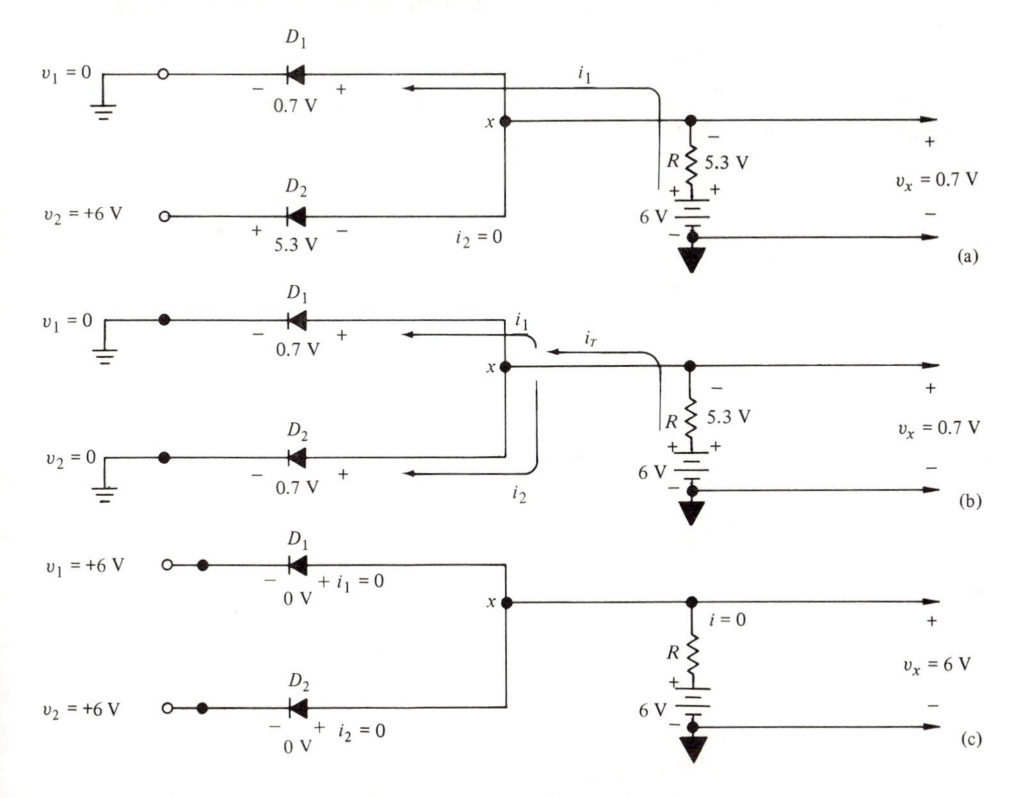

Figure 7–7 **AND gate under various input conditions**

Case II: $v_1 = 6$ V, $v_2 = 0$ V

This case is obviously the same as the first one except that D_2 will be ON and D_1 will be OFF. The output (v_x) is again equal to 0.7 V.

Case III: $v_1 = v_2 = 0$ V

This case is illustrated in Figure 7–7(b). Since both diodes have their cathodes at ground potential, both will be in the ON state and conducting current. Thus, there will be 5.3 V across R, which gives a *total* current, $i_T = 5.3\ \text{V}/R$, that splits between the two diodes; that is, $i_1 = i_2 = i_T/2$, assuming identical diodes. Once again the output (v_x) is equal to 0.7 V.

Case IV: $v_1 = v_2 = 6$ V

This final case is shown in Figure 7–7(c). Neither diode is forward-biased since cathode and anode are at the same potential. Thus, neither will conduct current, so no current will flow through R. With no current there will be zero voltage across R, so $v_x = +6$ V, which is the HIGH level.

To summarize, the AND gate produces a HIGH output level only when *all* the inputs are at a HIGH level.

Example 7.2 Determine the output of the two-input AND gate for the input waveforms shown in Figure 7–8(a).

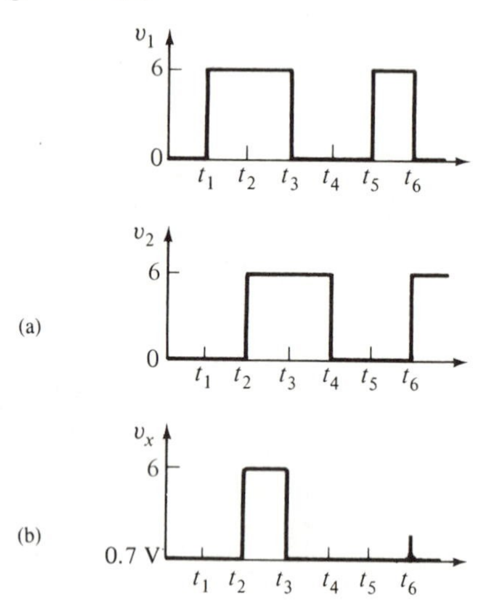

Figure 7–8

Solution: To determine the output (v_x), it is only necessary to look at the two inputs; and whenever *both* are at 6 V, the output will be 6 V, otherwise, $v_x = 0.7$ V. The result is shown in Figure 7–8(b). The only time v_1 and v_2 are both HIGH is during the interval between t_2 and t_3. Note again the glitch at t_6 caused by the simultaneous opposite transitions of the inputs.

Example 7.3 If $R = 100\ \Omega$, what is the minimum R_L that the AND gate can drive without causing the output to drop below 5 V when $v_1 = v_2 = 6$ V?

Solution: With both inputs at 6 V, neither diode conducts. With no load on the output, the output will be 6 V. However, with a load R_L on the output, current will flow from the 6 V source through R and through R_L. The output will be some portion of 6 V determined by the voltage divider ratio between R_L and R (see Figure 7–9).

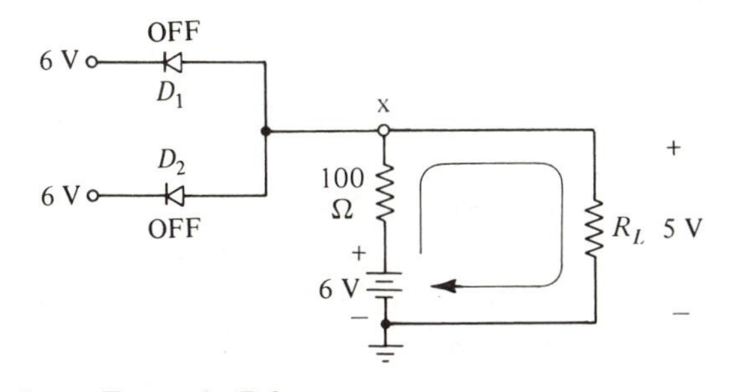

Figure 7–9 **Example 7.3**

Since v_x is not allowed to drop below 5 V, the minimum R_L will be the value that produces $v_x = 5$ V. Thus, using the voltage divider rule, we have

$$5\text{ V} = \left(\frac{R_L}{R_L + 100}\right) \times 6\text{ V}$$

or

$$R_L(\text{min}) = \mathbf{500\ \Omega}$$

Any value of R_L below this will drag the output down below 5 V.

Like OR gates, AND gates with more than two inputs can be constructed by adding a diode for each input. Regardless of how many inputs, the output of the AND gate will be HIGH only if *all* the inputs are HIGH.

Diode logic gates are not too useful in themselves in large-scale digital systems or circuits because they have no inherent mechanism for keeping the output levels from going into the undefined zone. For example, if a 6 V input is fed to an OR gate, the output is only 5.3 V due to the diode drop. If this output is fed to another OR gate input, the output of the second OR gate will only be 4.6 V, and so on. As such, diode gates are usually used in conjunction with transistor amplifier circuits, which help to bring the output levels back to the desired values.

7.4 NOT, NOR and NAND Gates

The NOT circuit The fundamental transistor logic circuit is the NOT circuit, shown in Figure 7–10. It should be recognized as the basic common-emitter inverter circuit

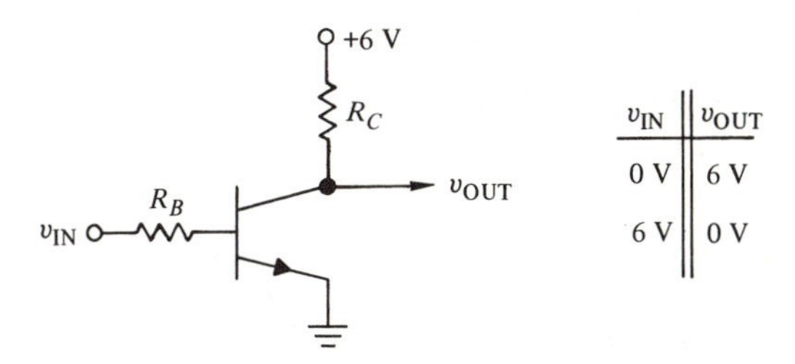

Figure 7–10 **NOT circuit and truth table**

studied in the last chapter. The truth table in the figure reflects what is already known about the inverter circuit; namely, that the output voltage is at the *opposite* level as the input. A NOT circuit always has only *one* input.

The NOR gate One way to construct a two-input NOR gate is shown in Figure 7–11. Here, a diode OR gate is combined with a transistor NOT circuit (inverter). The result is a circuit that produces an output opposite to what an OR gate would produce. This is reflected in the truth table, which indicates a LOW output when any input is HIGH. Recall that an OR gate produces a HIGH output when any input is HIGH.

The circuit operation is relatively simple. If either or both inputs is 6 V (HIGH), the transistor will be turned ON (saturated), so v_{OUT} will be approximately 0 V. If both inputs are LOW, the transistor will be OFF, and v_{OUT} will be 6 V.

The NOR gate is an OR gate followed by a NOT circuit and derives its name from a contraction of NOT-OR. Although the NOR gate can have any number of inputs, in practice the number of inputs rarely exceeds eight. The NOR-gate logic is the same regardless of the number of inputs; that is, the output will be HIGH only when *all* inputs are LOW.

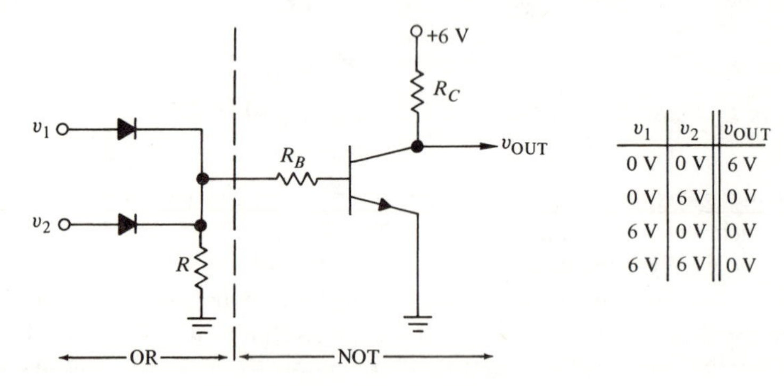

v_1	v_2	v_{OUT}
0 V	0 V	6 V
0 V	6 V	0 V
6 V	0 V	0 V
6 V	6 V	0 V

Figure 7–11 **NOR gate and truth table**

The NAND gate One means of obtaining a NAND gate is shown in Figure 7–12, which combines a diode AND gate with a NOT circuit to produce an output that is opposite what an AND gate would produce. This is reflected in the truth table, which indicates a LOW output only when *both* inputs are HIGH. Recall that an AND gate produces a HIGH output only when *both* inputs are HIGH.

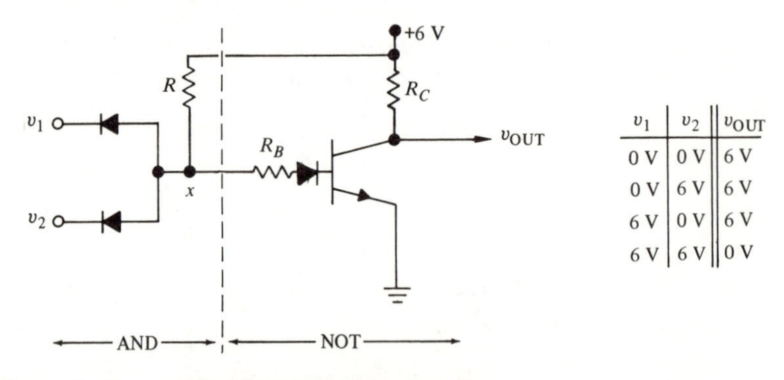

v_1	v_2	v_{OUT}
0 V	0 V	6 V
0 V	6 V	6 V
6 V	0 V	6 V
6 V	6 V	0 V

Figure 7–12 **NAND gate and truth table**

The diode connected to the base of the transistor ensures that the transistor will be OFF when the AND gate output (x) is LOW.

The NAND gate is so called because it is an AND gate followed by a NOT circuit (NOT-AND contracts to NAND). It can have more than two inputs but is usually limited to eight. The NAND-gate logic will be the same regardless of the number of inputs; the output will be LOW only when *all* inputs are HIGH.

Now that we have seen examples of the circuits used for the basic logic gates, we will begin a general study of logic circuits using the standard logic symbols for each type of gate.

7.5 Boolean Algebra

We have seen the five basic digital logic gates—AND, OR, NOT, NAND, and NOR. Each of these gates and all other logic circuits share the same characteristic, namely, their input and output voltages are at either a LOW level or a HIGH level. This characteristic makes it possible to use Boolean algebra as a tool in the analysis and design of digital circuits and systems. Boolean algebra uses only *two* values or numbers—0 and 1. The 0 and 1 can be used to represent, respectively, the LOW voltage and HIGH voltage levels present in digital circuits. We shall now see how digital logic circuits can be mathematically described and analyzed beginning with the basics of Boolean algebra.

Boolean constants and variables A major difference between Boolean algebra and ordinary algebra is that in Boolean algebra the constants and variables are allowed to have either of only *two* possible values—0 or 1. A Boolean variable is a quantity that may, at different times, be equal to either 0 or 1. The Boolean variables are often used to represent the voltage level present on a wire or at the input/output terminals of a circuit. For example, in a certain digital system the Boolean value of 0 might be assigned to any voltage in the 0–0.8 V range, and the Boolean value of 1 might be assigned to any voltage in the 2–5 V range.*

Thus, Boolean 0 and 1 do not represent actual numbers but instead represent the state of a voltage variable, or what is called its *logic level*. A voltage in a digital circuit is said to be at the logic level 0 or the logic level 1 depending on its actual numerical value. In the digital logic field there are several other terms that are used synonymously with 0 and 1. Some of the more common ones are shown in Table 7–1.

Logic 0	Logic 1
False	True
OFF	ON
LOW	HIGH
No	Yes
Open	Closed

TABLE 7–1

*Voltages between 0.8 V–2 V are undefined (neither 0 nor 1) and under normal circumstances should not occur. This undefined range is necessary to separate the 0 and 1 ranges.

Boolean algebra is used to express the effects that various digital circuits have on logic inputs and to manipulate logic variables for the purpose of determining the best method for performing a given circuit function. In all our work to follow we will use letter symbols to represent logic variables. For example, A might represent a certain digital circuit output, and at any time we must have either $A = 0$ or $A = 1$.

The fact that only two values are possible in Boolean algebra makes it relatively easy to work with as compared to ordinary algebra. In Boolean algebra there are no fractions, decimals, negative numbers, square roots, cube roots, logarithms, imaginary numbers, and so on. In fact, in Boolean algebra there are only three basic operations.

Boolean operations The three basic logic operations in Boolean algebra are the following:

1. *Logical addition,* which is also called *OR addition,* or simply the *OR* operation. The common symbol for this operation is the plus sign (+).
2. *Logical multiplication,* which is also called *AND multiplication,* or simply the *AND* operation. The common symbol for this operation is the multiplication sign (·).
3. *Logical complementation* or *inversion,* which is also called the *NOT operation.* The common symbol for this operation is a bar ($\overline{\ }$).

7.6 The OR Operation

Let A and B represent two independent logic variables. When A and B are combined using OR addition, the result, x, can be expressed as

$$x = A + B$$

In this expression the plus sign does not stand for ordinary addition; it stands for OR addition, whose rules are given in the table shown in Figure 7–13(a).

It should be apparent from the table that except for the case where $A = B = 1$ the OR operation is the same as ordinary addition. However, for $A = B = 1$ the OR sum is 1 (not 2, as in ordinary addition). This is easy to remember if we recall that only 0 and 1 are possible values in Boolean algebra, so 1 is the largest value we can get. This same result is true if we have $x = A + B + C$ for the case where $A = B = C = 1$. That is,

$$x = 1 + 1 + 1 = 1$$

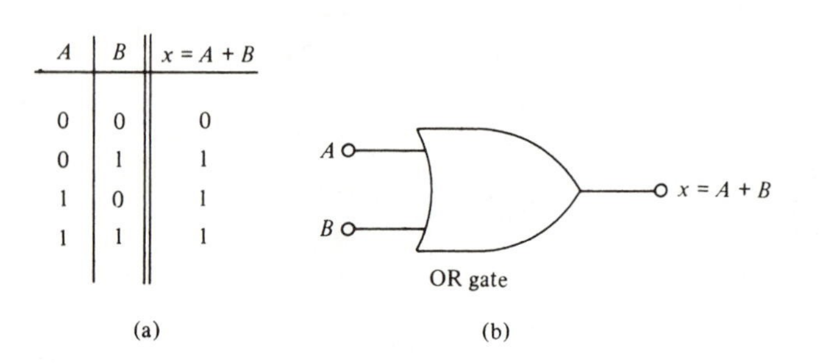

Figure 7–13 **(a) Truth table defining the OR operation (b) Circuit symbol for a two-input OR gate**

We can therefore say that in the OR operation the results will be 1 if any one *or* more variables is a 1. We can also see this in the table in Figure 7–13.

The expression $x = A + B$ can either be read "x equals A *plus* B" or "x equals A *or* B." Both expressions are in common use. The key is that the plus sign stands for the OR operation as defined by the truth table in Figure 7–13, and not for ordinary addition.

The OR gate In digital circuitry an *OR gate* is a circuit that has two or more inputs and whose output is equal to the OR sum of the inputs. Figure 7–13(b) shows the symbol for a two-input OR gate. The inputs, A and B, are logic voltage levels, and the output, x, is a logic voltage level whose value is the result of the OR combination of A and B; that is, $x = A + B$. In other words, the OR gate operates such that its output is HIGH (logic 1) if either input A *or* B *or* both are at a logic-1 level. The OR gate output will be LOW (logic 0) only if all its inputs are at logic 0.

This same idea can be extended to OR gates with more than two inputs. For example, Figure 7–14 shows a three-input OR gate and its accompanying truth table. Note that there are eight cases shown in the table, whereas there were four cases for the two-input gate. In general, if N is the number of input variables, there will be 2^N possible cases, since each variable can take on either of two values. So, for four inputs there will be $2^4 = 16$ possible cases to consider, and so on.

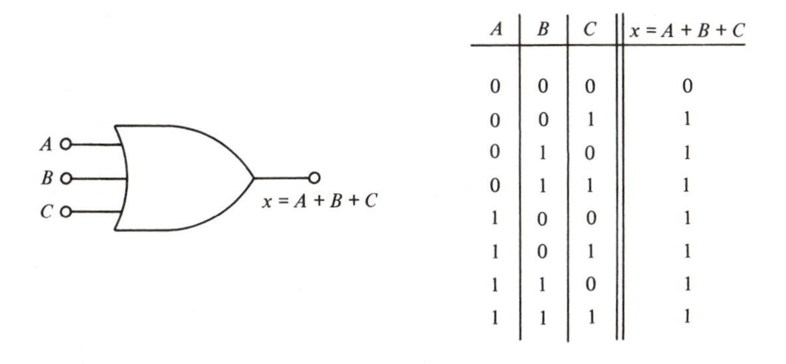

A	B	C	$x = A + B + C$
0	0	0	0
0	0	1	1
0	1	0	1
0	1	1	1
1	0	0	1
1	0	1	1
1	1	0	1
1	1	1	1

Figure 7–14 **Symbol and truth table for three-input OR gate**

Examination of the truth table of Figure 7–14 shows again that the output is 1 for every case where one or more inputs is 1. This general principle of the OR-gate operation is the same for any number of inputs.

Using the language of Boolean algebra, the output, x, can be expressed as $x = A + B + C$, where again it must be emphasized that the plus sign means OR addition. The output of any OR gate, then, can be expressed as the OR sum of its various inputs. We will put this knowledge to use when we subsequently analyze logic circuits.

Summary of the OR operation The important points concerning the OR operation and OR gates are the following:

1. The OR operation produces a result of 1 when *any* of the input variables is 1.
2. The OR operation produces a result of 0 *only* when *all* the input variables are 0.
3. In OR addition $1 + 1 = 1$; $1 + 1 + 1 = 1$; and so on.

Example 7.4 In many industrial control systems an output function must be activated whenever *any one* of several inputs is activated. For example, in a chemical process it may be desirable that an alarm be activated whenever the process temperature exceeds a maximum value *or* whenever the pressure goes above a certain limit. Figure 7–15 is a block diagram of a circuit for this situation. The temperature transducer circuit produces an output voltage proportional to the process temperature. This voltage (V_T) is compared with a temperature reference voltage (V_{TR}) in a voltage comparator circuit. The comparator output is normally a LOW voltage (logic 0), but it switches to a HIGH voltage (logic 1) whenever V_T exceeds V_{TR}, indicating that process temperature is excessive. A similar arrangement is used for the pressure measurement, so its associated comparator output goes from LOW to HIGH when the pressure becomes excessive.

Since we want the alarm to be activated when either temperature *or* pressure is too high, it should be apparent that the two comparator outputs can be fed to a two-input OR gate. The OR gate output thus goes HIGH (logic 1) for either alarm condition and will activate the alarm. This same idea can easily be extended to situations with more than two process variables.

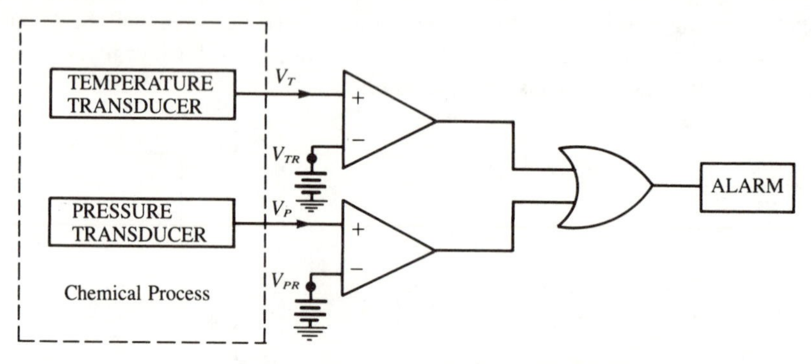

Figure 7–15 **Example of the use of OR gate in an alarm system**

Example 7.5 For the conditions depicted in Figure 7–16, determine the waveform at the OR gate output.

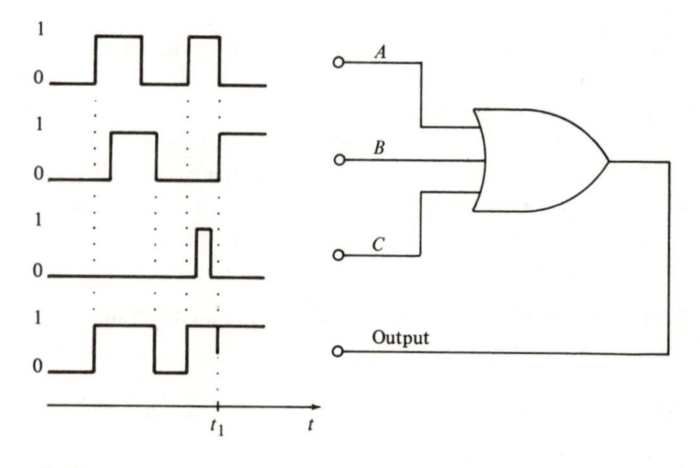

Figure 7–16

Solution: The three OR gate inputs—A, B, and C—are varying, as shown by their waveform diagrams. The OR gate output is determined by realizing that it will be HIGH whenever *any* of the three inputs are at a HIGH level. Using this reasoning, the OR output waveform is as shown in the figure. Particular attention should be paid to what occurs at time t_1. The diagram shows that at that instant of time input A is going from HIGH to LOW, and input B is going from LOW to HIGH. Since these inputs are making their transitions at approximately the same time, and since these transitions take a certain amount of time, there is a short interval when both these OR gate inputs are in the undefined range between 0 and 1. When this occurs the OR-gate output also becomes a value in this undefined range, as evidenced by the glitch, on the output waveform at t_1.

7.7 The AND Operation

If two logic variables A and B are combined using the AND operation, the result, x, can be expressed as

$$x = A \cdot B$$

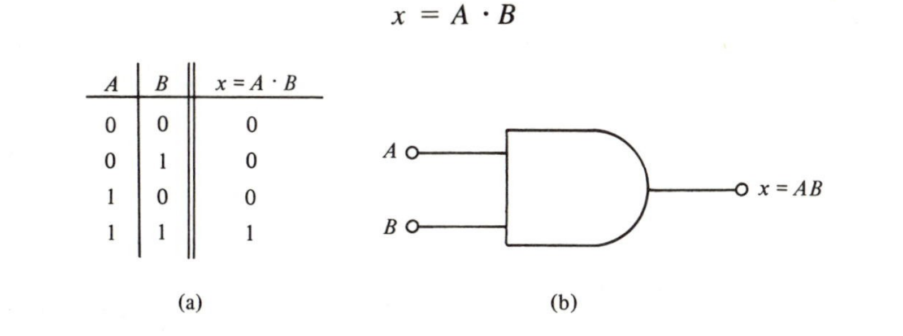

A	B	$x = A \cdot B$
0	0	0
0	1	0
1	0	0
1	1	1

Figure 7–17 **(a) Truth table for the AND operation (b) AND gate symbol**

In this expression the multiplication sign stands for the Boolean operation of AND multiplication, whose rules are given in the truth table shown in Figure 7–17.

It should be apparent from this table that the AND operation is *exactly* the same as ordinary multiplication. Whenever A or B are 0 their product is zero; when both A *and* B are 1 their product is 1. We can therefore say that in the AND operation the result will be 1 *only if* all the inputs are 1; for all other cases the result is 0.

The expression $x = A \cdot B$ is read "x equals A *and* B." The multiplication (AND) sign is usually omitted, as it is in ordinary algebra, so the expression becomes $x = AB$. The key thing to remember is that the AND operation is the same as ordinary multiplication where the variables can be either 0 or 1.

The AND gate A two-input AND gate is shown symbolically in Figure 7–17(b). The AND-gate output is equal to the AND product of the logic inputs; that is, $x = AB$. In other words, the AND gate is a circuit that operates such that its output is HIGH only when all its inputs are HIGH. For all other cases the AND gate output is LOW.

This same operation is characteristic of AND gates with more than two inputs. As an example, a three-input AND gate and its accompanying truth table are shown in Figure 7–18. Once again, note that the gate output is 1 only for the case where $A = B = C = 1$. The expression for the output is $x = ABC$. For a four-input AND gate the output is $x = ABCD$, and so on.

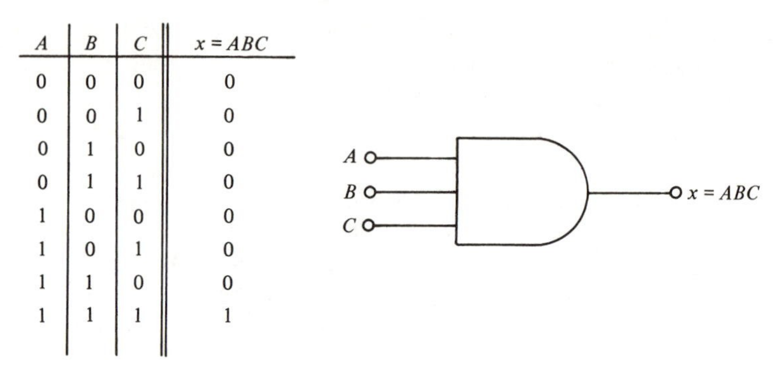

A	B	C	x = ABC
0	0	0	0
0	0	1	0
0	1	0	0
0	1	1	0
1	0	0	0
1	0	1	0
1	1	0	0
1	1	1	1

Figure 7–18 **Truth table and symbol for a three-input AND gate**

Summary of the AND operation

1. The AND operation is performed exactly like ordinary multiplication of 1's and 0's.
2. An output equal to 1 occurs only for the single case where all inputs are 1.
3. The output is 0 for any case where one or more inputs are 0.

Example 7.6 Determine the output waveform for the AND gate shown in Figure 7–19.

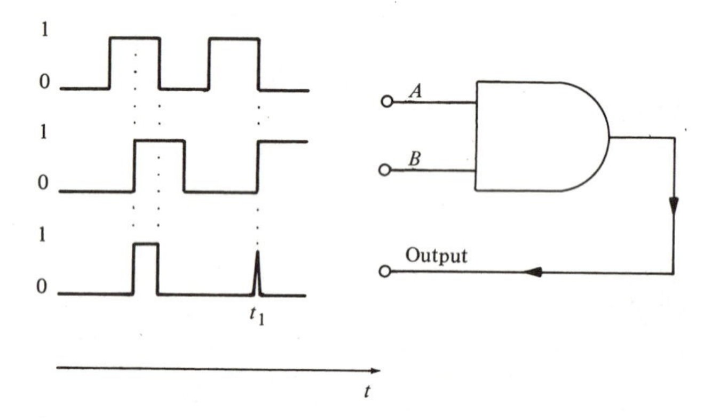

Figure 7–19

Solution: The gate output will be at 0 at all times except when the inputs are simultaneously HIGH. The waveform at the output is easily determined using this principle. Note again the presence of an output glitch at time t_1 due to the simultaneous opposite transitions occurring at the inputs.

Example 7.7 Determine the waveform for output x in Figure 7–20.

Solution: The output x will be at 1 only when A and B are both high at the same time. Using this fact the x waveform can be determined as shown in the figure.

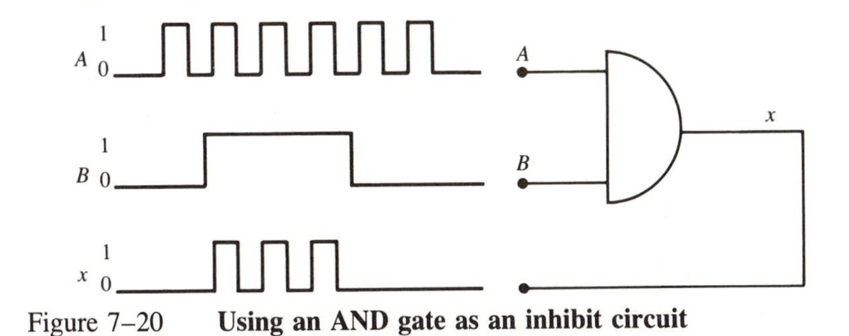

Figure 7–20 **Using an AND gate as an inhibit circuit**

Notice that the x waveform is 0 whenever B is 0, regardless of the signal at A. Also notice that the x waveform is the same as A whenever B is 1. Thus, we can think of the B input as a *control* input whose logic level determines whether the A waveform gets through to the x output or not. In this situation, the AND gate is used as an *inhibit circuit*. We can say that $B = 0$ is the *inhibit* condition producing a 0 output. Conversely, $B = 1$ is the *enable* condition enabling A to reach the output. This inhibit operation is an important application of AND gates.

7.8 The NOT Operation

The NOT operation is unlike the OR or AND operations in that it can be performed on a single variable. For example, if the variable A is subject to the NOT operation, the result, x, can be expressed as

$$x = \overline{A}$$

where the overhead bar represents the NOT operation. This expression is read "x equals NOT A," "x equals the *inverse* of A," or "x equals the *complement* of A." Each of these forms is in common usage, and they all indicate that the logic value of $x = \overline{A}$ is *opposite* the logic value of A. The truth table in Figure 7–21(a) clarifies this for the two cases $A = 0$ and $A = 1$. That is,

$$\overline{1} = 0 \text{ because NOT 1 is 0}$$

and

$$\overline{0} = 1 \text{ because NOT 0 is 1}$$

A	$x = \overline{A}$
0	1
1	0

(a)

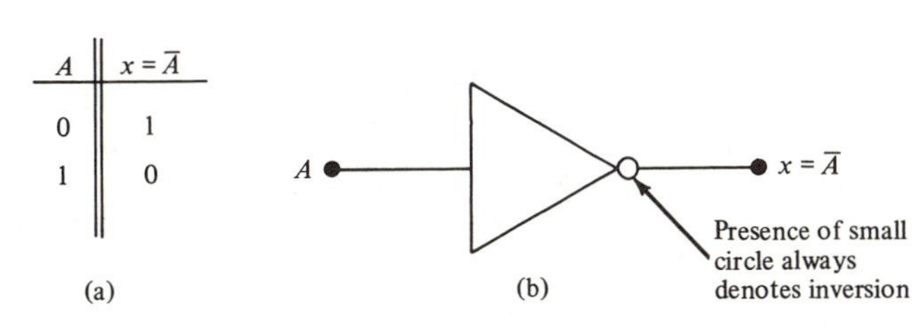

Figure 7–21 **(a) Truth table for the NOT circuit (inverter) (b) Symbol for the NOT circuit**

The NOT operation is also referred to as *inversion*, or *complementation*, and these terms will be used interchangeably throughout the text. Although we will use the bar symbol (‾) exclusively to represent inversion, it is important to mention that another commonly used symbol for inversion is the prime (′). That is,

$$A' = \overline{A}$$

Both symbols should be recognized as NOT symbols.

The NOT circuit Figure 7–21(b) shows the symbol for a NOT circuit, which is also called an *inverter*. This circuit always has a single input, and its output logic level is always opposite the logic level of this input. There is no such thing as an inverter with more than one input.

Summary of Boolean operations The rules for the OR, AND, and NOT operations are summarized as follows:

OR	AND	NOT
$0 + 0 = 0$	$0 \cdot 0 = 0$	$\overline{0} = 1$
$0 + 1 = 1$	$0 \cdot 1 = 0$	$\overline{1} = 0$
$1 + 0 = 1$	$1 \cdot 0 = 0$	
$1 + 1 = 1$	$1 \cdot 1 = 1$	

7.9 Describing Logic Circuits Algebraically

Any logic circuit, no matter how complex, can be completely described using the Boolean operations previously defined, since the OR gate, AND gate, and NOT circuit are the building blocks of digital systems. For example, consider the circuit in Figure 7–22. This circuit has three inputs, A, B, and C, and a single output, x. Using the Boolean expression for each gate, we can easily determine the expression for the output.

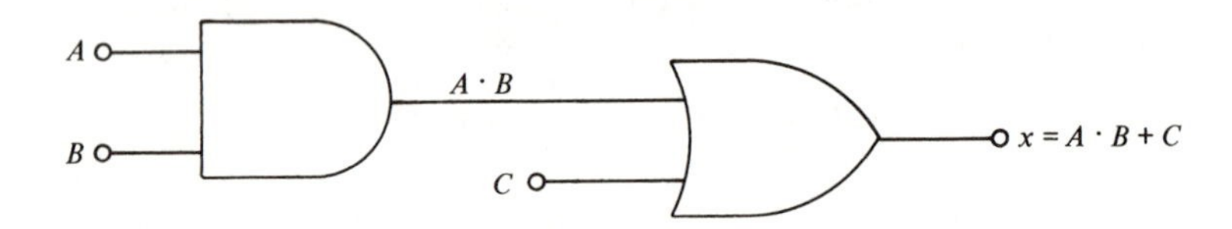

Figure 7–22 **Logic circuit with its Boolean expression**

The expression for the AND-gate output is written $A \cdot B$. This AND output is connected as an input to the OR gate along with C, another input. The OR gate operates on its inputs so that its output is the OR sum of the inputs. Thus, we can express the OR output as $x = A \cdot B + C$. (This final expression could also be written $x = C + A \cdot B$ since it does not matter which term of the OR sum is written first.)

Occasionally there may be confusion about which operation in an expression is performed first. The expression $A \cdot B + C$ can be interpreted in two different ways—that $A \cdot B$ is ORed with C or that A is ANDed with the term $(B + C)$. To avoid this confusion, it is understood that if an expression contains both AND and OR operations, the AND operations are performed first, unless there are *parentheses* in the expression, in which case the operation(s) inside the parentheses is to be performed first. This is the same rule used in ordinary algebra to determine the hierarchy of operations.

To further illustrate, consider the circuit in Figure 7–23. The expression for the OR-gate output is simply $A + B$. This output serves as an input to the AND gate along with another input, C. Thus, we express the output of the AND gate as $x = (A + B) \cdot C$. Note the use of parentheses here to indicate that A and B are ORed *before* their OR sum is ANDed with C. Without the parentheses it would be interpreted *incorrectly*, since $A + B \cdot C$ means A is ORed with the product $B \cdot C$.

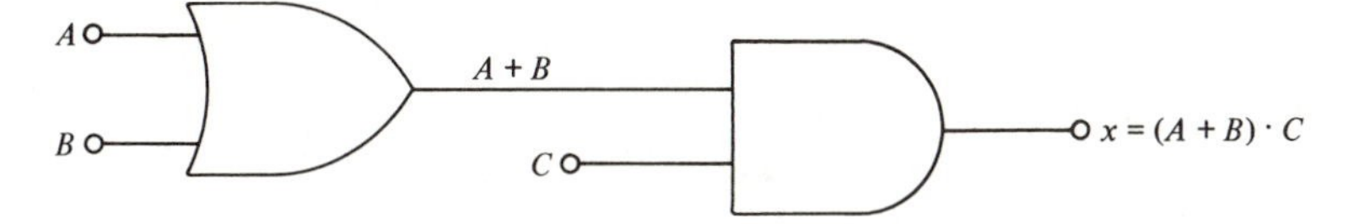

Figure 7–23 **Logic circuit whose output expression requires parentheses**

Circuits containing INVERTERs Whenever an INVERTER (NOT circuit) is present in a logic-circuit diagram, its output expression is simply equal to the input expression with a bar over it. Figure 7–24 shows two examples using INVERTERs. In Figure 7–24(a) the input A is fed through an INVERTER, whose output is therefore $\overline{A}$. The INVERTER output is fed to an OR gate along with B, so the OR output is equal to $\overline{A} + B$. Note that the bar is over only the A, indicating that A is first inverted and then ORed with B.

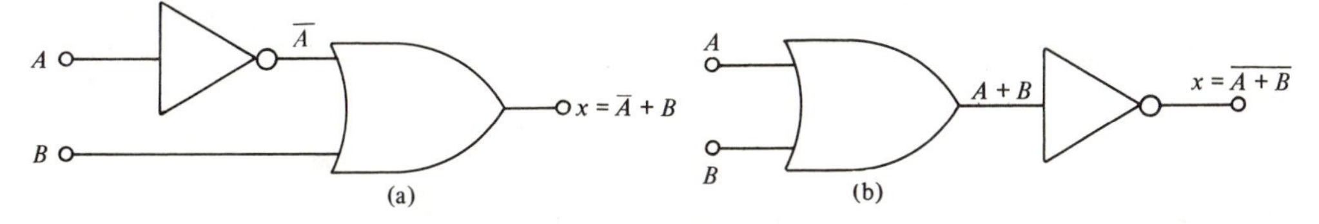

Figure 7–24 **Circuits using INVERTERs**

In Figure 7–24(b) the output of the OR gate is equal to $A + B$ and is fed through an INVERTER. The INVERTER output is therefore equal to $(\overline{A + B})$ since it inverts the *complete* input expression. Note that the bar covers the whole expression $(A + B)$. This is important because, as we shall see later, the expressions $(\overline{A + B})$ and $(\overline{A} + \overline{B})$ are *not* equivalent. The expression $(\overline{A + B})$ means that A is ORed with B and then their OR sum is inverted, whereas the expression $(\overline{A} + \overline{B})$ indicates the A is inverted, B is inverted, and then the results are ORed together.

Figure 7–25 shows two additional examples that should be studied carefully. Note especially the use of *two* separate sets of parentheses in Figure 7–25(b). Also notice in Figure 7–25(a) that the input variable C is connected as an input to two different gates, which is common in digital circuits.

7.10 Evaluating Logic-Circuit Outputs

Once the Boolean expression for a circuit is obtained, the logic level of the output can be determined for any values of circuit inputs. For example, suppose we wanted to know the logic level of the output x for the circuit in Figure 7–25(a) for the case where $A = 0$,

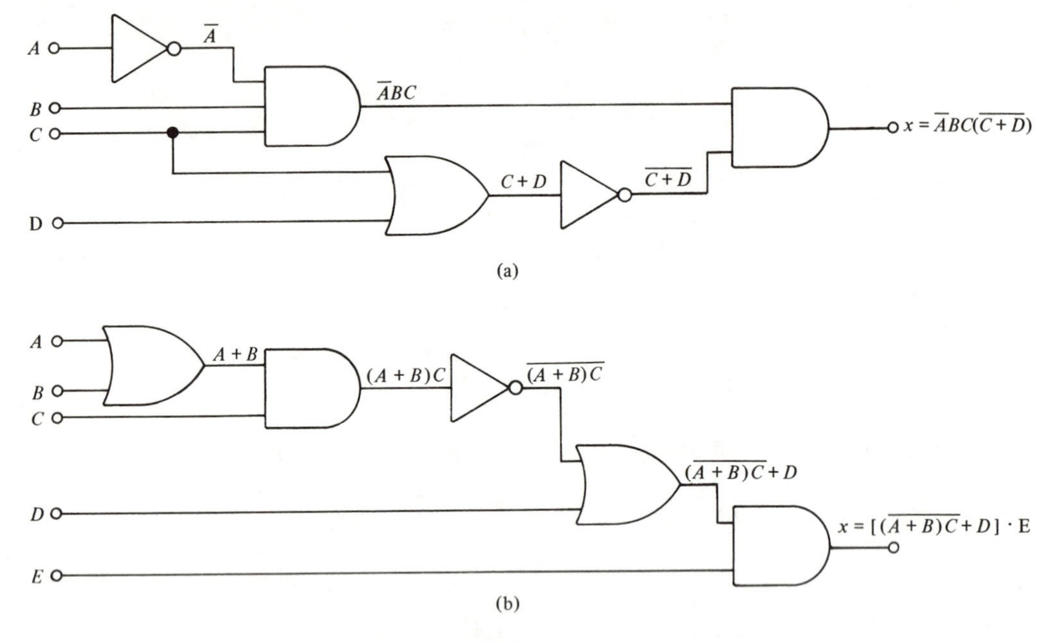

Figure 7–25 **More examples**

$B = 1$, $C = 1$ and $D = 0$. As in ordinary algebra, the value of x can be found by plugging the values of the variables into the expression for x and performing the indicated operations as follows:

$$\begin{aligned} x &= \overline{A}BC(\overline{C + D}) \\ &= \overline{0} \cdot 1 \cdot 1 \cdot (\overline{1 + 0}) \\ &= 1 \cdot 1 \cdot 1 \cdot (\overline{1 + 0}) \\ &= 1 \cdot 1 \cdot 1 \cdot (\overline{1}) \\ &= 1 \cdot 1 \cdot 1 \cdot 0 \\ x &= 0 \end{aligned}$$

As another illustration, let us evaluate the output of the circuit in Figure 7–25(b) for $A = 0$, $B = 0$, $C = 1$, $D = 1$, $E = 1$.

$$\begin{aligned} x &= [D + \overline{(A + B)C}] \cdot E \\ &= [1 + \overline{(0 + 0) \cdot 1}] \cdot 1 \\ &= (1 + \overline{0 \cdot 1}) \cdot 1 \\ &= (1 + \overline{0}) \cdot 1 \\ &= (1 + 1) \cdot 1 \\ &= 1 \cdot 1 \\ x &= 1 \end{aligned}$$

In general, the following rules must be followed when evaluating expressions like the foregoing:

1. First, perform all inversions of single variables, that is, $\overline{0} = 1$, or $\overline{1} = 0$.
2. Then, perform all operations within parentheses.
3. Perform an AND operation before an OR operation unless parentheses indicate otherwise.
4. If an expression has a bar (¯) over it, perform the operations of the expression first and then invert the result.

Determining the output level from a diagram The output logic level for given input levels can also be determined directly from the circuit diagram without using the Boolean expression. This technique is often used by a technician during the troubleshooting or testing of a logic system since it tells him what each gate output is supposed to be as well as gives him the final output. To illustrate, the circuit of Figure 7–25(a) is redrawn in Figure 7–26 with the input levels $A = 0$, $B = 1$, $C = 1$, and $D = 0$. The procedure is to start from the inputs and proceed through each inverter and gate, writing down each of the outputs in the process until the final output is reached.

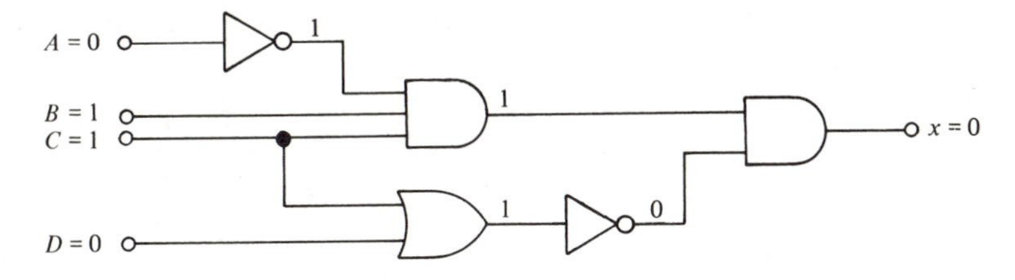

Figure 7–26 **Determining the output level from a circuit diagram**

In Figure 7–26 the first AND gate has *all* three inputs at the 1 level because the INVERTER changes the $A = 0$ to an $\overline{A} = 1$. This condition produces a 1 at the AND output since $1 \cdot 1 \cdot 1 = 1$. The OR gate has inputs of 1 and 0, which produce a 1 output since $1 + 0 = 1$. This 1 is inverted to 0 and applied to the final AND gate along with the 1 from the first AND output. The 0 and 1 inputs to the final AND gate produce an output of 0 since $0 \cdot 1 = 0$.

7.11 Implementing Circuits from Boolean Expressions

If the operation of a circuit is defined by a Boolean expression, a logic-circuit diagram can be implemented directly from that expression. For example, if we needed a circuit that was defined by $x = A \cdot B \cdot C$, we would immediately know that all that was needed was a three-input AND gate. If we needed a circuit that was defined by $x = A + \overline{B}$, we would use a two-input OR gate with an INVERTER for one of the inputs. The same reasoning used for these simple cases can be extended to more complex circuits.

Suppose we wanted to construct a circuit whose output is $y = AC + B\overline{C} + \overline{A}BC$. This Boolean expression contains three terms (AC, $B\overline{C}$, $\overline{A}BC$) that are ORed together. This tells us that a three-input OR gate is required with inputs that are respectively equal to AC, $B\overline{C}$, and $\overline{A}BC$. This situation is illustrated in Figure 7–27(a).

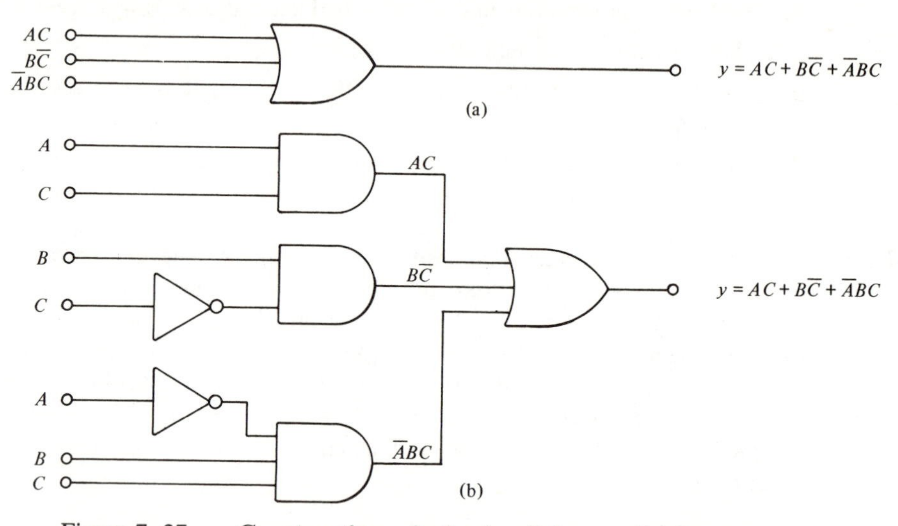

Figure 7–27 **Constructing a logic circuit from a Boolean expression**

Each OR-gate input is an AND product term, which means that an AND gate with appropriate inputs can be used to generate each of these terms. This is shown in Figure 7–27(b), which is the final circuit diagram. Note the use of INVERTERs to produce the $\overline{A}$ and $\overline{C}$ terms required in the expression.

This same general approach can always be followed, although we shall find that there are some clever, more efficient techniques that can be employed. For now, however, this straightforward method will be used to minimize the number of new things to be learned.

7.12 NOR Gates and NAND Gates

Two other types of logic gates—NOR gates and NAND gates—are used extensively in digital circuitry. Because these gates actually combine the basic AND, OR, and NOT operations it is relatively easy to describe them using the Boolean algebra operations previously studied.

The NOR gate Figure 7–28(a) shows the common symbol for a two-input NOR gate. The operation of the NOR gate is equivalent to the OR gate followed by an INVERTER, as shown in Figure 7–28(b), so the output expression for each is $x = \overline{A + B}$. In other

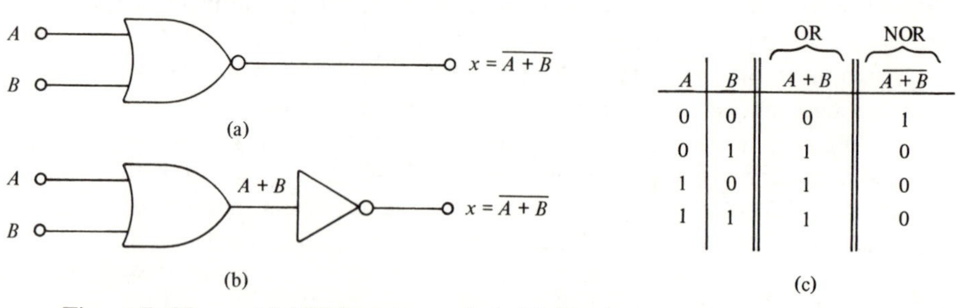

A	B	OR $A + B$	NOR $\overline{A + B}$
0	0	0	1
0	1	1	0
1	0	1	0
1	1	1	0

Figure 7–28 **(a) NOR gate symbol (b) Equivalent circuit (c) Truth table**

words, the NOR gate first performs the OR operation on the inputs and then performs the NOT operation on the OR output. This is easy to remember because the NOR gate symbol is just the OR symbol with a small circle on the output. This small circle represents the inversion operation.

The truth table in Figure 7–28(c) shows that the NOR output in each case is the inverse of the OR output. Whereas the OR output is HIGH when *any* input is HIGH, the NOR gate output is LOW when any input is HIGH. This same operation can be extended to NOR gates with more than two inputs.

Example 7.8 Determine the Boolean expression for a three-input NOR gate followed by an INVERTER.

Solution: Refer to Figure 7–29, where the circuit diagram is shown. The expression at the NOR output is $(\overline{A + B + C})$, which is then fed through an INVERTER to produce

$$x = \overline{(\overline{A + B + C})}$$

The double inversion signs (=) indicate that the quantity $(A + B + C)$ has been inverted and then inverted again. It should be clear that this simply results in the expression $(A + B + C)$ being unchanged. That is,

$$x = \overline{(\overline{A + B + C})} = A + B + C$$

Whenever two inversion bars are over the same variable or expression, they cancel each other out as in this example. However, in cases such as $\overline{\overline{A} + \overline{B}}$, the inverter bars do not cancel (in the same way). This is because the small inverter bars are inverting the single variables, A and B, while the longer bar inverts the expression $(\overline{A} + \overline{B})$. Thus, $\overline{\overline{A} + \overline{B}} \neq A + B$.

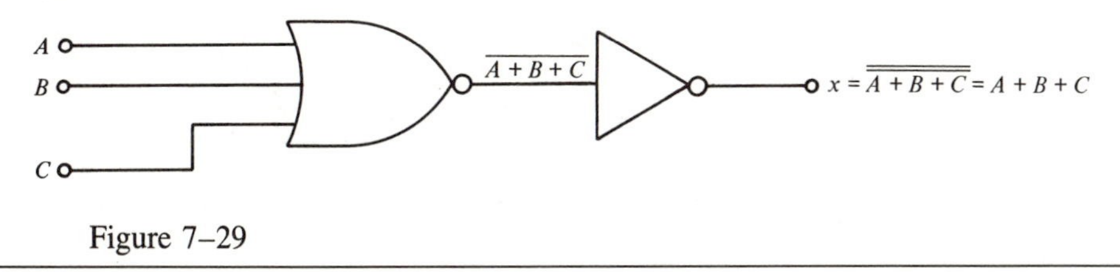

Figure 7–29

The NAND gate Figure 7–30(a) shows the common symbol for a two-input NAND gate. The operation of the NAND gate is equivalent to the AND gate followed by an INVERTER, as shown in Figure 7–30(b) so that the output expression for each is $x = \overline{A \cdot B}$. Thus, the NAND gate first performs the AND operation on the inputs and then

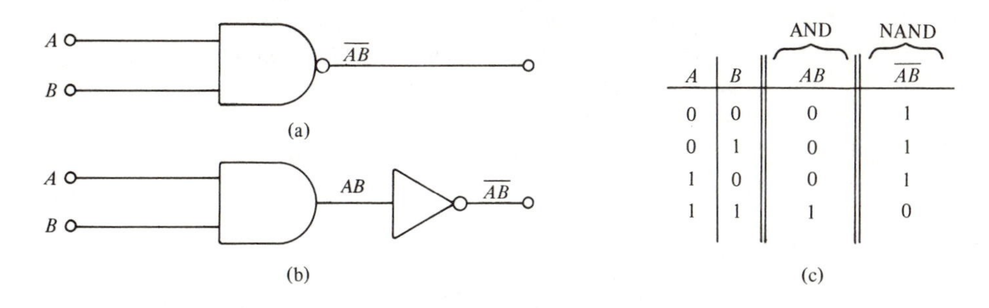

A	B	AND AB	NAND $\overline{AB}$
0	0	0	1
0	1	0	1
1	0	0	1
1	1	1	0

Figure 7–30 **(a) NAND symbol (b) Equivalent circuit (c) Truth table**

performs the NOT operation on the AND product. Once again, the order of these operations should be clear from the NAND symbol.

The truth table in Figure 7–30(c) shows that the NAND-gate output in each case is the inverse of the AND output. Whereas the AND output is HIGH only when *all* inputs are HIGH, the NAND output is LOW only when *all* inputs are HIGH. This same type of operation can be extended to NAND gates with more than two inputs.

The NAND gate and NOR gate are much more versatile than the AND and OR gates, since each can be used to produce any desired Boolean expression. We will be able to investigate this property later after we have studied some Boolean rules and theorems.

7.13 The Power of Boolean Algebra

We have seen that Boolean algebra can be used to analyze a logic circuit and express its operation mathematically. This brings us to the most important application of Boolean algebra—*the simplification of logic circuits*. Once the Boolean expression for a logic circuit has been obtained, it may often be reduced to a simpler form by the use of certain Boolean theorems. Since it is equivalent to the original, the simpler Boolean expression can be used in place of the original expression. This procedure often considerably reduces the number of gates and connections in the logic circuit.

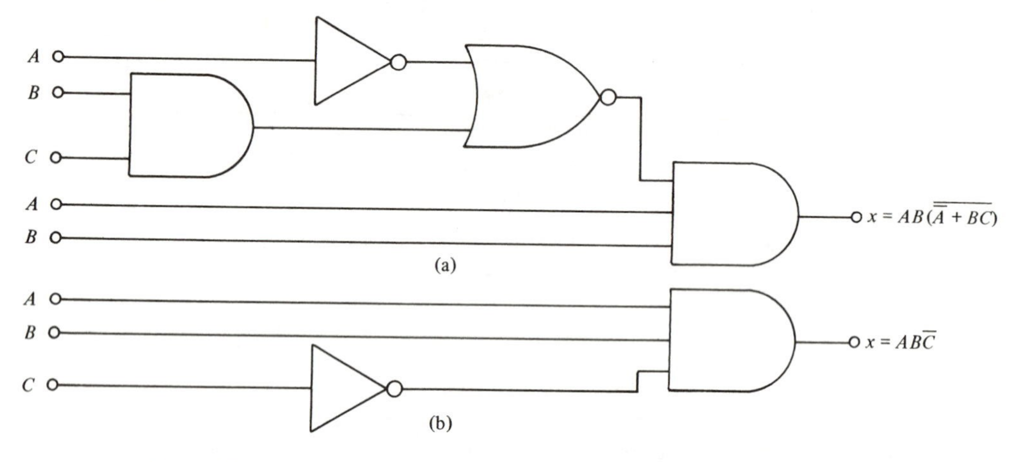

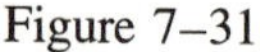
Figure 7–31

For example, Figure 7–31(a) shows a logic circuit whose output expression, $x = AB(\overline{\overline{A} + BC})$, can be obtained by the methods we used earlier. By using Boolean theorems, we can simplify this expression to its equivalent $x = AB\overline{C}$, as shown in Figure 7–31(b). It should be apparent that if both circuits are equivalent in the logic that they perform, then the simpler circuit is more desirable from the standpoints of cost and size.

The following sections will introduce a set of Boolean theorems and show how they can be applied to realize simplifications like those in this example.

7.14 Boolean Theorems

The first group of Boolean theorems is given in Figure 7–32. In each theorem x represents a logic variable which can be either 0 or 1. Each theorem is accompanied by its equivalent logic diagram, which will help to validate the theorem. Any one of these theorems, and in fact any Boolean theorem, may be proved simply by checking it for all possible values of the variables. The theorems in Figure 7–32 are single-variable theorems with x the only variable, so we have to check for the cases $x = 0$ and $x = 1$.

Theorem (1) states that if any variable is ANDed with 0, the result has to be 0. This is easy to remember, since the AND operation is just like ordinary multiplication, where we know that anything multiplied by 0 is 0. We also know that the output of an AND gate will be 0 whenever any input is 0, regardless of the level on the other input.

Theorem (2) is also obvious from its comparison with ordinary multiplication.

Theorem (3) can be proven by trying each case. If $x = 0$, then $0 \cdot 0 = 0$; if $x = 1$, then $1 \cdot 1 = 1$.

Theorem (4) can be proven in the same manner. However, it can also be reasoned that at any time either x or its inverse, $\bar{x}$, has to be at the 0 level, so their AND product is always 0.

Theorem (5) is straightforward since *adding* 0 to anything does not affect its value either in regular addition or in OR addition.

Theorem (6) states that if any variable is ORed with 1, the result will always be 1. Checking this for both values of x, we obtain $0 + 1 = 1$ and $1 + 1 = 1$. Equivalently, remember that an OR-gate output will be 1 when *any* input is 1, regardless of the value of the other input.

Theorem (7) can be proven by checking for both values of x: $0 + 0 = 0$ and $1 + 1 = 1$.

Theorem (8) can be proven similarly, or we can just reason that at any time either x or $\bar{x}$ has to be at the 1 level, so we are always ORing a 0 and a 1, the result of which is always 1.

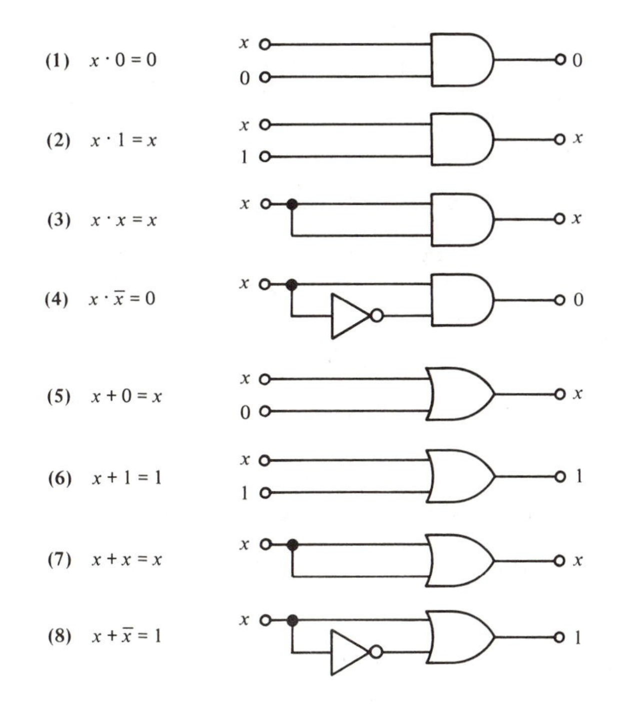

Figure 7–32 **Single-variable theorems**

Before introducing any more theorems, it should be pointed out that in applying theorems (1)–(8), the variable x may actually represent an expression containing more than one variable. For example, if we have $A\overline{B}(\overline{A\overline{B}})$, we can invoke Theorem (4) by letting $x = A\overline{B}$. Thus, we can say that $A\overline{B}(\overline{A\overline{B}}) = 0$. This same idea can be applied to the use of any of the theorems presented.

Multivariable theorems The following theorems involve more than one variable:

(9) $x + y = y + x$

(10) $x \cdot y = y \cdot x$

(11) $x + (y + z) = (x + y) + z = x + y + z$

(12) $x(yz) = (xy)z = xyz$

(13) $x(y + z) = xy + xz$

(14) $x + xy = x$

(15) $x + \overline{x}y = x + y$

Theorems (9) and (10) are called the *commutative laws*. These laws indicate that the order in which we add or multiply two variables is unimportant, because the results are the same.

Theorems (11) and (12) are the *associative laws*, which state that we can group the terms of a sum or a product any way we want. In other words, given the product $x \cdot y \cdot z$, we can first multiply x and y, then multiply the result by z; or we can first multiply y and z, then multiply the result by x. A similar idea applies to addition (the OR operation).

Theorem (13) is the *distributive law*, which states that an expression can be expanded by multiplying term by term, just as in ordinary algebra. This theorem also indicates that we can factor an expression. That is, if we have a sum of two (or more) terms, each of which contains a common variable, the common variable can be factored out just like in ordinary algebra. For example, if we have the expression $A\overline{B}C + \overline{A}\,\overline{B}\,\overline{C}$, we can factor out the $\overline{B}$ variable, which gives us

$$A\overline{B}C + \overline{A}\,\overline{B}\,\overline{C} = \overline{B}(AC + \overline{A}\,\overline{C})$$

As another example, consider the expression $ABC + ABD$. The two terms of the expression have the variables A and B in common, so $A \cdot B$ can be factored out of both terms. That is,

$$ABC + ABD = AB(C + D)$$

This *distributive* law is very useful in simplifying expressions, as the following examples illustrate.

Example 7.9 Simplify the expression $y = A\overline{B}D + A\overline{B}\,\overline{D}$.

Solution: Factor out the common variables $A\overline{B}$ using Theorem (13).

$$y = A\overline{B}(D + \overline{D})$$

The term in the parentheses is equivalent to 1 using Theorem (8). Thus,

$$y = A\overline{B} \cdot (1)$$

$$= \boldsymbol{A\overline{B}} \qquad \text{[Using Theorem (2)]}$$

Example 7.10 Simplify $z = (\overline{A} + B)(A + \overline{B})$.

Solution: The expression can be expanded by multiplying the terms out.

$$z = \overline{A} \cdot A + \overline{A} \cdot \overline{B} + B \cdot A + B \cdot \overline{B}$$

Invoking Theorem (4), the terms $\overline{A} \cdot A$ and $B \cdot \overline{B}$ are both equal to 0. Thus,

$$\begin{aligned} z &= 0 + \overline{A} \cdot \overline{B} + B \cdot A + 0 \\ &= \mathbf{\overline{A} \cdot \overline{B} + B \cdot A} \qquad \text{[Using Theorem (5)]} \end{aligned}$$

Theorems (9)–(13) are easy to remember and use since they are identical to theorems of ordinary algebra. Theorems (14) and (15), on the other hand, do not have any counterparts in ordinary algebra. Both of these are also useful in simplification techniques. Each can be proven by trying all possible cases for x and y. This is illustrated for Theorem (14) as follows:

Case 1: For $x = 0$, $y = 0$, we have

$$\begin{aligned} x + xy &= x \\ 0 + 0 \cdot 0 &= 0 \\ 0 &= 0 \end{aligned}$$

Case 2: For $x = 0$, $y = 1$, we have

$$\begin{aligned} x + xy &= x \\ 0 + 0 \cdot 1 &= 0 \\ 0 + 0 &= 0 \\ 0 &= 0 \end{aligned}$$

Case 3: For $x = 1$, $y = 0$, we have

$$\begin{aligned} x + xy &= x \\ 1 + 1 \cdot 0 &= 1 \\ 1 + 0 &= 1 \\ 1 &= 1 \end{aligned}$$

Case 4: For $x = 1$, $y = 1$, we have

$$\begin{aligned} x + xy &= x \\ 1 + 1 \cdot 1 &= 1 \\ 1 + 1 &= 1 \\ 1 &= 1 \end{aligned}$$

Theorem (14) can also be proven by factoring and using Theorem (6) as follows:

$$\begin{aligned} x + xy &= x(1 + y) \\ &= x \cdot 1 \qquad \text{[Using Theorem (6)]} \\ &= x \qquad \text{[Using Theorem (2)]} \end{aligned}$$

Example 7.11 Simplify $x = ACD + \overline{A}BCD$.

Solution: Factoring out the common expression, CD, we have

$$x = CD(A + \overline{A}B)$$

Using Theorem (15), we can replace $A + \overline{A}B$ by $A + B$, so

$$\begin{aligned} x &= CD(A + B) \\ &= \boldsymbol{ACD + BCD} \end{aligned}$$

Example 7.12 Simplify $x = AB + C(A + C)$.

Solution: Multiply out the term in parentheses.

$$x = AB + CA + CC$$

Invoking Theorem (3) we have $CC = C$. Thus,

$$x = AB + CA + C$$

Invoking Theorem (14) we have $CA + C = C$. Thus,

$$x = AB + C$$

De Morgan's theorems Two of the most important theorems of Boolean algebra were contributed by the great English mathematician Augustus De Morgan (1806–1871). These theorems are extremely useful in simplifying expressions in which a product or sum of variables is complemented (inverted).

The two theorems are the following:

(16) $\overline{(x + y)} = \overline{x} \cdot \overline{y}$

(17) $\overline{(x \cdot y)} = \overline{x} + \overline{y}$

Theorem (16) states that when the OR sum of two variables $(x + y)$ is complemented, it is the same as if the two variables were individually complemented and then ANDed together. Stated differently, *the complement of an OR sum equals the AND product of the complements*.

Theorem (17) shows that when the product of two variables $(x \cdot y)$ is complemented, the result is equivalent to complementing the individual variables and then ORing the results. In other words, *the complement of an AND product is equal to the OR sum of the complements*. These two theorems can be proven easily by checking each one for all values of x and y and is left as an exercise for the student.

De Morgan's theorems are used whenever we want to modify an expression containing large inversion signs (inversion over more than one variable). When applying these theorems, remember that x and y can represent complicated expressions. For example, let us apply them to the expression $\overline{(A\overline{B} + C)}$. First, note that we have the OR sum of $A\overline{B}$ and C, which is inverted. Invoking Theorem (16), we can write this as

$$\overline{(A\overline{B} + C)} = \overline{(A\overline{B})} \cdot \overline{C}$$

Note that here we treated $A\overline{B}$ as x and C as y. The result can be further simplified since we have a product $A\overline{B}$ that is complemented. Using Theorem (17), the expression becomes

$$\overline{A\overline{B}} \cdot \overline{C} = (\overline{A} + \overline{\overline{B}}) \cdot \overline{C}$$

Notice that we can replace $\overline{\overline{B}}$ by B, so we have

$$(\overline{A} + B) \cdot \overline{C} = \overline{A}\,\overline{C} + B\overline{C}$$

This final result contains only inversion signs that invert a single variable.

De Morgan's theorems are easily extended to more than two variables. For example, it can be proven that

$$\overline{x + y + z} = \bar{x} \cdot \bar{y} \cdot \bar{z}$$

and

$$\overline{x \cdot y \cdot z} = \bar{x} + \bar{y} + \bar{z}$$

and so on for more variables.

Implications of De Morgan's theorems Let us examine De Morgan's two theorems from the standpoint of logic circuits. First, consider Theorem (16).

$$\overline{x + y} = \bar{x} \cdot \bar{y}$$

The left-hand side of the equation can be viewed as the output of a NOR gate whose inputs are x and y. The right-hand side of the equation, on the other hand, is the result of first inverting both x and y and then putting them through an AND gate. These two representations are equivalent and are illustrated in Figure 7–33(a). This means that an AND gate with INVERTERs on each of its inputs is equivalent to a NOR gate. In fact, both representations are used to represent the NOR function. When the AND gate with inverted inputs is used to represent the NOR function it is usually drawn as shown in Figure 7–33(b), where the small circles on the inputs represent the INVERTERs.

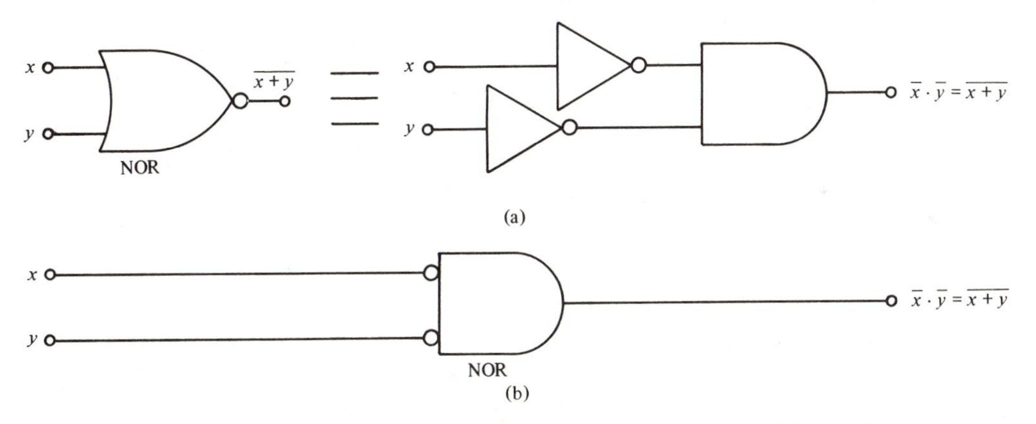

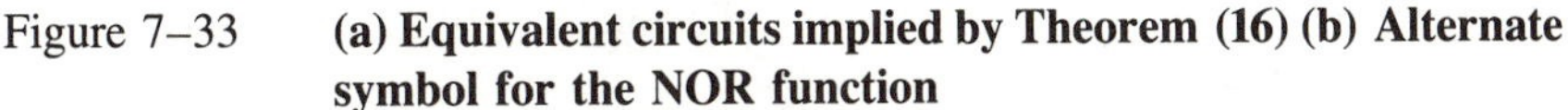

Figure 7–33 **(a) Equivalent circuits implied by Theorem (16) (b) Alternate symbol for the NOR function**

Now, consider Theorem (17).

$$\overline{x \cdot y} = \bar{x} + \bar{y}$$

The left side of the equation can be implemented by a NAND gate with inputs x and y. The right side can be implemented by first inverting inputs x and y and then putting them through an OR gate. These two equivalent representations are shown in Figure 7–34(a). The OR gate with INVERTERs on each of its inputs is equivalent to the NAND gate. In fact, both representations are used to represent the NAND function. When the OR gate with inverted inputs is used to represent the NAND function it is usually drawn as shown in Figure 7–34(b), where again the circles represent inversion.

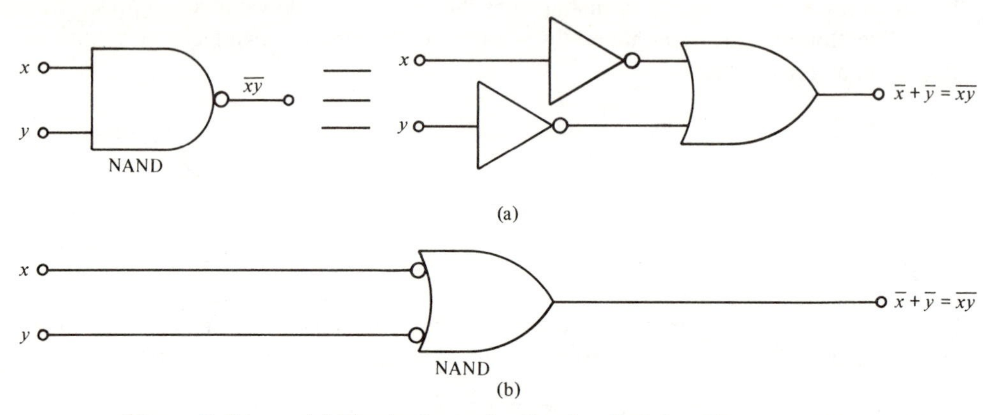

Figure 7–34 **(a) Equivalent circuits implied by Theorem (17) (b) Alternate symbol for the NAND function**

Example 7.13 Simplify the expression $z = \overline{(\bar{A} + C) \cdot (B + \bar{D})}$.

Solution: Using Theorem (17) we can rewrite this expression as

$$z = (\overline{\bar{A} + C}) + (\overline{B + \bar{D}})$$

We can think of this as breaking the large inverter sign down the middle and changing the AND sign (·) to an OR sign (+). Now, the term $(\overline{\bar{A} + C})$ can be simplified by applying Theorem (16). Likewise, $(\overline{B + \bar{D}})$ can be simplified.

$$\begin{aligned} z &= (\overline{\bar{A} + C}) + (\overline{B + \bar{D}}) \\ &= (\bar{\bar{A}} \cdot \bar{C}) + (\bar{B} \cdot \bar{\bar{D}}) \end{aligned}$$

Here we have broken the larger inversion signs down the middle and replaced the + with a ·. Cancelling out the double inversions gives us the final expression,

$$\boldsymbol{z = A\bar{C} + \bar{B}D}$$

Example 7.13 points out that when using De Morgan's theorems to reduce an expression, an inversion sign may be broken at any point in the expression, and the operator at that point in the expression is changed to its opposite (+ is changed to ·, and vice versa). This procedure is continued until the expression is reduced to one in which only single variables are inverted. Two additional examples follow. Study these carefully and note the procedure.

(1)

$$\begin{aligned} z &= \overline{A + \bar{B} \cdot C} \\ &= \bar{A} \cdot (\overline{\bar{B} \cdot C}) \\ &= \bar{A} \cdot (\bar{\bar{B}} + \bar{C}) \\ z &= \bar{A} \cdot (B + \bar{C}) \end{aligned}$$

(2)

$$\begin{aligned} q &= \overline{(A + BC) \cdot (D + EF)} \\ &= (\overline{A + BC}) + (\overline{D + EF}) \\ &= (\bar{A} \cdot \overline{BC}) + (\bar{D} \cdot \overline{EF}) \\ &= [\bar{A} \cdot (\bar{B} + \bar{C})] + [\bar{D} \cdot (\bar{E} + \bar{F})] \\ q &= \bar{A}\bar{B} + \bar{A}\bar{C} + \bar{D}\bar{E} + \bar{D}\bar{F} \end{aligned}$$

7.15 Simplifying Logic Circuits

The theorems that have just been presented can be used to simplify any Boolean expression. Unfortunately, it is not always obvious which theorems should be applied and in what order. Usually, if followed correctly, any path will lead to the simplest result. Sometimes, two different paths may lead to different results that are equally simple. The following examples should be studied carefully to see some of the reasoning used in the simplification of logic circuits. Even though there is no set procedure in applying these theorems, by studying these and the preceding examples, and doing the problems at the end of the chapter, you will eventually develop a facility for arriving at a reasonably good solution.

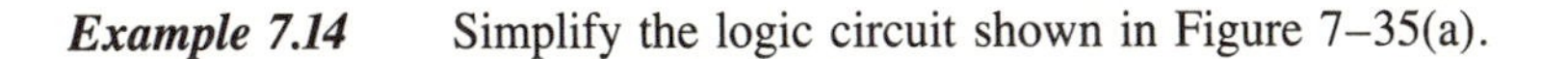

Example 7.14 Simplify the logic circuit shown in Figure 7–35(a).

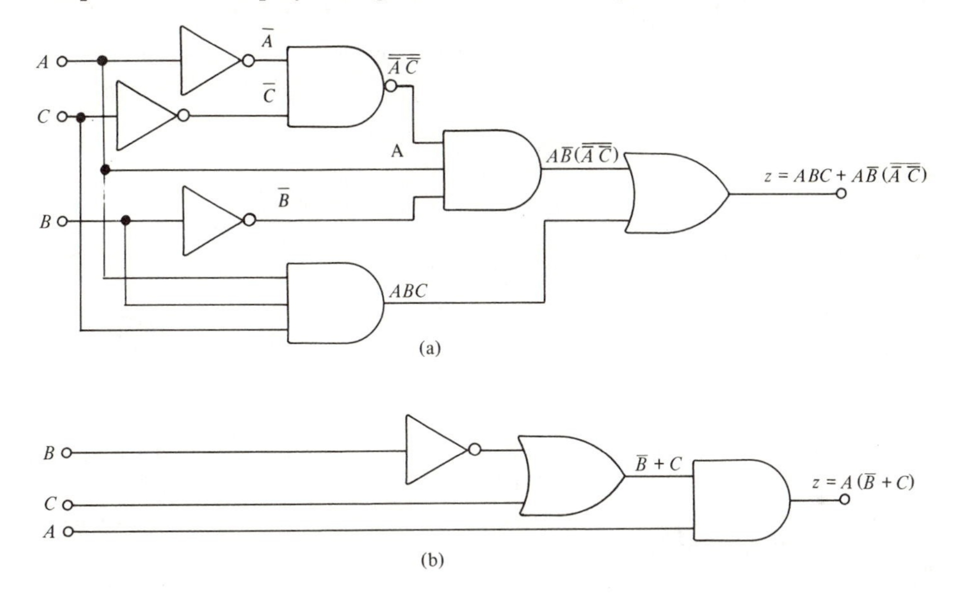

Figure 7–35

Solution: The first step is to determine the expression for the output. The result is

$$z = ABC + A\overline{B}(\overline{\overline{A}\,\overline{C}})$$

Once the expression is determined it is usually a good idea to break down all large inversion signs using De Morgan's theorems and then multiply out all terms.

$$\begin{aligned} z &= ABC + A\overline{B}(\overline{\overline{A}} + \overline{\overline{C}}) && \text{[Theorem (17)]} \\ &= ABC + A\overline{B}(A + C) \\ &= ABC + A\overline{B}A + A\overline{B}C \\ z &= ABC + A\overline{B} + A\overline{B}C && \text{[Theorem (3)]} \end{aligned}$$

Once the expression is in this form, we should look for common variables among the various terms with the intention of factoring. In this example, AC is found in the first and third terms and can be factored out.

$$z = AC(B + \overline{B}) + A\overline{B}$$

Since $B + \overline{B} = 1$, then

$$z = AC(1) + A\overline{B}$$
$$= AC + A\overline{B}$$

We can now factor out A, which results in

$$\mathbf{z = A(C + \overline{B})}$$

This last result cannot be simplified further. Its circuit implementation is shown in Figure 7–35(b). It is obvious that the circuit in part (b) is a great deal simpler than the original circuit in part (a).

Example 7.15 Simplify the expression $z = ABC + AB\overline{C} + A\overline{B}C$.

Solution: We will look at two different ways to arrive at the same result.

(*Method 1*) The first two terms in the expression have AB in common. Thus,

$$z = AB(C + \overline{C}) + A\overline{B}C$$
$$= AB(1) + A\overline{B}C$$
$$z = AB + A\overline{B}C$$

We can then factor the variable A from both terms.

$$z = A(B + \overline{B}C)$$

Invoking Theorem (15), we have

$$\mathbf{z = A(B + C)}$$

(*Method 2*) The original expression is $z = ABC + AB\overline{C} + A\overline{B}C$. The first two terms have AB in common. The first and last terms have AC in common. How do we know whether to factor AB from the first two terms or AC from the two end terms? Actually, we can do both by using the ABC term *twice*. In other words, we can rewrite the expression as

$$z = ABC + AB\overline{C} + A\overline{B}C + ABC$$

where we have added an extra term, ABC. This procedure is valid and will not change the value of the expression since $ABC + ABC = ABC$ (Theorem 7). Now, we can factor AB from the first two terms and AC from the last two terms.

$$z = AB(C + \overline{C}) + AC(\overline{B} + B)$$
$$= AB \cdot 1 + AC \cdot 1$$
$$z = AB + AC = \mathbf{A(B + C)}$$

This is, of course, the same result as that calculated for method 1. This trick of using the same term twice can always be used. In fact, the same term can be used more than twice if necessary.

Example 7.16 Simplify $z = \overline{A}C(\overline{\overline{A}BD}) + \overline{A}B\overline{C}\overline{D} + A\overline{B}C$.

Solution:

$$z = \overline{A}C(A + \overline{B} + \overline{D}) + \overline{A}B\overline{C}\overline{D} + A\overline{B}C \qquad \text{[Theorem (17)]}$$

Multiplying out, we have

$$z = \overline{A}CA + \overline{A}C\overline{B} + \overline{A}C\overline{D} + \overline{A}\overline{B}\overline{C}\overline{D} + A\overline{B}C$$

Since $\overline{A} \cdot A = 0$, the first term is eliminated.

$$z = \overline{A}\overline{B}C + \overline{A}C\overline{D} + \overline{A}\overline{B}\overline{C}\overline{D} + A\overline{B}C$$

Factoring $\overline{B}C$ from the first and last terms and $\overline{A}\overline{D}$ from the second and third terms gives us

$$z = \overline{B}C(\overline{A} + A) + \overline{A}\overline{D}(C + \overline{B}\overline{C})$$
$$= \mathbf{\overline{B}C + \overline{A}\overline{D}(C + B)} \qquad \text{[Theorems (8) and (15)]}$$

This expression cannot be simplified further*.

These examples have illustrated how Boolean algebra rules and theorems are used to simplify a logic circuit. It is *not* expected that the reader will become expert at simplifying logic circuits; logic design textbooks usually devote several chapters to this topic. However, by reviewing these examples and by trying the exercises at the end of the chapter, you should develop a reasonably good understanding of how to apply Boolean algebra principles to logic circuit analysis.

7.16 EXCLUSIVE-OR and EXCLUSIVE-NOR Circuits

EXCLUSIVE-OR and *EXCLUSIVE-NOR* circuits are special logic circuits that occur quite often in digital systems.

The EXCLUSIVE-OR circuit Consider the logic circuit of Figure 7–36(a). The output expression of this circuit is

$$x = \overline{A}B + A\overline{B}$$

The accompanying truth table shows that $x = 1$ for two cases—$A = 0$, $B = 1$ (the $\overline{A}B$ term) and $A = 1$, $B = 0$ (the $A\overline{B}$ term). In other words, *this circuit produces a HIGH output only whenever the two inputs are at opposite levels.* This is the EXCLUSIVE-OR circuit, which will hereafter be abbreviated EX-OR.

This particular combination of logic gates occurs quite often and is very useful in certain applications. In fact, the EX-OR circuit has been given a symbol of its own, shown in Figure 7–36(b). This symbol is assumed to contain all of the logic contained in the EX-OR circuit and therefore has the same logic expression and truth table. This EX-OR symbol is commonly referred to as an *EX-OR gate,* and we can consider it as another type of logic gate.

An EX-OR gate has only *two* inputs; there are no three-input or four-input EX-OR gates. The two inputs are combined such that $x = \overline{A}B + A\overline{B}$. A shorthand notation used to indicate the EX-OR output expression is

$$x = A \oplus B$$

where the symbol $\oplus$ represents the EX-OR gate operation.

---

*In truth, it can be simplified further using techniques that are beyond the scope of this text.

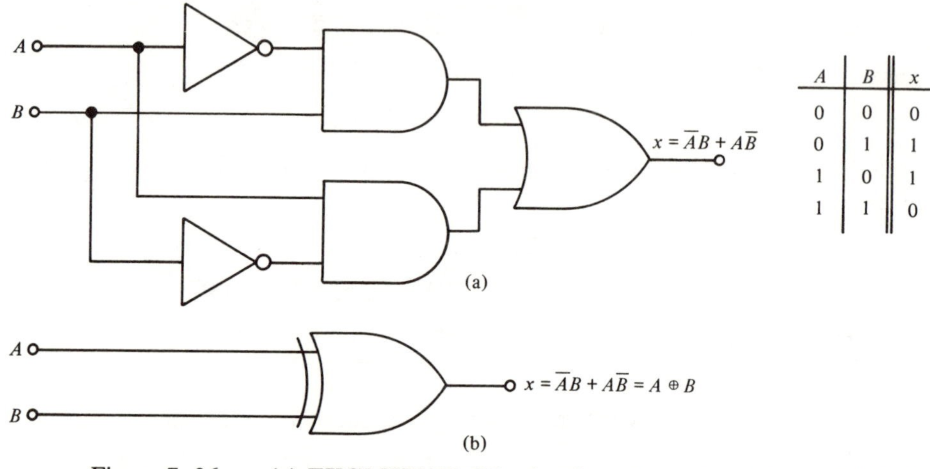

A	B	x
0	0	0
0	1	1
1	0	1
1	1	0

Figure 7–36 **(a) EXCLUSIVE-OR circuit and truth table (b) EXCLUSIVE-OR gate symbol**

The characteristics of an EX-OR gate are summarized as follows:

1. It has only two inputs, and its output is $x = \overline{A}B + A\overline{B} = A \oplus B$
2. Its output is HIGH only when the two inputs are at *opposite levels*.

The EXCLUSIVE-NOR circuit The EXCLUSIVE-NOR circuit (abbreviated EX-NOR) operates completely opposite to the EX-OR circuit. Figure 7–37(a) shows an EX-NOR circuit and its accompanying truth table. The output expression is

$$x = AB + \overline{A}\overline{B}$$

which indicates along with the truth table that x will be 1 for two cases—$A = B = 1$ (the AB term) and $A = B = 0$ (the $\overline{A}\overline{B}$ term). In other words, *this circuit produces a HIGH output only whenever the two inputs are at the same level*. It should be apparent that the output of the EX-NOR circuit is the exact inverse of the output of the EX-OR circuit. As such, the symbol for an EX-NOR gate is simply obtained by adding a small circle at the output of the EX-OR symbol. The EX-NOR gate symbol is shown in Figure 7–37(b).

The EX-NOR gate also has only *two* inputs, and it combines them such that its output is

$$x = AB + \overline{A}\overline{B}$$

A shorthand way to indicate the output expression of the EX-NOR is

$$x = \overline{A \oplus B}$$

which is simply the inverse of the EX-OR operation. The EX-NOR gate is summarized as follows:

1. It has only two inputs, and its output is

$$x = AB + \overline{A}\overline{B} = \overline{A \oplus B}$$

2. Its output is HIGH only when the two inputs are at the *same* level.

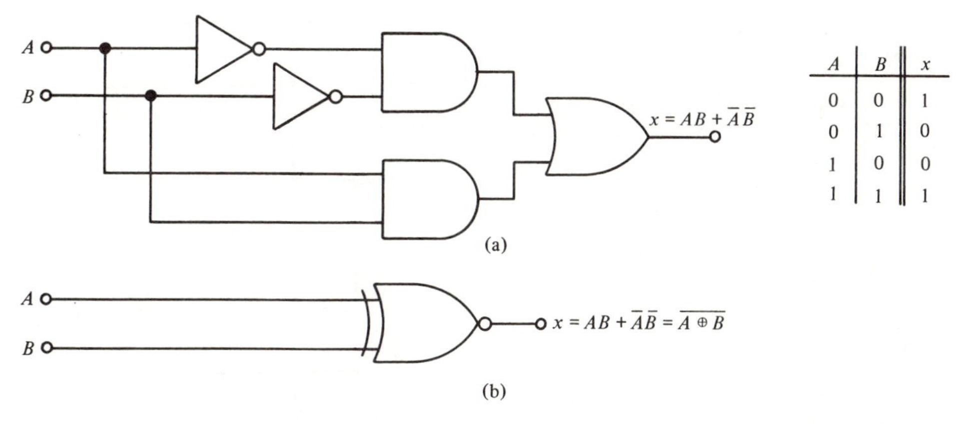

Figure 7–37 **(a) EXCLUSIVE-NOR circuit (b) EXCLUSIVE-NOR gate symbol**

Example 7.17 An EX-NOR circuit is often used as a *digital comparator* circuit, which detects whether two digital signals are exactly coincident or not. For example, Figure 7–38 shows two logic signals, A and B, applied to an EX-NOR circuit. The EX-NOR output, x, will remain HIGH as long as the A and B signals are exactly the same, but will go LOW whenever A and B become different.

The EX-NOR is the digital counterpart of the op-amp comparator. The EX-NOR compares the *logic levels* of the two inputs, while the op-amp comparator compares the actual voltages of the two inputs.

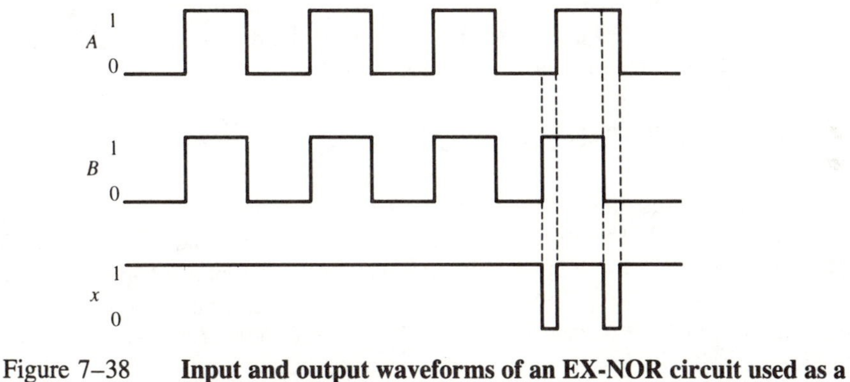

Figure 7–38 **Input and output waveforms of an EX-NOR circuit used as a digital comparator**

Example 7.18 An EX-OR circuit is often used as a *controlled inverter,* whereby one of its inputs is used to control whether a logic signal at the other input is inverted or not. This strategy is illustrated in Figure 7–39(a), where input A is a time-varying logic signal, and input B is the CONTROL input.

Consider first the case where the CONTROL input is kept at the 0 level, as shown in Figure 7–39(b). The output x is determined by using the EX-OR truth table or by recalling that the EX-OR output will be HIGH only when the two inputs are different. With $B = 0$, x will go HIGH only when A is HIGH. The resulting x waveform is clearly the same as the A input; that is, $x = A$.

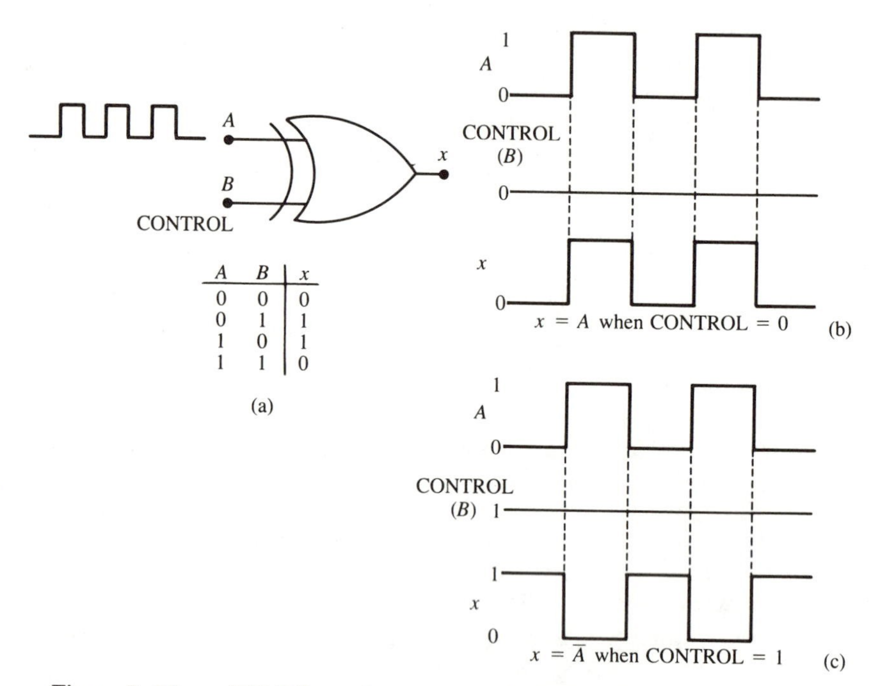

Figure 7–39 **EX-OR used as a controlled inverter**

Now consider what happens when the CONTROL input is kept at the 1 level. Since $B = 1$, x will go HIGH only when A is LOW. The resulting output waveform is clearly the inverse of A; that is, $x = \overline{A}$.

Thus, the EX-OR output will be the same as the A signal when CONTROL = 0 and will be the inverse of A when CONTROL = 1.

7.17 Universality of NAND Gates and NOR Gates

All Boolean expressions consist of various combinations of the basic operations OR, AND, and INVERT. Therefore, any expression can be implemented using OR gates, AND gates, and INVERTERs. It is possible, however, to implement any logic expression using *only* NAND gates, because NAND gates, in the proper combination, can be used to perform each of the Boolean operations OR, AND, and INVERT. This is demonstrated in Figure 7–40.

First, in part (a) of the figure we have a two-input NAND gate whose inputs are purposely connected together, so the variable A is applied to both. In this configuration, the NAND simply acts as an INVERTER since its output is $x = \overline{A \cdot A} = \overline{A}$.

In part (b) of the figure we have two NAND gates connected so that the AND operation is performed. NAND gate 2 is used as an INVERTER to change $\overline{AB}$ to $\overline{\overline{AB}} = AB$, which is the desired AND function.

The OR operation can be implemented using NAND gates connected as in part (c). Here, NAND gates 1 and 2 are used as INVERTERs to invert the inputs, so the final output is $x = \overline{\overline{A} \cdot \overline{B}}$, which can be simplified to $x = A + B$ using De Morgan's theorem [Theorem (17)].

In a similar manner, it can be shown that NOR gates can be arranged to implement any of the Boolean operations, as illustrated in Figure 7–41. Part (a) of the figure shows

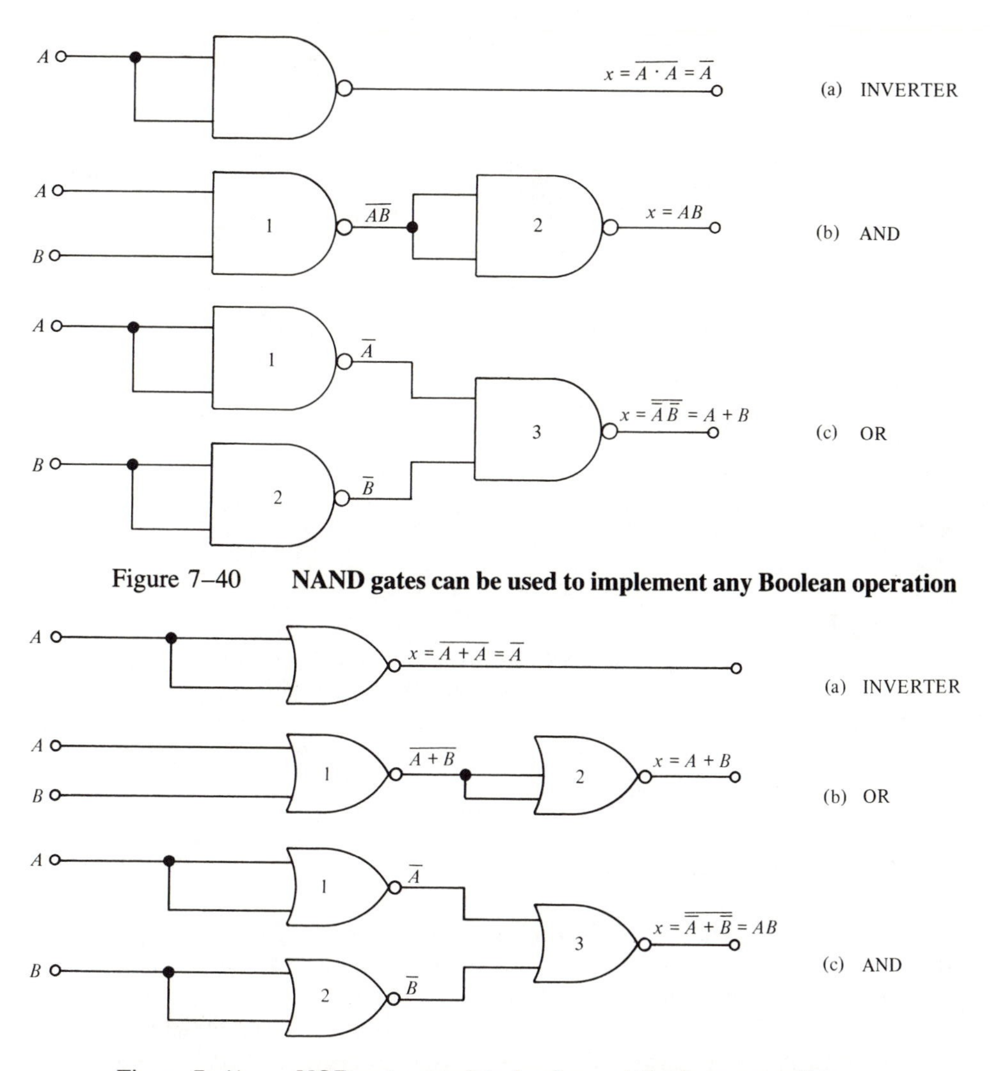

Figure 7–40 **NAND gates can be used to implement any Boolean operation**

Figure 7–41 **NOR gates used to implement Boolean operations**

that a NOR gate with its inputs connected together behaves as an INVERTER since the output is $x = \overline{A + A} = \overline{A}$.

In part (b) two NOR gates are arranged so that the OR operation is performed. NOR gate 2 is used as an INVERTER to change $\overline{A + B}$ to $\overline{\overline{A + B}} = A + B$, which is the desired OR function.

The AND operation can be implemented with NOR gates as shown in part (c) of the figure. Here, NOR gates 1 and 2 are used as INVERTERs to invert the inputs, so the final output is $x = \overline{\overline{A} + \overline{B}}$, which can be simplified to $x = A \cdot B$ using De Morgan's theorem [Theorem (16)].

Because NAND gates can be used to implement *all* of the Boolean operations, any logic circuit, no matter how complex, can be implemented using *only* NAND gates. The same is true for NOR gates. This characteristic of NAND and NOR gates is important. For example, it means that a company designing and producing a piece of digital equipment needs to stock only one type of logic gate, which can greatly reduce its costs.

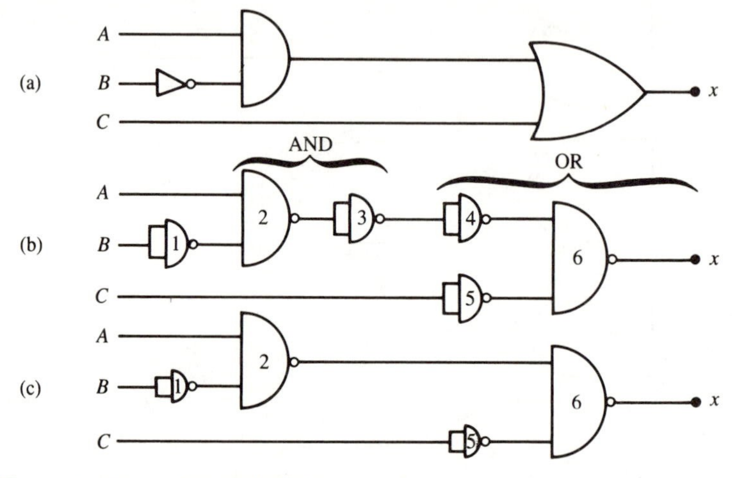

Figure 7–42 **Example 7.19**

Example 7.19 Convert the circuit of Figure 7–42(a) to one that uses only NAND gates.

Solution: Replace each gate by its equivalent NAND-gate implementation (from Figure 7–40). Figure 7–42(b) shows the result. Note that NAND gates 3 and 4 form a double inverter that will have no effect on the logic level coming from the output of NAND gate 2 into the input of NAND gate 6. Thus, we can delete NAND gates 3 and 4 to produce the final result shown in Figure 7–42(c).

7.18 Equivalent Logic Representations

As we have seen, there is more than one way to implement a given logic operation. To illustrate, Figure 7–43(a) is a standard NAND gate, while Figure 7–43(b) shows the same NAND operation implemented as an OR gate with inverted inputs. Both of these circuits provide the NAND operation and are therefore equivalent. Either representation can be used on logic circuit schematics; the choice depends on how the NAND operation is to be interpreted.

The NAND operation can be stated as:

(a) The output goes LOW only when *all* inputs are HIGH.

Or as

(b) The output goes HIGH when any input is LOW.

Obviously, these two statements are simply two ways of saying the same thing. When deciding which to use for a particular application, we must determine the *normal* output state and the *active* output state for that situation.

To illustrate, suppose the output is normally resting in the HIGH state and is activated to the LOW state only at certain times. For this situation, the representation of Figure 7–43(a) should be used. Stated another way, when the LOW output state is the active state that causes other events (e.g., triggers another circuit, turns on a device), then the standard NAND symbol should be used. The small circle on the output tells us that the LOW state is the active state; i.e., the output is active-LOW.

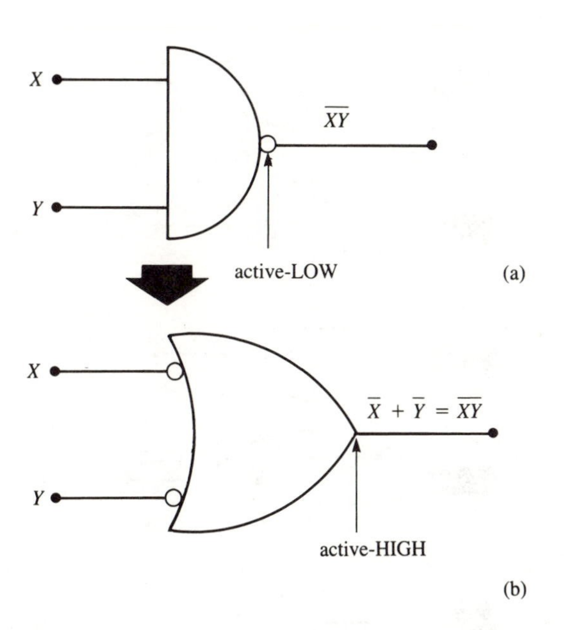

Interpretation:
Output goes LOW only when all inputs are HIGH.

Interpretation:
Output goes HIGH when any input is LOW.

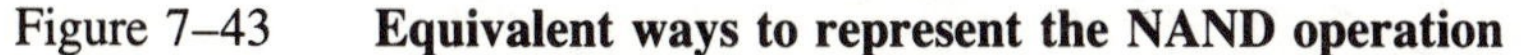

Figure 7–43 **Equivalent ways to represent the NAND operation**

On the other hand, suppose the output is normally resting in the LOW state and is activated to the HIGH state only at certain times. The representation of Figure 7–43(b) should be used for this situation. In other words, when the HIGH output state is the active state that causes other events, the NAND operation should be represented by the OR symbol with inverted inputs as in Figure 7–43(b). The absence of a circle on the output indicates that the output is active-HIGH.

This same idea can be applied to the other logic operations, such as the OR operation (Figure 7–44). Here again both representations are equivalent to the OR operation. The representation in (a) should be used in applications where the active output state is the HIGH state. The representation in (b) is used where the output is active-LOW.

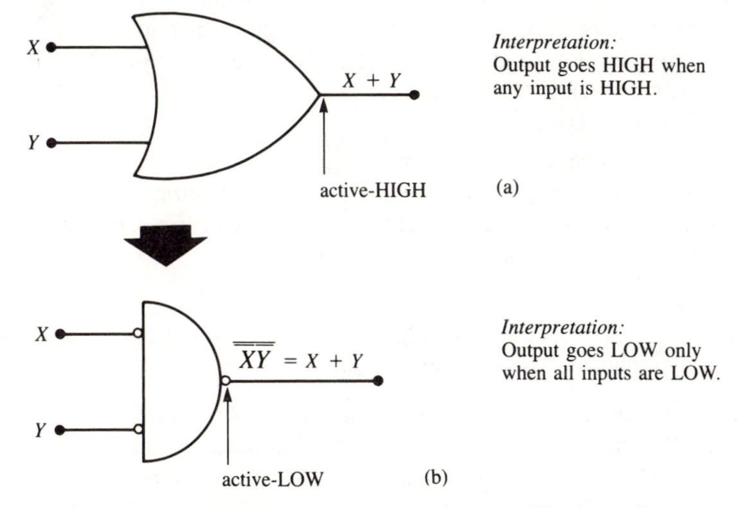

Figure 7–44 **Equivalent ways to represent the OR operation**

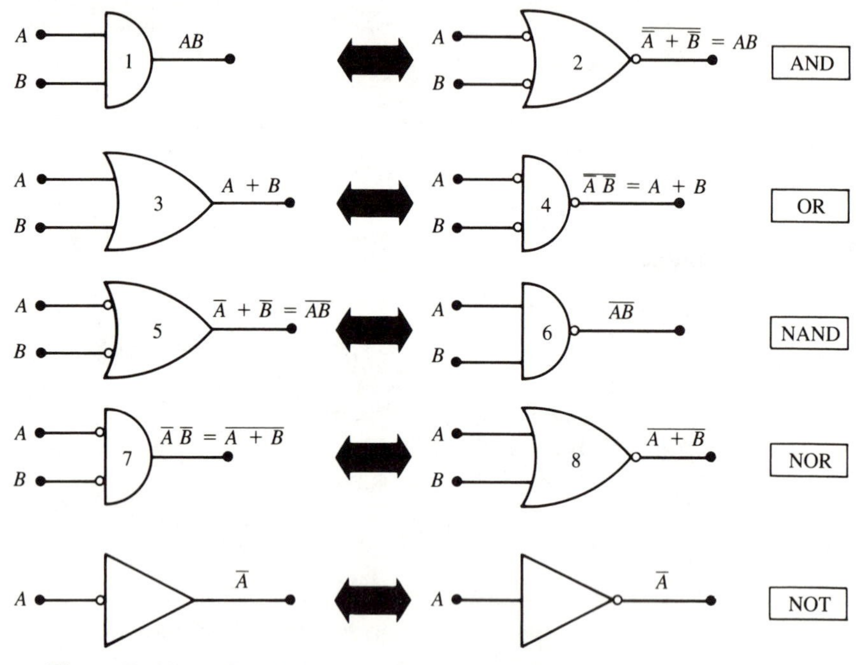

Figure 7–45 **Equivalent gate representations**

Figure 7–45 shows the equivalent logic representations for each of the basic logic gates. The representations on the left have active-HIGH outputs, and those on the right have active-LOW outputs. Again, the choice as to which representation should be used in a schematic is determined by how the output is being used (i.e., its active state).

Figure 7–45 points out another important characteristic of the equivalent logic representations. Gates 1, 4, 6, and 7 all use AND logic, which means that *all* inputs have to be at their active levels to produce the output active level. For example, the operation of gate 7 can be interpreted as "The output goes HIGH only if *all* inputs are LOW." The output is active HIGH (no circle) and the inputs are active LOW (circles).

Similarly, gates 2, 3, 5, and 8 all use OR logic, which means that *any* input has to be at its active level to produce an active output level. For example, the operation of gate 2 can be interpreted as "Output goes LOW when *any* input is LOW." The output and inputs are all active LOW.

Use of this convention is helpful in troubleshooting or testing logic circuitry because it shows which output level of a given circuit is significant; that is, which output level will cause something to happen in the circuits being driven. This information makes it easier to follow a complicated logic schematic.

Example 7.20 For the circuit of Figure 7–46, determine the input conditions necessary to cause output z to go to its active state.

Solution: Output z is active-LOW, and it will go LOW when either x or y is LOW. x will go LOW when both A and B are HIGH. Likewise, y will go LOW when both C and D are HIGH. Thus we can say that z will go LOW when both A and B are HIGH, or when both C and D are HIGH.

Note that gate 3 is the active-LOW equivalent of an AND gate.

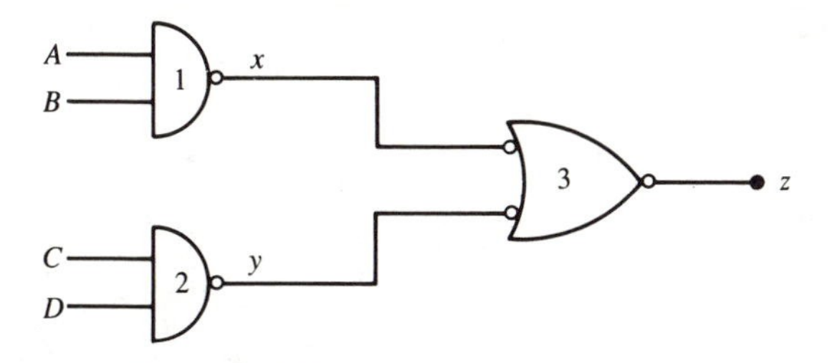

Figure 7–46 **Example 7.20**

Questions

Sections 7.1–7.4

7.1 In the circuit of Figure 7–47, the inputs v_1–v_3 can be either 0 V or +10 V. The diodes are silicon. Determine the truth table for the circuit. What kind of circuit is it?

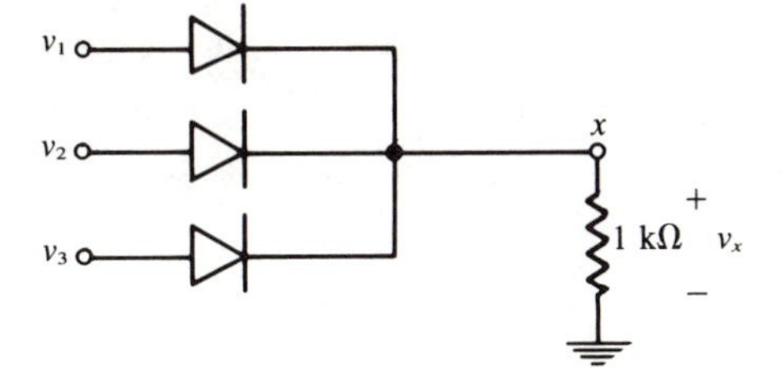

Figure 7–47

7.2 In a certain manufacturing process, ten detector circuits are used to detect various process malfunctions such as over-temperature, oversized parts, etc. The detector circuits, which use photocells and thermistors, each put out a 100 ms-wide pulse of +10 V amplitude when the particular malfunction occurs. Design a simple circuit that will take the ten detector outputs and will produce a single output pulse to energize a *latching relay* whenever *any* of the detector circuits produces a pulse. [A latching relay is a relay that will energize and stay energized (latch) upon receipt of an input pulse until it is mechanically unlatched. The relay is used to sound an alarm or shut off the process.]

7.3 In the circuit of Figure 7–48 the inputs v_1, v_2, and v_3 can be either 0 V or +12 V. Under what input conditions will the output be HIGH? What is the maximum current rating that each diode should have?

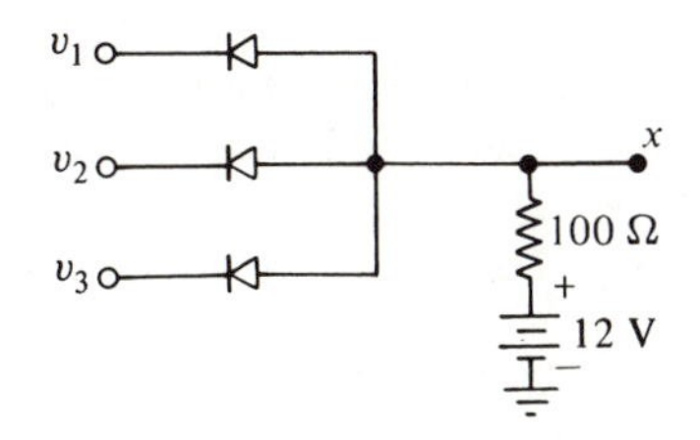

Figure 7–48

7.4 Consider the circuit of Figure 7–49. The diodes are all silicon. The inputs v_1, v_2, and v_3 are logic inputs that can be either 0 V or 6 V. Determine what various input conditions will produce a HIGH level at v_{OUT}.

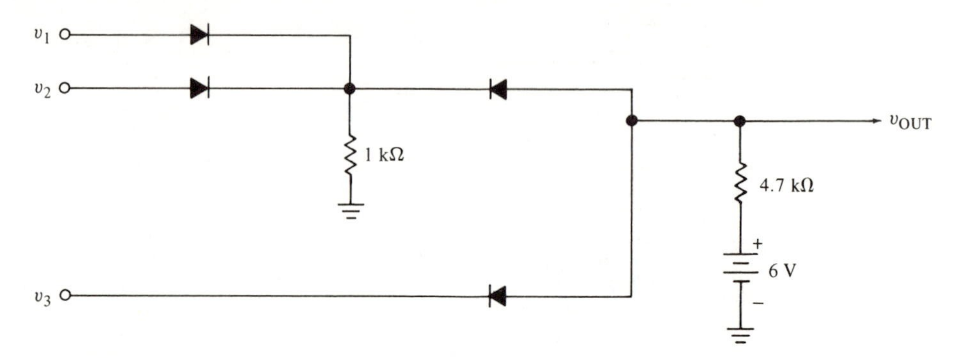

Figure 7–49

7.5 Determine v_{OUT} in Figure 7–49 for each of the eight possible input conditions. One set of input voltages will produce an output that does not fall into either the LOW range (0–1 V) or the HIGH range (5–6 V).

7.6 Apply the v_1 and v_2 waveforms of Figure 7–50 to the circuit of Figure 7–11 and determine the output waveform. Repeat for the circuit of Figure 7–12.

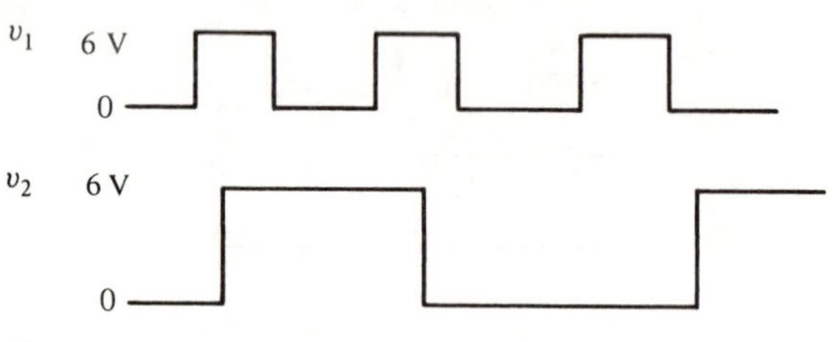

Figure 7–50

Sections 7.5–7.8

7.7 In Section 7.6 we learned that if a digital circuit has N logic input variables, then there will be 2^N different possible cases. There is a methodical way to construct a truth table that contains all the cases. Consider the truth table for the four variables M, N, O, and P shown in Figure 7–51. Since $N = 4$, there are $2^4 = 16$ cases. All sixteen cases can be listed if we momentarily let the variables M, N, O, P represent a 4-bit binary number with P the LSB and M the MSB. If we now simply count in binary from 0000 to 1111 as described in Appendix A, we will have all sixteen cases. The same procedure is followed for any number of inputs. Fill in the outputs for each case in the truth table if it represents a four-input OR gate.

7.8 What is the only condition in which an OR-gate output will be 0?

7.9 Determine the waveform at the output of the OR gate in Figure 7–52.

7.10 Write the Boolean expression for a six-input OR gate.

7.11 Under what conditions will the output of an AND gate be 0?

7.12 Write the expression for a four-input AND gate. Construct the complete truth table, showing the output for all possible cases.

7.13 Change the OR gate in Figure 7–52 to an AND gate, and determine the output waveform.

7.14 Change the OR gate in Figure 7–16 to an AND gate; determine the output.

7.15 Refer to Example 7.4. Redesign the circuit so that the alarm is activated only if the temperature and pressure both exceed their maximum limits.

Sections 7.9–7.11

7.16 Write the Boolean expression for the output (x) in Figure 7–53(a). Determine the value of x for all possible input conditions, and list them in a truth table.

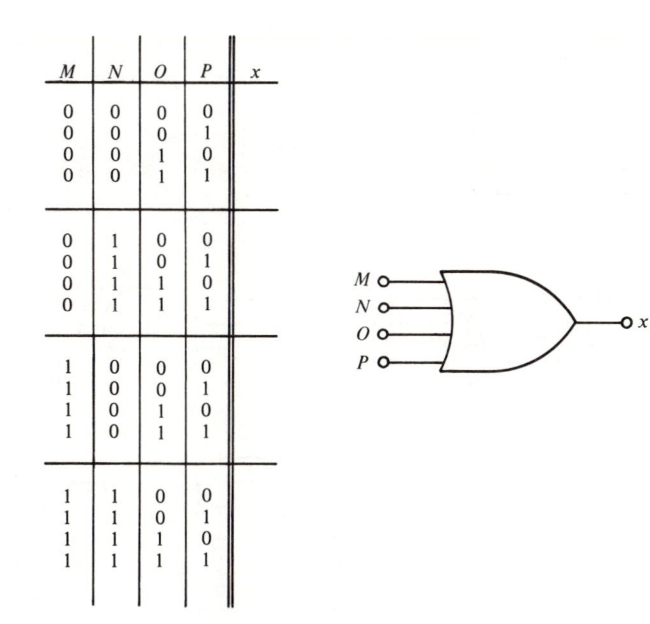

M	N	O	P	x
0	0	0	0	
0	0	0	1	
0	0	1	0	
0	0	1	1	
0	1	0	0	
0	1	0	1	
0	1	1	0	
0	1	1	1	
1	0	0	0	
1	0	0	1	
1	0	1	0	
1	0	1	1	
1	1	0	0	
1	1	0	1	
1	1	1	0	
1	1	1	1	

Figure 7–51

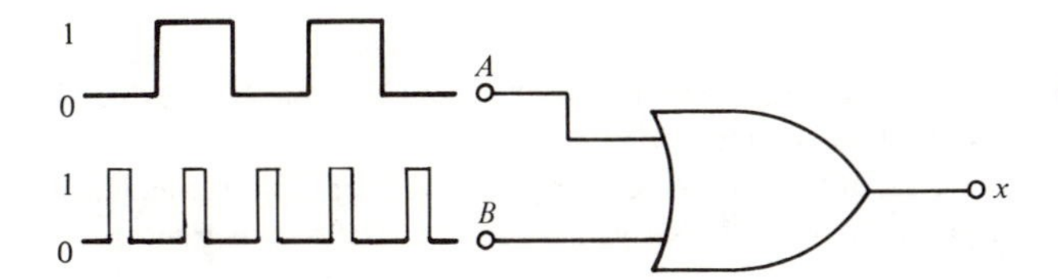

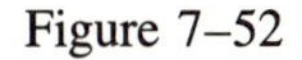

Figure 7–52

7.17 Repeat Question 7.16 for the circuit in Figure 7–53(b).

7.18 Determine the output level of the circuit in Figure 7–25(b) for the following cases:

(a) $A = B = C = 1, D = 0, E = 1$

(b) $A = B = 0, C = 1, D = E = 0$

7.19 For each of the following expressions, construct the corresponding logic circuit using AND and OR gates and INVERTERs:

(a) $x = \overline{A\overline{B}(C + D)}$

(b) $z = \overline{(A + B + \overline{C}D\overline{E})} + \overline{B}C\overline{D}$

(c) $y = \overline{(M + N + \overline{P}Q)}$

Section 7.12

7.20 Change the OR gate in 7–16 to a NOR gate; determine the output.

7.21 Construct a truth table for a three-input NOR gate.

7.22 Under what conditions is the output of a NOR gate HIGH?

7.23 Construct a truth table for a three-input NAND gate.

7.24 Change the OR gate in 7–16 to a NAND gate; determine the output.

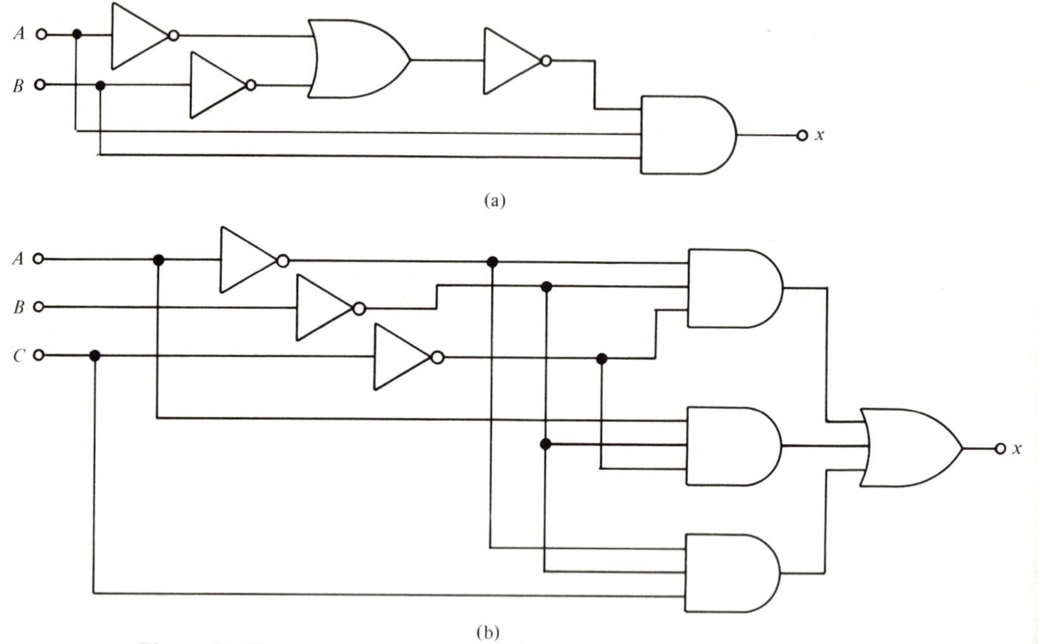

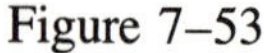
Figure 7–53

7.25 Modify the circuits that were constructed for Question 7.19 so that NAND and NOR gates are used where appropriate.

7.26 Consider each of the following statements, and for each one indicate for which logic gate or gates (AND, OR, NAND, NOR) the statement is true:

(a) Output is HIGH *only* if all inputs are LOW.

(b) Output is LOW if the inputs are at different levels.

(c) Output is HIGH when both inputs are HIGH.

(d) Output is LOW *only* if all inputs are HIGH.

(e) All LOW inputs produce a HIGH output.

7.27 In each of the circuits of Figure 7–53, change each OR to a NOR, and each AND to a NAND. Then determine the expression and truth table for the output of each circuit.

Sections 7.13–7.15

7.28 Prove Theorem (15) by trying all possible values for x and y.

7.29 Prove De Morgan's theorems by trying all values for x and y.

7.30 Simplify the following terms using De Morgan's theorems:

(a) $\overline{\overline{A}B\overline{C}}$

(b) $\overline{A + B\overline{C}}$

(c) $\overline{AB\overline{CD}}$

(d) $\overline{\overline{AB} + C\overline{D} + AC}$

7.31 Use Boolean algebra rules and theorems to simplify the circuits of Figure 7–53. Then prove that your simplified circuits are logically equivalent to the original ones by determining the truth tables and comparing to the results of Problems 7.16 and 7.17.

7.32 Simplify the expressions obtained in Problem 7.27.

7.33 Simplify each of the following expressions:

(a) $x = RST + RS(\overline{\overline{T}} + V)$
(b) $y = \overline{A}\overline{B}\overline{C} + \overline{A}B\overline{C} + \overline{A}\overline{B}C + A\overline{B}C$
(c) $z = (M + N)(\overline{M} + P)(\overline{N} + \overline{P})$
(d) $q = AB(\overline{\overline{B} + C}) + C$
(e) $x = \overline{AB + \overline{A}\overline{B}}$

Section 7.16

7.34 Show how the EX-NOR circuit (by itself) can be used as a controlled inverter.
7.35 Use Boolean algebra to prove that the expression for the EX-NOR circuit ($\overline{A}\overline{B} + AB$) is the inverse of the expression for the EX-OR circuit ($\overline{A}B + A\overline{B}$).
7.36 Determine and simplify the expression for the output in Figure 7–54.

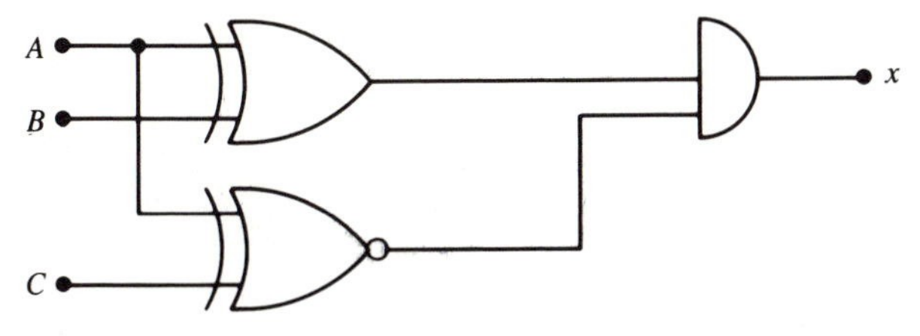

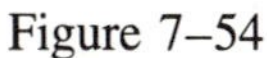
Figure 7–54

Section 7.17

7.37 (a) Construct the logic circuit corresponding to the expression $x = AB + \overline{C}D + E\overline{F}$ using AND gates, OR gates, and INVERTERs.
(b) Replace each AND and OR gate in the circuit of part (a) with its equivalent NAND-gate implementation.
(c) Write the expression for the revised circuit. Simplify it, and compare it with the original expression.

7.38 (a) Construct the logic circuit for $y = (A + \overline{B}) \cdot (C + \overline{D})$ using ANDs, ORs, and INVERTERs.
(b) Replace each AND and OR gate in the circuit of part (a) by its equivalent NOR-gate implementation.
(c) Write the expression for the revised circuit, simplify it, and compare it with the original.

7.39 Prove that the circuit of Figure 7–42(c) is equivalent to that of Figure 7–42(a).
7.40 Show how to implement an EX-OR circuit, using all NAND gates.
7.41 Repeat using all NOR gates.

Section 7.18

7.42 Which AND-gate representation should be used when the output is driving a circuit that is activated by a LOW input level?
7.43 Which NOR-gate representation should be used in the circuit of Figure 7–55?

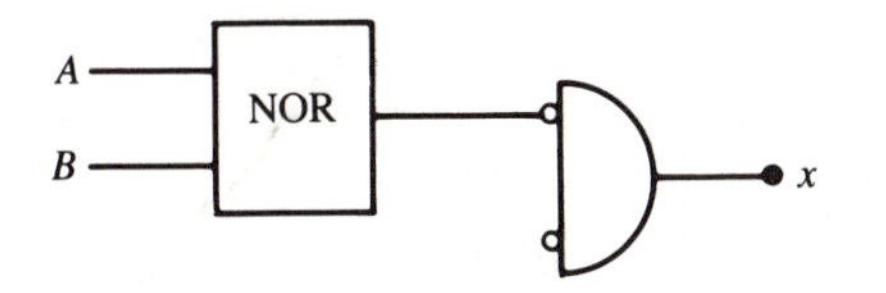

Figure 7–55

7.44 Determine the input conditions necessary to cause the output z to go to its active state in Figure 7–56(a).

7.45 Repeat for the circuit in Figure 7–56(b).

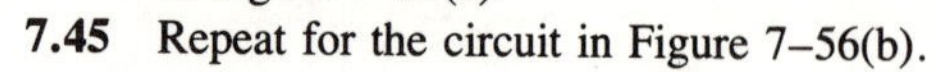

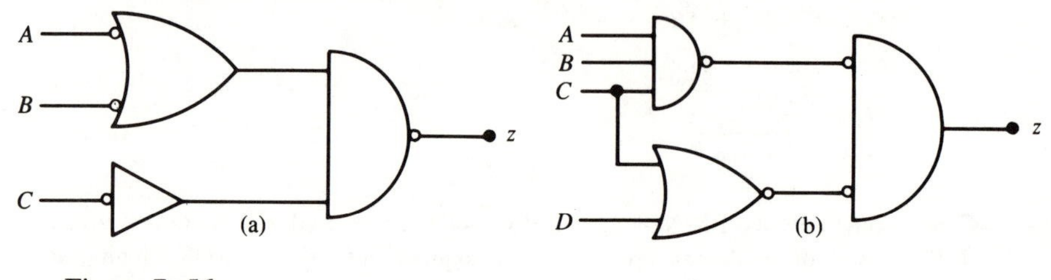

Figure 7–56

8 Flip-Flops

8.1 Introduction

The logic circuits considered thus far have been combinatorial circuits whose output levels at any instant of time are dependent upon the levels present at the inputs at that time.* Any prior input level conditions have no effect on the present outputs because combinatorial logic circuits have no memory. Most digital systems are made up of both combinatorial circuits and memory elements.

The most widely used memory element is the *flip-flop* (FF), which we shall study thoroughly in this chapter. The FF is a logic circuit that has two outputs that are the logical inverse of each other. Figure 8–1 indicates these outputs as Q and $\overline{Q}$ (actually, any letter could be used, though Q is the most common). The Q output is called the *normal* FF output, and $\overline{Q}$ is the *inverted* FF output. When a FF is said to be in the HIGH (1) state or the LOW (0) state, this is the condition at the Q output. Of course, the $\overline{Q}$ output is always the inverse of Q.

There are only two possible operating states for the FF: $Q = 0$, $\overline{Q} = 1$; and $Q = 1$; $\overline{Q} = 0$. The FF has one or more inputs which are used to cause the FF to switch back and forth between these two states. As we shall see, once an input signal causes a FF to go to a given state, the FF will remain in that state even after that input signal is terminated. This is its *memory* characteristic.

**Combinatorial* refers to combinations of logic gates with no feedback.

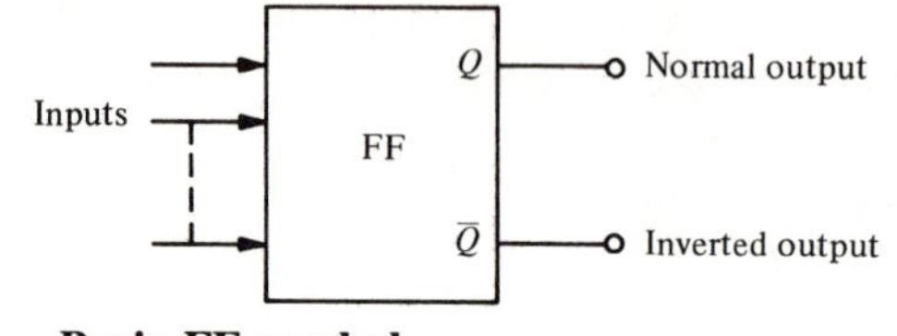

Figure 8–1 **Basic FF symbol**

Incidentally, the flip-flop is known by several other names, including *bistable multivibrator, latch,* and *binary.* We will usually use the term *flip-flop* since it is the most frequently used designation in the digital field.

Other memory elements are used in digital systems. However, FFs are the most versatile of the memory elements because of their high speed of operation, the ease with which information can be stored into and read out of them, and the ease with which they can be interconnected with logic gates.

Flip-flops can be constructed from discrete components. However, the availability of all types of FFs in integrated circuit form has virtually eliminated the need for design and analysis of discrete FF circuits. Digital IC manufacturers offer a wide variety of FFs compatible with logic gates so that FFs and gates can be easily interconnected to perform more complex logic functions.

We will begin our study by analyzing two simple FF circuits constructed from *cross-coupled* pairs of inverting logic gates. These simple FFs form the basis for the complex ones.

8.2 The NOR-Gate FF

Figure 8–2 shows two cross-coupled NOR gates where the output of NOR-1 serves as an input to NOR-2, and vice versa. The other gate inputs are labelled "SET" and "CLEAR," respectively. The gate outputs are the FFs outputs Q and $\overline{Q}$. These outputs (under normal circumstances) will always be the inverse of each other.

The SET and CLEAR inputs are *normally LOW* inputs; that is, they are normally resting in the 0 state, and they have to be pulsed to the 1 state to perform their function. In other words, the SET and CLEAR inputs are active-HIGH inputs. The truth table in Figure 8–2 summarizes the input/output relationship. We will examine each case separately.

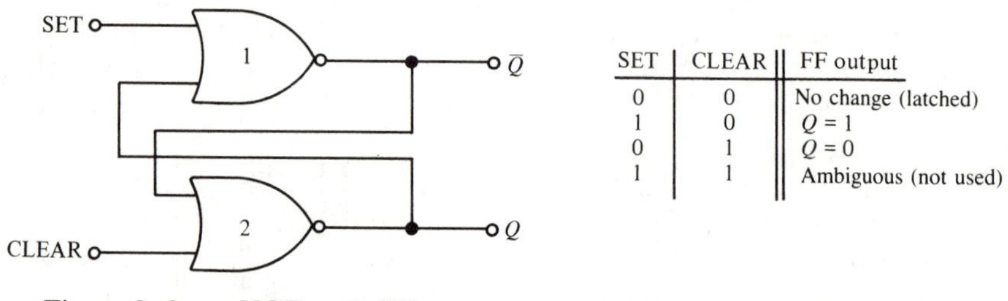

SET	CLEAR	FF output
0	0	No change (latched)
1	0	$Q = 1$
0	1	$Q = 0$
1	1	Ambiguous (not used)

Figure 8–2 **NOR-gate FF**

SET = 0/CLEAR = 0 As stated previously, SET = 0/CLEAR = 0 is the normal inactive state to which the inputs are always returned. Whenever this condition is applied to the inputs, the FF output Q does not change. (We say that the output is *latched* in its current state.) In other words, Q will stay at the level it had prior to the occurrence of the SET = CLEAR = 0 condition. Of course, the same is true of the $\overline{Q}$ output.

SET = 1/CLEAR = 0 This activates the SET input. The 1 at the SET input causes the NOR-1 output to go to 0 ($\overline{Q} = 0$). This 0 is applied to the top input of NOR-2, so that NOR-2 has 0's at both its inputs, which produces $Q = 1$. Thus, the SET = 1/CLEAR = 0 condition always causes Q to go HIGH. This is called *setting* the FF.

In practice, the SET input would be momentarily pulsed to the 1 state and then returned to the 0 state. The FF outputs, however, will remain latched in the $Q = 1/\overline{Q} = 0$ state because the 1 from Q will keep $\overline{Q} = 0$, and the 0's from $\overline{Q}$ and the CLEAR input will, in turn, keep $Q = 1$.

SET = 0/CLEAR = 1 This activates the CLEAR input. The 1 at the CLEAR input causes the NOR-2 output to go to 0 ($Q = 0$). This 0 is applied to the bottom input of NOR-1, so that NOR-1 has 0's at both its inputs, which produces $\overline{Q} = 1$. Thus, SET = 0/CLEAR = 1 always causes Q to go LOW. This is called *clearing* the FF.

If the CLEAR input is momentarily pulsed to the 1 state and then returned to the 0 state, the FF output will remain latched in the $Q = 0/\overline{Q} = 1$ state. The 1 from $\overline{Q}$ will keep $Q = 0$, and the 0's from Q and the SET input will, in turn, keep $\overline{Q} = 1$.

SET = 1/CLEAR = 1 This condition will produce a LOW at each NOR output so that $Q = 0$ and $\overline{Q} = 0$. This state is undesirable because Q and $\overline{Q}$ are supposed to be inverses of each other. Furthermore, when the inputs are simultaneously returned to the SET = CLEAR = 0 condition, it is impossible to predict in which state the outputs will end up. For these reasons, the SET = CLEAR = 1 condition should never be used.

Simplified symbol Figure 8–3 shows the simplified logic symbol we will often use to represent the NOR-gate FF. The S and C, of course, represent the SET and CLEAR inputs of Figure 8–2. As we have seen, these are active-HIGH inputs.

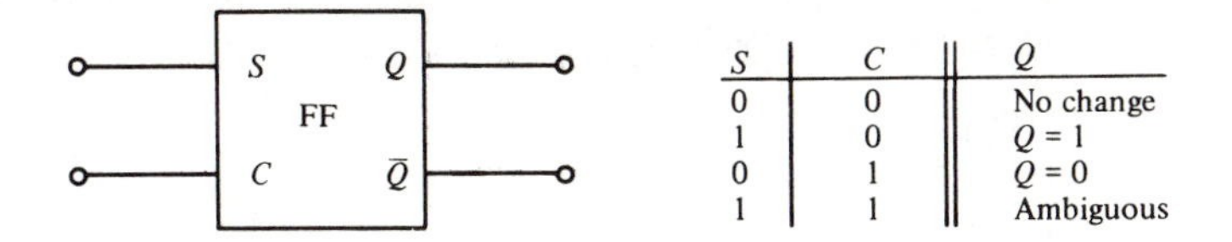

S	C	Q
0	0	No change
1	0	$Q = 1$
0	1	$Q = 0$
1	1	Ambiguous

Figure 8–3 **Symbol for NOR-gate *SC* flip-flop**

Example 8.1 The waveforms of Figure 8–4 are applied to the inputs of a NOR-gate *SC* flip-flop. Determine the Q output waveform. Assume Q is LOW prior to $t = t_1$.

Solution: Q will stay at 0 until the SET input goes HIGH at t_1. At that time the FF will be set to the $Q = 1$ state. It will remain in this state when SET returns LOW at t_2.

At t_3 the CLEAR input goes HIGH, clearing Q to the 0 state. It will remain in this state when CLEAR returns LOW at t_4.

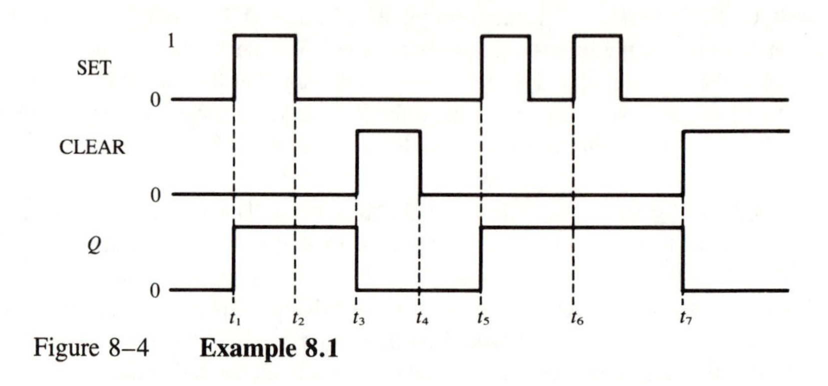

Figure 8–4 **Example 8.1**

At t_5 the SET input will set $Q = 1$, where it will remain until t_7, when the HIGH at CLEAR will make $Q = 0$. Note that the SET pulse at t_6 has no effect since the FF is already in the 1 state.

Of course, the $\overline{Q}$ output waveform would be the exact inverse of the Q waveform.

Example 8.2 The *SC* FF is useful as a memory element to store information. Figure 8–5 shows a simple burglar alarm circuit using the *SC* FF. (Remember, the FF symbol represents the cross-coupled NOR-gate circuit of Figure 8–2.)

Assume that light is initially focused on the photocell and describe the circuit operation.

Solution: With light on the photocell, transistor Q_1 will be ON so that $V_x \approx 0$ V, which makes $S = 0$. If the C input is momentarily switched to +5 V, this will guarantee that $Q = 0$, thereby keeping transistor Q_2 OFF and the alarm de-activated.

When the light on the photocell is interrupted, its resistance becomes very large. This large resistance will turn Q_1 OFF, producing $V_x = 5$ V. This HIGH at the FF *S* input will set $Q = 1$. This turns ON Q_2 and activates the alarm.

Once the FF has been set to the $Q = 1$ state, it will remain there even if the photocell goes light again and brings *S* back to 0. Thus, the alarm will persist until the FF is manually cleared back to $Q = 0$ by momentarily switching *C* to +5 V.

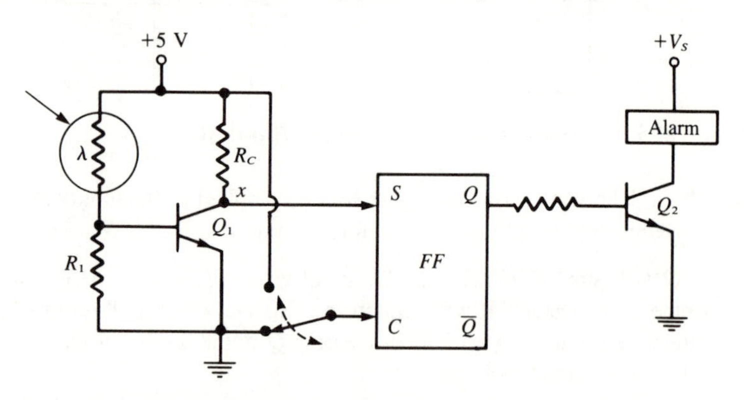

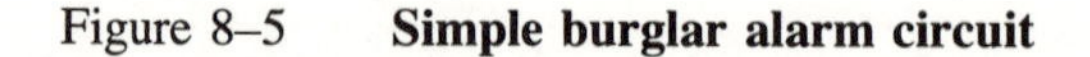
Figure 8–5 **Simple burglar alarm circuit**

8.3 The NAND-Gate FF

Figure 8–6(a) shows how two cross-coupled NAND gates can be used as a *SC* flip-flop. The arrangement is similar to the NOR-gate FF, except that the Q and $\overline{Q}$ outputs are reversed.

The SET and CLEAR inputs of the NAND-gate FF are normally HIGH inputs; that is, they are normally resting in the 1 state and have to be pulsed to the 0 state to perform their function. In other words, they are active-LOW inputs. The truth table in Figure 8–6 shows how these inputs affect the FF output.

SET = CLEAR = 1 In this normal resting state, the FF outputs will not change; that is, they are latched in their current state.

SET = 0/CLEAR = 1 This activates the SET input and sets Q to the 1 state, where it will stay even when SET is returned to the 1 state.

SET = 1/CLEAR = 0 This activates the CLEAR input and clears Q to the 0 state, where it will stay even when CLEAR is returned to the 1 state.

SET = 0/CLEAR = 0 This produces a 1 at both Q and $\overline{Q}$ and is therefore an undesirable condition. Furthermore, it is impossible to predict in which state the outputs will end up when SET and CLEAR are simultaneously returned to the 1 level. Thus, the SET = CLEAR = 0 condition is *never* used for this FF.

Comparison with NOR-gate FF The NAND-gate FF operates basically the same as the NOR-gate FF except for one major difference: the NOR FF inputs are normally 0 and must be pulsed to the 1 level to change the state of the output, while the NAND FF inputs are normally 1 and must be pulsed to the 0 level to change the state of the output.

Since the NAND FF has active-LOW inputs, the representation of Figure 8–6(b) is often used. Here each NAND gate has been replaced by its equivalent OR-gate representation.

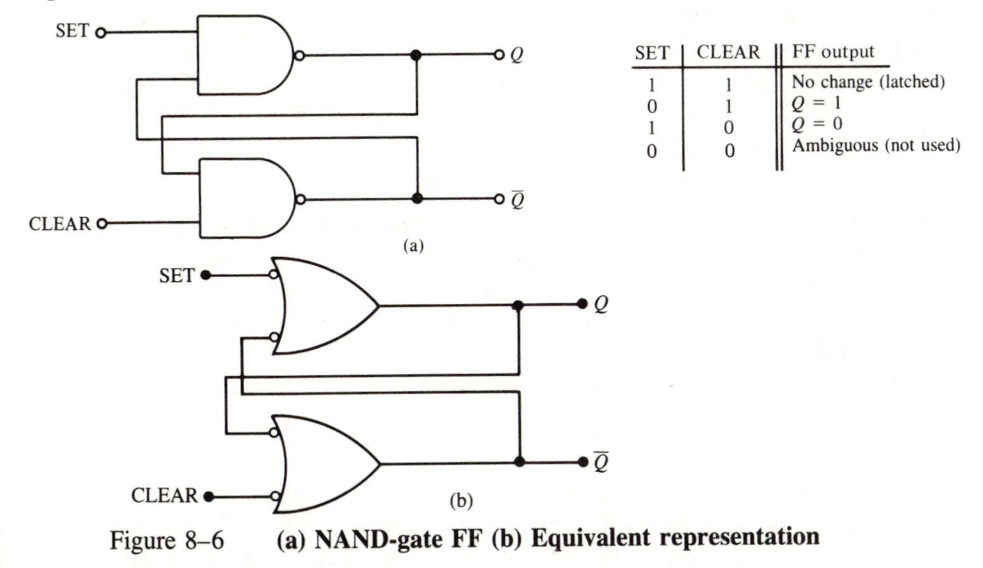

SET	CLEAR	FF output
1	1	No change (latched)
0	1	$Q = 1$
1	0	$Q = 0$
0	0	Ambiguous (not used)

Figure 8–6 **(a) NAND-gate FF (b) Equivalent representation**

Simplified symbol Figure 8–7 shows the simplified logic symbol we will often use to represent the NAND-gate FF. Note the circles on the S and C inputs to indicate that they are both active-LOW.

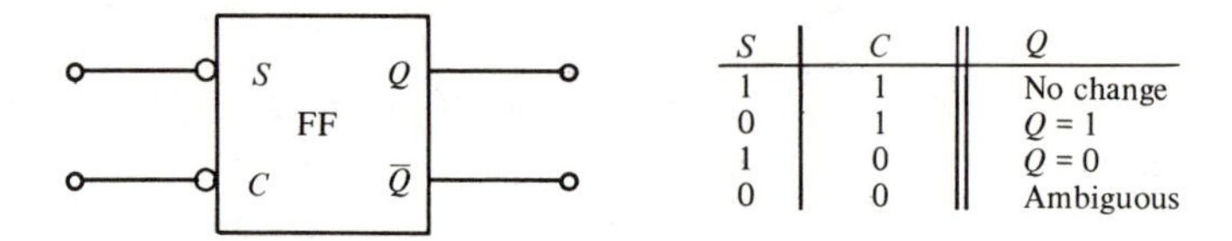

S	C	Q
1	1	No change
0	1	$Q = 1$
1	0	$Q = 0$
0	0	Ambiguous

Figure 8–7 **Symbol for NAND-gate FF**

Example 8.3 The waveforms of Figure 8–8 are applied to the inputs of a NAND-gate FF. Assume $Q = 0$ initially and determine the Q waveform.

Solution: The output waveform can be obtained by realizing that Q will go HIGH whenever SET goes LOW, and Q will stay HIGH until the CLEAR input goes LOW.

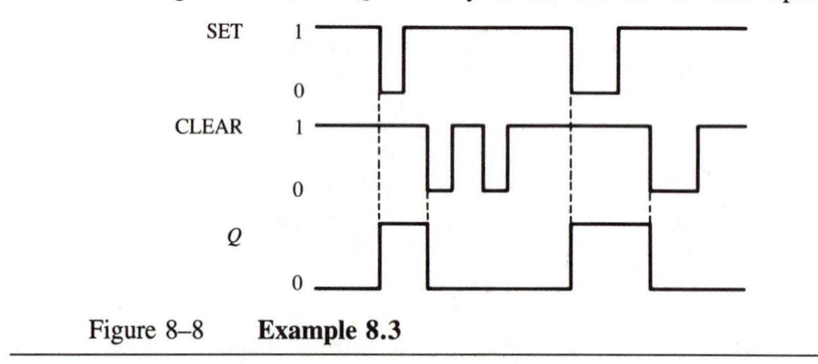

Figure 8–8 **Example 8.3**

Example 8.4 When a mechanical switch is closed, the metal contacts will alternately collide and bounce apart several times before they finally settle after a few milliseconds in the closed position. This phenomenon is called *contact bounce*. Because of contact bounce, a mechanical switch will not be able to produce a single, clean voltage transition when it is actuated; instead, multiple transitions occur as the switch bounces open and closed, illustrated in Figure 8–9(a).

A NAND-gate FF can be used to "debounce" a mechanical switch as shown in Figure 8–9(b). The FF will produce a single output transition each time the switch is moved to the opposite position.

Explain the operation of the "debounce" circuit.

Solution: When the switch is in the lower position, the FF is cleared (to the $Q = 0$ state). If the switch is moved to its upper position, the FF will set (to the $Q = 1$ state) and it will stay there even as the switch momentarily bounces away from the S input. Likewise, when the switch is moved to the lower position, the FF will clear (to the $Q = 0$ state) and will remain there as the switch bounces away from the C input. Thus, as the switch is moved from one position to the other, the Q output switches from one logic level to the other without multiple transitions.

Resetting a FF We have talked about *setting* a FF ($Q = 1$) and *clearing* a FF ($Q = 0$). *Resetting* a FF means the same thing as clearing a FF. SET/CLEAR FFs are often referred to as RESET/SET FFs (RS FFs) where the RESET input is the same as the

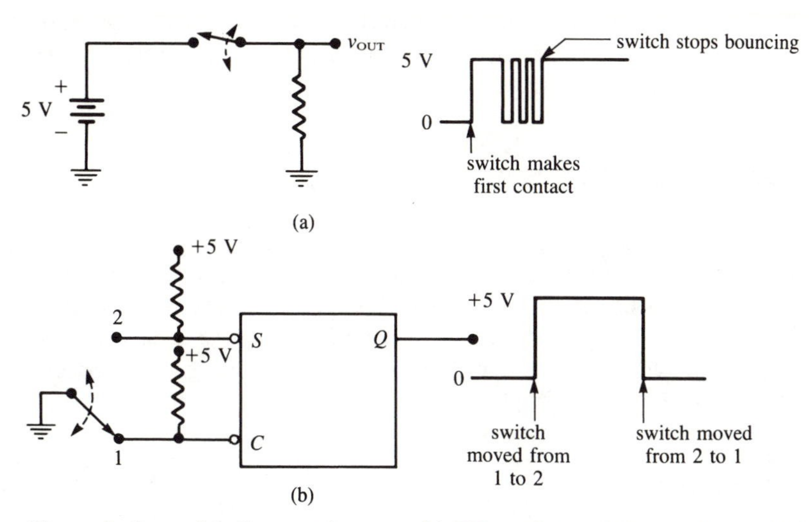

Figure 8–9 **(a) Contact bounce (b) FF used to "debounce" a switch**

CLEAR input. We will use the terms *clear* and *reset* alternatively throughout the remainder of the text since both are in common use in the digital field.

8.4 Clock Signals

Most digital systems operate as *synchronous sequential systems*. What this means is that the sequence of operations that takes place is synchronized by a "master clock" signal that generates periodic pulses which are distributed to all parts of the system. This *clock* signal is usually one of the forms shown in Figure 8–10; most often, it is a square wave (50 percent duty cycle) like that in Figure 8–10(b).

The clock signal is the signal that causes things to happen at regularly spaced time intervals. In particular, operations in the system are made to take place at times when the clock signal is making a transition either from 0 to 1 or from 1 to 0. These transition times are indicated in Figure 8–10. The transition from 0-to-1 is called the *rising edge,* or *positive-going edge,* of the clock signal; the transition from 1-to-0 is called the *falling edge,* or *negative-going edge,* of the clock signal.

The synchronizing action of the clock is obtained by using *clocked FFs,* which are designed to change states on either (but not on both) the rising edge or falling edge of the clock signal. In other words, the clocked FFs will change states at the appropriate clock transition and will rest between successive clock pulses. The maximum frequency of the clock pulses is generally determined by how long it takes the FFs and gates in the circuit

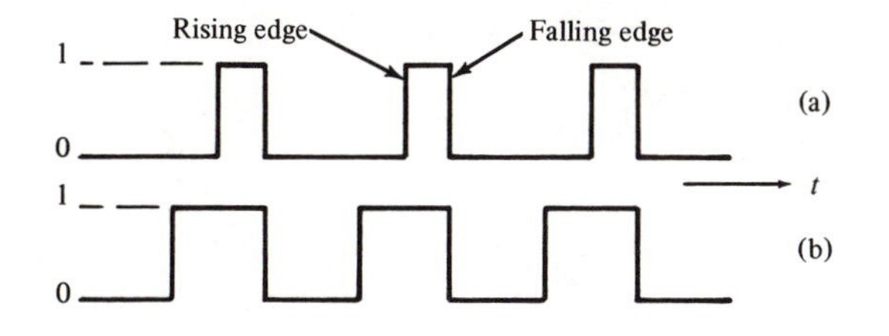

Figure 8–10 **Clock signals**

to respond to the level changes initiated by the clock pulse, that is, the propagation delays of the various logic circuits. In the next section we will begin the study of the various types of clocked FFs that are being used extensively in most digital systems.

8.5 The Clocked SET-CLEAR Flip-Flop

Figure 8–11(a) shows the symbol for a clocked *SC* FF that is triggered by the positive-going edge of the clock signal. This means that the FF will change states *only* when a signal applied to its clock input (abbreviated CLK) makes a transition from 0 to 1. The *S* and *C* inputs control the state of the FF in the manner described earlier for the basic (unclocked) *SC* FF, but the FF does not respond to these inputs until the occurrence of the rising edge of the clock signal. This situation is illustrated by the waveforms in Figure 8–11(b) and can be analyzed as follows:

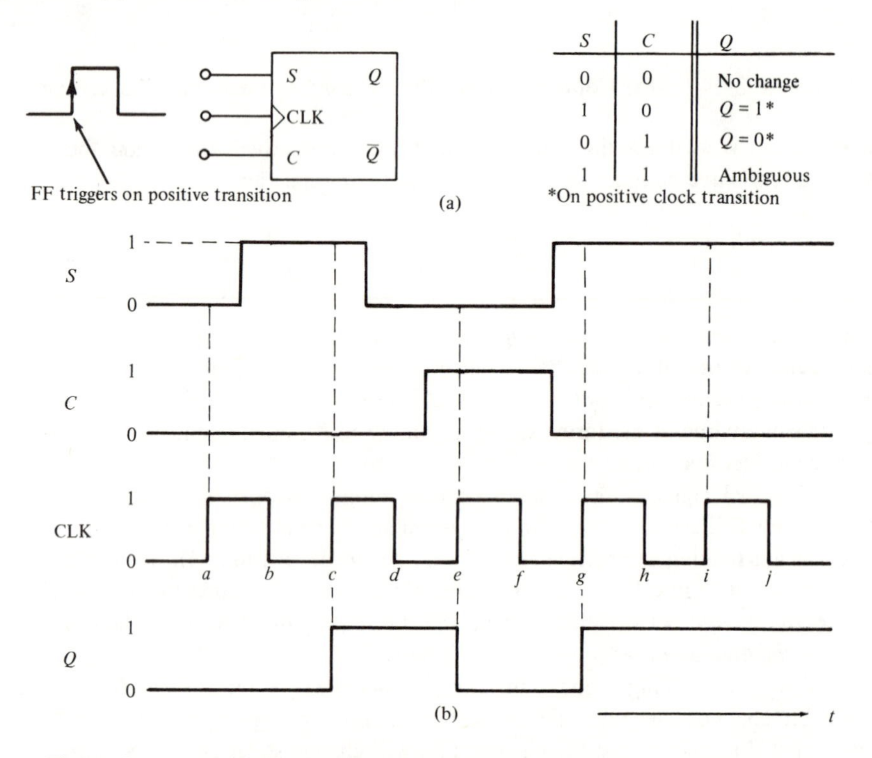

Figure 8–11 **(a) Clocked *SC* FF that responds to the positive-going edge of the clock pulse (b) Waveforms**

1. Initially, all inputs are 0, and the Q output is 0.
2. When the rising edge of the first clock pulse occurs (point *a*), the *S* and *C* inputs are both 0. Therefore, the FF is unaffected and remains in the $Q = 0$ state.
3. At the occurrence of the rising edge of the second clock pulse (point *c*), the *S* input is HIGH with *C* still LOW. Thus, the FF sets to the 1 state at the rising edge of this clock pulse.

4. When the third clock pulse makes its positive transition (point e), it finds $S = 0$ and $C = 1$. Therefore, the FF clears to the 0 state.
5. The fourth pulse sets the FF once again to the $Q = 1$ state because $S = 1$ and $C = 0$ when it occurs.
6. The fifth pulse also finds $S = 1$ and $C = 0$ when it makes its positive-going transition. However, Q is already HIGH, so it remains in that state.
7. The $S = C = 1$ condition should not be used since it results in an ambiguous condition.

Note from these waveforms that this FF is not affected by the negative-going edge of the clock pulses. Also note that the S and C levels have no effect on the FF except upon the occurrence of a positive-going transition of the clock signal. The S and C inputs are essentially *control* inputs that control what state the FF will go to when the clock pulse occurs; the CLK input is the *trigger* input that actually causes the FF to change states according to the levels on the S and C inputs when the clock edge occurs.

The symbol for the edge-triggered SC flip-flop shows a small triangle on the CLK input. This triangle is used to indicate that the CLK input only responds to *transitions* and is not affected by a constant HIGH or LOW level. This convention is not used by all digital circuit manufacturers; but as it is gaining wider acceptance, we will use it throughout the text.

Figure 8–12 shows the symbol for a clocked SC flip-flop that triggers on negative-going transitions at its CLK input. The small circle on the CLK input indicates that this FF will respond when the CLK input goes from 1 to 0. This FF operates in the same manner as the positive-edge FF of Figure 8–11, except that the output can change states *only* on the falling edge of the clock pulses (points b, d, f, h, j), and is not affected by the positive-going transitions.

Both positive-edge and negative-edge triggering FFs are used in digital systems.

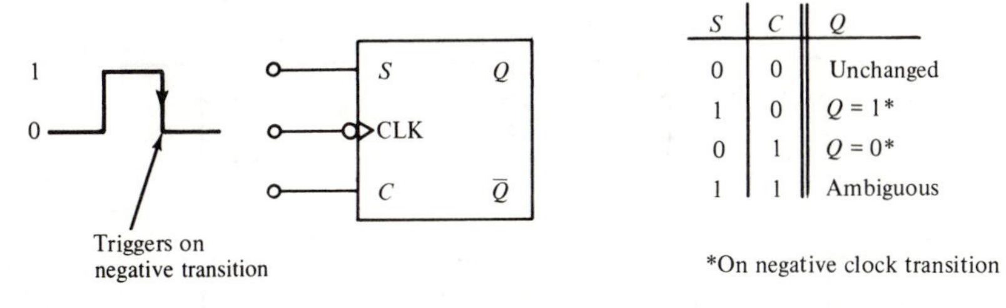

S	C	Q
0	0	Unchanged
1	0	$Q = 1$*
0	1	$Q = 0$*
1	1	Ambiguous

Figure 8–12 **Clocked *SC* FF that triggers on negative-going transitions**

8.6 The Clocked *JK* Flip-Flop

Figure 8–13(a) shows a clocked JK FF that is triggered by the positive-going edge of the clock signal. The J and K inputs control the state of the FF in the same way as the S and C inputs do for the clocked SC FF except for one major difference: *The $J = K = 1$ condition does not result in an ambiguous output*. For this 1-1 condition, the FF will always go to the opposite state upon the positive transition of the clock signal. This is called the *toggle* mode of operation. In this mode, if both J and K are left high, this FF will change states (toggle) on each positive-going clock edge. The operation of this FF is illustrated by the waveforms in Figure 8–13(b) which can be analyzed as follows:

1. Initially, all inputs are 0, and the Q output is 1.
2. When the positive-going edge of the first clock pulse occurs (point a) the $J = 0$, $K = 1$ condition exists. Thus, the FF will reset (clear) to the $Q = 0$ state.
3. The second clock pulse finds $J = K = 1$ when it makes its positive transition (point c). This causes the FF to toggle to its opposite state, $Q = 1$.
4. At point e on the clock waveform J and K are both 0, so the FF does not change states on this transition.

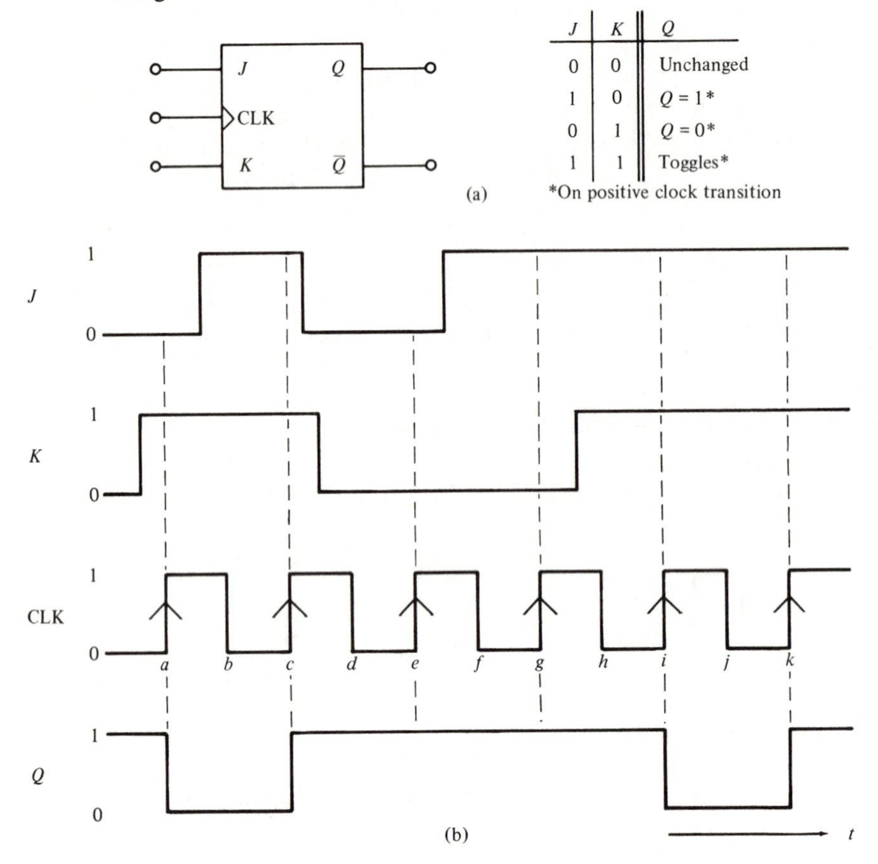

Figure 8–13 **(a) Clocked *JK* FF that responds to the positive edge of the clock (b) Waveforms**

5. At point g, $J = 1$, and $K = 0$. This is the condition which sets Q (to the 1 state). However, it is already 1, so it will remain there.
6. At point i, $J = K = 1$, so the FF toggles to its opposite state. The same thing occurs at point k.

It should be noted from these waveforms that the FF is not affected by the negative-going edge of the clock pulses. Also, the J and K input levels have no effect except upon the occurrence of the positive-going edge of the clock signal. The *J* and *K* inputs by themselves cannot cause the FF to change states.

Figure 8–14 shows the symbol for a clocked *JK* FF that triggers on the negative-going clock signal transitions. The small circle on the CLK indicates that this FF will trigger when the CLK goes from 1 to 0. This FF operates in the same manner as the positive-edge FF of Figure 8–13 except that the output can change states only on negative-going clock signal transitions (points *b, d, f, h,* and *j*). Both polarities of edge-triggered *JK* FFs are in common usage.

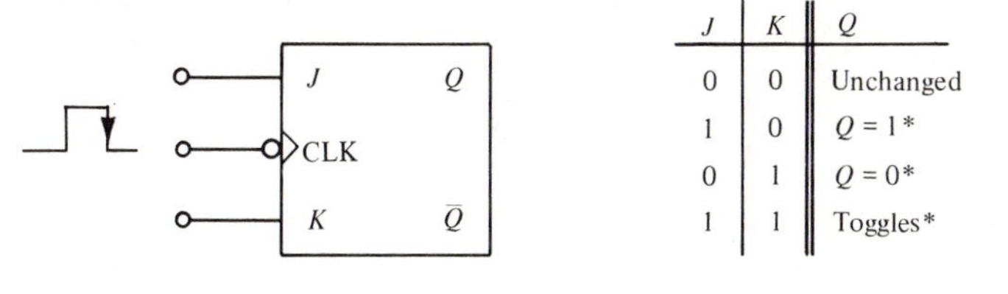

J	*K*	*Q*
0	0	Unchanged
1	0	*Q* = 1*
0	1	*Q* = 0*
1	1	Toggles*

*On negative clock transition

Figure 8–14 **_JK_ FF that triggers on negative-going transitions**

The *JK* FF is much more versatile than the *SC* FF because it has no ambiguous states. The $J = K = 1$ condition that produces the toggling operation is used extensively in all types of binary counters. In essence, the *JK* FF can do anything the *SC* FF can do *plus* it has the useful toggle mode of operation. As such, the *JK* FF presently enjoys widespread use in almost all modern digital systems.

8.7 The Clocked *D* Flip-Flop

Figure 8–15(a) shows the symbol for a clocked *D*-type FF that triggers on positive transitions at the CLK input. The *D* input is a single control input that determines the state

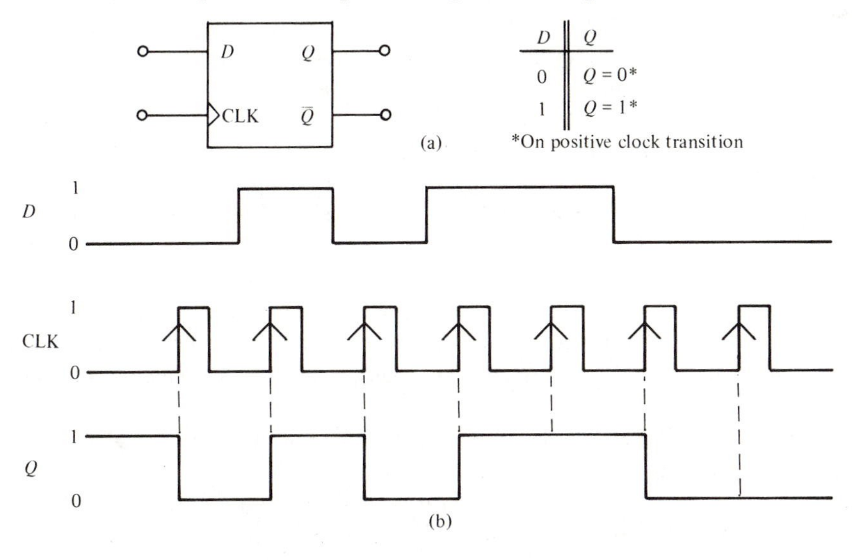

D	*Q*
0	*Q* = 0*
1	*Q* = 1*

*On positive clock transition

Figure 8–15 **(a) _D_ FF that triggers on positive-going transitions**
(b) Waveforms

of the FF according to the accompanying truth table. Essentially, whenever a positive transition occurs at the CLK input, the FF output Q will go to the same state that is present on the D input. This situation is illustrated by the waveforms in Figure 8–15(b).

Note that each time a positive transition occurs on the CLK input, the Q output takes on the same value as the level present at the D input. The negative transitions at the CLK input have no effect. The levels present at the D input have no effect until a positive clock transition occurs.

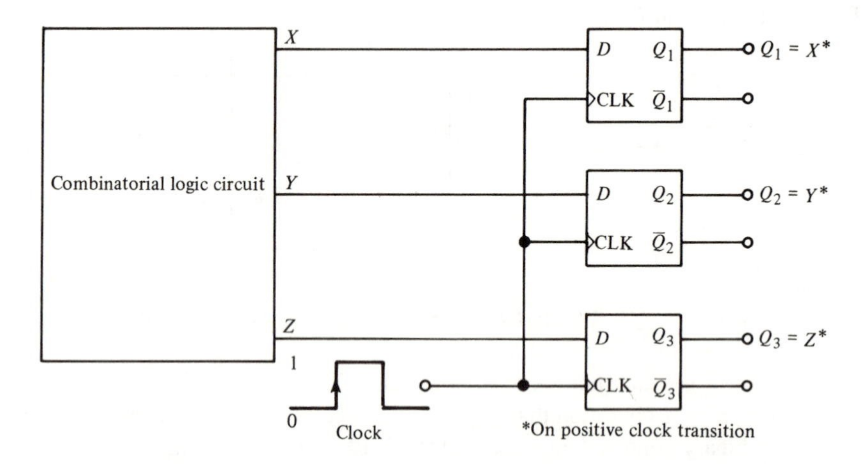

Figure 8–16 **Parallel transfer of binary data using *D* FFs**

Negative-edge triggered D FFs are also available and operate similarly except that they trigger on the negative-going clock transitions. The symbol for the negative-edge triggered D FF has a small circle on the CLK input.

The D FF is used principally in the transfer of binary data, as illustrated in Figure 8–16. Here, the outputs X, Y, and Z of a combinatorial logic circuit are to be transferred to FFs Q_1, Q_2, and Q_3 for storage. Using the D FFs, the levels present at X, Y, and Z will be transferred to Q_1, Q_2, and Q_3, respectively, upon application of a pulse to the common CLK inputs. The FFs can store these values for subsequent processing. This is an example of *parallel* transfer of binary data because the three bits X, Y, and Z are transferred simultaneously.

SET-CLEAR and JK FFs can easily be modified to operate as D FFs, as illustrated in Figure 8–17. Either of these arrangements may be used to perform the data transfer operation when D FFs are not available. The student should verify that these arrangements will act the same as D FFs.

D-type latch As we have seen, edge-triggered FFs will change states *only* when the CLK input receives the proper transition. Another type of FF called a *latch* operates somewhat differently in that the output can change while the CLK input is at a fixed logic level.

Figure 8–18(a) shows the symbol and truth table for a D-type latch, and Figure 8–18(b) illustrates the device's operation. When the CLK input is HIGH, output Q will follow the D input; that is, $Q = D$ even if D changes. When CLK goes LOW, Q will store (latch) the last value it had while CLK was HIGH. The D input has no effect on Q while CLK is LOW.

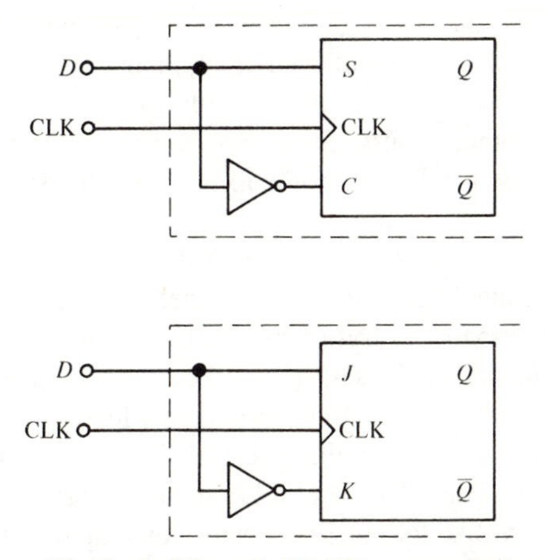

Figure 8–17 **Clocked *SC* and *JK* FFs arranged to operate as *D* FFs**

The small circle on the CLK input indicates that the *latching* function occurs when CLK goes LOW. Note that there is no small triangle on the CLK input because the latch is not edge-triggered.

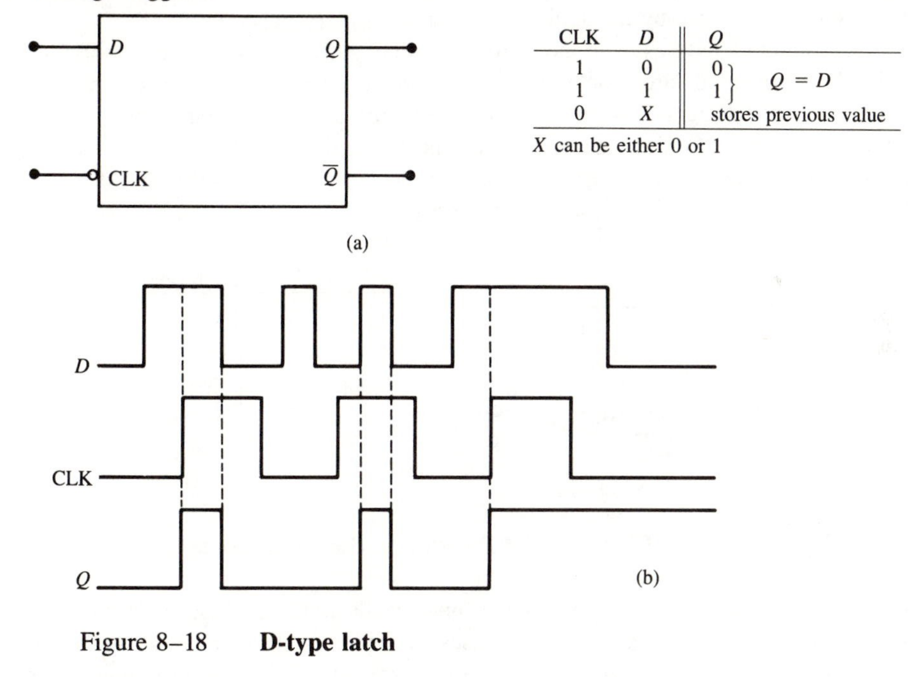

CLK	D	Q
1	0	0 } $Q = D$
1	1	1 } $Q = D$
0	X	stores previous value

X can be either 0 or 1

Figure 8–18 **D-type latch**

8.8 Synchronous and Asynchronous Flip-Flop Inputs

For the clocked FFs that we have been studying, the *S, C, J, K,* and *D* inputs have been referred to as control inputs. These inputs are also called *synchronous* inputs because their effect on the FF output is synchronized with the CLK input. As we have seen, the synchronous inputs must be used in conjunction with a clock signal to trigger the FF.

Most clocked FFs also have one or more *asynchronous* inputs which operate independently of the synchronous inputs and clock input. These asynchronous inputs can be used to set the FF (to the 1 state) or clear the FF (to the 0 state) at any time regardless of the conditions at the other inputs. Stated another way, the asynchronous inputs are *override* inputs that can be used to override all the other inputs in order to place the FF in one state or the other.

Figure 8–19 shows a clocked *JK* FF with dc SET and dc CLEAR inputs. Either of these asynchronous inputs is activated by a 0 level as indicated by the small circles on the FF symbol. The accompanying truth table indicates how these inputs operate. A low on the dc SET input *immediately* sets Q to the 1 state. A low on the dc CLEAR *immediately* clears Q to the 0 state. Simultaneous low levels on dc SET and dc CLEAR are forbidden since an ambiguous condition results. When neither of these inputs is low, the FF is free to respond to the *J, K,* and *CLK* inputs as previously described.

It is important to realize that these asynchronous inputs respond to dc levels. This means that if a constant 0 is held on the dc SET input (with dc CLEAR $= 1$) the FF will remain in the $Q = 1$ state regardless of what is occurring at the other inputs. Similarly, a constant low on the dc CLEAR input (with dc SET $= 1$) holds the FF in the $Q = 0$ state. Thus, the asynchronous inputs can be used to hold the FF in a particular state for any desired interval. Most often, however, the asynchronous inputs are used to set or clear the FF to the desired state by the application of a momentary pulse.

Some FFs have asynchronous inputs that are activated by 1's rather than 0's. For these FFs the small circle on the dc SET and dc CLEAR inputs is omitted. Other commonly used designations for asynchronous inputs are *PRESET* (same as *dc SET*) and *RESET* (same as *dc CLEAR*). In many cases, the "dc" is dropped, and the inputs are simply referred to as *SET* and *CLEAR*. It is important to remember, however, that these inputs respond to dc levels and are not edge-triggered.

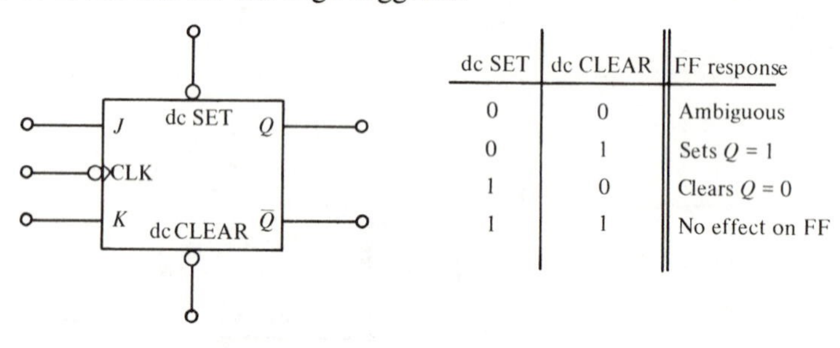

dc SET	dc CLEAR	FF response
0	0	Ambiguous
0	1	Sets $Q = 1$
1	0	Clears $Q = 0$
1	1	No effect on FF

Figure 8–19 **Clocked *JK* FF with asynchronous inputs**

The operation of the asynchronous inputs is illustrated in Figure 8–20 for a *JK* FF. Here we are keeping $J = K = 1$ while pulses are applied to the CLK input, so that output Q will normally toggle on each negative-going transition. This happens at points *a, c, d,* and *g*. However, the clock edge at point *f* will not cause Q to toggle, because there is a LOW at the CLEAR input that forces Q to stay LOW; this is the override feature mentioned previously.

Note that Q goes HIGH at point *b* in response to the LOW at the PRESET input. This event happens independently from the CLK input. Similarly, Q goes LOW at point *e* in immediate response to the LOW on the CLEAR input.

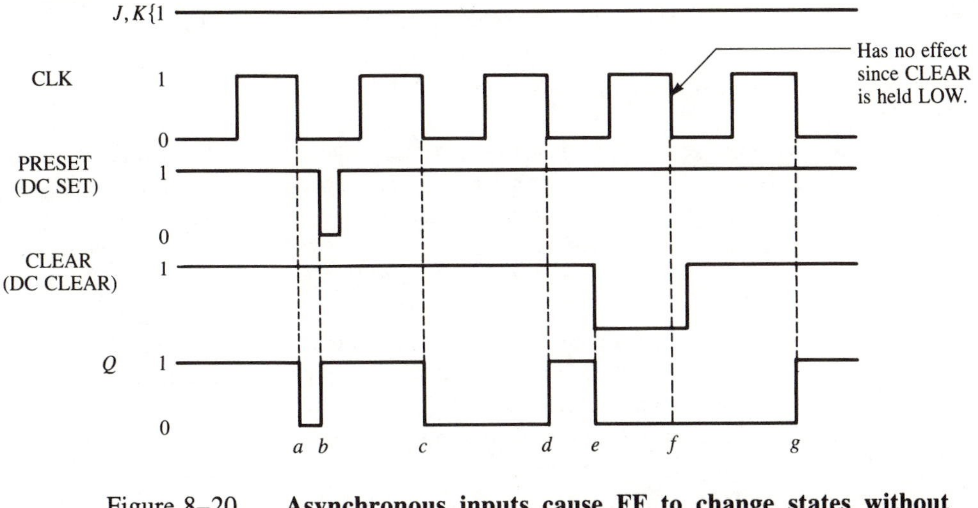

Figure 8–20 **Asynchronous inputs cause FF to change states without waiting for CLK transition, and they can override the effect of CLK, *J* and *K***

8.9 FF Operating Characteristics

There are several operating characteristics of flip-flops that are usually specified on manufacturers' data sheets. These characteristics or parameters are often needed by the logic circuit designer to determine whether a particular FF will be satisfactory for a given application.

Propagation delays Whenever a signal is to change the state of a FF's output, there is a delay from the time the signal is applied to the time the output makes its change. Figure 8–21 illustrates the propagation delays that occur in response to a positive transition

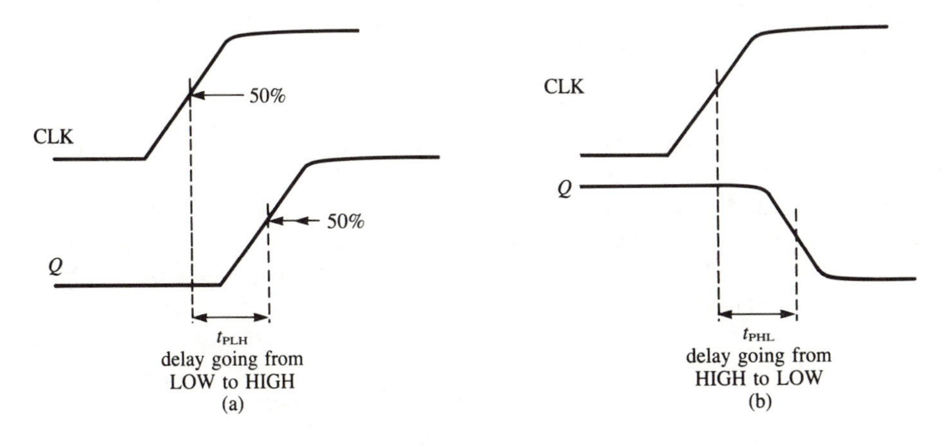

Figure 8–21 **FF propagation delays**

on the CLK input. Note that these delays are measured between the 50 percent points on the input and output waveforms. The same types of delays occur in response to signals on a FF's asynchronous inputs (PRESET and CLEAR). The manufacturer's data sheets usually specify propagation delays in response to all FF inputs.

Modern IC flip-flops have propagation delays that range from a few nanoseconds to around one microsecond. The values of t_{PLH} and t_{PHL} are generally not the same and they increase in direct proportion to the number of loads being driven by the Q output. FF propagation delays play an important part in many situations that you will encounter in later work.

Set-up and hold times The *set-up time*, t_s, is the minimum amount of time that the control inputs (S, C, J, K, and D) must be held stable prior to the occurrence of the triggering edge of the CLK input in order for the FF to respond reliably. This concept is illustrated in Figure 8–22 for a D flip-flop. Here the D input has to be in the 1 state for a time, t_s, before the negative-going CLK transition occurs, in order for the FF to go to the 1 state when the CLK edge occurs.

The *hold time*, t_H, is the minimum amount of time that the control inputs have to be held stable after the triggering edge of the CLK input. Figure 8–22 shows the D input being held high for a time, t_H, after the CLK goes from 1 to 0. If this requirement is not met, proper FF triggering cannot be assured.

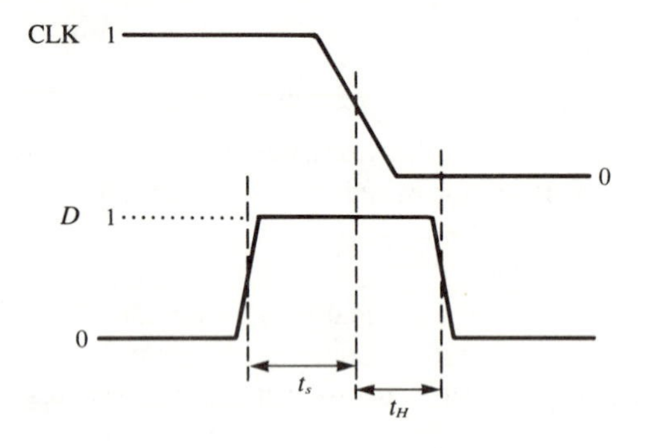

Figure 8–22 **Set-up and hold times**

In modern integrated circuit FFs, the values of t_s and t_H are typically in the nanosecond region. Set-up times generally fall in the range 5–50 ns, and hold times are usually less than 10 ns. Many edge-triggered FFs in the TTL IC family have $t_H = 0$, which means that the control inputs can change state at the same time as the CLK input transition. The *master/slave* type of FF, discussed in the next section, actually has *negative* t_H values, indicating that the control inputs can change state *before* the clock transition occurs.

The t_s and t_H parameters are particularly important in applications where the control inputs are changing at approximately the same time as the CLK input. These applications include certain types of counters and shift registers. The t_s and t_H parameters are characteristics of all clocked FFs, including the S-C and the J-K types.

Maximum frequency A flip-flop can have clock pulses applied to its CLK input at a rate that must not exceed f_{MAX}, its maximum frequency limit. At clock frequencies above f_{MAX}, the FF would be unable to respond quickly enough and its output would be distorted. Modern IC FFs have f_{MAX} values ranging from 1 MHz up to hundreds of MHz.

8.10 Master/Slave FFs

In digital systems the outputs of FFs are often connected (directly or through logic gates) to the inputs of other FFs. Figure 8–23 is an illustration of this. The output of FF Q_1 acts as the J input of FF Q_2, and both FFs are triggered by the same clock pulse at their CLK inputs. Let us assume that initially $Q_1 = 1$ and $Q_2 = 0$. Since Q_1 has both its J and K inputs HIGH, it will toggle to the 0 state on the negative transition of the clock pulse. Q_2 has a 1 on its J input via Q_1 prior to the negative transition of the clock pulse. However, when the clock pulse goes LOW, Q_1 will also go LOW, so the J input of Q_2 will be changing from 1 to 0 while Q_2 is being clocked. This is called a *race condition* and can sometimes lead to unpredictable triggering of Q_2.

Recall from our discussion of set-up and hold times (t_s and t_H) that edge-triggered FFs require that the control inputs (S, C or J, K or D) remain stable for a time t_H after the clock makes its transition. If the t_H requirement for the FF is very small or zero, the race situation in Figure 8–23 might not cause a problem, because Q_1 will actually go LOW a short time after the clock goes LOW due to the propagation delay of FF Q_1 in responding to the clock pulse. Thus, Q_2 will respond properly by going to the 1 state on the falling edge of the clock.

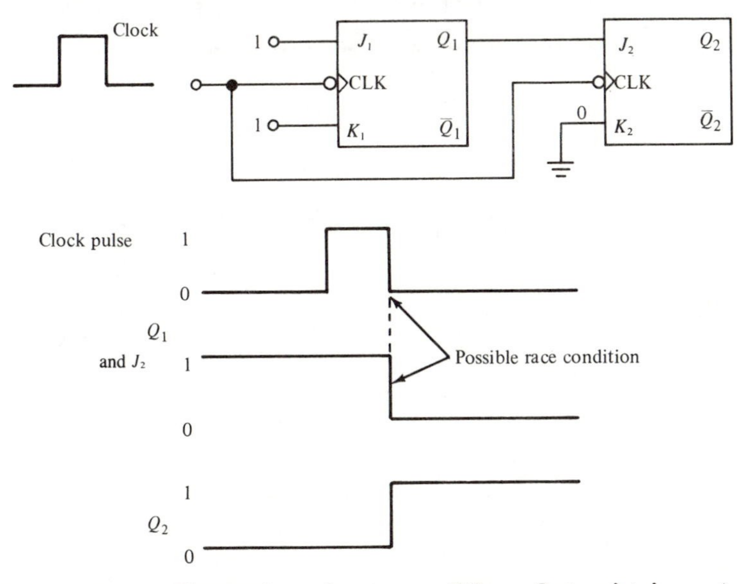

Figure 8–23 **Illustration of race conditions that exist in certain applications**

What this means is that for the *edge-triggered SC, JK,* and *D* FFs, the race condition is not a problem as long as the control inputs meet the hold time requirement. In some applications, however, either this cannot be reliably guaranteed, or the situation is marginal. For these reasons another group of clocked FFs, called *master/slave* (*M/S*) *FF*s, have been developed. The *SC*-, *JK*-, and *D*-type FFs are available in both edge-triggered and M/S versions. All FFs that operate on the M/S principle will trigger reliably even for race conditions such as that in Figure 8–23.

Stated in simple terms, a M/S FF actually contains two FFs—a master and a slave. The master is used to store the conditions present on the control inputs prior to the clock transition, so when the clock makes its transition the slave can respond to those *stored* control-input values. Thus, even if the control inputs are changing when the clock transition occurs, the FF output will respond to the control input values that were present *just prior* to the clock transition. This action insures that the FF will respond predictably for situations such as that in Figure 8–23.

Difference between edge-triggered and master/slave flip-flops The major difference between the edge-triggered FFs examined earlier and the M/S FF can best be illustrated by considering the waveforms shown in Figure 8–24. Here, the outputs of an edge-triggered FF and a M/S FF are shown for the same J, K, and CLK inputs. The two FFs respond in the same manner at t_1 and t_2 when the CLK makes its negative-going transitions. At t_3 the J input and CLK input are both going from HIGH to LOW. The edge-triggered FF will respond unpredictably unless the fall of the J input is delayed beyond the falling edge of the CLK by at least an amount t_H. The M/S FF, however, will be set to the 1 state at t_3 because the 1 at the J input was stored in the master while the CLK was HIGH, and when the CLK went LOW this 1 was transferred to the slave output. In other words, the M/S FF responded to the $J = 1$ condition which was present *prior to* the CLK falling edge.

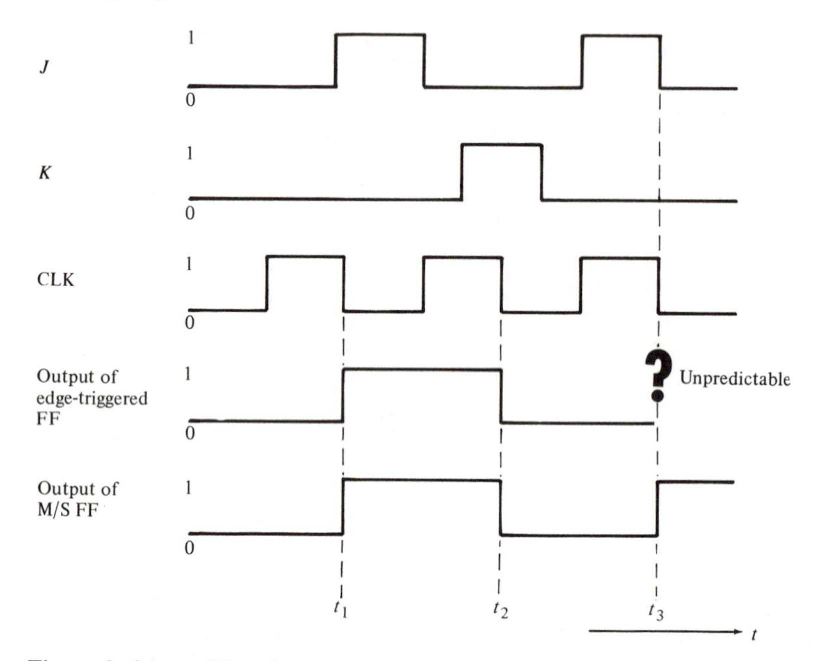

Figure 8–24 **Waveforms showing difference in response of edge-triggered and M/S FFs**

The situation at t_3 demonstrates the chief advantage of M/S FFs over the edge-triggered type. The M/S will provide reliable triggering even if the control inputs are changing at the same time the CLK transition occurs. For this reason, M/S FFs must be used in applications where race situations such as the ones in Figures 8–23 and 8–24 can occur. We will distinguish between the edge-triggered and M/S FFs by placing an M/S designation on the symbols for the M/S type.

8.11 Flip-Flop Operations—Parallel Transfer

An operation that occurs very frequently in digital systems is the transfer of information from one FF or group of FFs to another FF or group of FFs. Figure 8–25 illustrates how this operation is performed using *SC, JK,* and *D* FFs. In each case, the logic value stored in FFA is transferred to FFB upon the negative transition of the TRANSFER pulse. That is, the *B* output after the pulse occurs will be the same as what the *A* output was prior to the pulse.

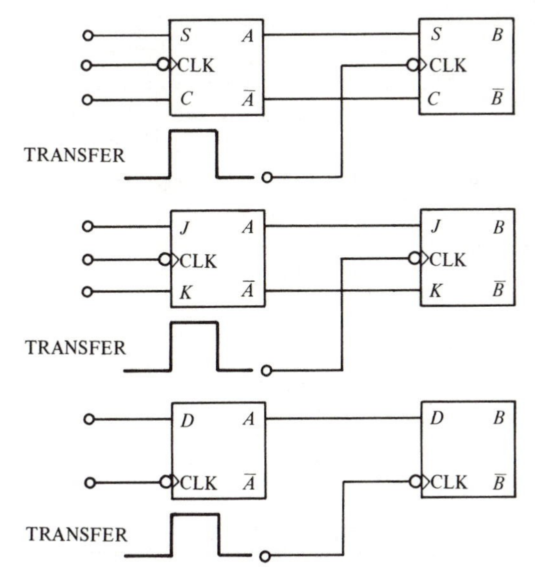

Figure 8–25 **Synchronous transfer operation performed by various types of FFs**

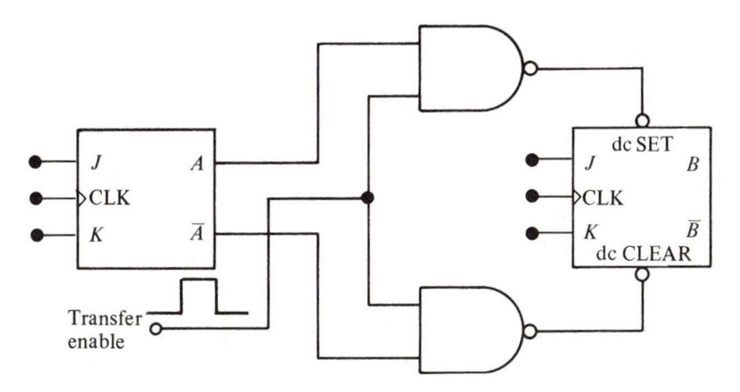

Figure 8–26 **Asynchronous transfer operation**

The transfer operations in Figure 8–25 are examples of *synchronous transfer* since the synchronous (control) and CLK inputs are used to perform the transfer. A transfer operation can also be performed using the asynchronous inputs of a FF. Figure 8–26 shows how an *asynchronous transfer* can be obtained using the dc SET and dc CLEAR inputs of any type of FF. Here, the asynchronous inputs respond to LOW levels (note small circles on symbol). When the TRANSFER ENABLE line is held LOW, the two NAND outputs are kept HIGH, with no effect on the FF inputs. When the TRANSFER ENABLE line is made HIGH, one of the NAND outputs will go LOW depending on the state of the A and $\overline{A}$ outputs. This LOW will either set or clear FFB to the same state as FFA. This asynchronous transfer is done independently of the synchronous and CLK inputs of the FF.

Parallel transfer of information Sets of FFs that are used to store some particular group of 0's and 1's are called *registers*. A common operation in digital systems involves transferring the information stored in one group of FFs to another group of FFs. Figure 8–27 illustrates this operation using D-type FFs. Register X consists of FFs X_1, X_2, and X_3; register Y consists of FFs Y_1, Y_2, and Y_3. On the occurrence of the positive transition of the TRANSFER pulse, the value stored in X_1 is transferred to Y_1, X_2 to Y_2, and X_3 to Y_3. The transfer of the contents of the X register into the Y register is a synchronous transfer. It is also referred to as a *parallel* transfer since the contents of X_1, X_2, and X_3 are transferred *simultaneously* into Y_1, Y_2, and Y_3, respectively. If a *serial* transfer were performed, the contents of the X register would be transferred to the Y register one bit at a time. This situation will be examined in the next section.

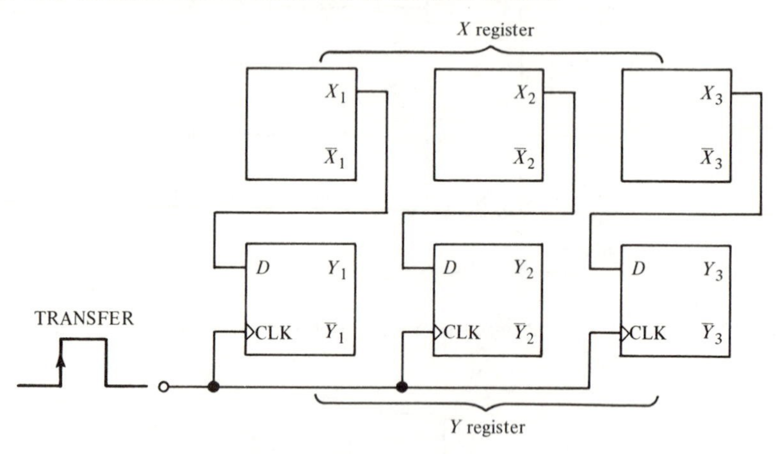

Figure 8–27 **Parallel transfer of contents of register *X* into *Y***

8.12 Flip-Flop Operations — Shift Registers

Shift registers are used to transfer the contents of one register into a second register one bit at a time (serially). Before examining this operation of serial transfer, let us first look at the operation of a basic shift register. Figure 8–28(a) shows four JK master/slave FFs wired as a four-bit shift register. Note that the FFs are connected so that the output of X_3 transfers into X_2, the output of X_2 into X_1, and X_1 into X_0. This means that upon the occurrence of the SHIFT pulse each FF takes on the value stored previously in the FF on its left. Flip-flop X_3 takes on a value determined by the conditions present on its J and K inputs when the SHIFT pulse occurs. For now, we will assume that X_3's J and K inputs are fed from the input waveforms shown in Figure 8–28(b). We will also assume that all FFs are initially in the 0 state before SHIFT pulses are applied.

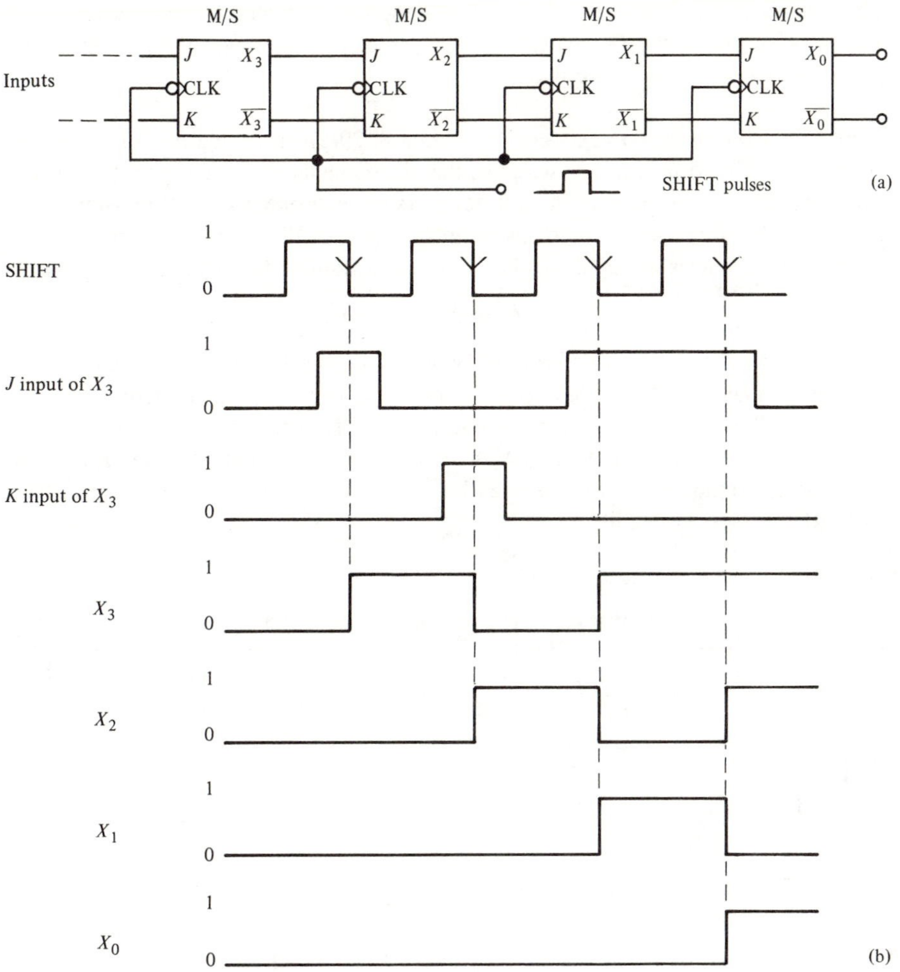

Figure 8–28 **Four-bit shift register**

The waveform diagrams show how the input information is essentially shifted from left to right from FF to FF. When the first SHIFT pulse occurs it finds all FFs have $J = 0$, $K = 1$ except for X_3, which has $J = 1, K = 0$. Thus, on the falling edge of this pulse only X_3 goes to 1. When the negative transition of the second SHIFT pulse occurs it finds X_2 has $J = 1, K = 0$ and all other FFs have $J = 0, K = 1$ so that X_2 goes HIGH and all other FFs go or stay LOW. Similar reasoning is used to determine the FF outputs for the succeeding SHIFT pulses.

Possible race problem This shift register exhibits the race condition discussed earlier because there are times when the JK inputs to a FF are changing at approximately the same time as its CLK input. For example, just prior to the falling edge of the second SHIFT pulse the J input of X_2 is 1 since $X_3 = 1$. On the falling edge X_3 will go low, so X_2's J input goes low at the same time as its CLK transition. However, since X_2 is a M/S FF it will trigger reliably according to what its J value was just prior to the CLK transition. Thus, the possible race problem is avoided. It is important to realize, then, that these simultaneous transitions on the *JK* and CLK inputs can be neglected when analyzing this type of circuit provided M/S FFs are used.

Serial transfer between registers Figure 8–29(a) shows the two three-bit registers once again. This time they are wired so that the contents of the *X* register are transferred serially (shifted) into register *Y*. Both registers are connected as shift registers. Flip-flop X_0, the last FF of register *X*, is connected to the inputs of Y_2, the first FF of register *Y*. Thus, as the SHIFT pulses are applied, the information transfer takes place as follows:

$$X_2 \rightarrow X_1 \rightarrow X_0 \rightarrow Y_2 \rightarrow Y_1 \rightarrow Y_0$$

The X_2 FF will go to states determined by its *JK* inputs. For now, $J = 0$ and $K = 1$ will be used so that X_2 will go LOW on the first pulse and will remain there.

To illustrate, let us assume before any SHIFT pulses are applied that the contents of the *X* register are 1-0-1 (that is, $X_2 = 1$, $X_1 = 0$, $X_0 = 1$) and that the *Y* register is at 0-0-0. Refer to the table in Figure 8–29(b), which shows how the states of each FF change as SHIFT pulses are applied.

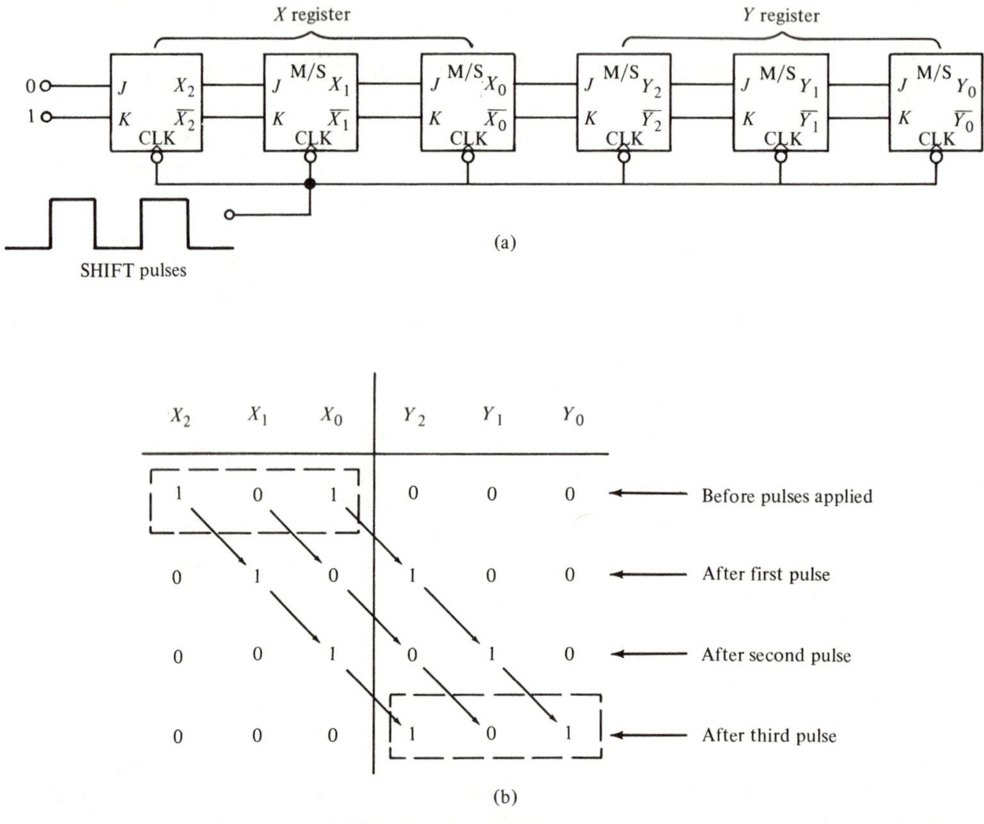

X_2	X_1	X_0	Y_2	Y_1	Y_0	
1	0	1	0	0	0	← Before pulses applied
0	1	0	1	0	0	← After first pulse
0	0	1	0	1	0	← After second pulse
0	0	0	1	0	1	← After third pulse

Figure 8–29 **Serial transfer of information from *X* register into *Y* register**

The following points should be noted:

1. On the falling edge of each pulse, each FF takes on the value that was stored in the FF on its left prior to the occurrence of the pulse.

2. After three pulses, the 1 that was initially in X_2 is in Y_2, the 0 initially in X_1 is in Y_1, and the 1 initially in X_0 is in Y_0. In other words, the 1-0-1 stored in the X register has now been shifted into the Y register.
3. The complete transfer of the three bits of data requires three SHIFT pulses.

Parallel versus serial transfer In parallel transfer all of the information is transferred simultaneously upon the occurrence of a single TRANSFER command pulse (Figure 8–27) no matter how many bits are being transferred. In serial transfer, as exemplified by Figure 8–29, the complete transfer of N bits of information requires N clock pulses (three bits require three pulses, four bits require four pulses, etc.). Parallel transfer, then, is obviously much faster than serial transfer using shift registers.

In parallel transfer the output of each FF in register X is connected to a corresponding FF input in register Y. In serial transfer only the last FF in register X is connected to register Y. In general, then, parallel transfer requires more interconnections between the sending register (X) and the receiving register (Y) than does serial transfer. This difference becomes more obvious when a greater number of bits of information are being transferred. This is an important consideration when the sending and receiving registers are remote from each other since it determines how many lines (wires) are needed in the transmission of the information.

The choice of either parallel or serial transmission depends on the particular system application and specifications. Often, a combination of the two types is used to take advantage of the speed of parallel transfer and the economy and simplicity of serial transfer.

8.13 Flip-Flop Operations—Counting*

Refer to Figure 8–30(a). Each FF has its J and K inputs at the 1 level, so it will change states (toggle) whenever the signal on its CLK input goes from HIGH to LOW. The train of clock pulses is applied only to the CLK input of FF X_0. Output X_0 is connected to the CLK input of FF X_1, and output X_1 is connected to the CLK input of FF X_2. The waveforms in Figure 8–30(b) show how the FFs change states as the pulses are applied. The following important points should be noted:

1. Flip-flop X_0 toggles on the negative-going transition of each input clock pulse. Thus, the X_0 output waveform has a frequency that is exactly ½ of the clock-pulse frequency.
2. Flip-flop X_1 toggles each time the X_0 output goes from HIGH to LOW. The X_1 waveform has a frequency equal to exactly ½ the frequency of the X_0 output, and therefore ¼ of the clock frequency.
3. Flip-flop X_2 toggles each time the X_1 output goes from HIGH to LOW. Thus, the X_2 waveform has ½ the frequency of X_1, and therefore ⅛ of the clock frequency.

As just described, each FF divides the frequency by *two*. Thus, if we were to add a fourth FF to the chain, it would have a frequency equal to 1/16 of the input clock frequency, and so on. Using the appropriate number of FFs, this circuit could divide a

*The student may wish to refer to Appendix A for a brief coverage of the binary number system.

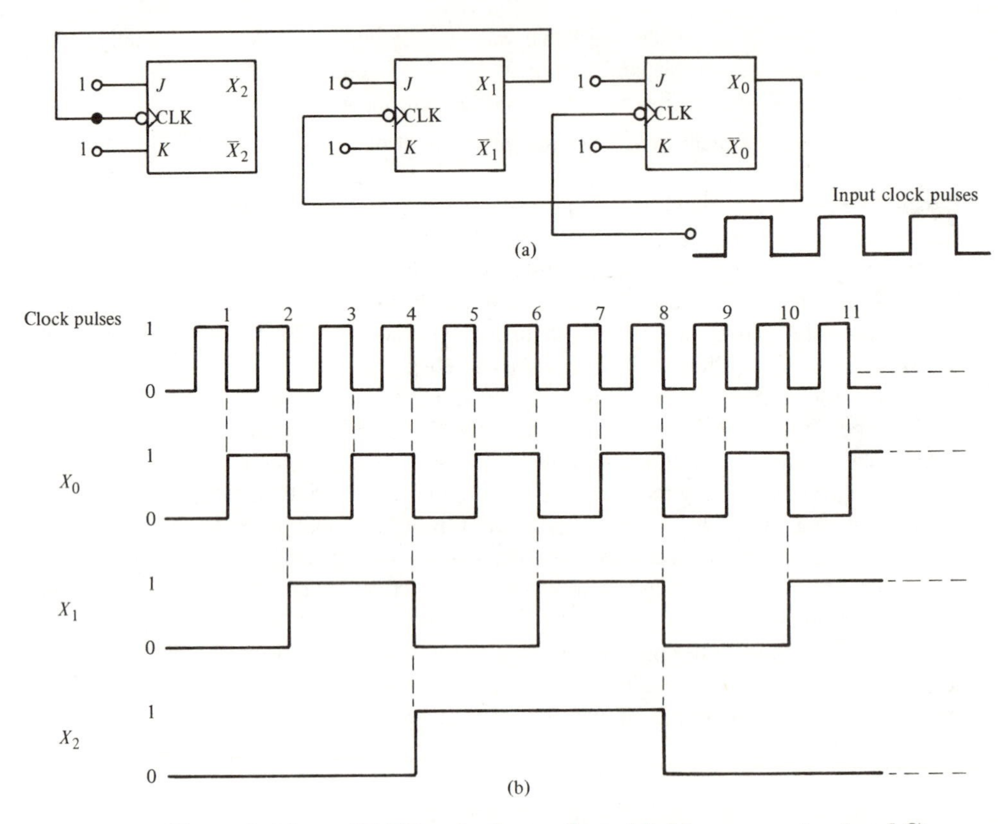

Figure 8–30 ***JK* FFs wired as a three-bit binary counter (mod-8)**

frequency by any power of two. Specifically, using N FFs would produce an output frequency from the last FF that is equal to $1/2^N$ of the input frequency.

Counting operation In addition to functioning as a frequency divider, the circuit of Figure 8–30 also operates as a *binary counter.* This can be demonstrated by examining the sequence of states of the FFs after the occurrence of each clock pulse. Figure 8–31 presents the results in tabular form. Let the $X_2X_1X_0$ values represent a binary number where X_2 is the 2^2 position, X_1 is the 2^1 position, and X_0 is the 2^0 position. The first eight $X_2X_1X_0$ states in the table should be recognized as the binary counting sequence from 000 to 111. After the first pulse the FFs are in the 001 state ($X_2 = 0$, $X_1 = 0$, $X_0 = 1$), which represents 001_2 (equivalent to decimal 1); after the second pulse the FFs represent 010_2, which is equivalent to 2_{10}; after three pulses $011_2 = 3_{10}$; after four pulses $100_2 = 4_{10}$; and so on, until after seven pulses $111_2 = 7_{10}$. On the eighth pulse the FFs return to the 000 state, and the binary sequence repeats itself for succeeding pulses.

Thus, for the first seven input pulses the circuit functions as a binary counter in which the states of the FFs represent a binary number equivalent to the number of pulses that have occurred. This counter can count as high as $111_2 = 7_{10}$, and then it returns to 000.

Mod number The counter of Figure 8–30 has $2^3 = 8$ different states (000 through 111). It would be referred to as a mod-8 counter, where the mod number indicates the number of states in the complete counting sequence. If a fourth FF were added, the

2^2 X_2	2^1 X_1	2^0 X_0	
0	0	0	Before applying clock pulses
0	0	1	After pulse 1
0	1	0	After pulse 2
0	1	1	After pulse 3
1	0	0	After pulse 4
1	0	1	After pulse 5
1	1	0	After pulse 6
1	1	1	After pulse 7
0	0	0	After pulse 8 recycles to 000
0	0	1	After pulse 9
0	1	0	After pulse 10
0	1	1	After pulse 11
⋮	⋮	⋮	⋮ ⋮

Figure 8–31 **Sequence of FF states shows binary counting sequence**

sequence of states would count in binary from 0000 to 1111, a total of 16 states. This would be called a mod-16 counter. In general, if N FFs are connected in the arrangement of Figure 8–30, the counter will have 2^N different states (including the zero state) and is a mod-2^N counter. It is capable of counting up to $2^N - 1$ before returning to its zero state.

The mod number of a counter also indicates the frequency division obtained from the last FF. For instance, a four-bit counter has four FFs, each of which represents one binary digit (bit), so it is a mod-2^4 = mod-16 counter. It can therefore count up to $2^4 - 1 = 15$. It can also be used to divide the input-pulse frequency by a factor of 16 (the mod number).

We have looked at the basic FF binary counter. There are many other types of counters used in digital applications. Unfortunately, their coverage is beyond the scope of this book. Many of these counters, however, operate on the same basic principle as the binary counter just examined.

Example 8.5 How many FFs are required to convert a 256 kpps clock signal to 500 pps?

Solution: The required frequency division factor is

$$\frac{256 \text{ kpps}}{500 \text{ pps}} = 512$$

Thus, we need a counter with a mod number of 512. Therefore

$$2^N = 512$$

so that $N = \mathbf{9}$ is the required number of FFs.

Example 8.6 A counter like the one in Figure 8–30 is to be used to count pulses coming from a photo-detector circuit, which detects items as they pass by on a conveyor belt. The counter has to be able to count up to 99_{10}. How many FFs does the counter need?

Solution: For N FFs, a counter will count up to $2^N - 1$ before recycling to zero. Thus, we want

$$2^N - 1 \geqslant 99$$

The smallest integer value of N that satisfies this requirement is **7**.

8.14 The One-Shot

A digital circuit that is somewhat related to the FF is the *one-shot* (abbreviated OS). Like the FF, the OS has two outputs Q and $\overline{Q}$ that are inverses of each other. Unlike the FF, the OS outputs have only *one* stable state (normally $Q = 0, \overline{Q} = 1$), where they remain until the OS is triggered by an input signal. Once triggered, the OS outputs switch to the opposite state ($Q = 1, \overline{Q} = 0$). The OS remains in this *quasi-stable* state for a fixed period of time t_p, which is usually determined by an RC time constant. After a time t_p, the OS outputs return to their stable resting state until triggered again.

Figure 8–32(a) shows the logic symbol for a OS. The value of t_p is usually indicated somewhere on the OS symbol. In practice, t_p can be made to vary from several nanoseconds up to several tens of seconds. The waveforms in the figure illustrate the OS operation. This particular OS is triggered by positive-going transitions at its T input. The important points to note are:

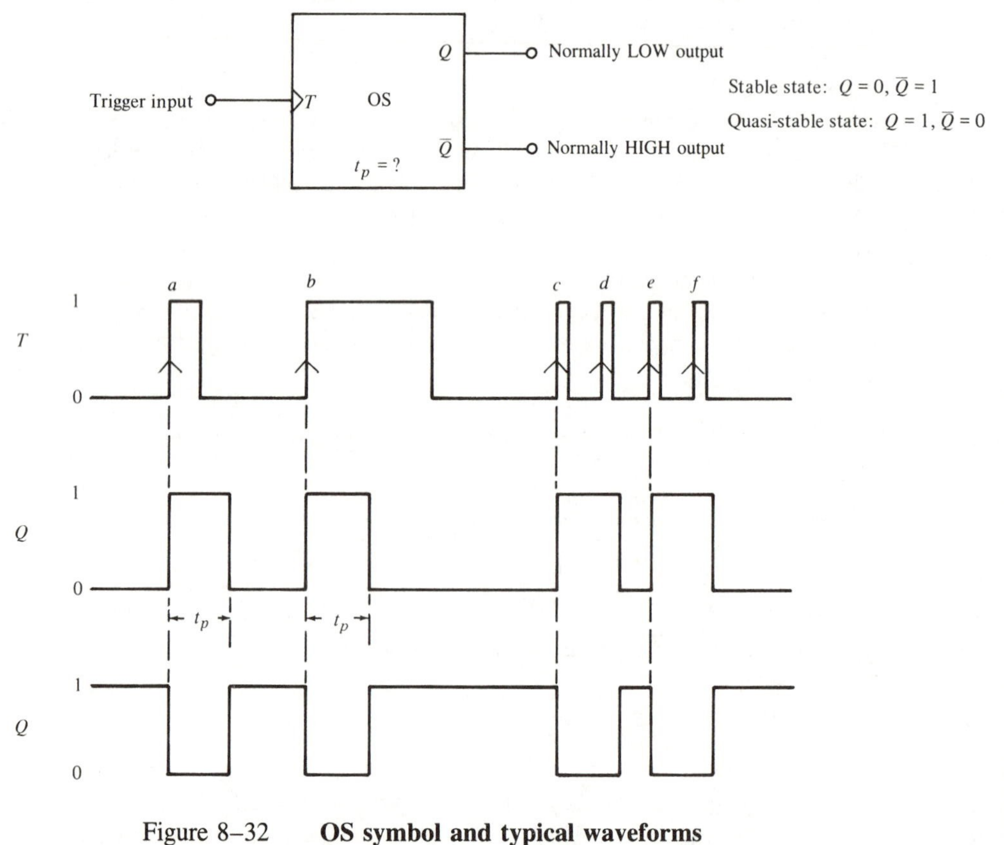

Figure 8–32 **OS symbol and typical waveforms**

1. The OS outputs go to the quasi-stable state for a time t_p whenever a positive-going trigger occurs (see points *a, b, c, e*).
2. The input-pulse duration has no effect on the OS operation; the OS responds only to the pulse's positive transition.
3. If a second trigger pulse occurs while the OS has already been triggered to its quasi-stable state, it will have no effect on the OS outputs (see points *d* and *f*). The OS must return to its stable state before it can be retriggered.

The OS is also referred to as a *monostable multivibrator,* or simply a *monostable,* because it has only one stable state. The major applications of the OS are as pulse generators, time-delay elements, and timing circuits. Some of these applications will be introduced in the following sections and in the problems at the end of the chapter.

8.15 One-Shot Circuit Operation

One-shots, like FFs, are readily and economically available in IC form and require only an external resistor and capacitor for producing the desired output-pulse duration. Even so, it would be instructive to look briefly at an actual OS circuit and analyze its operation. Figure 8–33 shows a OS circuit made from two NOR gates. Refer to the waveforms in the figure for the following analysis. We will use 0 V for logic 0 and +5 V for logic 1.

1. Assume that the trigger input T is initially LOW and the circuit is in steady state.
2. A HIGH voltage is applied to the upper input Z of NOR gate 2 through resistor R_T. This HIGH produces a LOW at output Y.
3. With $T = Y = 0$, the output X of NOR gate 1 will be HIGH. Since X and Z are both HIGH, there will be little or no voltage across capacitor C_T.
4. When the T input is pulsed HIGH, X immediately goes LOW. Since C_T cannot change its charge instantaneously, the voltage at Z must also go LOW (see waveforms). Capacitor C_T momentarily acts as a short circuit connecting Z to X.
5. The LOW at Z causes Y to go HIGH. This HIGH at Y holds X LOW even when T returns LOW.
6. The capacitor C_T now begins to charge through R_T, as illustrated in Figure 8–33(c). Remember, X is now virtually at ground potential. As C_T charges, the voltage at Z begins to rise.
7. When Z reaches a voltage high enough to activate the NOR gate, Y will go back LOW. This voltage is called the threshold voltage, V_{TH}. When Y goes LOW, X will go back HIGH.
8. The positive transition at X is coupled to Z through C_T, causing Z to increase. The voltage at Z then gradually returns to its steady-state level of +5 V as C_T discharges.
9. The important idea here is that the duration t_p of the pulses at X and Y is directly proportional to the R_TC_T time constant. That is, t_p is the number of time constants it takes for C_T to charge to V_{TH}. Normally, t_p is less than one

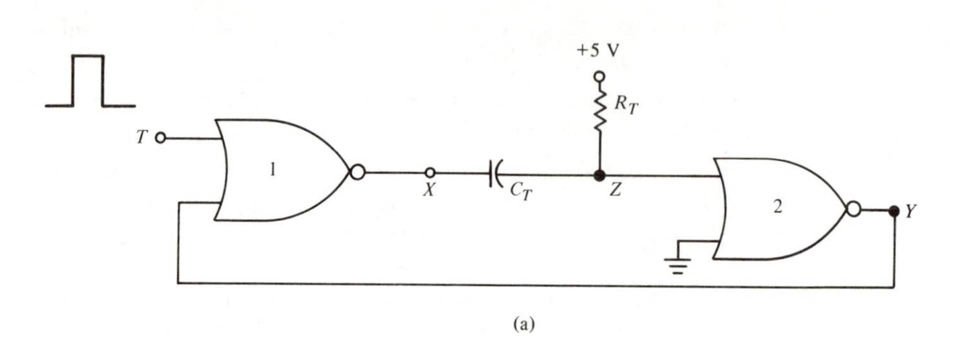

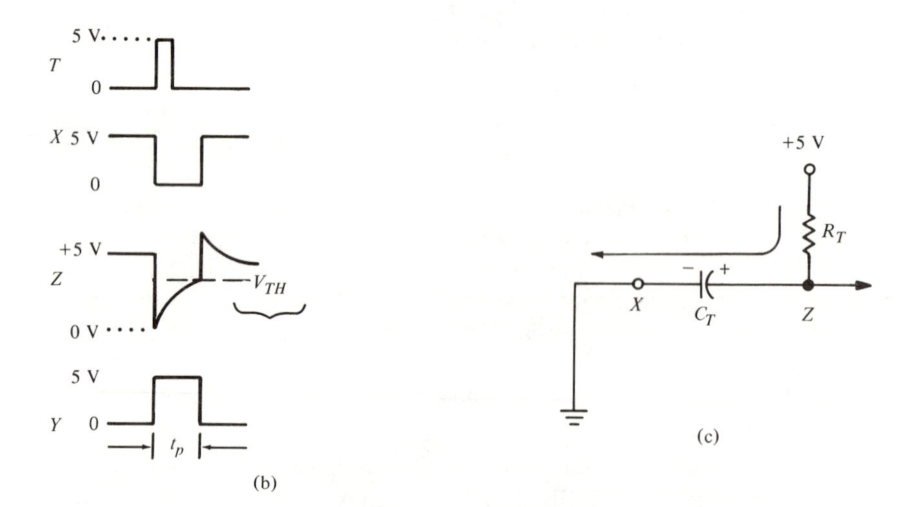

Figure 8–33 **(a) OS circuit (b) Waveforms (c) C_T charging path**

$R_T C_T$ time constant. The exact value depends on the values of V_{CC}, V_{TH}, and the gate characteristics.

10. This operation is repeated for each pulse at T; that is, pulses with duration t_p are produced at X and Y. It is important to note that the second pulse at T should not occur until after the recovery-time interval. Otherwise, C_T will not be completely discharged, and the second value of t_p will be shorter than that for the first input pulse.

As stated earlier, many one-shots are available as integrated circuit logic elements. The IC one-shot requires only the external components R_T and C_T. The user chooses these to provide the desired t_p. We will use the OS symbol of Figure 8–32 and will assume that R_T and C_T have been properly connected.

One-shot applications The following examples will illustrate how a OS can be used in various applications. Several other applications are covered in the questions at the end of the chapter.

Example 8.7 A certain binary counter is being used to count pulses that are continuously supplied by a signal source. It is required that the counter count these pulses for a total time duration of 1 s in response to the application of a START pulse. Design a circuit to do this.

Solution: Figure 8–34 shows one possible solution. The pulses from the signal source are fed to an AND gate along the Q output of a OS. Since Q is normally LOW the AND output will be kept LOW, and no pulses will reach the counter. When the START pulse occurs it will cause Q to go HIGH for a duration of $t_p = 1$ s. With $Q = 1$, the output of the AND gate will allow the input pulses into the counter to be counted. After 1 s the OS output returns to 0, and the counter will receive no further pulses.

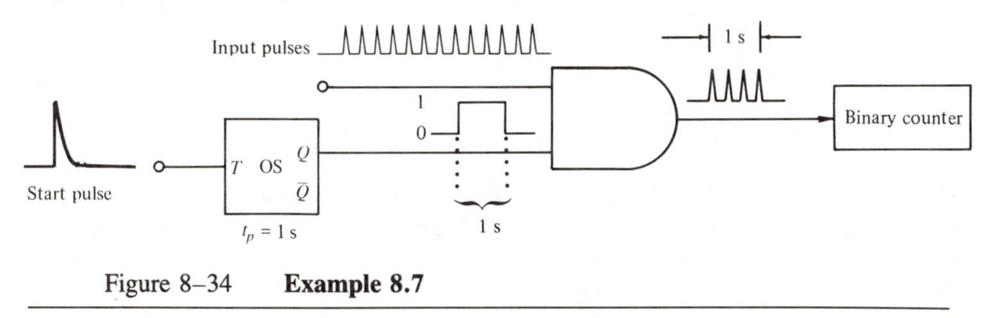

Figure 8–34 **Example 8.7**

Example 8.8 Design a circuit that will generate a positive 50 μs wide pulse whenever the two logic voltages A and B are both high. Use a OS that triggers on negative-going transitions.

Solution: Figure 8–35 shows one possible solution. The NAND-gate output is normally HIGH and will go LOW when both A and B are HIGH. The negative-going transition at the NAND output triggers the OS, producing a 50 μs positive pulse at the C output.

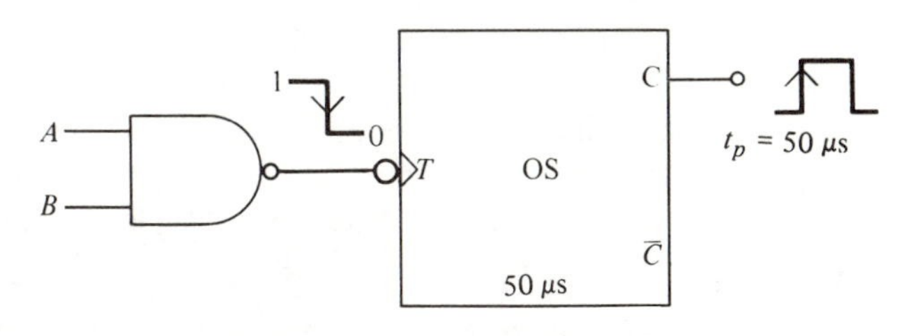

Figure 8–35 **Example 8.8**

Example 8.9 Design a circuit that will produce a positive 100 μs wide pulse 20 μs after the occurrence of a positive trigger spike.

Solution: Figure 8–36 shows one possible solution using two OSs that trigger on positive-going transitions. One-shot A is triggered by the trigger spike, resulting in a 20 μs negative-going pulse at $\overline{A}$. After 20 μs the $\overline{A}$ output goes positive; this positive transition triggers OS B, resulting in a positive 100 μs pulse at output B. Thus, the first OS provides the 20 μs delay, and the second OS produces the desired output pulse.

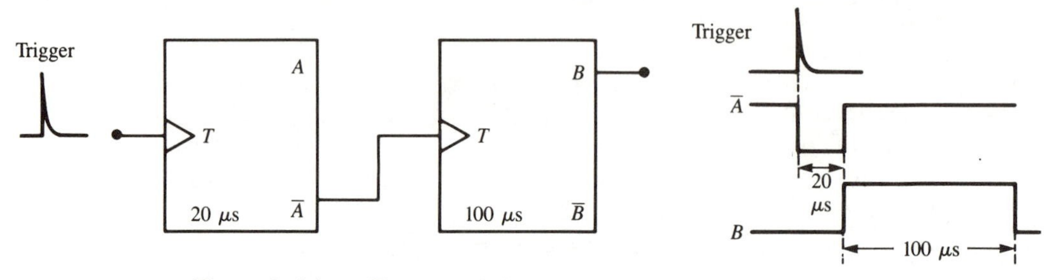

Figure 8–36 **Example 8.9**

8.16 Analyzing Flip-Flop Circuits

Once one has learned the basic operation of logic gates FFs, and OSs, the analysis of circuits that contain combinations of these elements is fairly straightforward. When the circuit is being controlled by a single master clock signal, the following procedure can be used to analyze the circuit operation.

1. Determine the initial logic levels present at each of circuit inputs and outputs *prior* to the occurrence of the first clock pulse.
2. Use these levels to determine the response of each FF and OS to the occurrence of the first clock pulse.
3. Determine the logic levels at each input and output prior to the occurrence of the second clock pulse.
4. Use these levels to determine how each FF and OS responds to the second clock pulse.
5. Repeat steps 3 and 4 for each successive clock pulse.

Example 8.10 Consider the circuit of Figure 8–37. Initially, all of the FFs are in the 0 state before the clock pulses are applied. These pulses have a repetition rate of 1 kHz. Determine the waveforms at X, Y, Z, W, $\overline{Q}$, A, and B for sixteen cycles of the clock input.

Solution: Initially, the FFs and the OS are in the 0 state, so $X = Y = Z = W = Q = 0$. The inputs to the NAND gate are $X = 0$, $\overline{Y} = 1$, and $\overline{Z} = 1$, so its output is $A = 1$. The inputs to the OR gate are $W = 0$ and $\overline{Q} = 1$, so its output is $B = 1$.

As long as B remains HIGH, FF Z will have its J and K inputs both HIGH, so it will operate in the toggle mode. Flip-flops X and Y are kept in the toggle mode since their J and K inputs are held at the 1 level permanently.

It should be clear, then, that FFs X, Y, and Z will operate as a three-bit counter as long as B stays HIGH. Thus, FF Z will toggle on the negative transition of each input clock pulse, FF Y will toggle on the negative transition of the Z output, and FF X will toggle on the negative transition of FF Y. This operation continues for the clock pulses 1 through 4.

When clock pulse 4 makes its negative transition, Z goes LOW, Y goes LOW, and X goes HIGH. Therefore, the inputs to the NAND gate become $X = 1$, $\overline{Y} = 1$, and $\overline{Z} = 1$, causing the NAND output A to go LOW. This negative transition at A triggers the OS to its quasi-stable state, so $\overline{Q}$ goes LOW for 3.5 ms. This negative transition of $\overline{Q}$ will not affect FF W since this D-type FF triggers on positive transitions at its CLK input. With $\overline{Q}$ and W both LOW, the OR gate output B goes LOW, so FF Z now has $J = 1$, $K = 0$. This means that Z will be *set* to the 1 state on the *next* clock pulse and will remain in this state for each successive clock pulse as long as $J = 1$, $K = 0$. Thus, Z goes HIGH on clock pulse 5 and stays HIGH for clock pulses 6 and 7. Note that A goes back HIGH when Z goes HIGH on pulse 5 because $\overline{Z} = 0$.

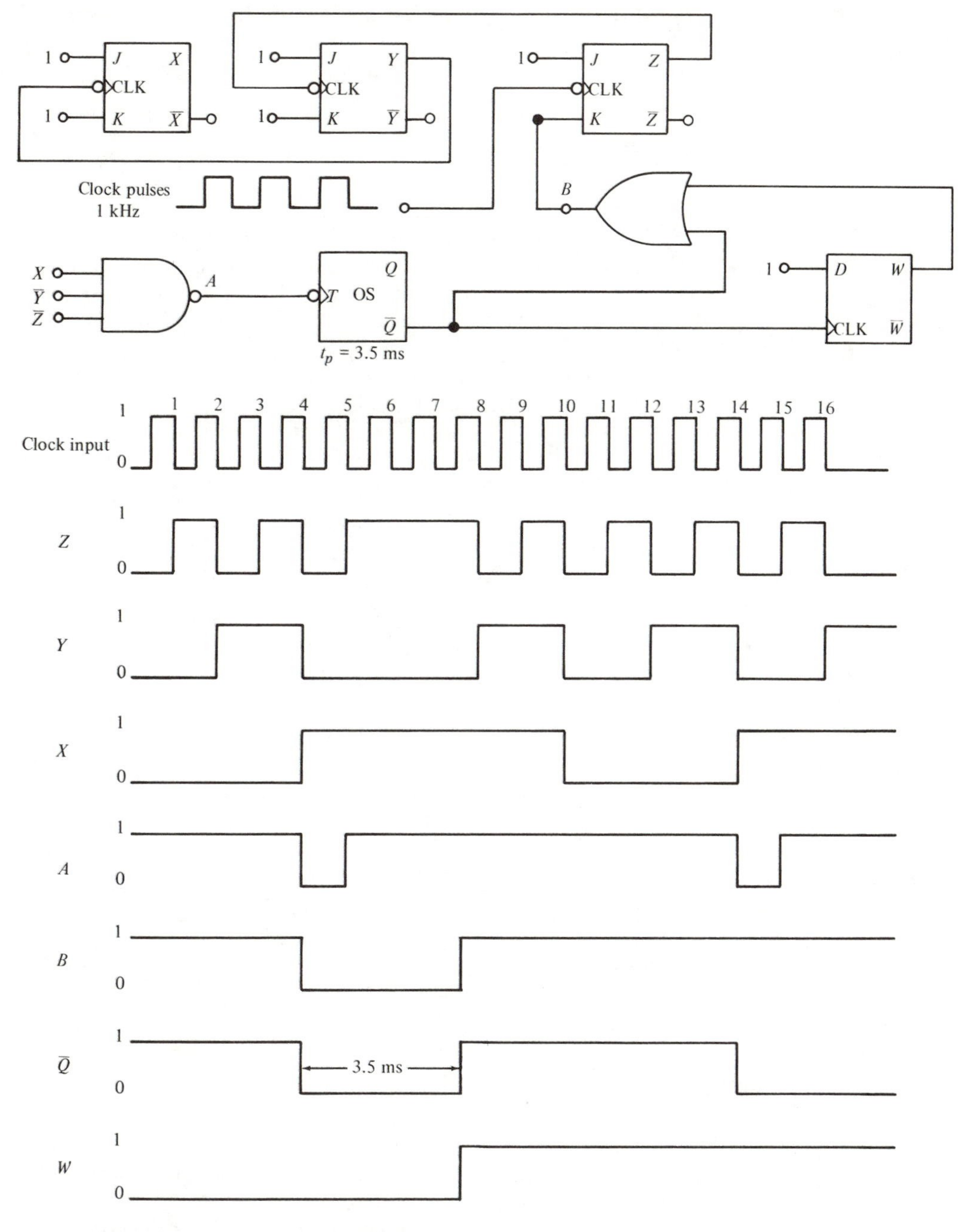

Figure 8–37 **Example 8.10**

After 3.5 ms the OS returns to its stable state with $Q = 0$, $\overline{Q} = 1$. This positive transition of $\overline{Q}$ causes FF W to go HIGH since $D = 1$. The 1 at W produces a 1 at B, so FF Z is back in the toggle mode with $J = K = 1$. The counter will now operate properly in response to all succeeding clock pulses.

On the negative transition of clock pulse 14 the NAND output A again goes LOW, thereby triggering the OS. However, the OS will not affect the K input of Z since W is keeping the OR output HIGH. Also, since its D input is HIGH, FF W will remain in the $W = 1$ state.

Questions

Sections 8.1–8.3

8.1 The NOR-gate FF of Figure 8–2 has waveforms w and x of Figure 8–38(a) applied to its SET and CLEAR inputs, respectively. Assume that initially $Q = 0$, and determine the Q and $\overline{Q}$ output waveforms in response to these inputs.

8.2 Waveforms w and y of Figure 8–38(a) are applied to the SET and CLEAR inputs of the FF of Figure 8–6. Assume that $Q = 1$ initially, and determine the Q and $\overline{Q}$ waveforms.

8.3 Why should the x and y waveforms in Figure 8–38(a) not be applied to the respective SET and CLEAR inputs of either of the FFs of Figures 8–2 or 8–6?

8.4 The x and y waveforms of Figure 8–38(a) are applied to the gates in Figure 8–38(b). The z_1 output signal is applied to the S input of the FF of Figure 8–3, and z_2 is applied to the C input. Determine the Q waveform. Assume that $Q = 0$ initially.

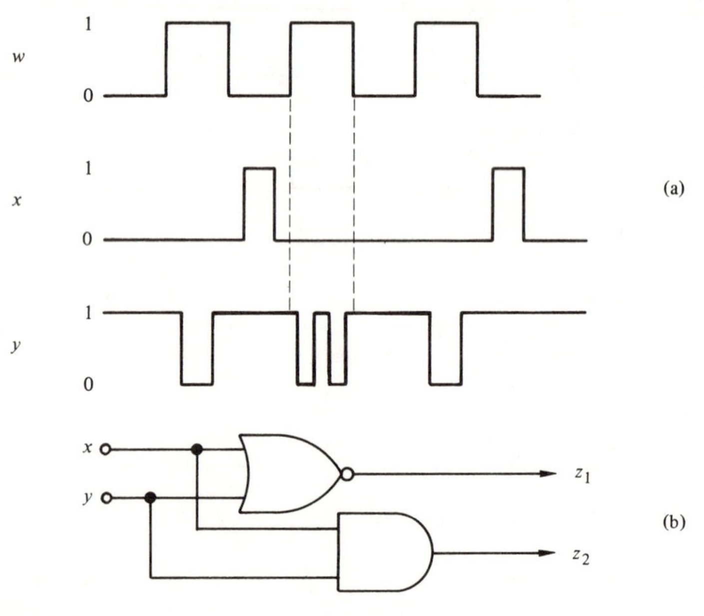

Figure 8–38

8.5 Change the NOR gate to an OR gate, and the AND gate to a NAND gate in Figure 8–38(b). Then apply z_1 and z_2 to the S and C inputs respectively of the FF in Figure 8–7, and determine Q.

8.6 Modify the circuit of Figure 8–5 so that the alarm is activated when the photocell goes from dark to light.

8.7 Modify the circuit of Figure 8–9 so that it uses a NOR-gate FF.

Sections 8.4–8.6

8.8 Apply the S, C, and CLK waveforms of Figure 8–11 to the FF of Figure 8–12, and determine the Q waveform.

8.9 A toggle FF operates such that the FF output changes state for each pulse applied to its input. The clocked *SC* FF can be wired to operate in the toggle mode as shown in Figure 8–39. The waveform applied to the CLK input is a 1 kHz square wave. Verify that this arrangement operates in the manner described, and then determine the Q output waveform. Assume that $Q = 0$ initially.

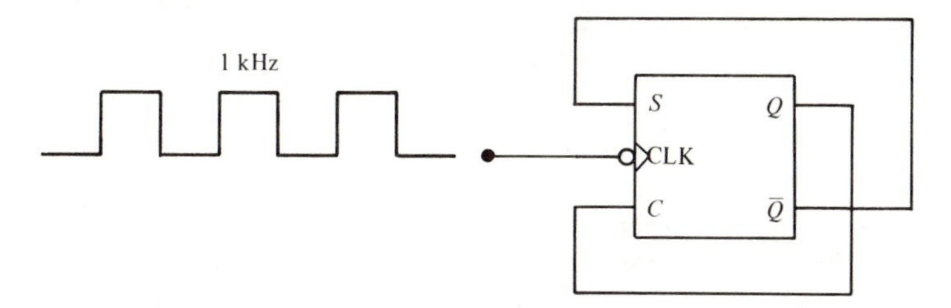

Figure 8–39

8.10 Apply the J, K, and CLK waveforms of Figure 8–13 to the FF of Figure 8–14. Assume that $Q = 1$ initially, and determine the Q waveform.

8.11 A 1 kpps square wave is applied to the CLK input of a negative-edge triggered *JK* FF.

(a) If $J = K = 1$, determine the duty cycle and frequency of the Q output.

(b) Repeat if the input is changed to a 1 kpps pulse waveform with a 20 percent duty cycle.

Section 8.7

8.12 One application of the D-type FF is as a *digital delay* circuit that delays a logic waveform by *one* clock cycle. Determine the output waveform in Figure 8–40. Then modify the circuit to delay the logic waveform by *two* clock cycles.

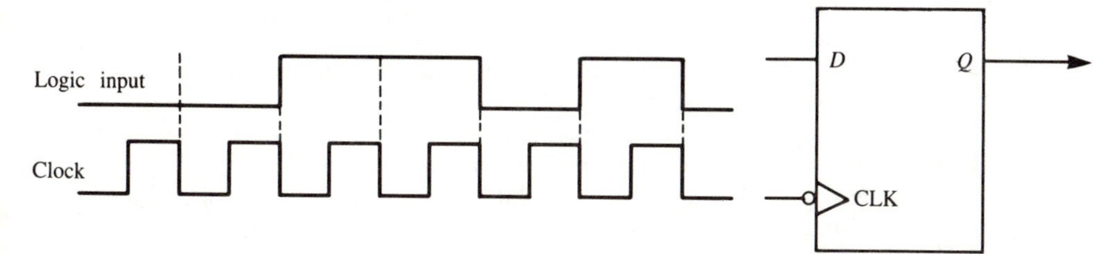

Figure 8–40

8.13 Change the D FF of Figure 8–15 to one that triggers on negative-going CLK transitions. Determine the Q waveform for the given D and CLK waveforms.

8.14 Prove that a D FF that has its $\overline{Q}$ output connected to its D input will toggle in response to the signal at CLK.

8.15 Explain how a D-type latch differs from an edge-triggered D FF.

Sections 8.8–8.10

8.16 Describe the differences between asynchronous and synchronous FF inputs.

8.17 Determine the Q waveform for the FF in Figure 8–41. Assume that $Q = 0$ initially.

8.18 Determine the Q output waveform if the input waveforms of Figure 8–20 are applied to a FF that responds to positive-going transitions at CLK.

8.19 A certain clocked *JF* FF has set-up and hold times specified typically as $t_s = 30$ ns and $t_H = 5$ ns.

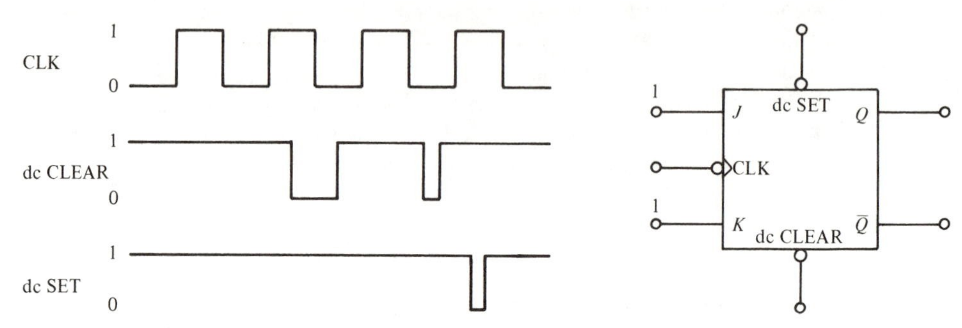

Figure 8–41

(a) What is the minimum amount of time that the J or K inputs have to be stable before the CLK input transition occurs? What is the maximum?

(b) How long must the J or K inputs remain HIGH once the CLK transition occurs?

8.20 What is the major difference in the operation of edge-triggered FFs and M/S FFs?

8.21 *True or false*

(a) The dc SET input will affect the FF output only while the CLK input is LOW.

(b) Master/slave FFs are sensitive only to what the synchronous control inputs are immediately prior to the CLK transition.

(c) In a clocked *SC* FF the Q output goes HIGH as soon as the S input goes HIGH.

(d) M/S FFs require stable control inputs when the CLK edge occurs.

(e) A FF's synchronous inputs should not be used without also using the CLK input.

(f) The dc SET or dc CLEAR inputs are effective only while a FF's CLK input and control inputs are HIGH.

Sections 8.11–8.12

8.22 Show the circuit arrangement required to perform parallel transfer from register X to register Y using four *JK* FFs in each register. Count the total number of connections between the two registers.

8.23 Repeat Problem 8.22, using serial transfer.

8.24 Show how to construct a shift register using four D FFs.

8.25 A *recirculating* shift register is a shift register that keeps the logic information circulating through the register as clock pulses are applied. The shift register of Figure 8–28 can be made into a circulating register by connecting X_0 to the J of FF X_3, and $\overline{X}_0$ to the K of FF X_3. No external J, K inputs are used. Assume that this circulating register starts out with 1011 stored in it (that is, $X_3 = 1, X_2 = 0, X_1 = 1$, and $X_0 = 1$). List the sequence of states that the register FFs go through as eight SHIFT pulses are applied.

8.26 Refer to Figure 8–29, where a three-bit number stored in register X is serially shifted into register Y. How could the circuit be modified so that at the end of the transfer operation the original number stored in X is present in both registers? *Hint:* See Question 8.25.

8.27 Compare the relative advantages of serial and parallel transfer.

Section 8.13

8.28 Refer to the binary counter of Figure 8–30. Change it by connecting $\overline{X}_0$ to the CLK of FF X_1, and $\overline{X}_1$ to the CLK of FF X_2. Start out with all FFs in the 1 state, and draw the various FF output waveforms (X_0, X_1, X_2) for sixteen input pulses. Then, list the sequence of FF states as was done in Figure 8–31. This counter is called a *down counter.* Why?

8.29 How many FFs would be required to count from 0 to 200?

8.30 How many FFs are in a mod-32 counter? If a 640 kHz clock signal is applied to a mod-32 counter, what is the frequency at the output of the last FF (the one farthest from the clock input signal)?

Section 8.14

8.31 Determine the waveforms at Q_1, Q_2, and Q_3 in response to the single input pulse in Figure 8–42. Each OS triggers on negative-going transitions.

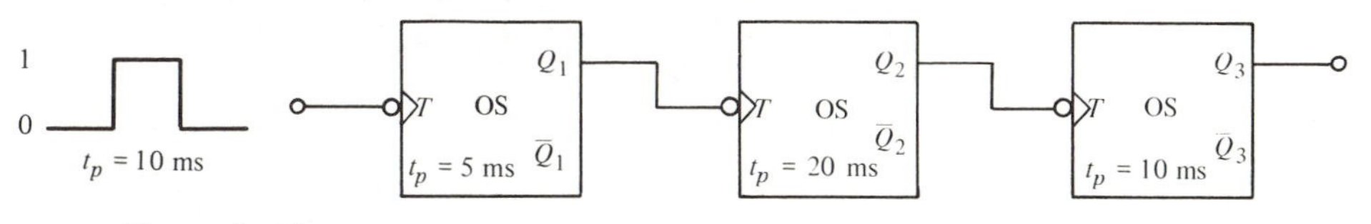

Figure 8–42

8.32 Repeat Example 8.9 using two OSs that trigger on negative-going transitions.

8.33 Consider the circuit in Figure 8–43. The input pulses are arriving at the rate of 100 pulses/s. Assume FF A is initially in the 0 state ($A = 0$). Determine the output waveform.

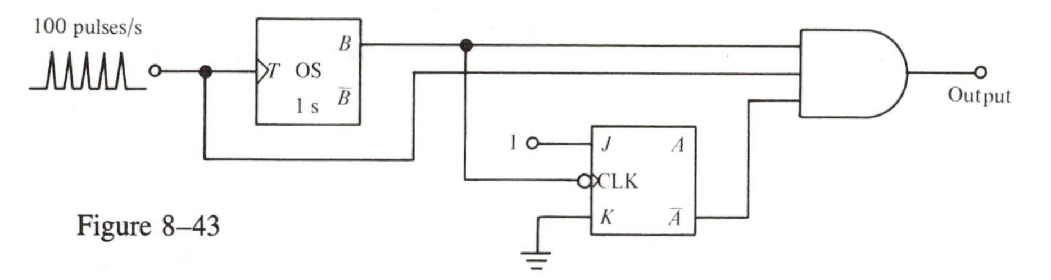

Figure 8–43

8.34 The circuit in Figure 8–44 uses a OS as a part of a *frequency-to-voltage converter.* The purpose of the circuit is to produce a dc output voltage that is proportional to the frequency of the input pulses. The OS converts the input pulses to rectangular pulses with this same repetition frequency and with $t_p = 100\ \mu s$. The RC network is used as a low-pass filter which is designed to pass only the dc (average) value of the OS output.

(a) Determine V_{OUT} for an input pulse frequency of 1 kHz. Assume the OS output levels are 0 V and 10 V.

(b) Repeat part (a) for 2 kHz.

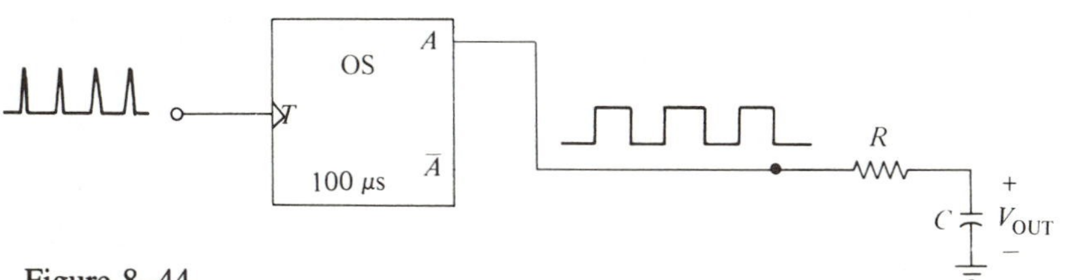

Figure 8–44

8.35 In the circuit of Figure 8–44 what factors determine the R and C values of the filter network? If the minimum input frequency is 500 Hz, determine appropriate values of R and C. Assume the output resistance of the OS output is 1 kΩ.

8.36 Figure 8–45(a) shows a situation that sometimes occurs in a digital system. A push-button switch is being used to provide a pulse at the output of the AND gate when $A = 1$. The gate output should normally be LOW (switch open) and goes HIGH for the period of time the switch is depressed. However, due to the phenomenon of *switch bounce* the gate output will jump back and forth between HIGH and LOW for several milliseconds, as shown in the figure. The contacts of many mechanical switches usually vibrate open and closed for a few milliseconds (<10 ms) before reaching a stable closed position.

In many cases the waveform at X is undesirable because the extra "bounce" pulses will cause erroneous operation of the circuit being driven. Several types of "debounce" circuits can be used to eliminate the effects of switch bounce. One such circuit is shown in Figure 8–45(b). Analyze and explain its operation, paying particular attention to the waveforms at Q and Z.

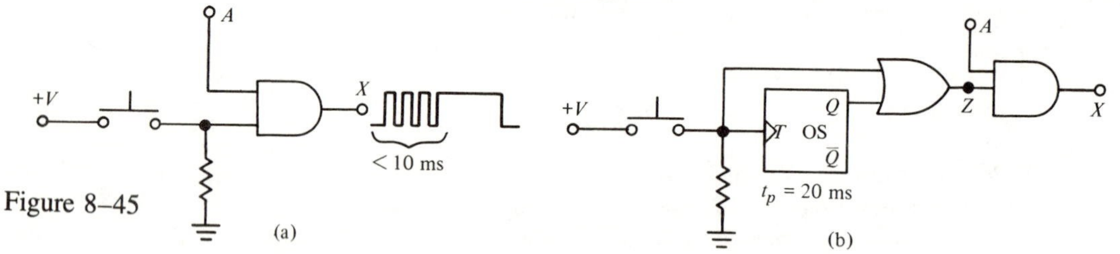

Figure 8–45

Section 8.15

8.37 Assume that both *JK* FFs in Figure 8–46 are initially in the 0 state. Plot the A and B waveforms in response to the input pulses.

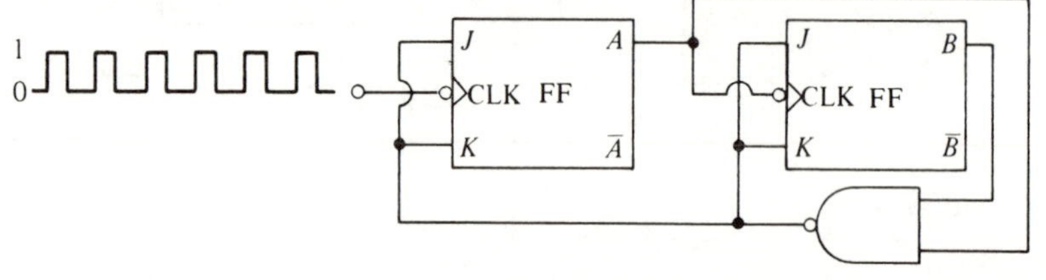

Figure 8–46

8.38 Consider the circuit of Figure 8–47. Initially, all FFs are in the 0 state. The circuit operation begins with a momentary START pulse applied to the dc SET inputs of FFs X and Y. Determine the waveforms at A, B, C, X, Y, Z, and W for twenty cycles of the clock pulses after the START pulse.

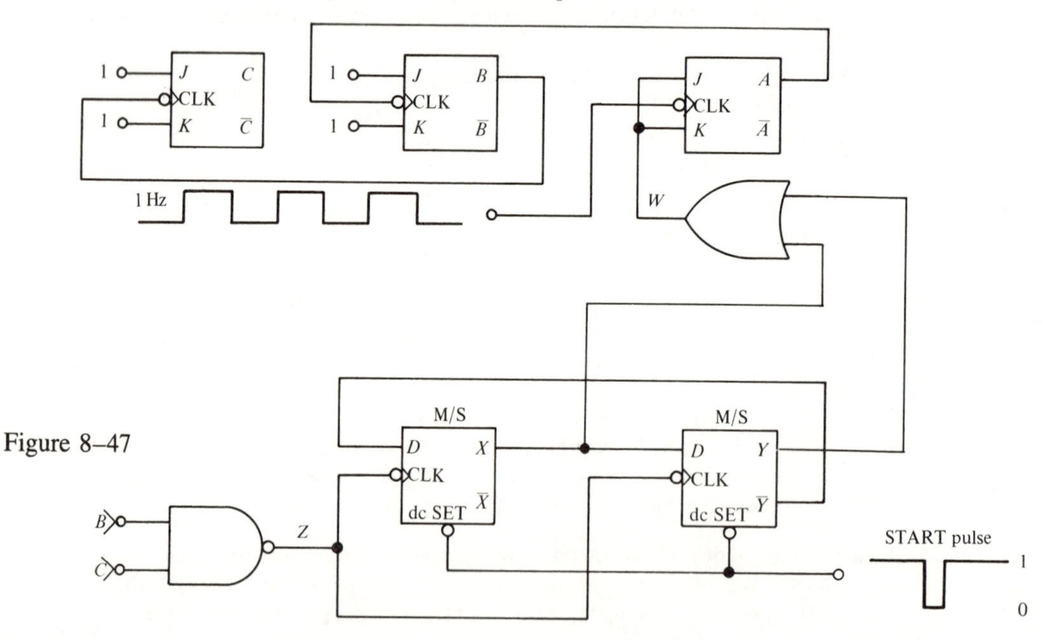

Figure 8–47

9

Integrated Circuit Logic Family Characteristics

9.1 Introduction

All modern digital systems use digital integrated circuits because they provide increased reliability along with a reduction in both size and cost. Digital IC technology has advanced rapidly from small-scale integration (SSI) with only a few logic gates per chip, to very-large-scale integration (VLSI) with thousands of gates per chip.

Anyone working in this field must know about the various *logic families*. A logic family is a specific class of logic circuits made using a specific manufacturing process. These families can be divided into two broad categories according to whether they use bipolar transistors or MOSFETs as their principal circuit element. The bipolar families include TTL, ECL, and I^2L. The MOS families include P-MOS, N-MOS, CMOS, and SOS.

In this chapter we will study the electrical characteristics of each of these families. Our emphasis will be on the IC's input and output characteristics and not its internal operation. Because TTL and CMOS currently dominate the digital field, we will concentrate most heavily on their characteristics.

9.2 Digital IC Terminology

Although there are many digital IC manufacturers, much of the nomenclature and terminology is fairly standardized. The most useful terms are defined and discussed in the following sections.

Input and output currents

High-level input voltage V_{IH} The voltage level required for a logic 1 at an *input*. Any voltage below this level will not be accepted as a HIGH (1) by the logic circuit.

Low-level input voltage V_{IL} The voltage level required for a logic 0 at an *input*. Any voltage above this level will not be accepted as a LOW (0) by the logic circuit.

High-level output voltage V_{OH} The voltage level at a logic circuit *output* in the logic-1 state. The minimum value of V_{OH} is usually specified.

Low-level output voltage V_{OL} The voltage level at a logic circuit *output* in the logic-0 state. The maximum value of V_{OL} is usually specified.

High-level input current I_{IH} The current that flows into an input when a specified high-level voltage is applied to that input.

Low-level input current I_{IL} The current that flows into an input when a specified low-level voltage is applied to that input.

High-level output current I_{OH} The current that flows from an output in the logic-1 state under specified load conditions.

Low-level output current I_{OL} The current that flows from an output in the logic-0 state under specified load conditions.

Fan-out In general, a logic-circuit output is required to drive several logic inputs. The *fan-out* (also called *loading factor*) is defined as the *maximum* number of standard logic inputs that an output can drive reliably. For example, a logic gate that is specified to have a fan-out of ten can drive ten standard inputs. If this number is exceeded, the output logic level voltages cannot be guaranteed.

Transition times Some digital circuits respond to logic levels at their inputs, but others are activated by the rate of change in voltage. In the latter type of circuit it is essential that the input signals have sufficiently fast transitions or the circuit may not respond properly. For this reason, the rise-time, t_R, and fall-time, t_F, of logic outputs are often specified. The values of t_R and t_F are not necessarily equal, and both are dependent on the amount of loading placed on a logic output.

Propagation delays A logic signal always experiences a delay in going through a circuit. The two propagation delay times are defined as

t_{PLH}—*Delay time in going from logic-0 to logic-1 state* (LOW-to-HIGH)

t_{PHL}—*Delay time in going from logic-1 to logic-0 state* (HIGH-to-LOW)

Figure 9–1 illustrates these propagation delays for an inverter. Note that t_{PHL} is the delay in the *output's* response as it goes to the 0 state; vice versa for t_{PLH}.

In general, t_{PLH} and t_{PHL} do not have the same value. Both will vary depending on loading conditions. The values of t_{PLH} and t_{PHL} are used as a measure of the relative speed of logic circuits. For example, a logic circuit with values of 10 ns is faster than one with values of 20 ns.

Power requirements The amount of power required by an IC is an important characteristic and is always specified on the manufacturer's data sheet. Sometimes it is given

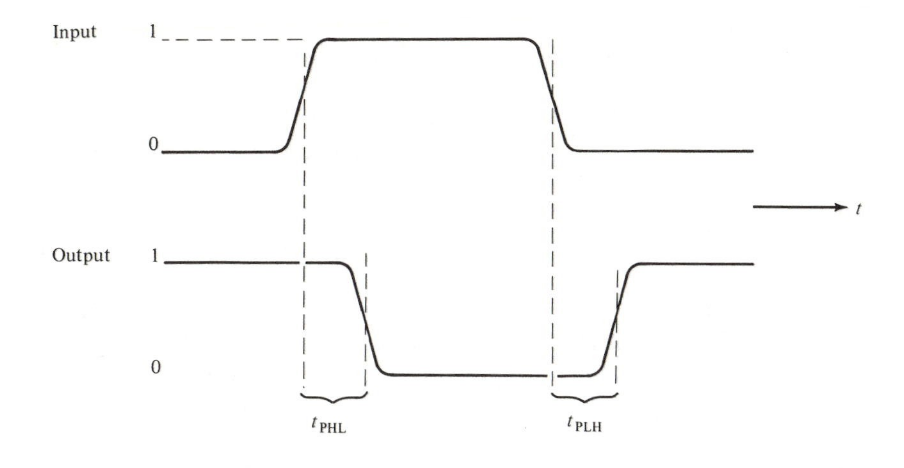

Figure 9–1 **Propagation delays**

directly as average power dissipation, P_D. More often it is indirectly specified in terms of the current drain from the IC power supply. This current is symbolized as I_{CC}. When the value for I_{CC} is known, the power drawn from the supply is obtained by multiplying I_{CC} by the power supply voltage.

For many ICs the value of supply current I_{CC} will be different for the two logic states. In such cases two values for I_{CC} are specified: I_{CCH} is the supply current when all *outputs* on the IC chip are HIGH; I_{CCL} is the supply current when all *outputs* are LOW.

Noise immunity Stray electrical and magnetic fields can induce voltages on the connecting wires between logic circuits. These unwanted, spurious signals are called *noise* and can sometimes cause the voltage at the input to a logic circuit to drop below V_{IH} or rise above V_{IL} and produce unreliable operation.

The *noise immunity* of a logic circuit refers to the circuit's ability to tolerate noise voltages on its inputs. A quantitative measure of noise immunity called *noise margin* is illustrated in Figure 9–2.

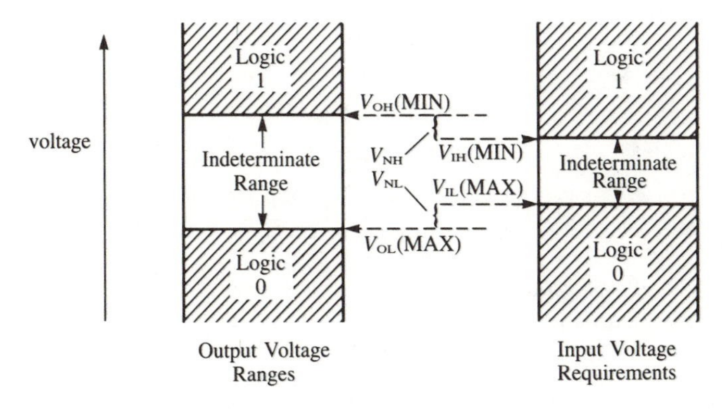

Figure 9–2 **Noise margins**

Figure 9–2(a) shows the range of voltages that can occur at a logic-circuit output. Any voltages greater than V_{OH}(MIN) are considered a logic-1, and any voltages lower than

V_{OL}(MAX) are considered a logic-0. Voltages in the indeterminate range should not appear at a logic-circuit output under normal conditions. Figure 9–2(b) shows the voltage requirements at a logic-circuit input. The logic circuit will respond to any input greater than V_{IH}(MIN) as a logic-1 and will respond to voltages lower than V_{IL}(MAX) as a logic-0. Voltages in the indeterminate range will produce an unpredictable response and should not be used.

The *high-state noise margin* V_{NH} is defined as

$$V_{NH} = V_{OH}(\text{MIN}) - V_{IH}(\text{MIN}) \tag{9–1}$$

As illustrated in Figure 9–2, V_{NH} is the difference between the lowest possible HIGH output and the minimum input voltage required for a HIGH. When a HIGH logic output is driving a logic-circuit input, any negative noise spikes smaller than V_{NH} appearing on the signal line will not cause the voltage to drop into the indeterminate range where unpredictable operation can occur.

The *low-state noise margin* V_{NL} is defined as

$$V_{NL} = V_{IL}(\text{MAX}) - V_{OL}(\text{MAX}) \tag{9–2}$$

and it is the difference between the largest possible LOW output and the maximum input voltage required for a LOW. When a LOW logic-output is driving a logic input, any positive noise spikes smaller than V_{NH} will not cause the voltage to rise into the indeterminate range.

Example 9.1 Here are the input/output voltage parameters for the standard TTL family. Calculate the noise margins.

Parameter	Min (V)	Typical (V)	Max (V)
V_{OH}	2.4	3.6	
V_{OL}		0.2	0.4
V_{IH}	2.0*		
V_{IL}			0.8*

*Normally only the minimum V_{IH} and maximum V_{IL} values are given.

Solution:

$$\begin{aligned} V_{NH} &= V_{OH}(\text{MIN}) - V_{IH}(\text{MIN}) \\ &= 2.4\text{ V} - 2.0\text{ V} = 0.4\text{ V} \\ V_{NL} &= V_{IL}(\text{MAX}) - V_{OL}(\text{MAX}) \\ &= 0.8\text{ V} - 0.4\text{ V} = 0.4\text{ V} \end{aligned}$$

These values are actually *worst-case guaranteed* noise margins. In typical operation, the noise margins might be around 1 V each.

Strictly speaking, the noise margins predicted by equations (9–1) and (9–2) are termed *dc noise margins*. The term *dc noise margin* might seem somewhat inappropriate when dealing with noise that is generally thought of as an ac signal of the transient variety. However, in today's high-speed ICs a pulse width of 1 μs is considered extremely long and may be treated as dc as far as the response of a logic circuit is concerned. As pulse widths decrease to the low nanosecond region, a point is reached where the *pulse duration* is too short for the circuit to respond. At this limit, the pulse amplitude would have to be increased to produce a change in the circuit output. This means that a logic circuit can tolerate a large noise amplitude if the noise pulse is of a very short duration. In other words, a logic circuit's *ac noise margins* are generally substantially greater than its dc noise margins [given by equations (9–1) and (9–2)]. Manufacturers usually supply information about ac noise margins in the form of a graph such as that in Figure 9–3. Note that the noise margins are constant for pulse widths greater than 10 ns but increase rapidly for narrower pulses.

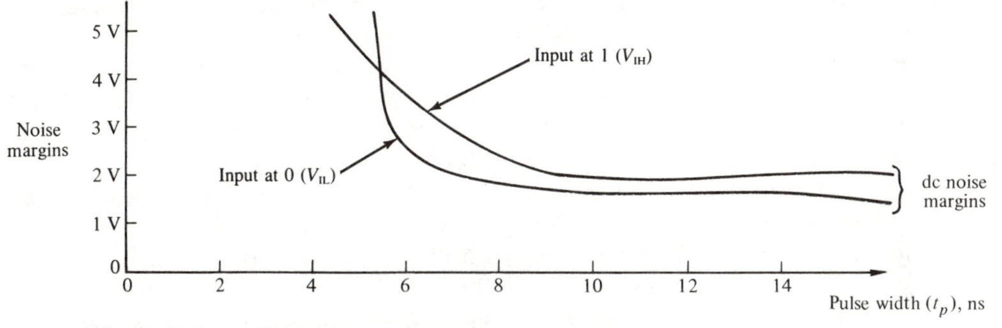

Figure 9–3 **Typical ac noise immunity graphs**

Current-sourcing and current-sinking logic Logic families can be categorized according to how current flows from the output of one logic circuit to the input of another. Figure 9–4(a) illustrates *current-sourcing* logic. When the output of gate 1 is in the HIGH state, it supplies a current I_{IH} to the input of gate 2, which acts essentially as a resistance to ground. Thus, the output of gate 1 is acting as a source of current for the gate 2 input.

Current-sinking logic is illustrated in Figure 9–4(b). Here the input circuitry of gate 2 is represented as a resistance tied to $+V_{CC}$, the positive terminal of a power supply. When the gate 1 output goes to its LOW state, current will flow in the direction shown from the input circuit of gate 2 back through the output resistance of gate 1 to ground. In other words, the circuit driving an input of gate 2 in the LOW state must be able to *sink* a current, I_{IL}, coming from that input.

The distinction between current-sourcing and current-sinking logic circuits is an important one, which will become more apparent as we examine the various logic families.

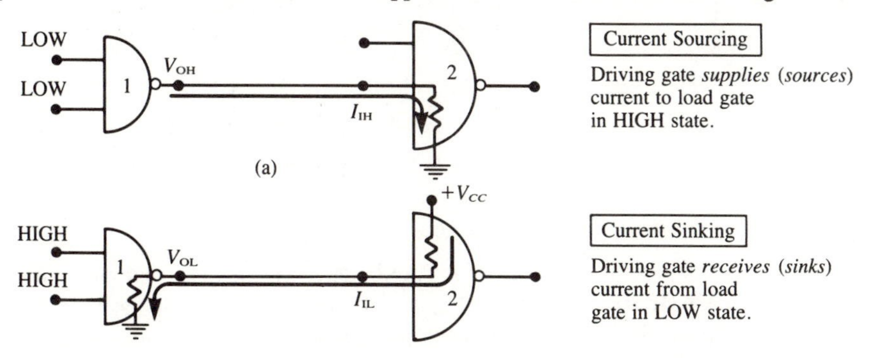

Figure 9–4 **Comparison of current-sourcing and current-sinking actions**

9.3 The TTL Logic Family

At this writing, the most widely used family of logic circuits is the transistor-transistor logic (TTL) family. Figure 9–5(a) is the circuit diagram for the basic TTL logic gate. Notice the *multiple-emitter* input transistor Q_1 and the *totem-pole* arrangement of output transistors Q_3 and Q_4. The TTL family uses *bipolar* transistors and so falls into the category of bipolar logic families.

Circuit operation The analysis can be simplified somewhat by using the diode equivalent of the multiple-emitter transistor Q_1 as in Figure 9–5(b). Diodes D_2 and D_3 represent the two base-emitter diodes, and D_4 is the collector-base diode. When inputs A and B are both HIGH, diodes D_2 and D_3 will not conduct. Thus, the current from the +5 V supply through R_1 will flow through D_4 into the base of Q_2, thereby turning Q_2 ON. Emitter current from Q_2 then turns Q_4 ON, while the low voltage at Q_2's collector keeps Q_3 OFF (this is ensured by diode D_1). With Q_4 ON, the voltage at output X will be very low. Thus, with both inputs HIGH, the output will be LOW.

If one (or both) of the inputs is at a LOW voltage, then its corresponding input diode will be forward-biased. Current will flow from the +5 V supply through R_1 and the input diode to ground. This will clamp point Y[see Figure 9–5(b)] to a voltage too small to turn ON D_4 and Q_2. Thus Q_2 will turn OFF, which in turn causes Q_4 to turn OFF. The high voltage at the collector of Q_2 now allows Q_3 to conduct. Then, Q_3 will act as an emitter follower, producing a high voltage at output terminal X, which will be typically around 3.6 V (5 V minus the 0.7 V voltage drops at the base-emitter of Q_3 and diode D_1) Thus, with *any* input LOW, the output will be HIGH.

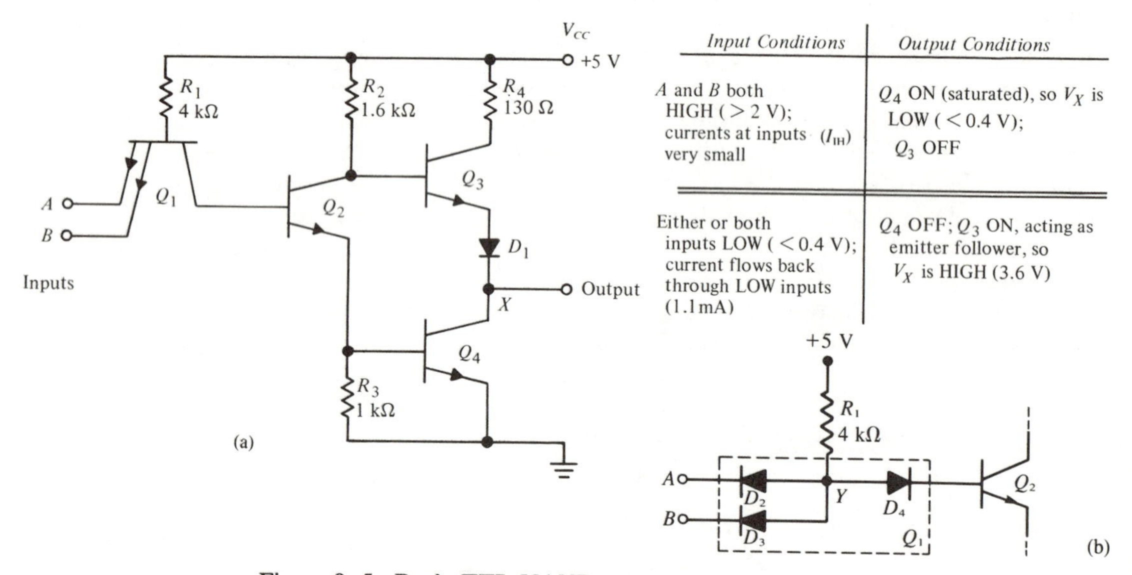

Input Conditions	Output Conditions
A and B both HIGH (> 2 V); currents at inputs (I_{IH}) very small	Q_4 ON (saturated), so V_X is LOW (< 0.4 V); Q_3 OFF
Either or both inputs LOW (< 0.4 V); current flows back through LOW inputs (1.1mA)	Q_4 OFF; Q_3 ON, acting as emitter follower, so V_X is HIGH (3.6 V)

Figure 9–5 **Basic TTL NAND gate**

Clearly, this circuit functions as a NAND gate since a LOW output occurs only when all inputs are HIGH. The table in Figure 9–5 summarizes the operation of the gate for both output states. This table also indicates that with a HIGH input voltage only a very small input current flows since the emitter-base junctions are reverse-biased. With a LOW input voltage, current will flow back through the corresponding emitter lead to the input. These are characteristics of current-sinking logic circuits.

TTL logic circuits are current-sinking circuits in which outputs *receive* current from the inputs they are driving in the LOW state. Figure 9–6 shows the output of one TTL gate driving the input of another. When the driving gate output is LOW, transistor Q_4 is saturated, and Q_3 is OFF. The LOW voltage at X forward-biases the emitter of Q_1, and current flows back through Q_4, as shown. The saturation collector current of Q_4 is provided by the gate being driven. Q_4 is acting as a current sink.

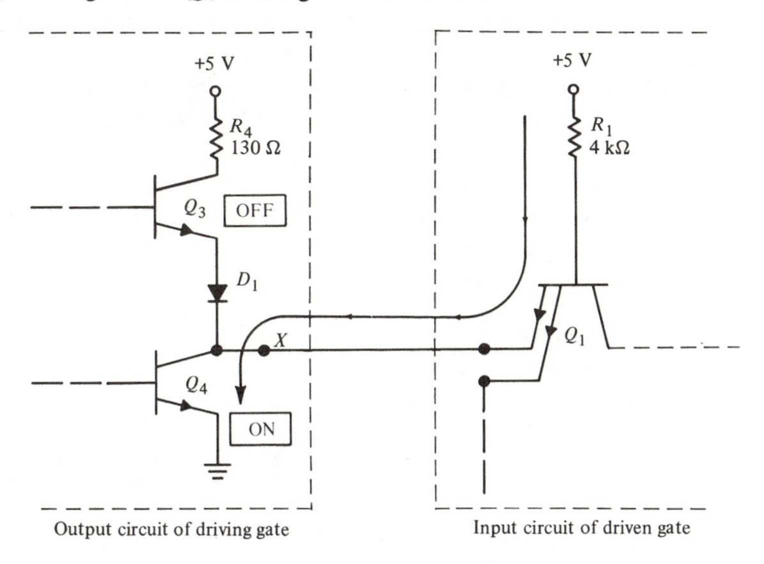

Figure 9–6 **Illustration of TTL current-sinking action when Q_4 is ON and Q_3 is OFF**

Totem-pole output circuit Several points should be mentioned concerning the totem-pole arrangement of the TTL output circuit since it is not readily apparent why it is used. The same logic could be accomplished by eliminating Q_3 and D_1 and connecting the bottom of R_4 to the collector of Q_4. However, this would mean that Q_4 would conduct a fairly heavy current in its saturation state (5 V/130 Ω ≈ 40 mA). With Q_3 present there will be no current through R_4 in the output LOW state. This is important because it reduces the circuit power dissipation.

Another advantage of this arrangement occurs in the output HIGH state. Here, Q_3 is acting as an emitter follower with its associated low output impedance (typically 70 Ω). This low output impedance provides a short time constant for charging up any capacitive load on the output. This action (commonly called *active pull-up*) provides very fast rise-time wave-forms at TTL outputs.

A disadvantage of the totem-pole output arrangement occurs during the transition from LOW to HIGH. Unfortunately, Q_4 turns OFF more slowly than Q_3 turns ON, so there is a period of a few nanoseconds during which both transistors are conducting, and a relatively large current (30 mA–40 mA) will be drawn from the 5 V supply. This situation can present a problem, which we shall examine later.

9.4 Standard TTL Series Characteristics

In 1964 Texas Instruments introduced the first standard product line of TTL circuits. The 5400/7400 series, as it is called, has been one of the most widely used families of IC logic. We will simply refer to it as the 7400 series since the only difference between the 5400 and 7400 versions is that the 5400 series is meant for military use and will operate over wider temperature and power supply ranges. Many IC manufacturers now produce the 7400 line of ICs although some use their own identification numbers. For example, Fairchild has a series of TTL ICs that uses numbers like 9N00, 9300, 9600, etc. However, on the Fairchild specification sheets the equivalent 7400 series number is usually indicated.

The 7400 series operates reliably over a temperature range of 0°C–70°C and with a supply voltage (V_{CC}) of from 4.75 V to 5.25 V. The 5400 series is somewhat more flexible since it can tolerate a temperature range of −55°C–+125°C and a supply variation of 4.5 V–5.5 V. Both series typically have a fan-out of 10, indicating that each output can reliably drive an additional 10 inputs.

7400 voltage levels Table 9–1 lists the input and output voltage levels for the standard 7400 series. The minimum and maximum values shown are for worst-case conditions of power supply, temperature, and loading.

Table 9–1 **Standard 7400 series voltage levels**

Voltage Level	Minimum	Typical	Maximum
V_{OL}		0.2 V	0.4 V
V_{OH}	2.4 V	3.6 V	
V_{IL}			0.8 V
V_{IH}	2.0 V		

Inspection of Table 9–1 reveals a guaranteed maximum logic-0 output $V_{OL} = 0.4$ V, which is 400 mV less than the logic-0 voltage needed at the input $V_{IL} = 0.8$ V. This means that the guaranteed LOW-state dc noise margin is 400 mV. That is,

$$V_{NL} = V_{IL}(\text{max}) - V_{OL}(\text{max}) = 0.8\text{ V} - 0.4\text{ V} = 0.4\text{ V} = 400\text{ mV} \qquad \textbf{(9–2)}$$

Similarly, the logic-1 output V_{OH} is a guaranteed minimum of 2.4 V, which is 400 mV greater than the logic-1 voltage needed at the input $V_{IH} = 2.0$ V. Thus, the HIGH-state dc noise margin is 400 mV.

$$V_{NH} = V_{OH}(\text{min}) - V_{IH}(\text{min}) = 2.4\text{ V} - 2.0\text{ V} = 0.4\text{ V} = 400\text{ mV} \qquad \textbf{(9–1)}$$

Thus, the *guaranteed worst-case* dc noise margins for the 7400 series are both 400 mV. In actual operation the *typical* dc noise margins are somewhat higher ($V_{NL} = 1$ V, and $V_{NH} = 1.6$ V).

Power dissipation The basic TTL logic circuit is the NAND gate of Figure 9–5. Typically, it draws an average supply current (I_{CC}) of 2 mA, resulting in a power dissipation of 2 mA × 5 V = 10 mW.

Propagation delay The basic TTL NAND gate has typical propagation delays of $t_{PLH} = 11$ ns and $t_{PHL} = 7$ ns, which give an *average* propagation delay of 9 ns.

Table 9–2 summarizes the basic characteristics of the standard 7400 series.

Table 9–2 **Standard 7400 series characteristics**

Noise margins (worst-case)	$V_{NL} = V_{NH} = 400$ mV
Average power dissipation (basic gate)	$P_D = 10$ mW
Average propagation delay	9 ns
Typical fan-out	10

Example 9.2 Refer to Appendix *B* for the data sheet for the 7420 NAND-gate IC. From the data sheet determine:

(a) The typical average power dissipation of *one* NAND gate.

(b) The typical average propagation delay of one NAND gate.

Solution:

(a) Under "electrical characteristics" we can find the typical values of I_{CC} for the two logic states

$$I_{CCH}(\text{typ}) = 2 \text{ mA}$$
$$I_{CCL}(\text{typ}) = 6 \text{ mA}$$

Note that the chip draws more current when the gate outputs are in the LOW state. This is generally true of most TTL ICs. The *average* I_{CC} is $(2 + 6)/2 = 4$ mA and represents the average current that the V_{CC} source has to supply to the *entire* chip. We can assume that this current will divide equally between the two NAND gates. Thus, one gate will draw an average current of 2 mA, which results in an average power drain of 2 mA × 5 V = **10 mW** per gate.

(b) The typical propagation delays are listed as

$$t_{PLH} = 12 \text{ ns}$$
$$t_{PHL} = 8 \text{ ns}$$

This gives an average delay of $(8 + 12)/2 =$ **10 ns.**

9.5 Other TTL Series

The standard 7400 series ICs offer a combination of speed and power dissipation suited for many applications. ICs offered in this series include a wide variety of gates, flip-flop circuits (flip-flops), and one-shots in the small-scale integration (SSI) line, and shift registers, counters, decoders, memories, and arithmetic circuits in the medium-scale integration (MSI) line.

Besides the standard 7400 series, several other TTL series have been developed to provide a wider choice of speed and power dissipation characteristics.

Low-power TTL, 74L00 series Low-power TTL circuits designated as the 74L00 series have essentially the same basic circuit as the standard 7400 series except that all the resistor values are *increased*. The larger resistors reduce the power requirements, but at the expense of longer propagation delays. A typical NAND gate in this series has an average power dissipation of 1 mW and an average propagation delay of 33 ns.

The 74L00 series is ideal for applications in which power dissipation is more critical than speed. Low frequency, battery-operated circuits such as calculators are well suited for this TTL series.

High-speed TTL, 74H00 series The 74H00 series is a high-speed TTL series. The basic circuitry for this series is essentially the same as that for the standard 7400 series except that *smaller* resistor values are used (for faster charging of junction capacitances) and the emitter-follower transistor Q_4 is replaced by a Darlington pair. These differences result in a much faster switching speed, with an average propagation delay of 6 ns. However, the increased speed is accomplished at the expense of increased power dissipation. The basic NAND gate in this series has an average P_D of 23 mW.

Schottky TTL, 74S00 series This series has the highest speed of all the TTL series. It achieves this high speed by using a Schottky barrier diode (SBD) connected as a clamp from base to collector of each circuit transistor [see Figure 9–7(a)]. The SBD has a low forward voltage (typically 0.25 V) and is faster than a normal PN diode. The presence of the SBD prevents the transistor's collector-base junction from becoming forward-biased by more than 0.25 V when the onset of saturation occurs. As a result, the transistor will not go as deeply into saturation, so its lower storage time, t_s, will produce a shorter turn-off time. The Schottky-clamped transistor is given the special symbol shown in Figure 9–7(b).

A typical 74S00 series NAND gate has a propagation delay of 3 ns. Figure 9–7(c) shows the circuit diagram for a 74S00 NAND gate. Note that it uses lower value resistors than the standard 7400 series (Figure 9–5) to help improve the switching times. The smaller resistors, however, produce an increase in the average power dissipation per gate to about 23 mW. Since it has essentially the same P_D as the 74H00 series, while operating at about twice the speed, the 74S00 series is generally chosen for applications where high speed is essential. In other words, the 74H00 series has become obsolete since the development of the 74S00 series, although it may still be found in circuits that were built several years ago.

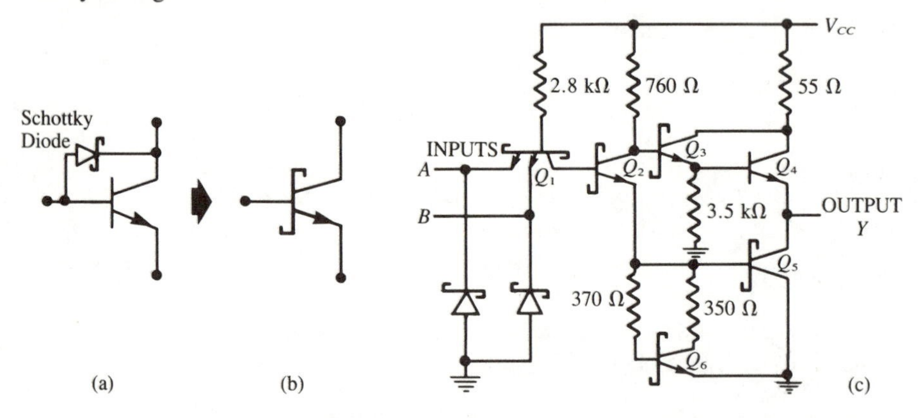

Figure 9–7 **(a) Schottky-clamped transistor (b) Equivalent symbol (c) 74S00 NAND-gate circuitry**

Low-power Schottky TTL series, 74LS00 This is becoming the most widely used TTL series because it combines the relatively high speed of the standard 7400 series with a very low P_D that is only slightly higher than that of the 74L00 series. This performance is achieved by using Schottky-clamped transistors, but with larger resistor values than in the 74S00 series. Figure 9–8 is a circuit for a 74LS00 NAND gate. The larger circuit resistances reduce the circuit power dissipation, but at the expense of increased switching times compared to the 74S00 series.

The 74LS00 NAND gate has a typical propagation delay of 9.5 ns, and an average P_D of only 2 mW. Since it has about the same speed as the 7400 series at a much lower P_D, the 74LS00 series has taken over many of the application areas previously dominated by the 7400 series.

Note that the 74LS00 circuit in Figure 9–8 does not use the multiple-emitter input transistor used by the other series. It uses input diodes instead (D_3 and D_4), but the basic circuit operation is the same as for the multiple-emitter inputs.

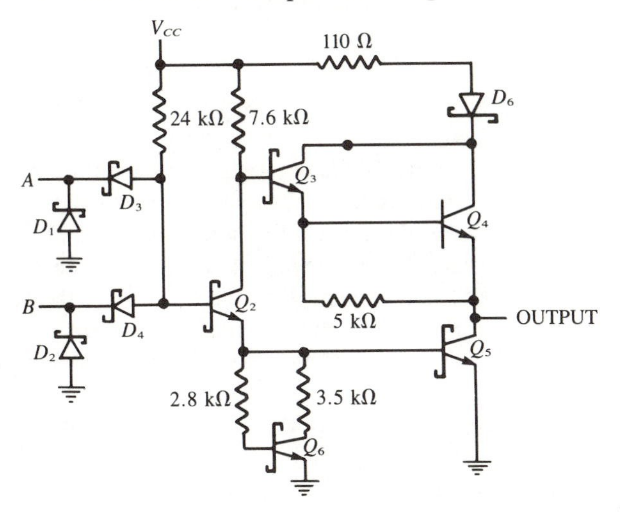

Figure 9–8 **74LS00 NAND-gate circuit**

Comparison of TTL series Table 9–3 is a comparison of the five TTL series as to typical propagation delays, power dissipation, and fan-out of a NAND gate. It also shows the maximum clock frequency for a *JK* flip-flop in each series.

Table 9–3

Series	Average Propagation Delay (ns)	Average Power Dissipation (mW)	Typical Fan-out	Maximum Clock Rate
7400	9	10	10	25 MHz
74L00	33	1	10	3 MHz
74H00	6	23	10	40 MHz
74S00	3	23	10	80 MHz
74LS00	9.5	2	10	30 MHz

Protective diodes Most TTL ICs are not designed to withstand any significant negative voltages at their inputs. For this reason, TTL manufacturers have connected shunt clipping diodes from each input to ground as shown in Figure 9–9 (also see Figures 9–7 and 9–8). These diodes limit the negative input voltages to the value of the diode forward voltage.

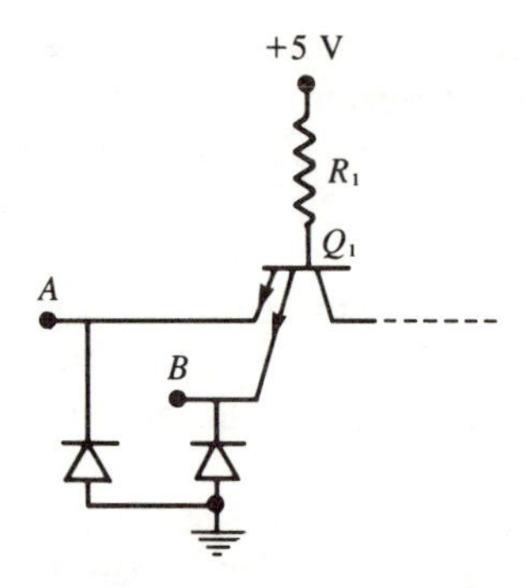

Figure 9–9 **Protective diodes on TTL inputs**

9.6 TTL Loading Rules

In designing digital systems using TTL devices, it is important to know how to determine and use the fan-out or drive capability of each circuit. Figure 9–10(a) shows a single TTL output in the LOW state connected to several TTL inputs. Transistor Q_4 is ON and is acting as a current sink for all of the currents flowing back from each input (I_{IL}). Although Q_4 is saturated, its ON-state resistance is some value other than 0, so the current I_{OL} produces an output voltage drop V_{OL}. Because V_{OL} must not exceed V_{OL}(MAX), which is 0.4 V for TTL (Table 9–1), the value of I_{OL}, and thus the number of loads that can be driven, is limited.

The HIGH-state situation is shown in Figure 9–10(b). Here, Q_3 is acting as an emitter-follower and is sourcing (supplying) current to each TTL input. These currents (I_{IH}) are just reverse-bias leakage currents since the TTL input emitter-base junctions are reverse-biased. However, if too many loads are driven, the total input current (I_{OH}) can become too large, causing larger drops across R_2, Q_3, and D_1, thereby lowering V_{OH} below V_{OH}(MIN), which is 2.4 V for TTL (Table 9–1).

Unit loads To simplify designing with TTL circuits, the manufacturers have established standardized input and output loading factors in terms of *current*. These currents are called units loads (U. L.) and are defined as follows:

$$\begin{aligned} 1 \text{ unit load (U. L.)} &= 40\ \mu\text{A in the HIGH state} \\ &= 1.6 \text{ mA in the LOW state} \end{aligned}$$

These unit-load factors are used to express the output drive capabilities and input requirements for TTL circuits in any of the five TTL series. The following examples illustrate how they are used.

Example 9.3 A certain TTL input is rated at 1 U. L., meaning the input draws a maximum current of 40 μA in the HIGH state; that is, $I_{IH}(\text{MAX}) = 40\ \mu\text{A}$. It also means that the maximum current this input will source back to the driving circuit in the LOW state is 1.6 mA; that is, $I_{IL}(\text{MAX}) = 1.6$ mA.

What are the input current limits for a TTL IC whose inputs are rated at 1.25 U. L.?

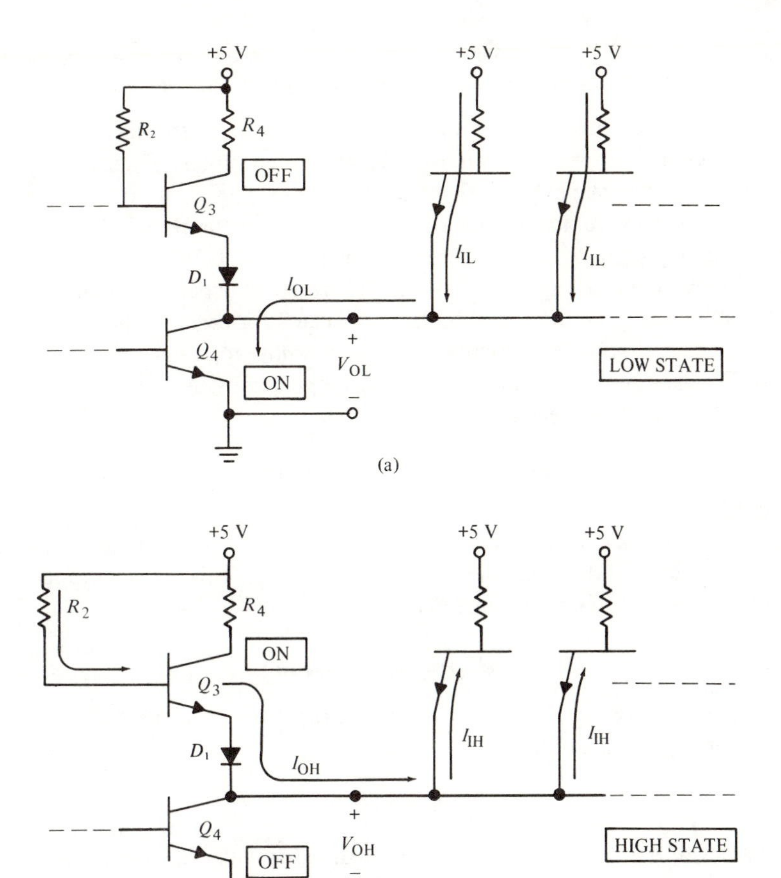

Figure 9–10 **TTL output drive capabilities**

Solution:

$$I_{IH}(\text{MAX}) = 1.25 \times 40\ \mu\text{A} = \mathbf{50\ \mu A}$$
$$I_{IL}(\text{MAX}) = 1.25 \times 1.6\ \text{mA} = \mathbf{2\ mA}$$

Example 9.4 A TTL IC has an output rated at 20 U. L. in both states. Determine its output current limitations.

Solution:

$$I_{OH}(\text{MAX}) = 20 \times 40\ \mu\text{A} = \mathbf{800\ \mu A}$$

This is the maximum current the output can *supply* to other loads in the HIGH state without V_{OH} dropping below $V_{OH}(\text{MIN}) = 2.4$ V.

$$I_{OL}(\text{MAX}) = 20 \times 1.6\ \text{mA} = \mathbf{32\ mA}$$

This is the maximum current the output can *sink* in the LOW state without the V_{OL} rising above $V_{OL}(\text{MAX}) = 0.4$ V.

Example 9.5 Refer to the data sheet for the 7400 quad NAND-gate IC in Appendix B. Determine its input and output loading factors in terms of unit loads.

Solution: The maximum input current parameters for the IC are listed as $I_{IH} = 40\ \mu A$* and $I_{IL} = -1.6$ mA (the negative sign simply indicates that the input current in the LOW state actually flows back to the driving circuit output). Thus, a 7400 NAND gate has an input loading factor of 1 U. L. in both the HIGH and LOW states. (In other words, any input to one of these gates acts as 1 U. L.)

The output drive capabilities of this IC are given under "recommended operating conditions," where a fan-out of 10 U. L. is indicated for each output, which means that each output can reliably drive a number of inputs whose total input loading factor equals 10 U. L. For example, one of the NAND-gate outputs (fan-out = 10 U. L.) can drive *ten* additional NAND inputs (each input = 1 U. L.).

Example 9.6 Determine the maximum output currents for the 7400 NAND gate in both states.

Solution: We saw in Example 9.5 that the 7400 NAND output has a fan-out of 10 U. L. In the HIGH state, 1 U. L. is 40 μA, so this gate output can *supply* $10 \times 40\ \mu A = 400\ \mu A = 0.4$ mA. Thus,

$$I_{OH}(\text{MAX}) = \mathbf{0.4\ mA}$$

In the LOW state, 1 U. L. = 1.6 mA; therefore this gate output can *sink* 10×1.6 mA = 16 mA. Thus,

$$I_{OL}(\text{MAX}) = \mathbf{16\ mA}$$

These values of I_{OH} and I_{OL} could also be found listed under the test conditions used for measuring V_{OH} and V_{OL} (refer to the 7400 data sheet).

Example 9.7

(a) Determine the fan-out of a 74S00 NAND gate.

(b) How many 74S00 inputs can a 74S00 output drive?

(c) How many 74S00 inputs can a 7400 output drive?

Solutions:

(a) The data sheet for the 74S00 IC in Appendix B does not specify the fan-out explicitly. However, the output current values can be obtained under the test conditions for V_{OH} and V_{OL}. They are given as

$$I_{OH} = 1\text{ mA} \qquad I_{OL} = 20\text{ mA}$$

Since $I_{OH} = 1$ mA, the output can supply 1 mA of current in the HIGH state. One U. L. in the HIGH state is 40 μA. Thus, one 74S00 output can drive 1 mA/40 μA = 25 U. L. That is,

$$\text{Fan-out (HIGH state)} = \frac{I_{OH}}{40\ \mu A} = \frac{1\text{ mA}}{40\ \mu A} = \mathbf{25\ U.\,L.}$$

---

*The second entry for I_{IH} (1 mA) is the input current when V_{IH} is at its maximum allowable value (5.5 V). This is not a normal operating condition.

Since $I_{OL} = 20$ mA, the 74S00 output can sink 20 mA in the LOW state. One U. L. in the LOW state is 1.6 mA. Thus,

$$\text{Fan-out (LOW state)} = \frac{I_{OL}}{1.6\text{ mA}} = \frac{20\text{ mA}}{1.6\text{ mA}} = \mathbf{12.5\ U.L.}$$

Notice that the fan-outs are different for the two output states.

(b) Each of the 74S00 inputs represents a load of 1.25 U. L., as indicated on the data sheet. Since the 74S00 has a fan-out of 12.5 U. L. in the LOW state, it can safely drive 12.5/1.25 = **10** 74S00 inputs.

(c) The 7400 NAND output has a fan-out of 10 U. L. Therefore, it can drive 10/1.25 = **8** 74S00 inputs.

Example 9.8 The output of a single 7400 NAND gate is providing clock pulses to a shift register made up of 7473 *JK* flip-flops. What is the maximum number of FFs the shift register can have?

Solution: The input loading factor for a 7473 clock input can be determined from the data sheet (Appendix B) by noting that $I_{IL} = 3.2$ mA at the clock input. This represents 3.2/1.6 = 2 U. L. Thus, one 7400 NAND gate (fan-out of 10 U. L.) can drive 10/2 = **5** clock inputs.

Table 9.4 gives the typical input and output loading factors for the five TTL series. These values are *typical* so there may be some variations depending on the particular device or manufacturer. The device data sheet should always be consulted before using the device in a circuit.

Table 9–4

TTL-Series	Input Loading HIGH	Input Loading LOW	Fan-out HIGH	Fan-out LOW
7400	1 U. L.	1 U. L.	10 U. L.	10 U. L.
74H00	1.25 U. L.	1.25 U. L.	12.5 U. L.	12.5 U. L.
74L00	0.5 U. L.	0.1 U. L.	10 U. L.	2.5 U. L.
74S00	1.25 U. L.	1.25 U. L.	25 U. L.	12.5 U. L.
74LS00	0.5 U. L.	0.25 U. L.	10 U. L.	5 U. L.

1 U. L. = { 40 μA (HIGH); 1.6 mA (LOW) }

Note that the higher speed series (74H00, 74S00) have larger input loading factors and fan-outs than the standard 7400 series. On the other hand, the lower power series (74L00, 74LS00) have lower input loading requirements and lower fan-outs. Also note that the HIGH-state and LOW-state values are not always the same.

Example 9.9 How many typical 7400 series inputs can be driven by a typical 74LS00 series output?

Solution: From Table 9–4, a typical 74LS00 device has a LOW-state fan-out of 5 U. L., and a typical 7400 input is rated at 1 U. L. in either state. Thus, the 74LS00 device can drive *five* 7400 inputs.

Example 9.10 Determine I_{IL}(MAX) for a 74LS04 INVERTER.

Solution: The 74LS04 data sheet (Appendix B) lists the 74LS04 input loading factor as 0.25 U. L. in the LOW state. Thus,

$$I_{IL}(\text{MAX}) = 0.25 \times 1.6 \text{ mA}$$
$$= \mathbf{0.4 \text{ mA}}$$

9.7 Other TTL Properties

There are several additional characteristics of TTL logic that must be understood if we are to intelligently use TTL circuits in a digital-system application.

Unconnected inputs Any input to a TTL circuit that is left disconnected (open) acts exactly like a logic 1 applied to that input, because in either case the emitter-base junction at the input will not be forward-biased. This means that on *any* TTL IC, an input will act as a logic 1 until it is tied to some logic signal or to ground.

Unused inputs Frequently, not all of the inputs on a TTL IC are used in a particular application, for example when all of the inputs to a logic gate are not needed for the required logic function. For instance, suppose we needed the logic operation $\overline{AB}$ and we were using a chip that had a three-input NAND gate. The various ways of accomplishing this are shown in Figure 9–11.

In Figure 9–11(a) the unused input is left disconnected, which means that it acts as a logic 1. The NAND-gate output is therefore $x = \overline{A \cdot B \cdot 1} = \overline{A \cdot B}$, which is the desired result. Although the logic is correct, it is usually undesirable to leave an input disconnected because it will act like an antenna that is likely to pick up stray radiated signals, which could cause the gate to operate improperly.

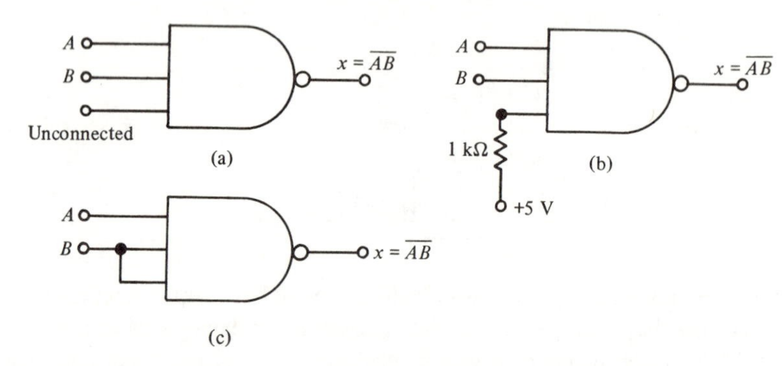

Figure 9–11 **Three ways to handle unused logic inputs**

A better technique is shown in (b). Here, the unused input is connected to +5 V through a 1 kΩ resistor, so the logic level is a 1. The 1 kΩ resistor is simply for current protection of the emitter-base junctions of the gate inputs in case of spikes on the power supply line. This same technique can be used for AND gates since a 1 on an unused input will not affect the output. As many as thirty unused inputs can share the same 1 kΩ resistor tied to V_{CC}.

A third possibility is shown in (c), where the unused input is tied to a used input. This is satisfactory, provided that the circuit driving input B is not going to have its fan-out exceeded. This technique can be used for *any* type of gate.

For OR gates and NOR gates the unused inputs cannot be left disconnected or tied to +5 V since these situations would produce a constant output logic level (1 for OR, 0 for NOR) regardless of the other inputs. Instead, for these gates the unused inputs must either be connected to ground (0 V) for a logic 0 or be tied to a used input as in Figure 9–11(c).

Tied-together inputs Sometimes two or more inputs to a TTL gate are connected together to act as a single input, as in Figure 9–11(c). In general, the tied-together inputs can be considered to have an input loading factor that is the *sum* of the individual input loading factors for each input. However, for AND and NAND gates, the LOW-state input loading factors do *not* add together, but instead act as a single input. For example, if two inputs of a NAND gate are tied together as a single input, they represent a loading factor of 1 U. L. + 1 U. L. = 2 U. L. (80 μA) in the HIGH state, but only 1 U. L. (1.6 mA) in the LOW state. Again, this is a characteristic of *only* ANDs and NANDs.

Example 9.11 Determine the number of U. L.s being driven by the X output in Figure 9–12.

Solution: The loading on X will be different for the two logic states, as shown in the figure. Note that in the HIGH state every gate input is counted as a separate 1 U. L., but in the LOW state the tied-together inputs are counted as a total of 1 U. L.

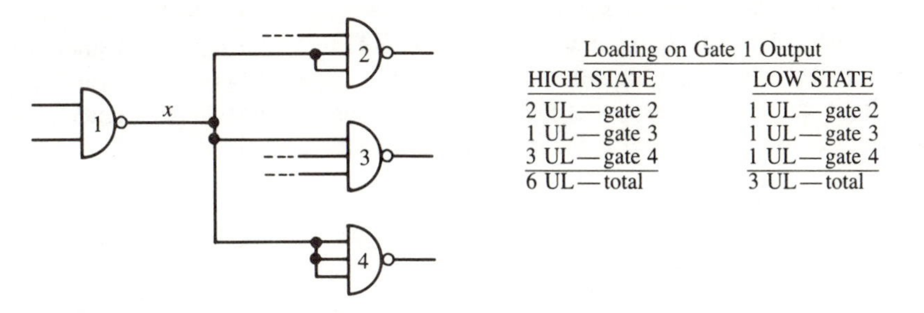

Loading on Gate 1 Output

HIGH STATE	LOW STATE
2 UL—gate 2	1 UL—gate 2
1 UL—gate 3	1 UL—gate 3
3 UL—gate 4	1 UL—gate 4
6 UL—total	3 UL—total

Figure 9–12 **Example 9.11**

Biasing TTL inputs to 0 Occasionally a TTL input must be held normally LOW and then caused to go HIGH by the actuation of a mechanical switch. This situation is illustrated in Figure 9–13 for the input to a flip-flop. This flip-flop triggers on a positive voltage transition which occurs when the switch is momentarily closed. The resistor, R, serves to hold the CLK input LOW while the switch is open. Care must be taken to keep the value of R low enough that the voltage developed across it by the current I_{IL} from the flip-flop input will not exceed 0.4 V, which is the maximum allowable LOW voltage, V_{IL}(MAX). Thus, the largest value of R is given by

$$I_{IL} \times R(\text{MAX}) = 0.4 \text{ V}$$

$$R(\text{MAX}) = \frac{0.4 \text{ V}}{I_{IL}} \qquad \textbf{(9–3)}$$

For example, if I_{IL} is 1.6 mA (1 U. L.), we have $R(\text{MAX}) = 250\ \Omega$. There is no limit on the smallest value of R except for the current drain on the 5 V supply when the switch is closed.

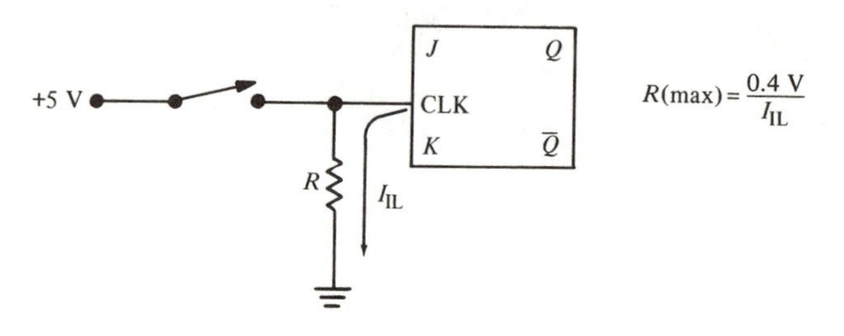

Figure 9–13

Transition times of inputs The input signals that drive TTL circuits must have relatively fast transitions for reliable operation. If the input rise times or fall times are greater than 100 ns, there is a possibility that oscillations might occur on the output as shown in Figure 9–14(a). These oscillations can cause serious problems if this output is being fed to circuits that are sensitive to pulses.

A slow signal can be sharpened up by passing it through a *Schmitt trigger*. The Schmitt trigger circuit produces fast transitions on the output (typically 10 ns) independent of the input rise and fall times. This output can then be fed to any TTL circuit.

Some TTL logic circuits are designed to include Schmitt triggers so they can handle slow input signal transitions with no problems. The 7414 IC contains six INVERTERs with Schmitt trigger inputs. Figure 9–14(b) shows how a 7414 INVERTER responds to a slow-changing input signal. Note that the INVERTER symbol contains a small representation of a Schmitt trigger input/output transfer curve (Chapter Six). Whenever you see this on a logic symbol, it means that the device has Schmitt trigger inputs and will respond properly to any input signal without regard to transition times.

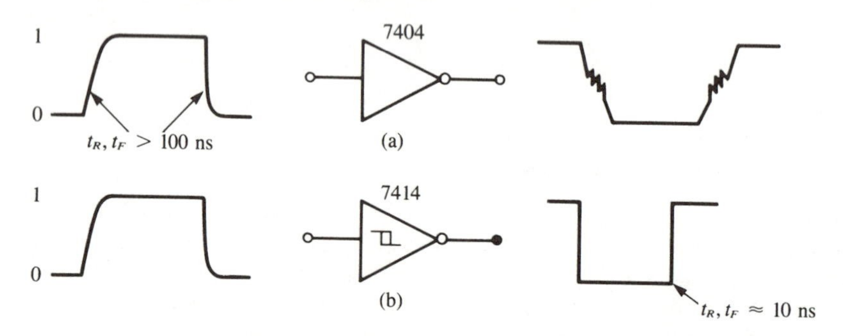

Figure 9–14 **(a) Slow t_R and t_F cause standard TTL output to momentarily oscillate (b) Schmitt trigger INVERTER responds properly**

Current spiking TTL logic circuits suffer from internally generated current transients or spikes because of the totem-pole output structure. When the output is switching from the LOW state to the HIGH state, the two output transistors are changing states—Q_3 OFF to ON and Q_4 ON to OFF. Since Q_4 is changing from the saturated condition, it will take longer than Q_3 to switch states. Thus, there is a short interval of time (around 2 ns) during the switching transition in which *both* transistors are conducting and a relatively large surge of current (30 mA–50 mA) is drawn from the +5 V supply (see Figure 9–15). The duration of this current transient is extended by the effects of any load capacitance on the circuit output. This capacitance consists of stray wiring capacitance and the input capacitance of any load circuits, and draws its charging current through Q_3. This overall effect can be summarized as follows: *Whenever a totem-pole TTL output goes from LOW to HIGH, a high-amplitude current spike is drawn from the V_{CC} supply.*

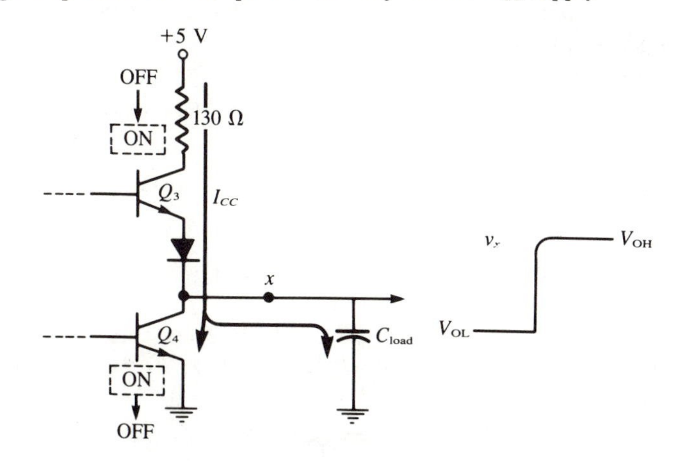

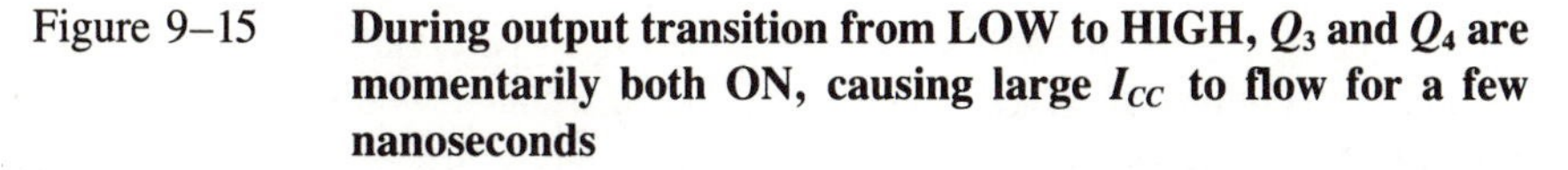

Figure 9–15 **During output transition from LOW to HIGH, Q_3 and Q_4 are momentarily both ON, causing large I_{CC} to flow for a few nanoseconds**

Figure 9–16 shows how the supply current, I_{CC}, will vary for a TTL NAND gate as the output switches levels. When the output is HIGH, a constant current, I_{CCH}, is drawn from the V_{CC} supply. Similarly, when the output is LOW, a constant I_{CCL} is flowing. Note the large-amplitude, narrow spikes that occur when the output switches from LOW to HIGH. The amplitude of these spikes is difficult to predict because both transistors are changing states.

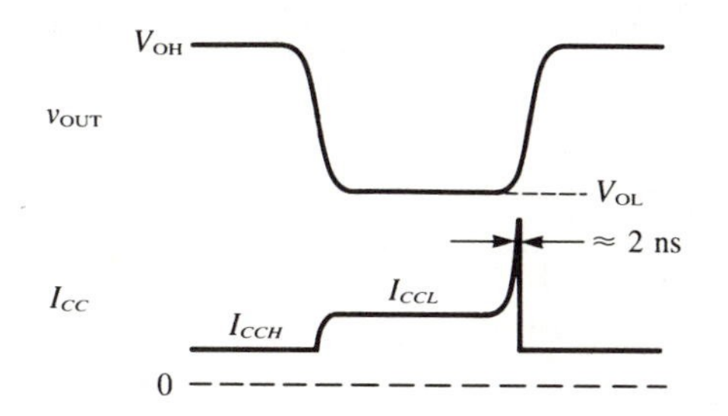

Figure 9–16 **Typical I_{CC} variation showing current spike**

In a complex digital circuit or system, many TTL outputs may be switching states at the same time, each one drawing a narrow spike of current from the power supply. The accumulative effect of all these current spikes is to produce a voltage spike on the common

V_{CC} line, mostly due to the distributed inductance on the supply line (remember, $V = L\ di/dt$ for inductance, and di/dt is very large for a 2 ns current spike). This voltage spike can cause serious malfunctions during switching transitions unless some type of filtering is used. The most common technique uses small *RF* capacitors connected from V_{CC} to ground to essentially "short out" these high-frequency spikes. This is called *power-supply decoupling*.

It is standard practice to connect a 0.01 μF or 0.1 μF low-inductance, ceramic disc capacitor between V_{CC} and ground near each TTL IC on a circuit board. The capacitor leads are kept very short to minimize inductance.

In addition, it is standard practice to connect a single large capacitor (2 − 20 μF) between V_{CC} and ground on each board to filter out possible variations in V_{CC} caused by the large changes in I_{CC} levels as outputs switch states.

9.8 TTL Open-Collector Outputs

Consider the logic circuit of Figure 9–17(a). NAND gates 4 and 5 provide the AND function which is ANDing the outputs of NAND gates 1, 2, and 3 so that the final output, x, has the expression

$$x = \overline{AB} \cdot \overline{CD} \cdot \overline{EF}$$

The circuit of Figure 9–17(b) shows the same logic operation obtained by simply tying together the outputs of NAND gates 1, 2, and 3. In other words, the AND operation is performed by tying the outputs together. This can be reasoned as follows: With all outputs tied together, when any one of the gate outputs goes to the LOW state, the common output point must go LOW due to the "shorting-to-ground" action of the Q_4 transistor of that gate. The common output point will be HIGH only when all gate outputs are in the HIGH state. Clearly, this is the AND operation.

The arrangement in Figure 9–17(b) has the advantage over the conventional arrangement of Figure 9–17(a) of requiring fewer logic gates to produce the desired output. This configuration is called a *wired-AND* because it produces the AND operation by connecting output wires together. It is sometimes misleadingly called the *wired-OR* operation.

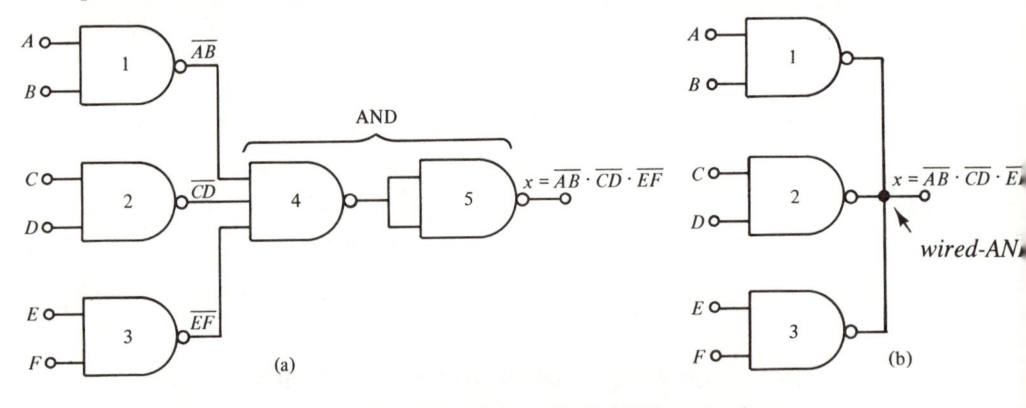

Figure 9–17 **Illustration of the wired-AND operation**

Totem-pole outputs and wired-ANDing To take advantage of the wired-AND configuration, the outputs of two or more gates must be tied together with no harmful effects.

Unfortunately, the totem-pole output circuitry of conventional TTL circuits prohibits tying outputs together. This restriction is illustrated in Figure 9–18, where the totem-pole outputs of two separate gates are connected together at point X. Suppose that the gate-A output is in the HIGH state (Q_{3A} ON, Q_{4A} OFF) and that the gate-B output is in the LOW state (Q_{3B} OFF, Q_{4B} ON). In this situation Q_{4B} is required to sink a large source current from Q_{3A}, which can go as high as 40 mA. This current can easily damage Q_{4B}, which is usually guaranteed to sink only 16 mA [I_{OL}(MAX)]. The situation is even worse when more than two TTL outputs are tied together. It is for this reason that some TTL circuits are designed with *open-collector* outputs instead of totem-pole outputs.

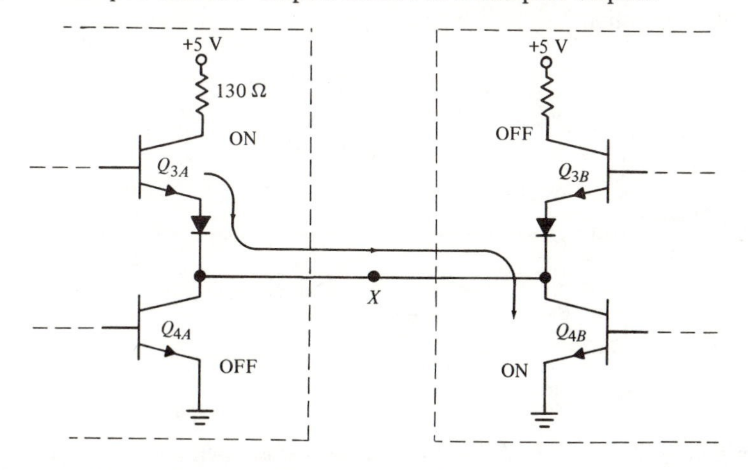

Figure 9–18 **Totem-pole outputs tied together can produce damaging current through Q_4 transistors**

Open-collector outputs As shown in Figure 9–19 (a), the open-collector circuit eliminates Q_3, D_1, and R_4. The output is taken at Q_4's collector, which is not connected to any bias source. When the output is in the LOW state, Q_4 is ON (has base current) and is able to sink collector current. When the output is in the HIGH state, Q_4 is OFF and is essentially an open circuit. There will be no voltage at the output terminal in the HIGH

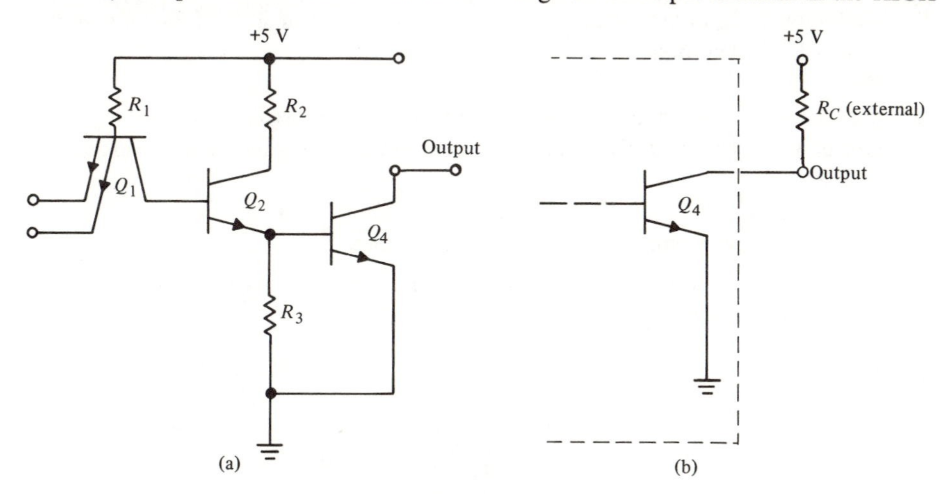

Figure 9–19 **(a) Open-collector TTL circuit (b) Open-collector TTL circuit with external pull-up resistor**

state unless an *external pull-up resistor,* R_C, is connected from the collector to +5 V as shown in Figure 9–19(b). Now when Q_4 is OFF, the output will be pulled up to +5 V.

Since open-collector outputs do not have the Q_3 transistor, they cannot supply current in the HIGH state. Therefore, open-collector outputs *can* be safely connected together in the wired-AND configuration as illustrated in Figure 9–20. Here three 7401 open-collector NAND gates are wired-ANDed. These open-collector gates have no special symbol, although sometimes the letters "O. C." are written on their symbols. Notice that only one pull-up resistor is used for the wired-AND connection; it is not necessary to use a separate R_C for each output. Also notice the some-times used symbol for the wired-AND connection.

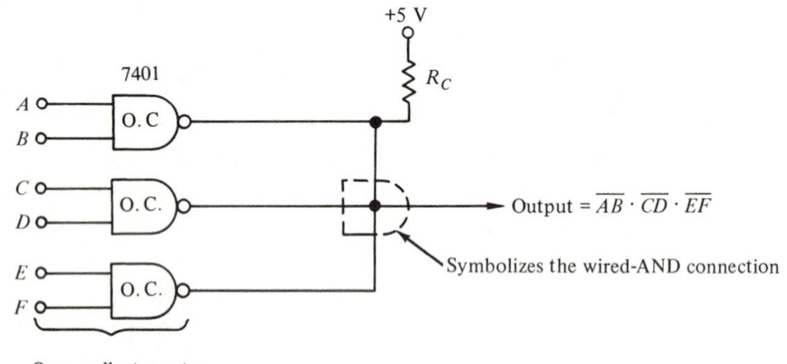

Figure 9–20 **Wired-AND operation using open-collector gates**

Value of R_C Clearly, the use of the wired-AND connection can save logic gates. The principal disadvantage of using open-collector outputs is that the signal rise time will be slowed down. This is because the Q_3 emitter-follower transistor is not there to provide rapid charging of any load capacitance.

Instead, the load capacitance has to charge up through R_C. This requires that R_C be made as small as possible. Unfortunately, its minimum value is limited by the I_{OL}(MAX) of Q_4. When Q_4 is ON, it has to sink the current from the +5 V supply through R_C, in addition to the I_{IL} currents from any TTL loads it is driving. The following example shows how R_C is determined.

Example 9.12 The 7405 IC (see Appendix B) contains six INVERTERs with open-collector outputs. These six INVERTERs are connected in a wired-AND arrangement in Figure 9–21(a).

(a) Determine the logic expression for output x.
(b) Determine a value for R_C assuming that output x is to drive other circuits with a total loading factor of 4 U. L.

Solution:

(a) Each INVERTER output is the inverse of its input. The wired-AND connection simply ANDs each INVERTER output. Thus,

$$x = \overline{A} \cdot \overline{B} \cdot \overline{C} \cdot \overline{D} \cdot \overline{E} \cdot \overline{F}$$

Using De Morgan's theorem, this expression is equivalent to

$$x = \overline{A + B + C + D + E + F}$$

which is the NOR operation.

(b) The minimum value for R_C is determined by considering the case where only one open-collector output is in the LOW state so that its output transistor has to sink all of the current from R_C and the total I_{IL} from the TTL loads. This case is illustrated in Figure 9–21(b).

Since the output is driving 4 U. L., the total I_{IL} is 4×1.6 mA = 6.4 mA. The 7405 has a rated fan-out of 10 U. L., so it can sink $I_{OL}(\text{MAX}) = 16$ mA. Thus,

$$I_{OL}(\text{MAX}) = I_{RC} + I_{IL}$$

$$\therefore \quad I_{RC} = I_{OL}(\text{MAX}) - I_{IL} = 16 \text{ mA} - 6.4 \text{ mA} = 9.6 \text{ mA}$$

The voltage across R_C when Q_4 is sinking its maximum current will be

$$V_{RC} = V_{CC} - V_{OL}(\text{MAX}) = 5 \text{ V} - 0.4 \text{ V} = 4.6 \text{ V}$$

This, then gives

$$R_C(\text{MIN}) = \frac{V_{RC}}{I_{RC}} = \frac{4.6 \text{ V}}{9.6 \text{ mA}} \approx \mathbf{480\ \Omega}$$

This is the smallest that we can make R_C. It can be made larger, but it is usually kept as small as possible in order to prevent deterioration of signal rise time.

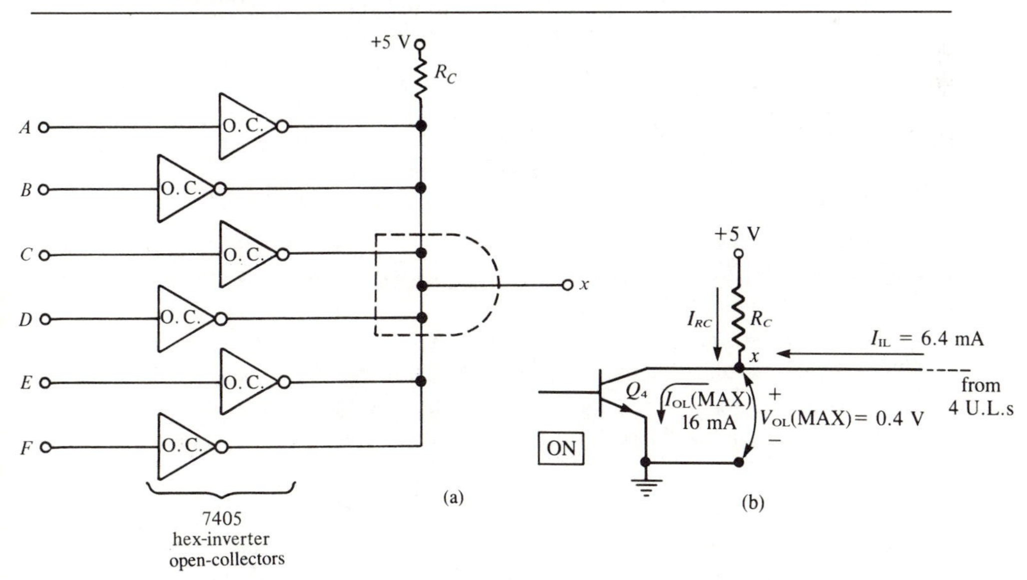

Figure 9–21 **Example 9.12**

9.9 TRI-STATE TTL

As we have seen, TTL devices with totem-pole outputs provide high-speed switching, but they cannot be wire-ANDed together. Open-collector devices can be wire-ANDed, but they are slower than totem-pole outputs even when the minimum R_C is used.

Tristate TTL is a third type of TTL output configuration that retains the high-speed characteristic of the totem-pole arrangement while permitting outputs to be tied together.

Tristate devices have *three* possible output states: HIGH, LOW, and HI-Z. In the HI-Z state, *both* Q_3 and Q_4 transistors in the totem-pole are turned OFF so that the output terminal is a high impedance to ground and V_{CC}. This situation is illustrated in Figure 9–21. Note that output terminal x is essentially an open or *floating* terminal. In practice, the resistance from x to ground and from x to V_{CC} is typically a few megohms.

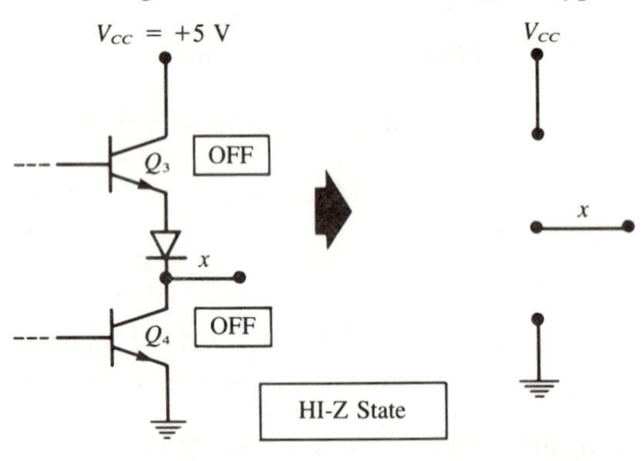

Figure 9–22 **In the HI-Z state the output terminal is essentially disconnected from the V_{CC} and ground terminals**

Figure 9–23(a) is a tristate NAND gate with two conventional logic inputs, A and B. In addition, it has a third input, the ENABLE input, which produces the HI-Z state. The table in Figure 9–23(b) describes the function of the ENABLE input. When ENABLE is HIGH, the NAND gate operates normally and its output will be at a HIGH or LOW logic level, depending on inputs A and B.

When ENABLE is LOW, the gate output is *disabled*; that is, it is in the high-impedance state (HI-Z) since Q_3 and Q_4 are both OFF. In this state, the output is not affected by the A and B inputs.

Note that the ENABLE input does *not* have a small circle where it enters the NAND-gate symbol. This indicates that the ENABLE input is active-HIGH; that is, a HIGH is required to *enable* the gate output to operate normally. If there was a circle on the ENABLE input, it would be active-LOW, indicating that a LOW *enables* the gate output.

Example 9.13 Figure 9–24 is a TTL device called a *tristate buffer*. It can be used to convert a totem-pole output to a tristate output. The 74125 does not change the logic level of the input A, but it controls whether or not A reaches the output X. When the ENABLE input is LOW, the tristate buffer is enabled and $X = A$. When ENABLE is HIGH, the buffer is disabled and output X is in its HI-Z state.

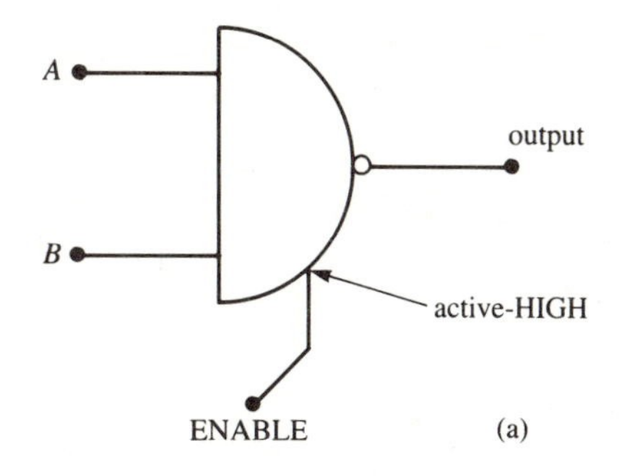

ENABLE	OUTPUT
HIGH	*Enabled*; so that $X = \overline{AB}$.
LOW	*Disabled*; output is in the HI-Z state. A and B have no effect.

(b)

Figure 9–23 **Tristate NAND gate**

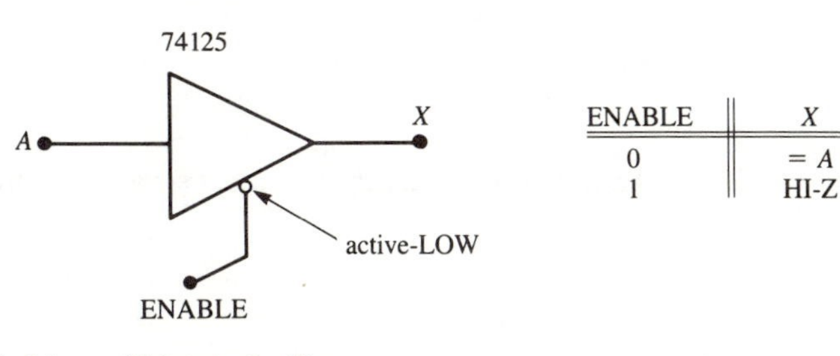

ENABLE	X
0	$= A$
1	HI-Z

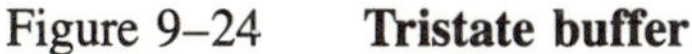
Figure 9–24 **Tristate buffer**

Figure 9–25 shows how tristate buffers such as the 74125 can be used to connect the outputs of three FFs to a common line called a *bus*. Suppose you want to transmit the output of each of the FFs, one at a time, over the bus line to other circuits. This can be done by enabling only one of the buffers at a time while disabling the other two. For example, with ENABLE $A = 0$ and ENABLE $B =$ ENABLE $C = 1$, only the output of FF A will be connected to the bus line because the other two buffers will be disabled (essentially disconnected from the bus).

Even though the three buffer outputs are connected together in Figure 9–25, no harmful currents will flow if only one buffer is enabled at one time. This requirement has to be met by the logic circuit designer.

Tristate outputs are superior to open-collector outputs because tristate outputs use the totem-pole arrangement. Thus, when the tristate output is enabled it has the low-impedance, high-speed characteristic of a totem-pole output.

Tristate outputs are also available in many MOS family devices. In fact, it was the development of MOS tristate operation that brought about the rapid development of microprocessor and memory chips. We will not discuss tristate operation in our subsequent study of MOS logic families, because it is basically the same as described here.

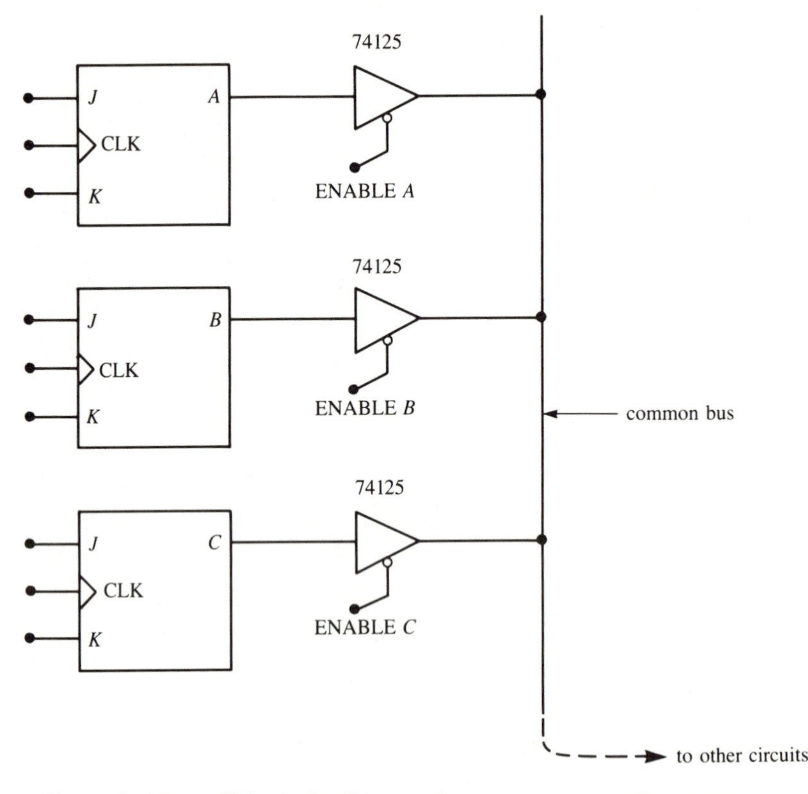

Figure 9–25 **Tristate buffers used to connect several outputs to a common bus**

9.10 The ECL Digital IC Family

The TTL logic family uses transistors that operate in the saturated mode. As such, TTL switching speed is limited by the storage delay time associated with a transistor that is driven into saturation. Another bipolar logic family has been developed that prevents transistor saturation, thereby increasing overall switching speed. This logic family is called *emitter-coupled logic* (ECL) and operates on the principle of current-switching whereby a fixed bias current less than I_C(SAT) is switched from one transistor's collector to another. Because of this current-mode operation, this logic form is also referred to as *current-mode logic* (CML).

Basic ECL circuit The basic circuit for emitter-coupled logic is essentially the differential amplifier configuration of Figure 9–26(a). The V_{EE} supply produces an essentially fixed current, I_E, which remains around 3 mA during normal operation. This current is allowed to flow through *either* Q_1 or Q_2, depending on the voltage level at V_{IN}. In other words, this current will switch between Q_1's collector and Q_2's collector as V_{IN} switches between its two logic levels of -1.7 V (logic 0 for ECL) and -0.8 V (logic 1 for ECL). The table in Figure 9–26(a) shows the resulting output voltages for these two conditions at V_{IN}. Two important points should be noted. First, V_{C1} and V_{C2} are the *complements* of each other. Second, the output voltage levels are not the same as the input logic levels.

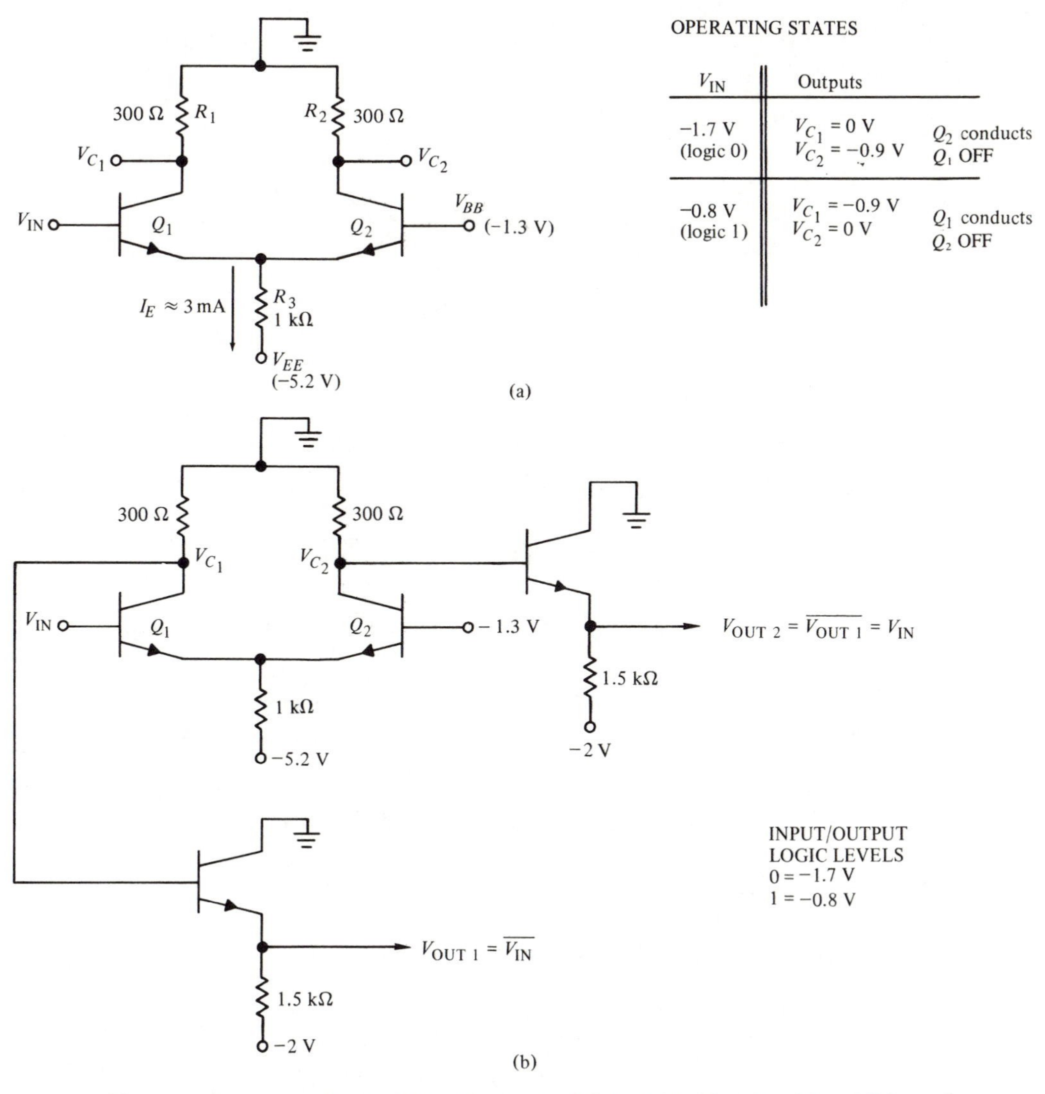

Figure 9–26 **(a) Basic ECL circuit (b) Basic ECL circuit with addition of emitter followers**

The second point just noted is easily taken care of by connecting V_{C1} and V_{C2} to emitter-follower stages (Q_3 and Q_4) as shown in Figure 9–26(b). The emitter-followers perform *two* functions. First, they *subtract* approximately 0.8 V from V_{C1} and V_{C2} to shift the output levels to the correct ECL logic levels. Second, they provide a very low output impedance (typically 7 Ω), which provides for large fan-out and fast charging of load capacitance. This circuit produces two complementary outputs: V_{OUT1}, which equals $\overline{V_{IN}}$, and V_{OUT2}, which equals V_{IN}.

ECL OR/NOR gate The basic ECL circuit of Figure 9–26(b) can be used as an INVERTER if the output is taken at V_{OUT1}. This basic circuit can be expanded to more than one input by paralleling transistor Q_1 with other transistors for the other inputs, as in Figure 9–27(a). Here, either Q_1 or Q_3 can cause the current to be switched out of Q_2, resulting in the two outputs, V_{OUT1} and V_{OUT2}, being the logical NOR and OR operations, respectively. This OR/NOR gate is symbolized in Figure 9–27(b) and is the fundamental ECL gate.

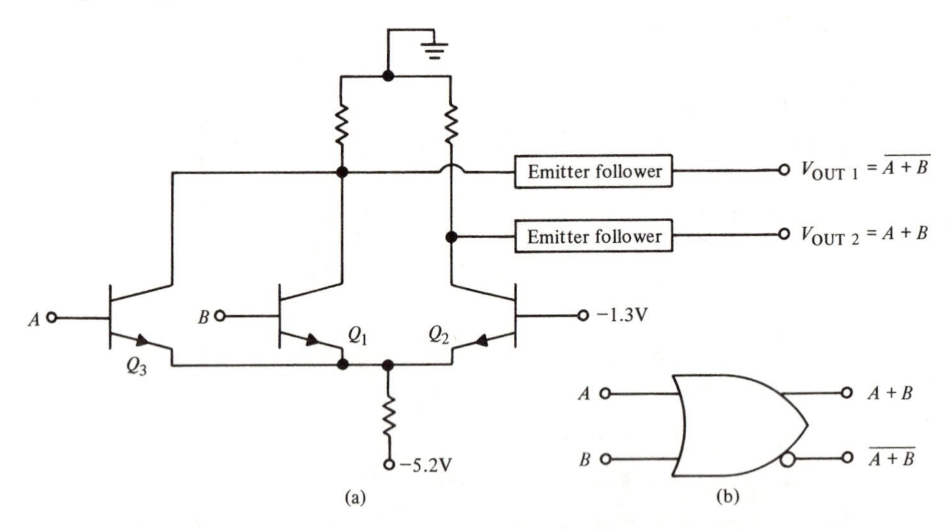

Figure 9–27 **(a) NOR/OR circuit (b) ECL NOR/OR logic symbol**

ECL characteristics The following are the most important characteristics of the ECL family of logic circuits:

1. The transistors never saturate, so switching speed is very high. Typical propagation delay time is 2 ns, which makes ECL a little faster than Schottky TTL (74S00 series). Although the 74S00 series is almost as fast as ECL, it requires a somewhat more complex fabrication process, so it is somewhat higher in cost.
2. The logic levels are nominally −0.8 V and −1.7 V for the logic 1 and 0, respectively.
3. Worst-case ECL noise margins are approximately 250 mV. These low noise margins make ECL somewhat unreliable for use in heavy industrial environments.
4. An ECL logic block usually produces an output and its complement, eliminating the need for INVERTERs.
5. Fan-outs are typically around 25 due to the low-impedance emitter-follower outputs.
6. Typical power dissipation for a basic ECL gate is 25 mW, just slightly higher than Schottky TTL.
7. The total current flow in an ECL circuit remains relatively constant regardless of its logic state. This helps to maintain an unvarying current drain on the circuit power supply even during switching transitions. Thus, no noise spikes will be internally generated like those produced by TTL totem-pole circuits.

Table 9–5 shows how ECL compares with the TTL-logic families using typical values.

Table 9–5

Logic Family	t_{pd} (ns)	P_D (mW)	Worst-case Noise Margin	Fan-out	Max. Clock Rate
7400 TTL	9	10	400 mV	10	25 MHz
74L00 TTL	33	1	400 mV	10	3 MHz
74H00 TTL	6	23	400 mV	10	40 MHz
74S00 TTL	3	23	300 mV	10	80 MHz
74LS00 TTL	9.5	2	300 mV	10	30 MHz
ECL	2	25	250 mV	25	120 MHz

9.11 MOS Digital Integrated Circuits

Metal-oxide-semiconductor (MOS) technology derives its name from the basic MOS structure of a **metal** electrode over an **oxide** insulator over a **semiconductor** substrate. Most of the MOS digital ICs are constructed entirely of enhancement-type MOSFETs.

The chief advantages of the MOSFET are that it is relatively simple and inexpensive to fabricate, it is small in size, and it consumes very little power. The fabrication of MOS ICs is approximately *one-third* as complex as the fabrication of bipolar ICs (TTL, ECL, etc.). In addition, MOS devices occupy much less space on a chip than bipolar transistors; typically, a MOSFET requires 1 square mil of chip area, while a bipolar transistor requires about 50 square mils. More importantly, MOS digital ICs normally do not use the IC resistor elements which take up so much of the chip area of bipolar ICs.

This means that MOS ICs can accommodate a much larger number of circuit elements on a single chip than bipolar ICs. Because of this, MOS ICs are surpassing bipolar ICs in the area of large-scale integration (LSI). The high packing density of MOS ICs results in a greater system reliability because of the reduction in the number of necessary external connections.

The principal disadvantage of MOS ICs is their relatively slow operating speed when compared with the bipolar IC families. In many applications speed is not a prime consideration, so MOS logic offers an often superior alternative to bipolar logic.

9.12 Digital MOSFET Circuits

Digital circuits that employ MOSFETs are broken down into three categories: (1) P-MOS, which uses *only* P-channel enhancement MOSFETs; (2) N-MOS, which uses *only* N-channel enhancement MOSFETs; and (3) CMOS (complementary MOS), which uses both P- and N-channel devices.

P-MOS and N-MOS digital ICs have a greater packing density (more transistors per chip) than CMOS and are therefore more economical. N-MOS has about twice the packing density of P-MOS. In addition to its greater packing density, N-MOS is also about twice as fast as P-MOS because free electrons are the current carrier in N-MOS while holes (slower-moving positive charges) are the current carriers for P-MOS. CMOS has the greatest complexity and lowest packing density of the MOS families, but it has the important advantages of higher speed and much lower power dissipation.

N-MOS and C-MOS are the most widely used MOS families, while P-MOS has become almost obsolete. We will study the basic circuitry and characteristics of N-MOS in this section, keeping in mind that P-MOS circuits would operate in exactly the same way except for the voltage polarities and current directions.

N-MOS inverter Figure 9–28(a) is the basic N-MOS inverter. It consists of two N-channel MOSFETs. Q_1's gate is connected permanently to +5 V so that it is always in the ON state; thus, it essentially acts as a resistor of value R_{ON}. Q_2 is a switching MOSFET whose state (ON/OFF) depends on E_{IN}. Typically Q_1 has R_{ON} = 100 kΩ while Q_2 is designed for R_{ON} = 1 kΩ and R_{OFF} = 10^{10}Ω.

Figure 9–28(b) summarizes the inverter's operation. We can consider each MOSFET as a resistance so that V_{OUT} is taken from a voltage divider. With E_{IN} = 0 V, Q_2 is OFF and is a very large resistance (10^{10}Ω), much larger than Q_1's resistance (100 kΩ). Thus, V_{OUT} will be approximately +5 V. With E_{IN} = +5 V, Q_2 is ON and its resistance is around 1 kΩ. The voltage divider is now 1 kΩ and 100 kΩ so that $V_{OUT} \approx 0.05$ V.

N-MOS NAND Gate The basic inverter can be modified to form NOR and NAND gates. Figure 9–29(a) shows a NAND gate that again uses Q_1 as a load resistance of 100 kΩ. Q_2 and Q_3 are MOSFET switches that are controlled by inputs A and B. Examination of the circuit shows that output X can only go LOW if *both* Q_2 and Q_3 are ON ($A = B$ = HIGH). If either one is OFF, there will essentially be an open circuit from X to ground, thereby producing an output of approximately 5 V.

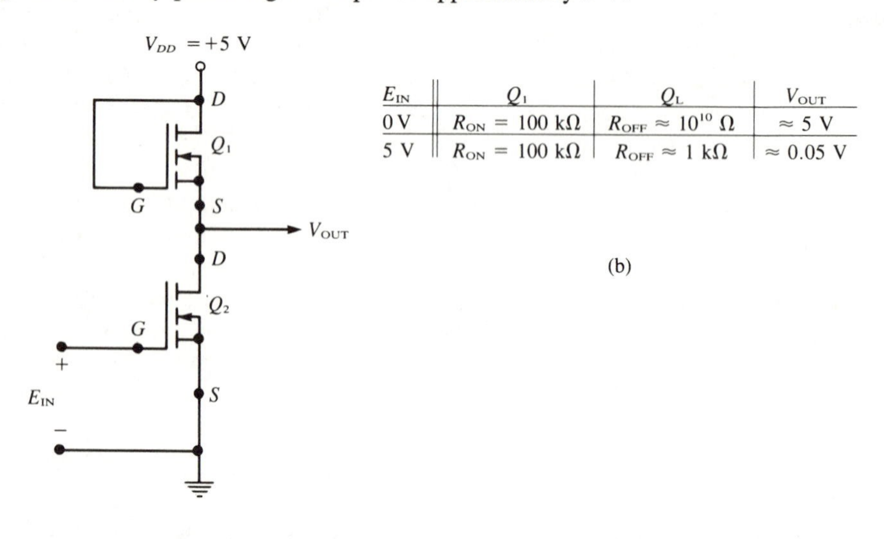

E_{IN}	Q_1	Q_L	V_{OUT}
0 V	R_{ON} = 100 kΩ	$R_{OFF} \approx 10^{10}$ Ω	≈ 5 V
5 V	R_{ON} = 100 kΩ	R_{OFF} ≈ 1 kΩ	≈ 0.05 V

(b)

Figure 9–28 **N-MOS inverter**

N-MOS NOR Gate Figure 9–29(b) is a NOR gate that uses Q_2 and Q_3 as parallel switches, with Q_1 again acting as a load resistance. The X output will go LOW when either Q_2 or Q_3 is ON, providing a low resistance from X to ground. X will be HIGH only when both Q_2 and Q_3 are OFF ($A = B$ = LOW).

Other N-MOS Logic Circuits OR gates, AND gates, and EX-OR gates can be easily formed by combining NOR and NAND gates with inverters. Flip-flops are constructed from cross-coupled NORs or NANDs (Chapter 8). The MOS FF is extremely important in MOS memory ICs that can store a large number (several thousand) of 1's and 0's on a single chip.

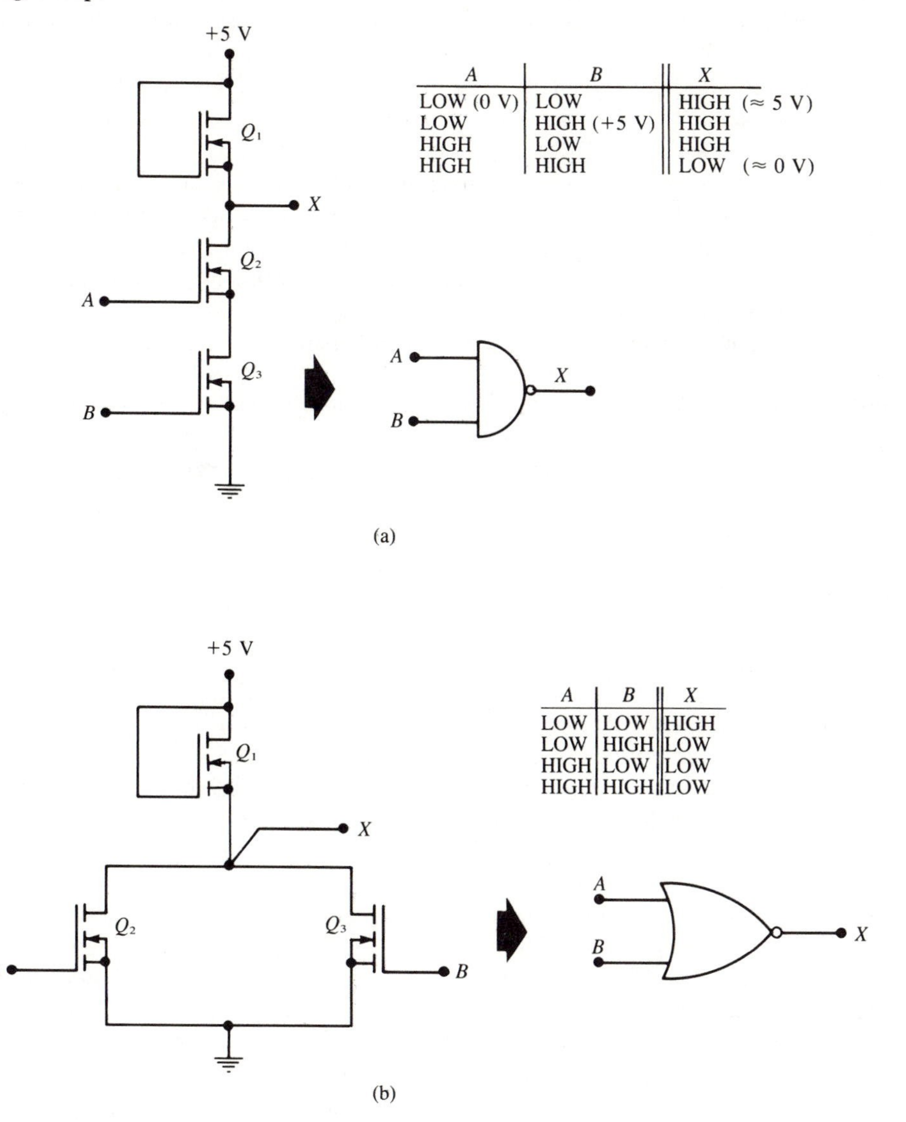

A	B	X
LOW (0 V)	LOW	HIGH (≈ 5 V)
LOW	HIGH (+5 V)	HIGH
HIGH	LOW	HIGH
HIGH	HIGH	LOW (≈ 0 V)

A	B	X
LOW	LOW	HIGH
LOW	HIGH	LOW
HIGH	LOW	LOW
HIGH	HIGH	LOW

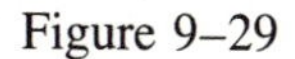
Figure 9–29 **(a) N-MOS NAND gate; (b) NOR gate**

9.13 Characteristics of MOS Logic

Compared with the bipolar logic families (TTL, ECL) the MOS logic families are slower in operating speed, require much less power, have a better noise margin and a higher fan-out, and, as was mentioned earlier, require much less "real-estate" (chip area).

Operating speed A typical N-MOS NAND gate has a propagation delay time of 50 ns. This is due to two factors—the relatively high output resistance (100 kΩ in the HIGH state) and the capacitive loading presented by the inputs of the logic circuits being driven. MOS logic inputs have very high input resistance ($> 10^{12}\ \Omega$), but they have a reasonably high gate capacitance (MOS capacitor), which is typically a few picofarads. This combination of large R_{OUT} and large C_{load} serves to increase switching time.

Noise margin Typically, MOS noise margins are around 1 V (when $V_{DD} = 5$ V), which is substantially higher than those for TTL or ECL.

Fan-out Because of the extremely high input resistance at each MOSFET input we would expect the fan-out capabilities of MOS logic to be virtually unlimited. This is essentially true for dc or low-frequency operation. However, for frequencies greater than around 100 kHz the gate input capacitances cause a deterioration in switching time that increases in proportion to the number of loads being driven. Even so, MOS logic can operate easily at a fan-out of 50, which is somewhat better than the bipolar families.

Power drain MOS logic circuits draw small amounts of power because of the relatively large resistances being used. To illustrate, we can calculate the power dissipation of the INVERTER of Figure 9–28 for its two operating states. For $V_{IN} = 0$ V,

$$R_{ON}(Q_1) = 100\ \text{k}\Omega \qquad R_{OFF}(Q_2) = 10^{10}\ \Omega$$

Therefore, I_D (current from V_{DD} supply) ≈ 0.05 nA, and $P_D = 5\ \text{V} \times 0.05\ \text{nA} = 0.25$ nW.

For $V_{IN} = 5$ V,

$$R_{ON}(Q_1) = 100\ \text{k}\Omega \qquad R_{ON}(Q_2) = 1\ \text{k}\Omega$$

Therefore, $I_D = 5\ \text{V}/101\ \text{k}\Omega \approx 50\ \mu\text{A}$, and $P_D = 5\ \text{V} \times 50\ \mu\text{A} = 250\ \mu\text{W}$.

This gives an *average* P_D of about 125 μW for the INVERTER. This low-power drain makes MOS logic suitable for large-scale integration, where many gates, flip-flops, etc., can be on one chip without causing overheating that would damage the chip, and for battery-operated devices.

Process complexity MOS logic is the simplest logic family to fabricate since it uses only one basic element: an N-MOS (or P-MOS) transistor. It requires no other elements such as resistors, diodes, etc. This characteristic along with its lower P_D makes it ideally suited for large-scale integration (large memories, calculator chips, microprocessors, etc.). It is in this area of the digital field that MOS logic has made its greatest impact. The operating speed of P-MOS and N-MOS are not comparable with TTL, so very little has been done with them in small-scale and medium-scale integration applications. In fact, there are very few MOS logic circuits in the small-scale or medium-scale integration categories (gates, flip-flops, counters, etc.). CMOS, however, is rapidly invading the medium-scale integration area which was heretofore dominated by TTL.

9.14 Complementary MOS Logic (CMOS)

The CMOS logic family uses *both* P- and N-channel MOSFETs in the same circuit and therefore has several advantages over the P-MOS and N-MOS families. Generally speaking, CMOS is faster and consumes even less power than MOS. These advantages are offset somewhat by the increased complexity of the CMOS IC fabrication process and a lower packing density. Thus, CMOS cannot compete with MOS in applications requiring the utmost in large-scale integration.

However, CMOS logic has undergone a constant growth in the medium-scale integration area mostly at the expense of TTL, with which it is directly competitive. The CMOS fabrication process is simpler than TTL and has a greater packing density, therefore permitting more circuitry in a given area and reducing the cost per function. CMOS uses only a fraction of the power needed for even the low power TTL series (74L00) and is thus ideally suited for applications using battery power or battery back-up power. CMOS operating speed is not yet comparable to the faster TTL series, but it is expected to keep on improving.

CMOS INVERTER The connection of the basic CMOS INVERTER is shown in Figure 9–30. For this diagram and those that follow, the standard symbols for the enhancement MOSFETs have been replaced by blocks labelled P and N to denote a P-MOSFET and N-MOSFET, respectively. This is done simply for convenience in analyzing the circuits.

The CMOS INVERTER has two MOSFETs in series such that the P-channel device has its *source* connected to $+V_{DD}$ (a positive voltage) and the N-channel device has its source connected to ground. The gates of the two devices are connected together as a common input. The drains of the two devices are connected together as the common output.

The logic levels for CMOS are essentially $+V_{DD}$ for logic 1 and 0 V for logic 0. First, consider the case where $V_{IN} = +V_{DD}$. In this situation the gate of Q_1 (P channel) is at 0 V relative to the source of Q_1. Thus, Q_1 will be in the OFF state with $R_{OFF} \approx 10^{10}\ \Omega$. The gate of Q_2 (N channel) will be at $+V_{DD}$ relative to its source. Thus, Q_2 will be ON with $R_{ON} \approx 1\ k\Omega$. The voltage divider between Q_1's R_{OFF} and Q_2's R_{ON} will give $V_{OUT} \approx 0$ V.

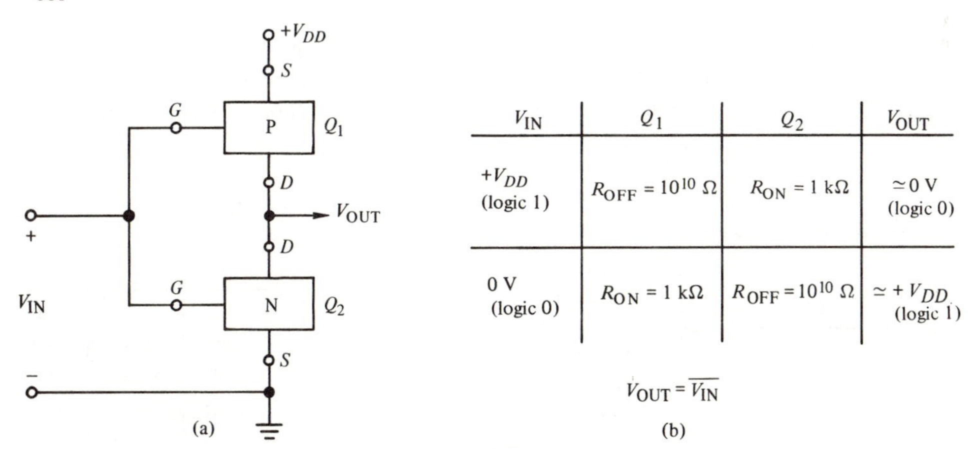

V_{IN}	Q_1	Q_2	V_{OUT}
$+V_{DD}$ (logic 1)	$R_{OFF} = 10^{10}\ \Omega$	$R_{ON} = 1\ k\Omega$	$\simeq 0$ V (logic 0)
0 V (logic 0)	$R_{ON} = 1\ k\Omega$	$R_{OFF} = 10^{10}\ \Omega$	$\simeq +V_{DD}$ (logic 1)

$V_{OUT} = \overline{V_{IN}}$

(b)

Figure 9–30 **Basic CMOS INVERTER**

Next, consider the case where $V_{IN} = 0$ V. In this case, Q_1 has its gate at a *negative* potential relative to its source, and Q_2 has $V_{GS} = 0$ V. Thus, Q_1 will be ON with $R_{ON} = 1\ k\Omega$, and Q_2 will be OFF with $R_{OFF} = 10^{10}\ \Omega$, producing a V_{OUT} of approximately

$+V_{DD}$. These two operating states are summarized in Figure 9–30(b), showing that the circuit does act as a logic INVERTER.

CMOS NAND gate Any logic function can be constructed by modifying the basic INVERTER. Figure 9–31 shows a NAND gate formed by adding a parallel P-channel MOSFET and a series N-channel MOSFET to the basic INVERTER. To analyze this circuit it helps to realize that a 0-V input turns ON its corresponding P-MOSFET and turns OFF its corresponding N-MOSFET, and vice versa for a $+V_{DD}$ input. Thus, it can be seen that the only time a LOW output will occur is when inputs A and B are both HIGH ($+V_{DD}$) to turn ON both N-MOSFETs. For any other combination at least one P-MOSFET will be ON and one N-MOSFET will be OFF, producing a HIGH output.

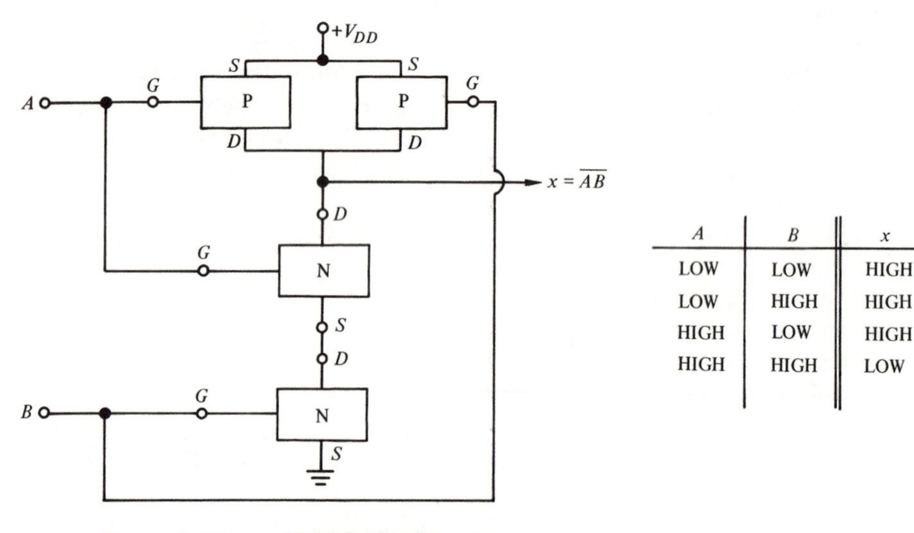

A	B	x
LOW	LOW	HIGH
LOW	HIGH	HIGH
HIGH	LOW	HIGH
HIGH	HIGH	LOW

Figure 9–31 **CMOS NAND gate**

CMOS NOR gate A CMOS NOR gate is formed by adding a series P-MOSFET and a parallel N-MOSFET to the basic INVERTER as shown in Figure 9–32. Once again, this circuit can be analyzed by realizing that a LOW at any input turns ON its corresponding

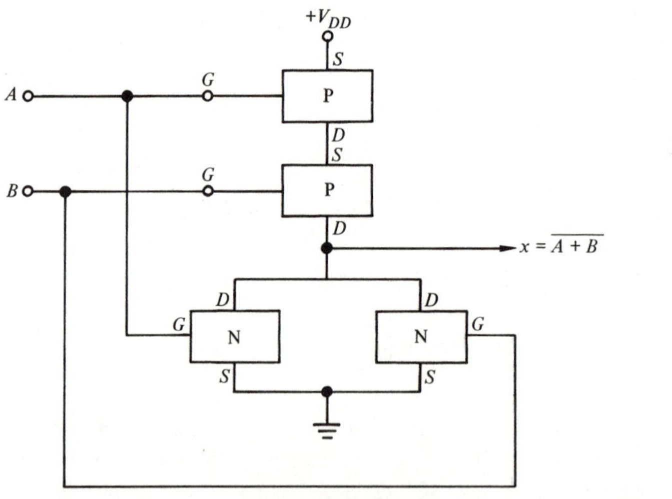

A	B	x
LOW	LOW	HIGH
LOW	HIGH	LOW
HIGH	LOW	LOW
HIGH	HIGH	LOW

Figure 9–32 **CMOS NOR gate**

P-MOSFET and turns OFF its corresponding N-MOSFET, and vice versa for a HIGH input. It is left to the student to verify that this circuit operates as a NOR gate.

CMOS AND and OR gates can be formed by combining NANDs and NORs with INVERTERs.

9.15 CMOS Family Characteristics

RCA developed the first CMOS logic series, known as the 4000 series, which was later manufactured by others. More recently, several firms have produced a CMOS series that is a direct pin-for-pin counterpart of TTL. This series, the 74C00, contains ICs with the same pin arrangements and logic operations as their TTL counterparts. For example, the 74C04 is a hex-inverter chip that is the equivalent of the 7404 TTL hex-inverter chip.

Voltage levels The CMOS logic levels are nominally 0 V for logic 0 and $+V_{DD}$ for logic 1. The $+V_{DD}$ supply can range from 3 V to 15 V, although some CMOS devices can use up to 20 V. This means that power-supply regulation is not a serious consideration for CMOS. When CMOS is being used with TTL, the V_{DD} supply voltage is made 5 V so that the voltage levels of the two families are almost the same. The choice of supply voltage often depends on other considerations, which will be discussed shortly.

The required CMOS input levels depend on V_{DD} as follows:

$$V_{IL}(\text{MAX}) = 30\% \times V_{DD}$$
$$V_{IH}(\text{MIN}) = 70\% \times V_{DD}$$

For example, with $V_{DD} = 5$ V, the value of V_{IL}(MAX) is 1.5 V, which is the highest input voltage that can be accepted as a LOW; and V_{IH}(MIN) = 3.5 V, which is the smallest voltage that can be accepted as a HIGH input.

Switching speed CMOS, like P-MOS and N-MOS, suffers from the relatively large load capacitances caused by the CMOS inputs being driven. Each CMOS input is typically a 5 pF load. CMOS circuits, however, have a faster switching rate than MOS circuits because of their lower output resistance in the HIGH state. Recall that for N-MOS any load capacitance is charged up in the HIGH state through the load MOSFET, which is about 100 kΩ. In the CMOS circuits the output resistance in the HIGH state is typically 1 kΩ(R_{ON} of a P-MOSFET), so the load capacitance is charged up more rapidly. A CMOS NAND gate typically has an average propagation delay of around 25 ns when a V_{DD} supply of 10 V is used, and 50 ns for $V_{DD} = 5$ V.

The switching speed of the CMOS family will vary with supply voltage. A large V_{DD} produces lower values of R_{ON}, which produce faster switching due to faster charging of load capacitances. This means that for higher frequency applications it is best to use a large value of V_{DD}. Of course, with a larger V_{DD} the power dissipation will increase although it will still be extremely low compared with those of other logic families.

Power dissipation The quiescent (dc) power dissipation of CMOS logic circuits is extremely low. This can be seen by re-examining the circuits of Figures 9–30—9–32. For these circuits, there is an extremely high resistance from V_{DD} to ground in either output state; that is, for any input condition there is always an OFF MOSFET in the current path. This situation results in typical CMOS dc power dissipations of 10 nW per gate using $V_{DD} = 10$ V, which is a significant improvement over even N-MOS logic. This extremely low P_D is the most important characteristic of CMOS logic.

Effect of frequency on P_D When CMOS logic circuits are in a static state for long periods of time or switching at very low frequencies the power dissipation will be extremely low. As the switching frequency of the CMOS circuits increases, however, the average power dissipation increases proportionally, because each time the CMOS output switches from LOW to HIGH a transient charging current must be supplied to any load capacitance. These momentary spikes of current come from the V_{DD} supply. Clearly, as frequency increases, the average current, and therefore average P_D, drawn from the V_{DD} supply will also increase (see Figure 9–33). For example, a CMOS NAND gate with a dc dissipation of 10 nW will have an average dissipation of 0.1 mW at a frequency of 100 kHz if it is driving two CMOS inputs. This increases to 1 mW at 1 MHz. Thus, CMOS loses some of its advantages at higher frequencies.

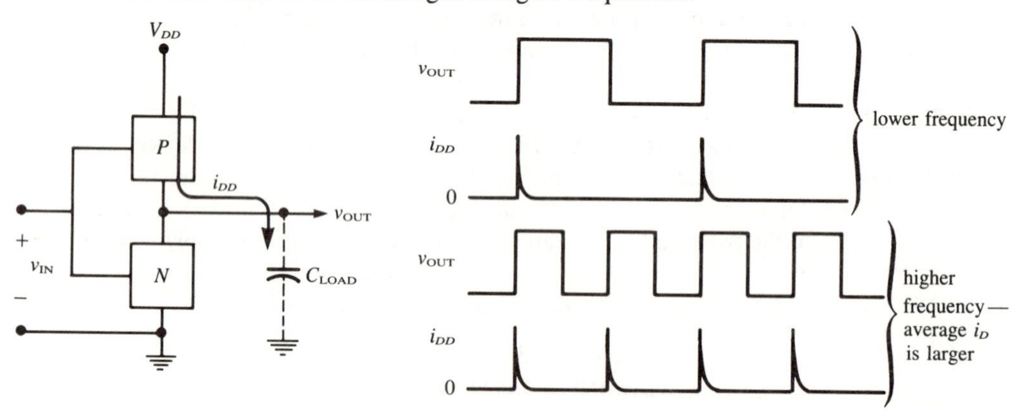

Figure 9–33 **Capacitive load draws charging current spikes from V_{DD} supply each time V_{OUT} goes HIGH; the average current will therefore increase with the frequency of switching**

Fan-out Like N-MOS and P-MOS, CMOS inputs have an extremely large resistance (10^{12} ohms) that draws essentially no current from the signal source. Each CMOS input, however, typically presents a 5 pF load to ground. This input capacitance limits the number of CMOS inputs that one CMOS output can drive (see Figure 9–34). The CMOS output has to charge and discharge the parallel combination of each input capacitance, so that the output switching time will be increased in proportion to the number of loads being driven. Typically, each CMOS load increases the driving circuit's propagation delay by 3 ns. For example, NAND gate 1 in Figure 9–34 might have a t_{PLH} of 30 ns if it were driving no loads; this would increase to 30 ns + 50(3 ns) = 180 ns if it were driving *fifty* loads.

Thus, CMOS fan-out depends on the permissible maximum propagation delay. Typically, CMOS outputs are limited to a fan-out of 50 for low-frequency operation ($\leq$1 MHz). Of course, for higher frequency operation the fan-out will be less.

Noise margin CMOS output levels are nominally 0 V and V_{DD}. These levels are not appreciably affected by CMOS loads. Thus,

$$V_{OH}(\text{MIN}) = +V_{DD}$$
$$V_{OL}(\text{MAX}) = 0\text{ V}$$

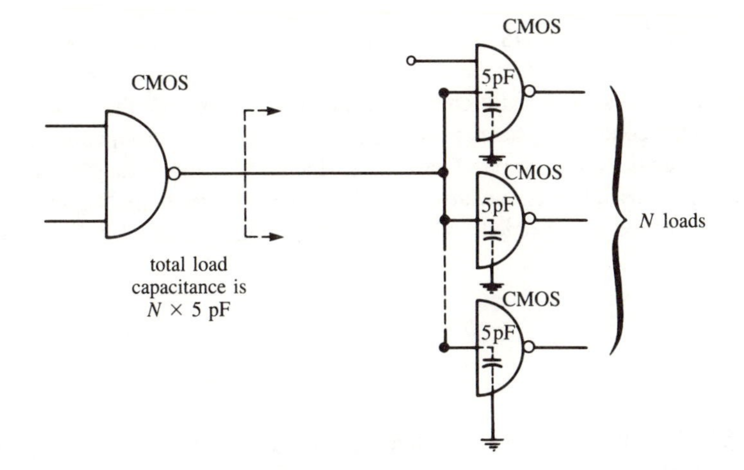

Figure 9–34 **Each CMOS input typically acts as a 5 pF load on the driving circuit's output**

As stated earlier, the CMOS input requirements are $V_{IL}(MAX) = 30\% \times V_{DD}$, and $V_{IH}(MIN) = 70\% \times V_{DD}$. Thus, the CMOS dc noise margins are

$$V_{NH} = V_{OH}(MIN) - V_{IH}(MIN)$$
$$= V_{DD} - 70\% V_{DD} = 30\% V_{DD}$$
$$V_{NL} = V_{IL}(MAX) - V_{OL}(MAX)$$
$$= 30\% V_{DD} - 0 = 30\% V_{DD}$$

These are *guaranteed* by the manufacturer, but in practice they are typically somewhat higher.

Example 9.14 Determine the guaranteed dc noise margins for CMOS operating at $V_{DD} = 5$ V. Repeat for $V_{DD} = 15$ V.

Solution: For $V_{DD} = 5$ V

$$V_{NH} = V_{NL} = 30\% \times 5 \text{ V} = \mathbf{1.5 \text{ V}}$$

For $V_{DD} = 15$ V

$$V_{NH} = V_{NL} = 30\% \times 15 \text{ V} = \mathbf{4.5 \text{ V}}$$

Both of these noise margins are much higher than for TTL. For this reason, CMOS is more suitable than TTL in applications that are exposed to a high-noise industrial environment, provided that its slower operating speed is acceptable.

The results above show that an improved noise margin can be obtained by using a higher V_{DD}. Of course, this improvement in noise immunity is obtained at the expense of a higher average P_D because of the larger supply voltage.

Susceptibility to static charge The high input resistance of CMOS inputs makes them especially susceptible to static charge buildup: voltages sufficient to cause electrical breakdown can result from simply handling the devices. Many of the newer CMOS devices have protective diodes on each input to guard against breakdown due to static charges.

Unused inputs All CMOS inputs *must* be tied to some voltage level, preferably ground or V_{DD}. Unused inputs cannot be left floating (disconnected) because they would be susceptible to noise and static charges, which could bias both the P- and N-channel MOSFETs in the conducting state, resulting in excessive power dissipation. Unused gate inputs can be tied directly to ground or $+V_{DD}$, whichever is appropriate for the particular logic function. Unused inputs may also be tied to one of the used inputs unless the fan-out of the signal source is exceeded, which is highly unlikely for CMOS because of its high fan-out.

9.16 Tristate CMOS

Standard CMOS outputs should not be tied together in the wired-AND configuration for the same reason stated for totem-pole TTL. CMOS manufacturers have produced several CMOS ICs with tristate outputs operating the same as TTL tristate devices. A CMOS tristate output can be tied to other tristate outputs provided that only one output is enabled at one time.

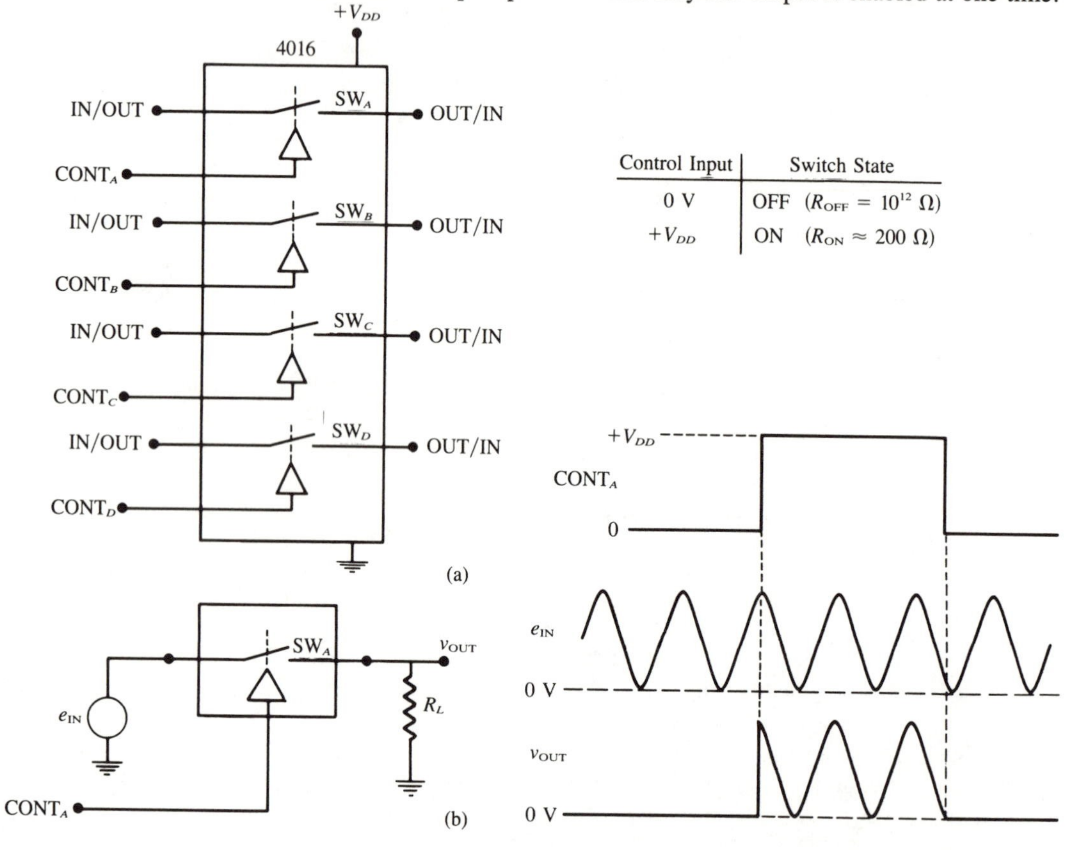

Figure 9–35 **(a) 4016 quad bilateral switch; (b) Bilateral switch used to switch an analog signal to a load**

9.17 CMOS Bilateral Switches

A special CMOS integrated circuit that has no TTL counterpart is the *bilateral switch*. As its name implies, this device will transmit signals in both directions, and it can be used to transmit both digital and analog signals. Its operation is similar to an electromechanical relay, and it is often referred to as a *solid-state relay*.

Figure 9–35(a) shows the functional diagram for a CMOS 4016 chip that contains four bilateral switches (SW_A, SW_B, SW_C, SW_D). The resistance across the terminals of each switch is controlled by the logic level at its control input ($CONT_A$, $CONT_B$, $CONT_C$, $CONT_D$). For example, when $CONT_A$ is at 0 V, SW_A is OFF and has a resistance of typically 10^{12} ohms. When $CONT_A = +V_{DD}$, SW_A will be ON, with a typical resistance of 200 Ω.

The switches are bilateral so that the signal to be switched can be applied to either terminal as the input, with the output taken at the other terminal. Figure 9–35(b) shows a typical application where a logic signal is used to control the transmission of an analog signal (e_{IN}) to a load R_L. When $CONT_A = 0$ V, the switch is open so that $V_{OUT} \approx 0$ V. When CONT $= +V_{DD}$, the switch is closed so that $v_{OUT} \approx e_{IN}$ as long as R_L is much greater than the switch's ON resistance, R_{ON}.

The analog signal has to be within the limits of 0 V to $+V_{DD}$. There are CMOS bilateral switches that will transmit bipolar analog signals, but they require a negative supply voltage in addition to $+V_{DD}$.

Questions

Refer to Appendix B for IC data sheets

Sections 9.1–9.2

9.1 Two different logic circuits have the characteristics tabulated below.

(a) Which circuit has the better LOW-state dc noise immunity? HIGH-state dc noise immunity?

(b) Which circuit can operate at higher frequencies?

(c) Which circuit draws more supply current?

(d) Which circuit is more likely to have a better ac noise margin? Explain.

	Circuit *A*	Circuit *B*
V_{supply}	6 V	5 V
V_{IH}	1.6 V–6 V	1.8 V–5 V
V_{IL}	0 V–0.9 V	0 V–0.7 V
V_{OH}	2.2 V–5 V	2.5 V–4.5 V
V_{OL}	0 V–0.4 V	0 V–0.3 V
t_{PLH}	10 ns	18 ns
t_{PHL}	8 ns	14 ns
P_D (MAX)	16 mW	10 mW

Section 9.3

9.2 Explain the difference between current-sourcing logic and current-sinking logic.

9.3 For the TTL NAND gate of Figure 9–5, calculate the current I_{IL} that flows when input A is connected to ground.

9.4 For the TTL NAND circuit of Figure 9–5, calculate the voltage at Q_2's collector when A and B are both at 5 V. Assume Q_2 has $h_{FE} \gtrsim 20$.
Can you now explain why the diode D_1 is needed?

9.5 Give two advantages and one disadvantage of the totem-pole output configuration.

9.6 Which totem-pole transistor conducts in the LOW state? In the HIGH state? Which one acts as a current sink?

Sections 9.4–9.5

9.7 Under normal conditions, what is the largest LOW-state voltage that should appear at the output of any 7400 series logic circuit?

9.8 Refer to the data sheet for the 7400 quad NAND gate. Determine the maximum power dissipation of a *single* NAND gate when all of its inputs are LOW.

9.9 (a) Which TTL series has the lowest power dissipation?

(b) Which TTL series has the longest propagation delay?

(c) Which TTL series can operate at the highest frequency?

(d) Which TTL series uses diode clamping to reduce storage time delay?

9.10 Determine and compare the average maximum P_D and t_{pd} for the 7404, 74H04, 74S04, and 74LS04 ICs.

Sections 9.6–9.7

9.11 Refer to the data sheet for a 74LS14. What is the most current it can *supply* to another logic circuit in the HIGH state? How much current can it *sink* in the LOW state?

9.12 Refer to the data sheet for the 7473 dual *JK* flip-flop (FF).

(a) Determine the input loading factor at the J, K inputs.

(b) Determine the input loading factor at the clock and clear inputs.

(c) How many other 7473 flip-flops can the output of one 7473 drive at the clock input?

9.13 Figure 9–36(a) shows a 7473 *JK* flip-flop (FF) whose output is required to drive a total of 14 U.L. Since this value exceeds the fan-out of the 7473, a *buffer* of some type is needed. Figure 9–36(b) shows one possibility using one of the NAND gates from the 7437 quad NAND buffer, which has a much higher fan-out than the 7473. Note that $\overline{Q}$ from the 7473 is used since the NAND is acting as an INVERTER. Refer to the data sheet for the 7437 and determine

(a) its fan-out

(b) its maximum sink current in the LOW state

9.14 Try to come up with a different way of solving the loading problem of Figure 9–36(a) using a 7400 quad NAND chip.

9.15 Refer to the logic diagram of Figure 9–37, where the 7486 EXCLUSIVE-OR output is driving several 7420 inputs. Determine whether the fan-out of the 7486 is being exceeded. Explain.

9.16 For the circuit of Figure 9–37, determine the *longest* time it will take for a change in the A input to be felt at output w. Use all worst-case conditions and maximum values of gate propagation delays. (*Hint:* Remember that NAND gates are *inverting* gates.)

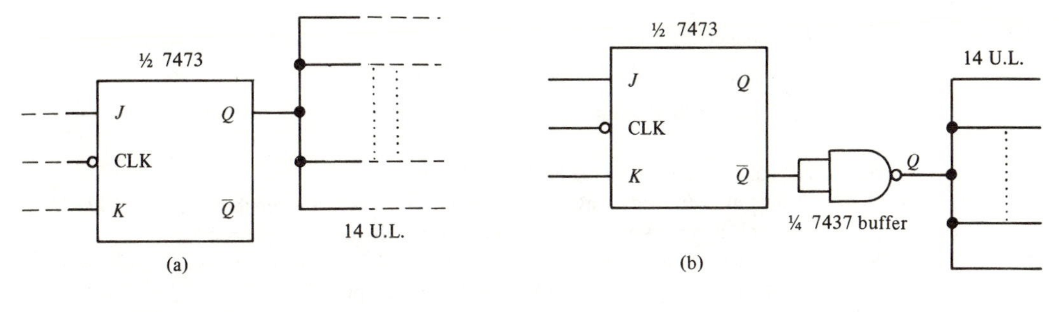

Figure 9–36

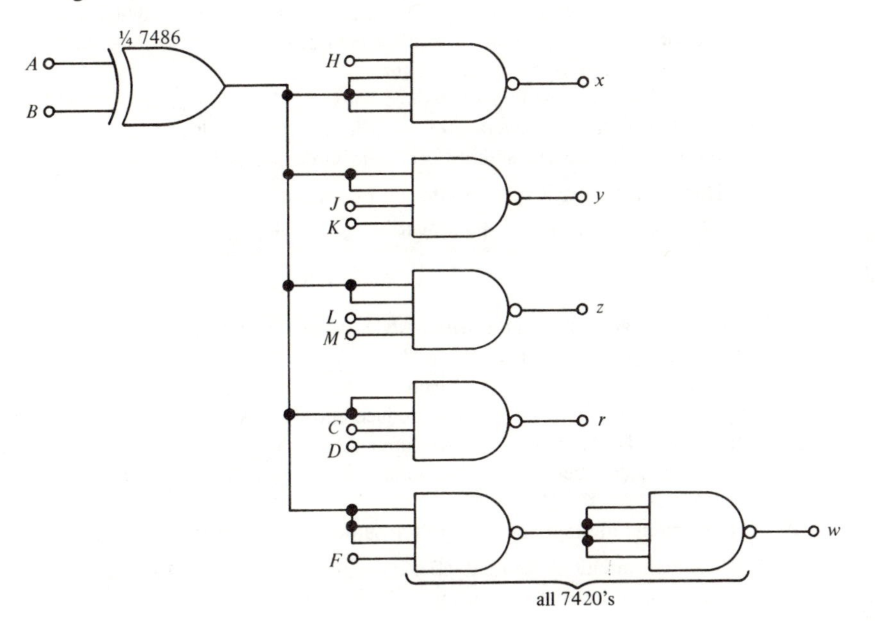

Figure 9–37

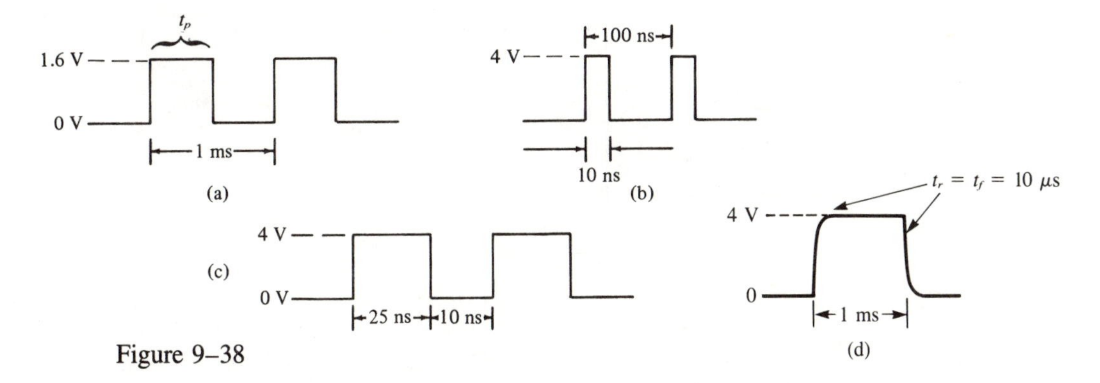

Figure 9–38

9.17 How many 7473 CLK inputs can be driven by a 74LS04 output?

9.18 Which of the following are acceptable ways of handling unused inputs of a TTL NAND or AND gate?

(a) Leave disconnected.

(b) Connect directly to ground.

(c) Connect to a used input.

(d) Connect directly to V_{CC}.

(e) Connect to V_{CC} through a 1-kΩ resistor.

9.19 Repeat Question 9.18 for NOR and OR gates.

9.20 Determine *why* each waveform in Figure 9–38 will *not* reliably trigger a 7473 FF at its clock input. (*Note:* t_p of a clock pulse is the *positive* portion.)

9.21 A 74121 one-shot is being triggered by a 1 MHz pulse waveform. Determine the $\overline{Q}$ output waveform if $R_x = 10$ kΩ, and $C_x = 68$ pF.

9.22 Figure 9–39 shows a 74121 one-shot being triggered by the closing of a switch. The one-shot is used to produce a *single* 20 ms pulse even though the switch contacts may bounce for a few milliseconds. Calculate the maximum value of R for proper operation. Allow for a 400 mV noise margin.

9.23 Determine the lower and upper threshold voltages (V_{LT} and V_{UT}) for a 74LS14 Schmitt trigger. (HINT: take the average of the MIN and MAX values.)

9.24 (a) Explain why current spiking occurs in circuits using TTL totem-pole outputs.

(b) How is this current spiking affected by capacitive loads?

(c) What is done to reduce the effects of current spiking on circuit operation?

Sections 9.8–9.9

9.25 Why is it unwise to form the wired-AND connection with conventional TTL NAND gates such as the 7400 quad NAND?

9.26 What are the two advantages of using the wired-AND connection?

9.27 Why don't open-collector outputs produce current spiking?

9.28 The 7409 TTL IC is a quad two-input AND with open collector outputs. Show how 7409's can be used to implement the operation $x = A \cdot B \cdot C \cdot D \cdot E \cdot F \cdot G \cdot H \cdot I \cdot J \cdot K \cdot M$.

9.29 (a) Determine the logic expression for output x in Figure 9–40.

(b) The logic circuit of Figure 9–40 is implemented using one 7401 chip. Can the same function be implemented using a single 7400 chip?

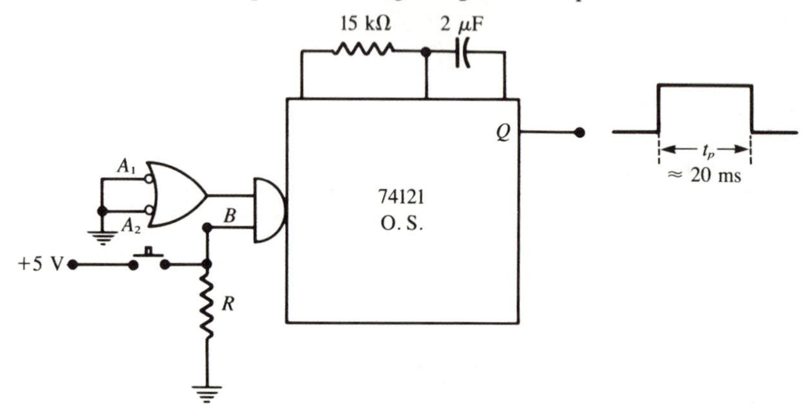

Figure 9–39

9.30 Determine a value for R_C in Figure 9–40 if output x is driving the *clear* inputs of *four* 7473 flip-flops (FFs).

9.31 Explain how tristate TTL overcomes the disadvantages of open-collector TTL.

9.32 Why should the open-collector pull-up resistor be kept as small as possible?

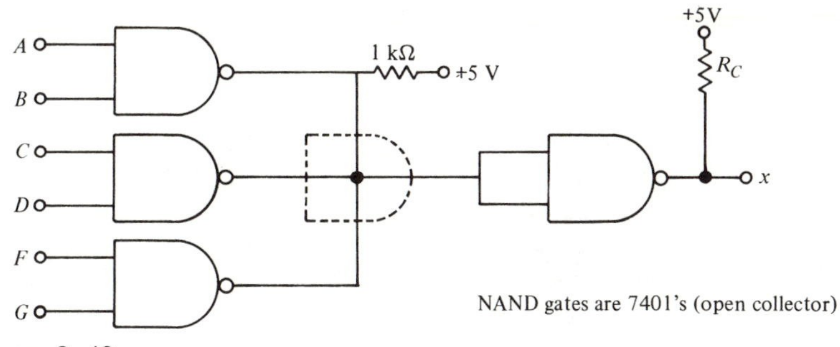

Figure 9–40

9.33 Figure 9–41 uses 74125 tristate buffers, described in Figure 9–24. Determine the output waveform for the input waveforms given.

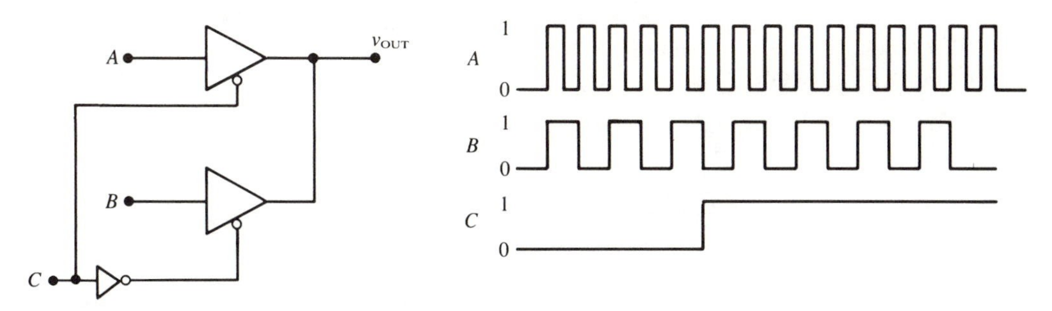

Figure 9–41

Section 9.10

9.34 What is the principal reason ECL logic is faster than TTL?

9.35 What functions do the emitter followers perform in the ECL circuits?

9.36 Which of the following are advantages that ECL has over conventional TTL (7400 series)?

(a) lower power dissipation

(b) shorter t_{pd}

(c) greater fan-out

(d) greater noise immunity

(e) complementary outputs

(f) no current spikes during switching

Sections 9.11–9.13

9.37 Draw the circuit diagram of an N-MOS INVERTER using a supply voltage of 12 V. Determine the output voltages for $V_{IN} = 0$ V and +12 V.

9.38 The circuit of Figure 9–42 is an N-MOS logic gate. Determine what type of gate it is. Assume +16 V = logic 1, 0 V = logic 0.

9.39 Which of the following are advantages of N-MOS logic over TTL?

(a) greater packing density

(b) greater operating speed

(c) greater fan-out

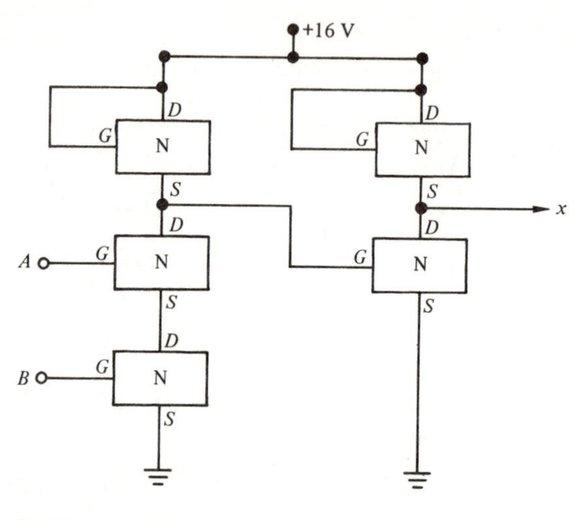

Figure 9–42

(d) greater suitability for large-scale integration

(e) lower P_D

(f) complementary outputs

(g) greater noise immunity

(h) simpler fabrication process

(i) more small-scale and medium-scale integration functions

Sections 9.14–9.17

9.40 Which of the following are advantages of CMOS over TTL?

(a) greater packing density (more circuits per chip)

(b) higher speed

(c) greater fan-out

(d) simpler fabrication process

(e) greater noise margins

(f) lower P_D

(g) uses wide range of supply voltages

(h) no supply-current spikes

9.41 Which of the following conditions would result in the lowest average P_D for a CMOS *JK* FF?

(a) V_{DD} = 10 V, clock input = 1 MHz

(b) V_{DD} = 5 V, clock input = 10 kHz

(c) V_{DD} = 10 V, clock input = 10 kHz

9.42 (a) Explain why the average P_D for a CMOS logic circuit is so low in either logic state.

(b) Explain why $P_{D'}$ increases with frequency.

9.43 What is the acceptable range of input voltages for each logic level for CMOS operating with V_{DD} = 12 V?

9.44 Refer to the 4011 data sheet in Appendix B and determine:

(a) current drawn from the supply when V_{DD} = 18 V at 25°C

(b) typical average propagation delay at V_{DD} = 5 V and at V_{DD} = 15 V

9.45 What limits the number of CMOS inputs a CMOS output can drive?

9.46 Determine the following values for a CMOS IC operating at V_{DD} = 12 V and driving only CMOS loads:

(a) V_{OH}(MIN)

(b) V_{OL}(MAX)

(c) V_{IH}(MIN)

(d) V_{IL}(MAX)

9.47 Determine the amplitude of the v_{OUT} waveform in Figure 9–35(b) when CONT$_A$ = +10 V, V_{DD} = 10 V, e_{IN} = 6 V*p-p*, and R_L=3.3 kΩ.

9.48 Figure 9–43 shows CMOS bilateral switches used to change the gain of an amplifier under the control of digital inputs *A* and *B*. The 4016s are operating at V_{DD} = 5 V.

Determine the amplitude of v_{OUT} for the four different conditions on the two logic inputs.

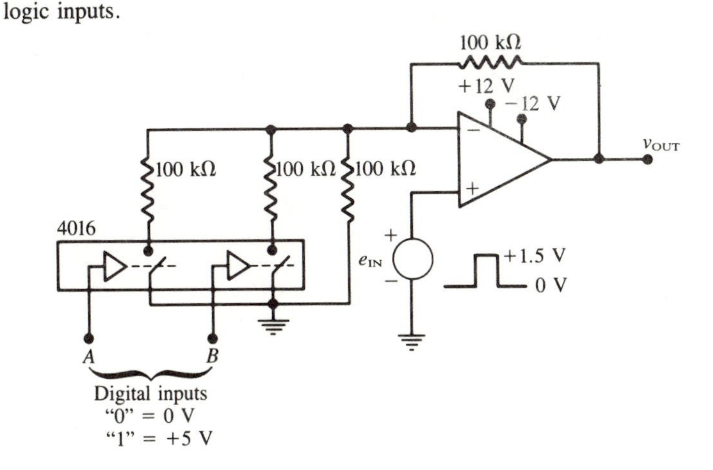

Figure 9–43

9.49 For each of the following statements, indicate which logic family (or families) is being described (TTL, ECL, N-MOS, CMOS).

(a) highest operating speed

(b) current-sinking logic

(c) does not use resistors in its basic circuit

(d) P_D increases dramatically with frequency

(e) operates at nominal supply voltage of 5 V

(f) has multiple-emitter inputs

(g) has lowest P_D

(h) slowest operating speed

(i) best family for large-scale integration

(j) extremely high input resistance

10

IC Interfacing

10.1 Introduction

There is no doubt that the availability of a wide variety of integrated circuits has made it possible to construct relatively complex electronic systems that would not have been feasible ten or fifteen years ago. While the use of ICs has greatly simplified the task of putting circuit functions together to form a working system, we must not be misled into thinking that we should be concerned only with the *logical* characteristics of the various circuit building blocks. It is true that the logical function of each IC has to be understood if we are to combine them to obtain the desired overall function, but it is just as important to understand the *electrical* characteristics of each IC.

The last chapter introduced each of the major IC families. We saw how their electrical characteristics determined the maximum number of different logic inputs, of the same family, that one logic output could drive. In this chapter we will address the more complex problem of connecting a logic output to inputs of different logic families (e.g., TTL to CMOS) or to loads that are not logic devices (e.g., relay, LED).

Interfacing means connecting electrically different circuits and devices. Sometimes a direct interconnection is not possible; rather, some type of *interface* circuit has to be used between the circuit providing the output signal (the *driver*) and the circuit receiving the signal (the *receiver*). The interface circuit thus takes the output signal from the driver and conditions this output so that it satisfies the input requirements of the receiver.

10.2 TTL Driving CMOS

The input impedance of a CMOS device is extremely high, so it will not load down the logic output voltages from a TTL device. It would appear, then, that a TTL driver can drive a CMOS receiver with no problems when the CMOS is operating with $V_{DD} = 5$ V. Unfortunately, this is not true for TTL totem-pole outputs because their HIGH-state output voltage, V_{OH}, is very close to the minimum V_{IH} of CMOS.

A high TTL output will typically be 3.6 V to 3.8 V, with no load. A CMOS input requires $V_{IH} = 70\% \times V_{DD} = 3.5$ V when operating from a 5 V supply. This leaves very little noise margin, and thus a small amount of noise could easily drop the CMOS input voltage into the ambiguous region. The situation is even worse when the TTL output is also driving other TTL loads that will reduce its V_{OH} to less than 3.5 V.

The standard solution to this interface problem is to connect the TTL output to +5 V through a pull-up resistor, as shown in Figure 10–1(a). This resistor will pull the TTL output voltage up to approximately 5 V in the HIGH state. This action can be explained by examining the totem-pole circuit with the added pull-up resistor [Figure 10–1(b).]

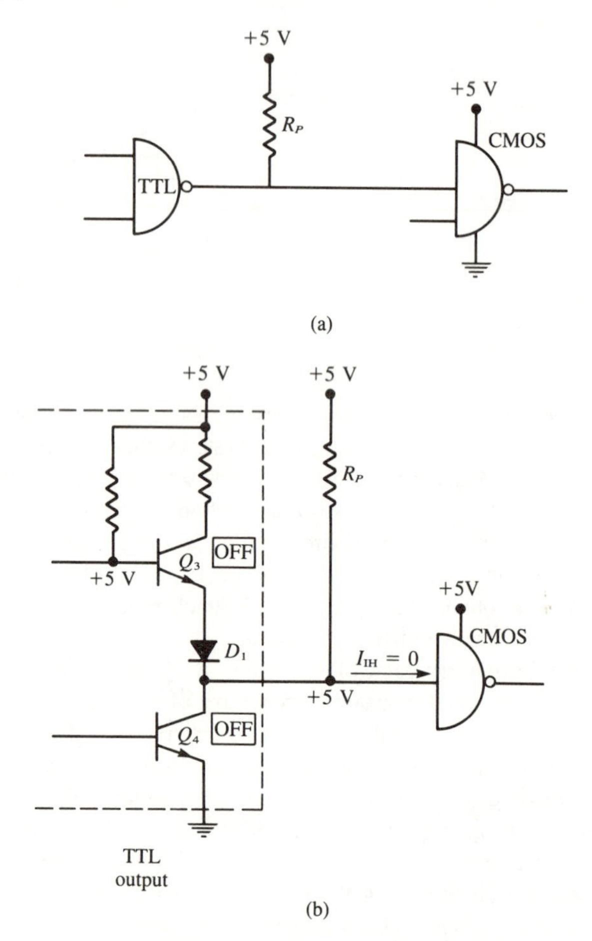

Figure 10–1 **TTL totem-pole outputs require a pull-up resistor to reliably drive CMOS inputs**

In the HIGH state, transistor Q_4 is OFF while Q_3 would normally be conducting as an emitter-follower. The pull-up resistor, however, prevents Q_3 from conducting because it connects the cathode of D_1 to +5 V. Thus, the +5 V at the base of Q_3 is not large enough to forward bias Q_3 and D_1. With both totem-pole transistors in the OFF state, there is no current through R_p (except leakage current), so that the voltage at the CMOS input will be +5 V.

The value of the pull-up resistor is chosen using the same considerations that we used for the open-collector pull-up. This strategy makes sense because the totem-pole output is essentially acting as an open-collector output since Q_3 is always OFF. If the TTL output is not driving any other loads, a 1 kΩ pull-up can be used.

TTL driving high-voltage CMOS When the CMOS receiver is being operated from a supply voltage greater than 5 V, the situation becomes complicated by the fact that *many* TTL outputs are not rated to withstand voltages as high as the V_{IH} required by the CMOS input. For example, when the CMOS is operating from V_{DD} = 10 V, it will have V_{IH} = 70% × 10 V = 7 V. The TTL output can be connected to +10 V through a resistor to pull up the voltage to +10 V *provided* that the TTL device has a maximum V_{OH} rating of 10 V or more. This rating should always be checked before using the pull-up resistor.

If the TTL output cannot handle the higher voltage, it will be necessary to connect an interface between the TTL driver and the CMOS receiver. Figure 10–2 shows one possible interface circuit using discrete transistors. The transistors form cascaded inverter circuits operating from +10 V. When the TTL output is LOW (0.0–0.4 V), Q_1 will be OFF and Q_2 will be ON, thereby producing ≈ 0 V at the CMOS input. When the TTL output is HIGH (2.4–3.6 V), Q_1 will be ON and Q_2 will be OFF. With Q_2 OFF, the CMOS input will be +10 V, as required.

This interface circuit is called a *level translator* since it changes the HIGH-state logic level from a TTL level to a +10 V level. Level translators are also available in IC form. The 40104 is a CMOS device designed to translate low-voltage logic levels to high-voltage levels.

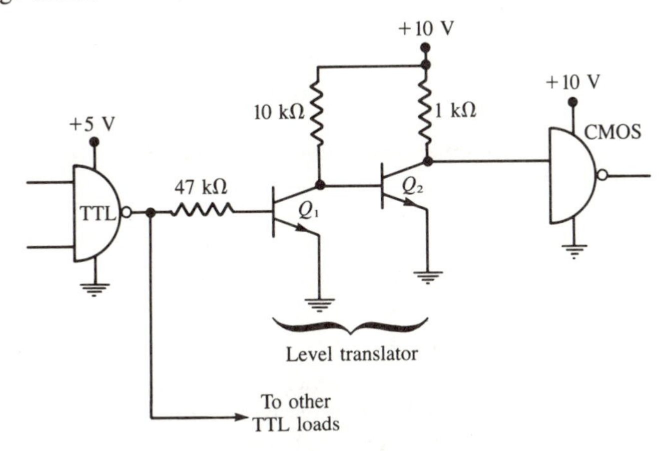

Figure 10.2 **TTL driving 10 V CMOS through a level translator interface circuit**

Example 10.1 The TTL output in Figure 10–2 is rated at 10 U. L. How many TTL loads can it drive in addition to the interface circuit?

Solution: The TTL HIGH output can source $10 \times 40\ \mu A = 400\ \mu A$ of current before it drops below the $V_{OH}(MIN)$ limit of 2.4 V. At 2.4 V, the TTL output has to supply a Q_1 base current of

$$I_B = \frac{2.4\text{ V} - 0.7\text{ V}}{47\text{ k}\Omega} \approx 36\ \mu\text{A}$$

This leaves $400 - 36 = 364\ \mu A$ available for the TTL loads. Since a standard TTL unit load is 40 μA, this means that **9 U. L.** can be driven.

10.3 TTL Driving NMOS and PMOS

The considerations for TTL driving NMOS are the same as for CMOS since NMOS uses a positive supply voltage. NMOS is used mainly in large-scale integration devices like memory chips and microprocessors, and typically operates from +5 V.

PMOS operates from a negative supply voltage and is therefore not capable of being directly driven by TTL outputs. Although there are not too many PMOS devices in current use, we will show the TTL-to-PMOS interface as another example of a level translator. A discrete version is shown in Figure 10–3. The circuit uses a P-channel JFET chosen to have a $V_{GS}(OFF)$ of no more than +2.4 V so that it will turn OFF when the TTL output is 2.4 V or greater.

This same circuit can also be used as an interface between CMOS and PMOS logic circuits, and between NMOS and PMOS logic circuits.

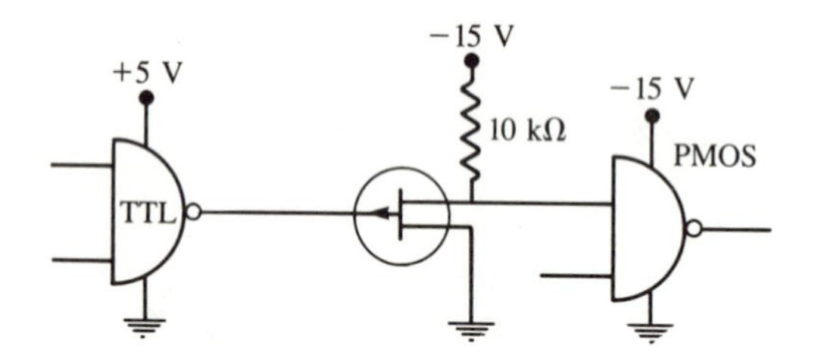

Figure 10–3 **A P-channel JFET used as an interface between TTL and PMOS**

10.4 CMOS Driving TTL

Figure 10–4 shows a CMOS output driving a TTL input. The HIGH state is generally not a problem since the TTL input current requirement is small ($I_{IH} = 40\ \mu A$ for standard TTL, and 20 μA for LS-TTL). The CMOS output has no trouble supplying this amount of current through the R_{ON} of its P-channel MOSFET.

In the LOW state, however, a problem can occur because the TTL input produces an I_{IL} that has to flow back through the CMOS output to ground. In other words, the CMOS device has to sink I_{IL} through the R_{ON} of its N-channel MOSFET, producing an output $V_{OL} = I_{IL} R_{ON}$. The value of R_{ON} will typically range from 100 Ω to 5 kΩ for different CMOS circuits. In many cases the V_{OL} produced by the $I_{IL} R_{ON}$ voltage will exceed the maximum allowable LOW state input voltage for the TTL device, $V_{IL}(MAX)$.

Recall that TTL has $V_{IL}(MAX) = 0.8$ V. However, in order to maintain a 0.4 V noise margin, V_{IL} should be made no greater than 0.4 V. Most CMOS data sheets will specify the amount of N-channel sink current that will produce $V_{OL} = 0.4$ V. This current value can then be used to determine how many TTL loads the CMOS output can drive.

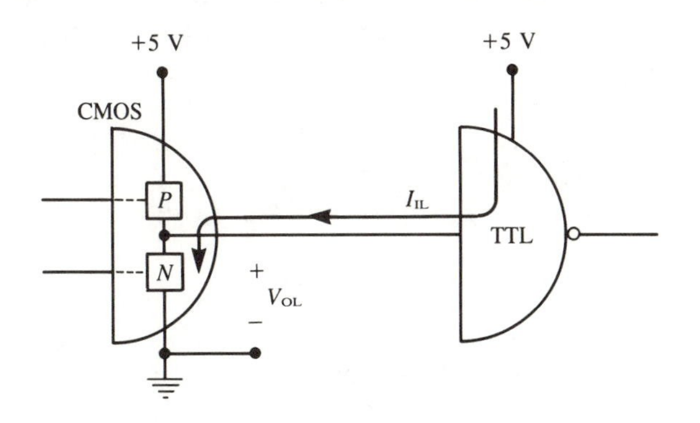

Figure 10–4 **In the LOW state, the CMOS output has to sink the TTL I_{IL} through its N-channel MOSFET. This produces a V_{OL} that may exceed the 0.4 V limit needed to drive TTL.**

Example 10.2 Refer to the data sheet for the CD4011 IC in Appendix B. Determine how many standard TTL loads it can reliably drive at 25° C. Repeat for LS-TTL loads.

Solution: Look under "static electrical characteristics" for the "Output Low (Sink) Current" entries. Find the $V_0 = 0.4$ V, $V_{DD} = 5$ V condition at 25° C and locate the minimum and typical values of I_{OL} as 0.51 mA and 1.0 mA, respectively. The minimum value is all that the manufacturer can guarantee, so we must use this value in our calculations.

Since the output can sink only 0.51 mA, and a standard TTL device has an I_{IL} of 1.6 mA, this output **cannot drive any** standard TTL inputs. An LS-TTL input typically has $I_{IL} = 0.4$ mA. Thus, this CMOS device can reliably drive only **one** LS-TTL input.

As the preceding example illustrates, a typical CMOS output *cannot* drive a standard TTL load, but it can drive an LS-TTL load. It can also easily drive L-TTL inputs since they have a smaller I_{IL} requirement than LS-TTL. Very often logic circuit designers use an LS-TTL device as a buffer between a CMOS driver and a standard TTL load because the LS-TTL device can readily drive standard TTL.

Not all CMOS devices have an I_{OL} limit as low as the 4011 device in the preceding example. There are several that can sink more than the 1.6 mA required to drive a standard TTL input. For instance, the 4050 is a non-inverting *buffer* that can sink 3.75 mA and so can reliably drive two standard TTL loads. It is called a buffer because it has a higher output current capability than the standard CMOS device.

High-voltage CMOS driving TTL Most standard TTL inputs cannot withstand input voltages greater than 5.5 V, so they cannot be driven directly by a CMOS device operating at $V_{DD} > 5.5$ V. A level translator is needed to change the CMOS output levels to the required TTL levels. Figure 10–5 shows one method for performing this function. Transistors Q_1 and Q_2 will both be OFF when the CMOS output is at +10 V, thereby producing

+5 V at the TTL input. Both transistors will turn ON when the CMOS output goes LOW, thereby producing ≈ 0 V at the TTL input.

CMOS-to-TTL level translators are also available as ICs. The 4049 and 4050 are CMOS devices that can perform high-to-low level translation. Each of these are also buffers that can drive *two* standard TTL loads. Figure 10–6 shows how the 4050 *non-inverting* buffer can be used as a level translator. The 4049 is an *inverting* buffer.

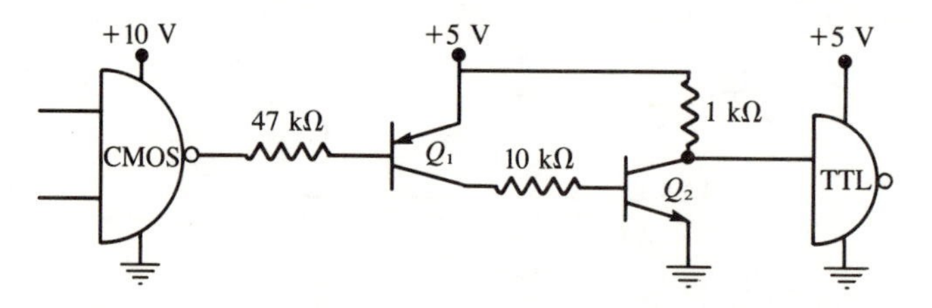

Figure 10–5 **A level translator is used as an interface between high-voltage CMOS and standard TTL**

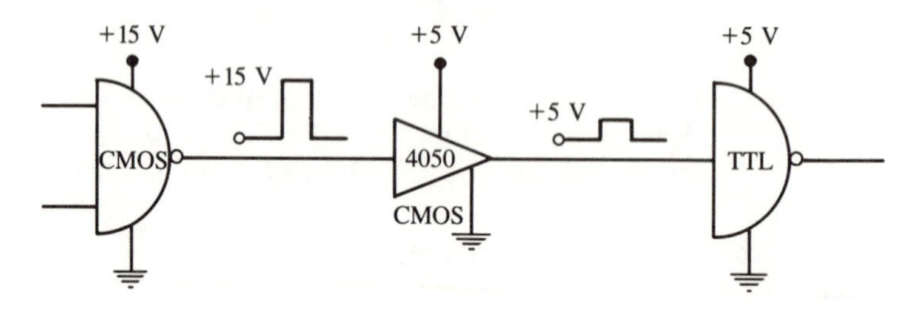

Figure 10–6 **The 4050 serves as a level translator and a buffer between high-voltage CMOS and TTL**

IC manufacturers have produced many LS-TTL devices that can withstand input voltages as high as +15 V. These devices, of course, can be driven directly by CMOS devices operating at supply voltages up to 15 V. The device data sheets should be consulted before connecting more than 5 V to any TTL input.

10.5 Interfacing Digital ICs to Non-Standard Devices

Thus far, we have considered only the problem of interfacing one digital IC to another digital IC. There is an equally important class of interfacing problems that involves connecting digital ICs to a wide variety of devices that do not have the standard characteristics of the digital IC family devices. For example, how can a TTL or CMOS output control a 220 V, 50 A industrial motor? Or how can a digital counter be used to count items that are interrupting a light beam?

In dealing with interface problems such as these, it becomes apparent that digital ICs can rarely be connected *directly* to any devices other than digital ICs. This is because the electrical requirements of these devices (i.e., current, voltage, power) are not compatible with those of the standard IC families. To perform the many different types of interfacing that are common to most digital systems, one has to know the characteristics and limitations of the ICs that are being used. With this knowledge, a suitable interface circuit can be designed to connect ICs to any type of device.

The following sections will examine some common interface problems and their solutions. These examples are chosen to illustrate the many considerations involved in connecting ICs to external devices and are in no way intended to provide an exhaustive survey of all types of interface problems. These examples will help you gain a general understanding of solution strategies useful in various interface problems. The end-of-chapter problems will provide some different interface situations to test your understanding.

10.6 Driving Light-Emitting Diodes (LEDs)

LEDs are often used to indicate the states of logic circuit outputs. These semiconductor light sources provide good visibility, do not burn out (as incandescent lights do) and require relatively low voltage and current for their operation. Discrete LEDs are available that emit the major visible colors (red, orange, yellow, and green) and the invisible infrared.

The brightness of the light emitted by an LED increases in almost direct proportion to the amount of forward current flowing through the device, until the LED saturates (no further increase in brightness). LEDs are typically operated at 10–30 mA.

LEDs are PN junctions, and so will have a relatively constant forward voltage drop, V_F, over a wide range of currents. The forward voltage, however, will depend on the semiconductor material used to fabricate the LED. For instance, a green gallium phosphide LED will have $V_F \approx 2.4$ V while a red gallium arsenide LED will have $V_F \approx 1.7$ V. Like any diode, once an LED is forward-biased into conduction, the current will be limited by series resistance in the circuit.

TTL driving an LED The following examples will illustrate how a TTL device can be used to drive an LED indicator.

Example 10.3 The LED in Figure 10–7(a) is to turn ON when the TTL FF output is HIGH ($Q = 1$). The LED is to be operated at $I_F = 10$ mA, and it has $V_F = 2.2$ V at 10 mA. Why won't this arrangement work?

Solution: A TTL output does not have a very large HIGH-state output current (I_{OH}). Even outputs having a fan-out of 20 U. L. will only have $I_{OH} = 20 \times 40\ \mu A = 800\ \mu A$. This is obviously not enough to turn on the LED.

Example 10.4 Figure 10–7(b) shows one way to have the TTL FF activate the LED. Here the LED will be forward-biased when $\overline{Q}$ goes LOW ($Q = 1$). The LED current will flow through the current-sinking transistor at $\overline{Q}$'s output. A standard TTL output will have $I_{OL}(MAX) = 16$ mA and so will be able to handle the 10 mA needed by this LED.

Calculate the value of R_s needed to produce 10 mA through the LED.

Solution: The voltage across R_s is determined by subtracting V_F and V_{OL} from +5 V. We will use $V_{OL} = 0.4$ V, to be conservative. Thus, there will be $5 - 2.2 - 0.4 = 2.4$ V across R_s, so that

$$R_s = \frac{2.4\ \text{V}}{10\ \text{mA}} = 240\ \Omega$$

In practice, a slightly smaller value might be used, perhaps 220 Ω, to allow for component tolerances.

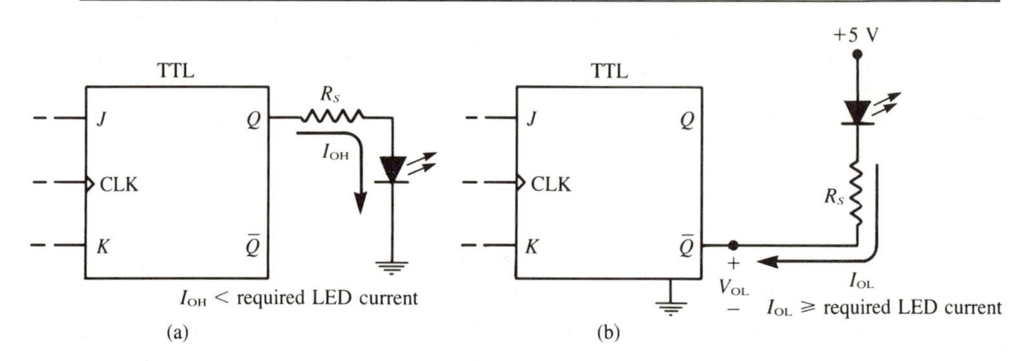

Figure 10–7 **(a) A HIGH TTL output cannot normally be used to turn ON an LED; (b) A LOW TTL output can be used to sink the LED current**

Although the arrangement in Figure 10–7(b) is often used, it does have some drawbacks. For one, the LED might not go completely dark when Q goes HIGH because V_{OH} could be as low as 2.4 V, especially if $\overline{Q}$ is driving other loads. For example, with $V_{OH} = 2.4$ V, there will be $5 - 2.4\text{ V} - 2.2\text{ V} = 0.4$ V across R_s. This will produce a small amount of current that could result in a dim, but not OFF, LED.

A second problem is that this standard TTL output is limited to $I_{OL} = 16$ mA and so cannot be used for the higher LED currents required for increased brightness.

Both of these drawbacks can be eliminated by using an *open-collector buffer,* which can sink more than 16 mA. One such device is the 7438 quad NAND-gate buffer with open-collector outputs (see Appendix B). Figure 10–8 shows its use.

The 7438 has a fan-out of 30 U. L. and can therefore sink up to 48 mA. Its open-collector output also ensures that the LED will be completely OFF when $Q = 0$.

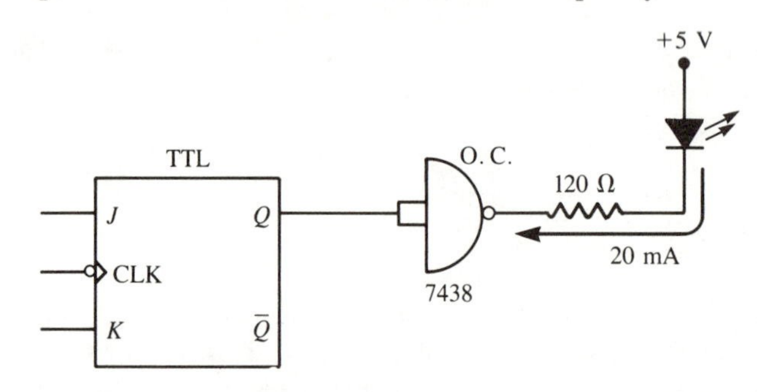

Figure 10–8 **Use of an open-collector buffer allows a larger LED current for more brightness (when Q = 1) and ensures LED will be OFF when $Q = 0$**

CMOS driving an LED Most CMOS devices have very low output current capabilities, so neither arrangement of Figure 10–7 would work. There are some CMOS buffers available that can handle the current required to turn on an LED. For example, the 40107 is a CMOS NAND-gate IC that can typically sink 32 mA when operated at $V_{DD} = 5$ V.

In addition, it has an *open-drain* output configuration (CMOS counterpart to the TTL open-collector) in which the upper P-channel MOSFET has been eliminated.

An LS-TTL device also can be used as a buffer between a CMOS output and an LED. This application is shown in Figure 10–9 where a 74LS38 NAND-gate open-collector buffer is used. The 74LS38 can be easily driven by any CMOS output, and it has a fan-out of 15 U. L. so that it can sink up to 24 mA.

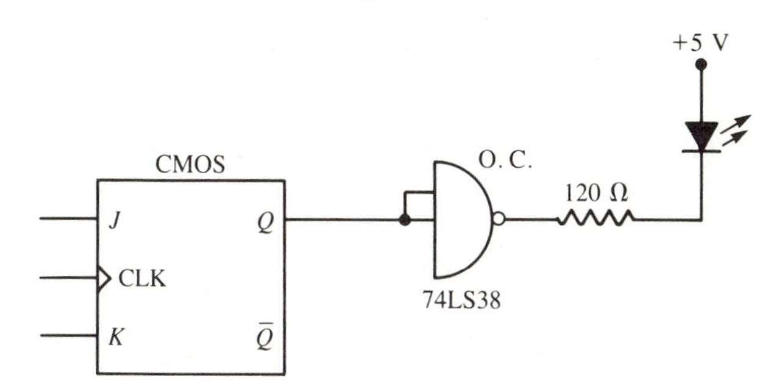

Figure 10–9 **An 74LS38 can act as an interface between a CMOS output and an LED**

10.7 Driving High-current and High-voltage Loads

A digital signal is often called on to activate a device that requires a current and/or voltage that cannot be delivered directly by an IC, even a buffer IC. For instance, a TTL or CMOS logic circuit could be controlling power to a motor, a solenoid, or some other heavy load. Clearly, some type of interface device is required between the IC and the load. Some of the most often used interface devices are power transistors, relays, opto-isolators, and a recent innovation, the VFET. We will now look at the way some of these are used.

Power transistors Figure 10–10(a) shows how a TTL output might supply base current to switch a power transistor ON. The power transistor will amplify this base current and supply a somewhat larger current to the load. A typical high-speed, medium-power switching transistor might have an h_{FE} somewhere in the 30–60 range. Clearly, since a TTL output has I_{OH} values that are usually $\leqslant$ 1 mA, the load current in this arrangement will not be very substantial.

There are several ways to modify this circuit to deliver heavier load currents. One method is given in Figure 10–10(b). Here the TTL output, X, is fed to an open-collector AND gate (e.g., 7409). When X is HIGH, the open-collector output will be in its OFF state so that current will flow from the +5 V through the 360 Ω resistor into the base of the power transistor. When X is LOW, the AND-gate output will become a low-resistance, thereby shunting this current to ground and turning OFF the transistor.

Example 10–5 The power transistor has the following characteristics and ratings:

$$BV_{CE} = 40 \text{ V}$$
$$I_C(\text{MAX}) = 1 \text{ A}$$
$$P_D\,(\text{MAX}) = 2 \text{ W}$$

$$V_{CE}(\text{SAT}) = 0.3\text{ V} @ I_C = 1\text{ A}$$
$$h_{FE} = 50 @ I_C = 1\text{ A}$$

The diode and transistor are both silicon.

(a) What is the largest load current that can be switched for the values given?
(b) What is the function of the diode?

Solution:

(a) When the AND-gate output is open, the base current will be

$$I_B = \frac{5\text{ V} - 0.7\text{ V} - 0.7\text{ V}}{360\ \Omega} = 10\text{ mA}$$

Thus, I_C can be no greater than $50 \times 10\text{ mA} = 500\text{ mA}$.

(b) The diode is there to ensure that the transistor will be completely OFF when the AND output is LOW. The V_{OL} from the AND gate could go as high as 0.4 V, which would slightly forward-bias the transistor *E-B* junction if the diode were not there.

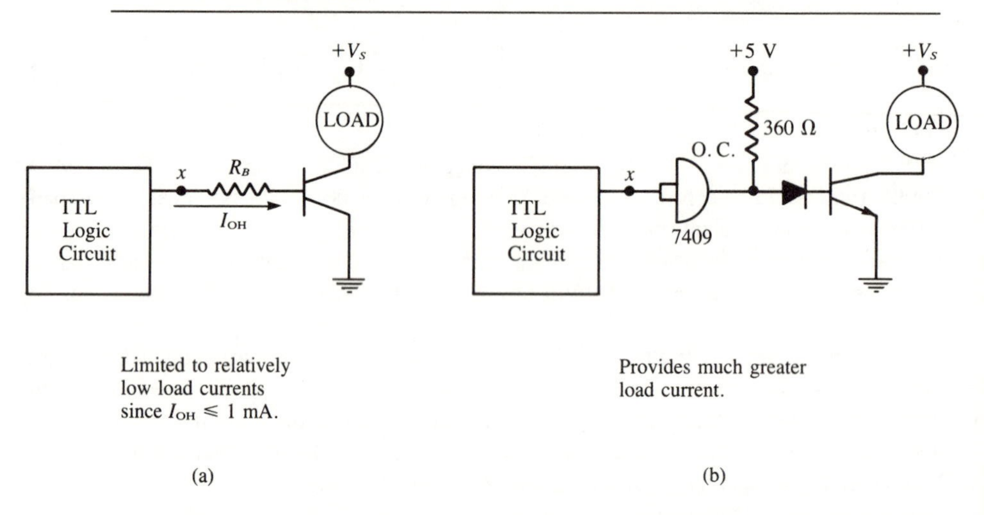

Figure 10–10 **TTL output driving a power transistor to switch current to a load**

Example 10.6 What can be done to increase the load current without changing the transistor?

Solution: The obvious answer is to increase the base current, by reducing the 360 Ω resistor. This, however, will increase the amount of current that the AND output has to sink in the LOW state. If the resistor is reduced, the AND gate would have to be replaced by one with a larger I_{OL} rating.

Power Darlington The preceding examples showed the limitation of using a single power transistor. A *power Darlington* transistor module can be used in place of the single power transistor, thereby providing a much higher current gain. Power Darlingtons are available with current gains of 10,000 (at currents of 1 A) and power ratings of 10 W. With this high current gain, TTL (and even CMOS) outputs can drive the base directly, as shown in Figure 10–11.

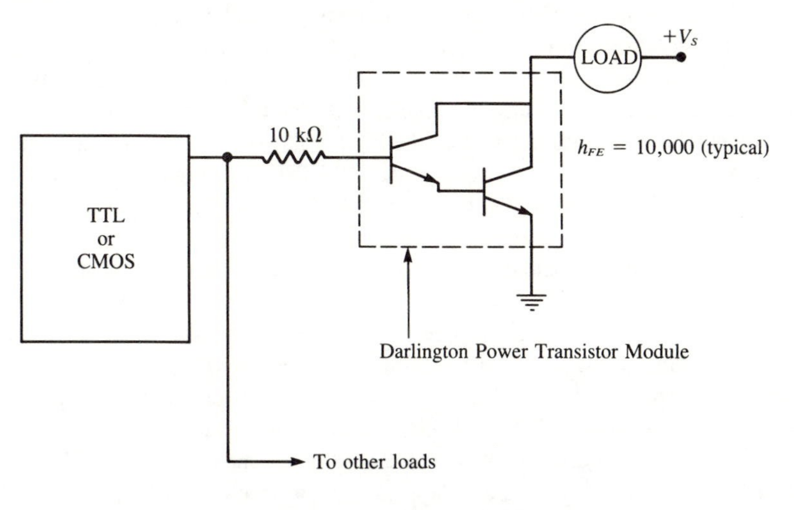

Figure 10–11 **Use of a high-gain Darlington power transistor allows TTL (or CMOS) outputs to drive the base directly and still provide heavy load current**

The V-MOSFET We have seen how field-effect transistors can operate as switches. Until recently these devices could not handle even moderate power levels. With the development of the V-MOSFET, there is now an FET counterpart to the bipolar power transistor. A V-MOSFET has a V-shaped conductive channel that allows operation at much higher current levels than standard MOSFETs.

Of course V-MOSFETs have the very high input impedance common to all MOSFET devices and this impedance makes them suitable as an interface between a digital IC and a high current load. One of the newer V-MOSFETs is the VN84GA (from Siliconix, Inc.), which can switch up to 12.5 A of current and can dissipate up to 80 W at low frequencies. Figure 10–12 shows a V-MOSFET used to switch power to a heavy load under the control of a digital IC output.

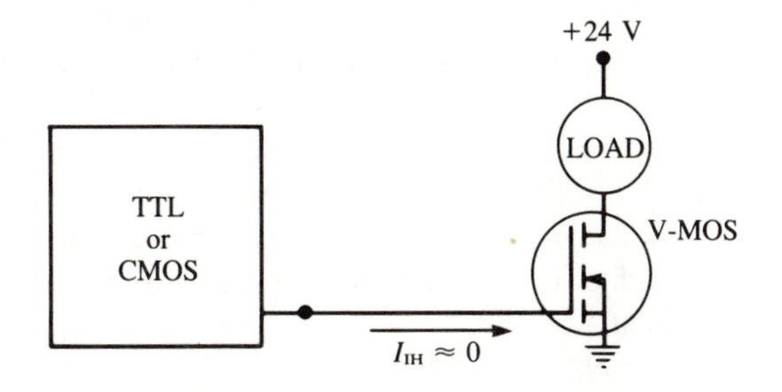

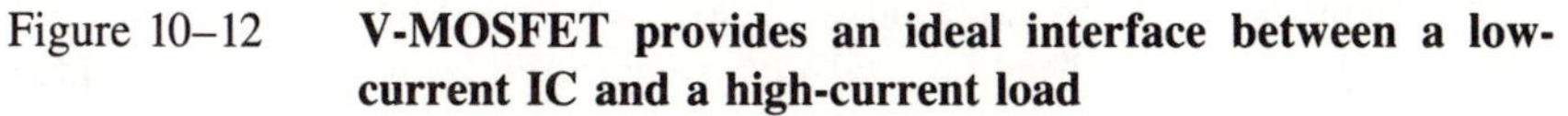
Figure 10–12 **V-MOSFET provides an ideal interface between a low-current IC and a high-current load**

The need for isolation Up to now we have shown the digital IC outputs connected either directly or indirectly to the device being driven, and there was always a common ground for the IC and for the load circuit. In many applications, it is more desirable to electrically *isolate* the low-power digital circuitry from the high-power load circuitry; that is, to have no electrical connections or common ground between them.

This isolation eliminates problems of *ground noise* caused by the heavy load currents flowing through the common ground connections. This ground noise can seriously effect the operation of any ICs connected to the common ground line. The isolation also protects the digital circuitry from the high voltages present in the load circuit in the event of a device failure. For example, if the V-MOSFET in Figure 10–12 were to develop a gate-to-drain short, the 24 V supply would be connected to the IC output through the low-resistance load. This connection would assuredly damage the driver IC and could easily produce a chain-reaction effect that could damage other ICs in the circuit.

Mechanical relays One of the oldest devices used to provide electrical isolation is the mechanical relay. This device consists of an inductive coil whose changing magnetic field controls the opening and closing of sets of contacts. The relay is controlled by switching the coil current ON and OFF. The amount of current needed to energize the relay will vary from a few mA to hundreds of mA, depending on the physical size of the relay and its contacts. Several types of smaller relays have been developed for use in digital circuits.

Reed relays consist of two small magnetic reeds that contact each other when the relay coil is energized. A reed relay requires very little coil current and can handle currents up to tens of amperes through its contacts. Figure 10–13 shows how a reed relay, designed for use with digital ICs, can be controlled by a digital signal. This particular relay has a coil resistance of 350 Ω and requires a coil current of 12 mA to be energized.

The diode is used to suppress the counter-emf that will be generated across the coil when the relay current is switched OFF. Without the diode, the counter-emf could easily damage the output transistor of the open-collector INVERTER.

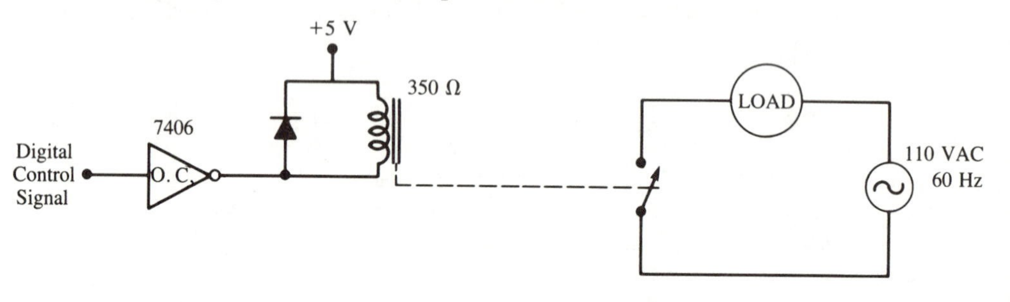

Figure 10–13 **Relay provides electrical isolation between digital circuitry and the high-voltage, high-current load circuit**

The relay provides nearly ideal electrical isolation, but it does have some serious disadvantages that could preclude its use in certain applications. For example, they have a shorter life than semiconductor devices because their contacts do corrode or wear out, and they are slow compared to semiconductor devices, typically requiring *milliseconds* to respond to a change in input.

Opto-isolators These are semiconductor counterparts of the mechanical relay, often called *semiconductor relays* or *solid-state relays*. While relays use electromagnetic energy, opto-isolators use *light* energy to couple the control signal to the load. An opto-isolator consists of a light source (usually an infrared LED), a light-sensitive device (such as a phototransistor), and a switching device. In many cases, the light sensor and the switching device are one and the same.

Figure 10–14 shows some of the common opto-isolator devices. In each case, the devices inside the dotted lines are encapsulated into a single light-tight package with only input terminals *x-y* and output terminals *a-b* accessible to the user. For each of these opto-isolators the input circuit consists of an LED that emits radiation when it conducts forward current. This radiation is focused on the light-sensitive device, which will be switched into conduction when sufficient current flows through the LED.

The various opto-isolators have different switching devices across their output terminals *a-b*. The opto-isolators in Figure 10–14(a), (b), and (c) can be used to switch DC power to a load, since transistors and SCRs conduct in only one direction. The devices in (d) and (e) can also switch AC power since the TRIACs are bilateral devices. Note that in each of these opto-isolators there is no electrical connection between the LED and the output device.

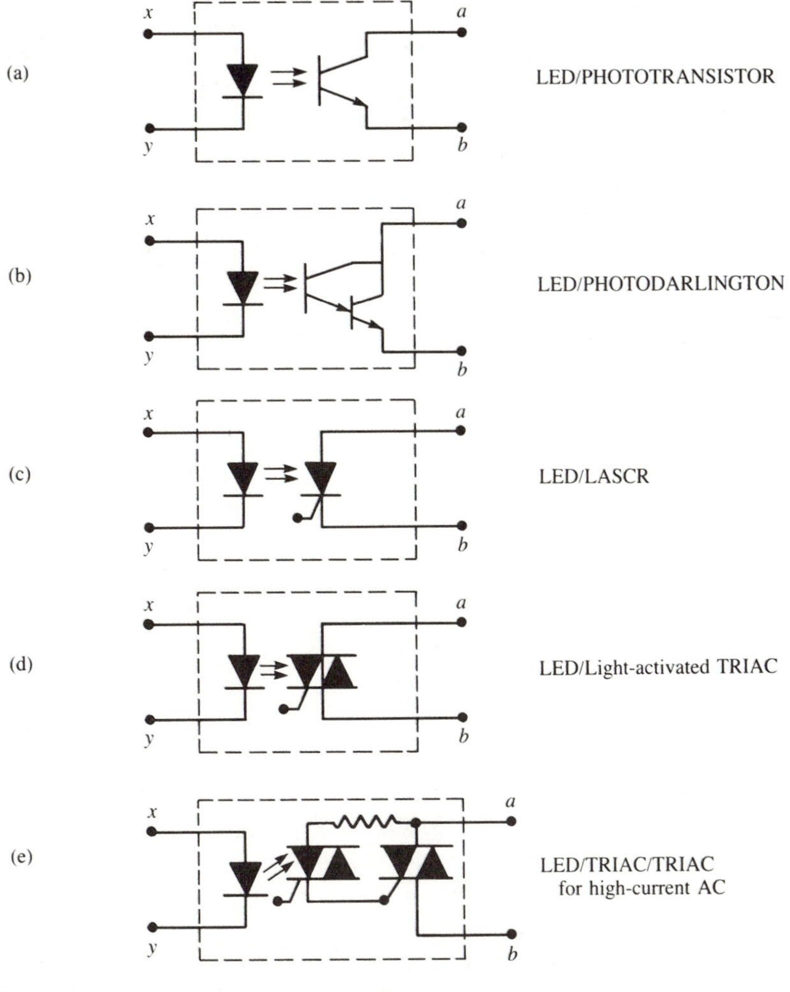

Figure 10–14 **Common opto-isolators**

Since the opto-isolator inputs are LEDs, these devices can be driven by digital ICs using the same considerations we discussed in Section 10.6. Figure 10–15 shows the opto-isolator counterpart to the relay circuit of Figure 10–13.

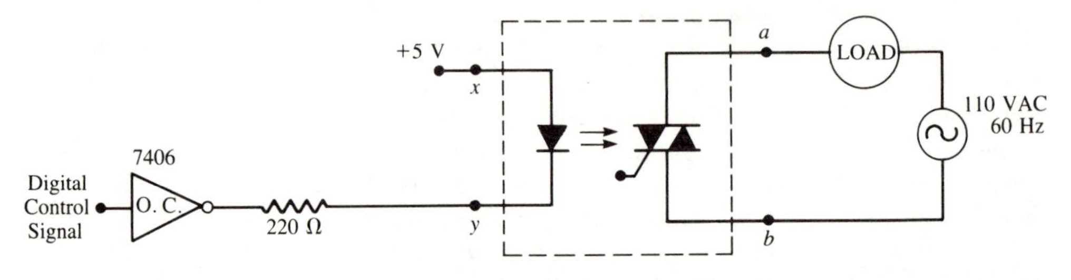

Figure 10–15 **Opto-isolator provides electrical isolation between digital input and the load circuit**

Example 10.7 Figure 10–16(a) shows the waveforms for the digital input and the 60 Hz AC voltage used in the circuit of Figure 10–15. Sketch the approximate load voltage waveform.

Solution: When the digital input is LOW, the open-collector output will be open, so that no LED current will flow, the TRIAC will be OFF, and the load voltage will be zero. When the digital input goes HIGH, the LED will conduct and its radiation will turn ON the TRIAC. If we neglect the small voltage dropped across the TRIAC, the load voltage will follow the AC voltage.

The load voltage waveform is sketched in Figure 10–16(b). Note that when the digital input switches LOW, the load voltage (and current) continues following the AC voltage until it reaches zero. This occurs because the TRIAC current has to drop below its holding current value before the TRIAC will turn OFF.

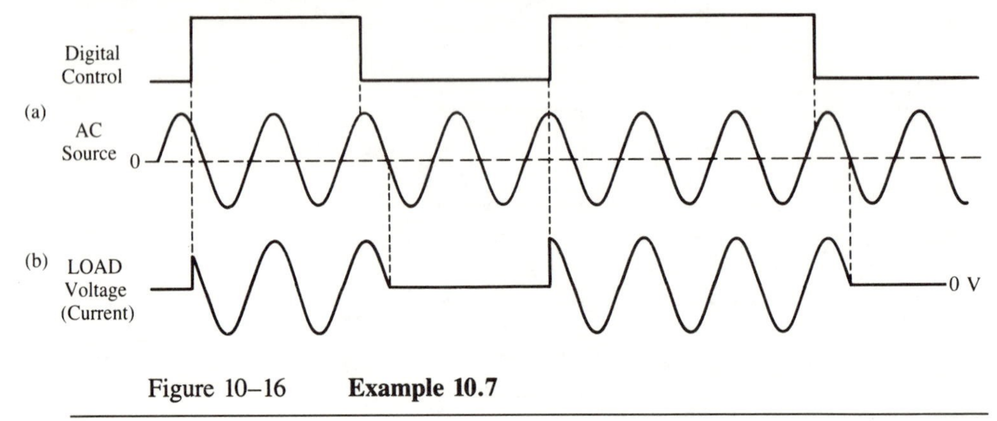

Figure 10–16 **Example 10.7**

10.8 Input Interfacing

Up to now, we have been concerned with how a digital output can be interfaced to various loads. We will now consider some of the problems that are encountered when a digital input has to be driven by a signal from a source other than another digital output.

AC input Figure 10–17 shows how an AC input can be converted to the sharp rectangular pulses required by digital circuits. A circuit like this one is used in the circuitry for a digital household clock to convert the 60 Hz line voltage to positive pulses.

The diode is used to eliminate the negative half-cycle of the AC voltage because most digital ICs cannot handle negative voltages. The half-wave rectified sinewave at point x cannot be used to drive standard TTL or CMOS inputs since these inputs generally require fast signal transitions for proper operation. The Schmitt trigger INVERTER will convert the half-wave signal to sharp pulses that can drive any digital IC.

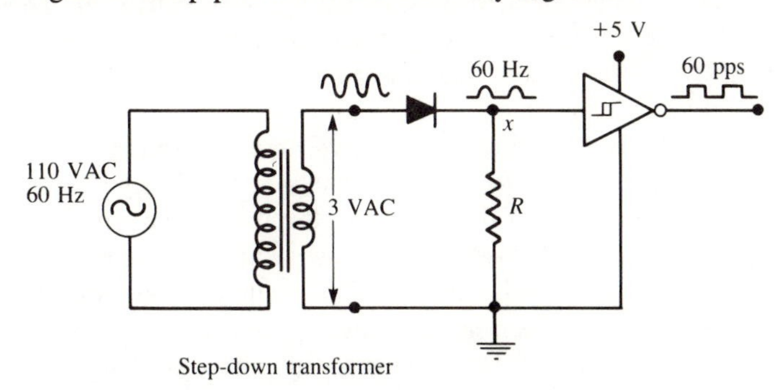

Figure 10–17 **Method for converting a 60 Hz AC signal to fast rectangular pulses using a Schmitt trigger inverter**

Both TTL and CMOS Schmitt trigger ICs are readily available. If a CMOS device is used, the value of R is not very critical. Almost any value can be used since its only function is to provide a discharge path for the CMOS input capacitance, C_{IN}. This capacitance will charge up through the diode during the positive half-cycle of the AC. It needs a path to discharge through when the diode turns OFF during the negative half-cycle. Otherwise it would stay charged and would keep the INVERTER output LOW. C_{IN} is typically 5 pF for CMOS, and R should be chosen such that C_{IN} will discharge completely during the negative half-cycle of the AC input.

Example 10.8 Determine the maximum value for R if the Schmitt trigger is CMOS.

Solution:

$$5\tau = 5 \times R \times C_{IN} \leqslant 8.33 \text{ ms}$$

$$\therefore \quad R \leqslant \frac{8.33 \text{ ms}}{5\, C_{IN}}$$

$$R \leqslant 333.3 \text{ M}\Omega$$

Thus, almost any value of R can be used.

If the Schmitt trigger INVERTER is TTL, the value of R has to be chosen more carefully. This is because R has to keep the INVERTER input LOW when the diode is OFF, and it must sink the I_{IL} of the TTL input without bringing the voltage above 0.4 V. For example, if the INVERTER is standard TTL, it will have I_{IL} = 1.6 mA. This will require that

$$R \times 1.6 \text{ mA} \leqslant 0.4 \text{ V}$$

or

$$R \leqslant 250\ \Omega$$

If LS-TTL is used, R can be larger since I_{IL} will be smaller.

Photocell input Figure 10–18(a) shows a circuit used for counting the number of times a light beam is interrupted. It uses a CdS photoconductive cell whose resistance switches from 1 kΩ to 1 MΩ as the cell goes from light to dark. This change in resistance produces a negative-going voltage change at point x that is fed to the clock input of the first *JK* flip-flop in a TTL counter.

There are two reasons why this circuit will not work properly; they can best be illustrated by looking at the waveform for the voltage at x as the light beam is repetitively turned on and off, [Figure 10–18(b)]. First, notice that the voltage waveform has relatively slow transition times. This is caused by the slow switching speeds of a CdS photocell (10 ms). There are faster photocells, such as CdSe, but they exhibit a much smaller change in resistance with light level. Even so, these faster photocells would still produce a signal at x with transition times that are not fast enough for reliable clocking of the *JK* flip-flop. Recall that TTL devices generally require input transition times that are less than 100 ns for reliable operation. No photocell can meet that requirement.

A second problem with this circuit is that the v_x waveform only drops to 1.6 V in its LOW state. This is because the 1 kΩ resistor has to sink the I_{IL} of the TTL input. Clearly, this 1.6 V is not a satisfactory LOW since it is well above the V_{IL}(MAX) = 0.8 V. This, of course, would not be a problem if the counter was CMOS, since CMOS has essentially zero input current.

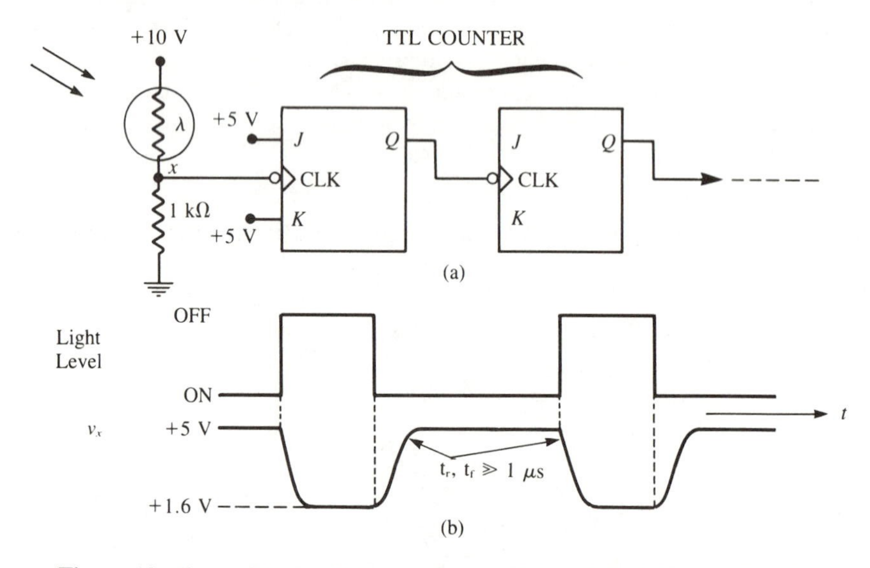

Figure 10–18 **Photocell is too slow to drive digital inputs directly**

Figure 10–19 shows one possible solution for both of these problems. A Schmitt trigger INVERTER has been connected between the photocell circuit and the *JK* flip-flop.

We are using a 74LS14 (Appendix A) because it has an I_{IL} of only 0.4 mA. The 1 kΩ resistor has been reduced to 820 Ω. Thus in the LOW state

$$v_x\,(\text{LOW}) = 820\ \Omega \times 0.4\ \text{mA} \approx .33\ \text{V}$$

This is acceptable since the 74LS14 has a guaranteed lower threshold (V_{LT}) of 0.6 V minimum. With 820 Ω, the HIGH-state voltage at v_x will be a little lower than before.

$$v_x\,(\text{HIGH}) = \frac{820}{1\ \text{k}\Omega + 820} \times 10\ \text{V} \approx 4.5\ \text{V}$$

This is more than adequate since the 74LS14 has a guaranteed upper threshold (V_{UT}) of 2.0 V maximum.

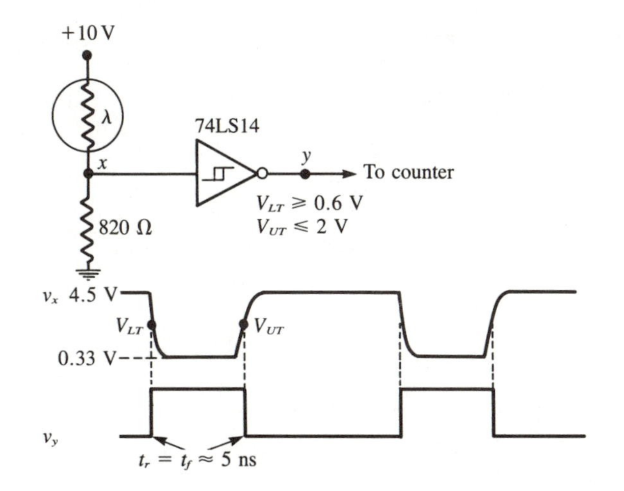

Figure 10–19 **An LS-TTL Schmitt trigger will sharpen up the photocell waveform**

Example 10.9 The circuit of Figure 10–19 uses a +10 V supply for the photocell circuit. Would the circuit function properly if the photocell circuit used the same +5 V supply as the digital circuits?

Solution: If the +10 V were replaced by +5 V, then the HIGH-state value of v_x would drop from 4.5 V to 2.25 V. The 74LS14 (Appendix B) has a guaranteed positive-going threshold (UTP) of 2.0 V. Thus, there would be a large enough voltage at v_x to trigger the 74LS14.

Questions

Sections 10.1–10.3

10.1 What is meant by the term *interfacing?*

10.2 A TTL output with a rated fan-out of 10 U. L. is driving a CMOS input. It is also driving four TTL inputs, each rated at 1 U. L. Calculate the range of values that can be used for the pull-up resistor.

10.3 The TTL gate in Figure 10–2 is driving 9 U. L. in addition to the interface circuit. Calculate the minimum h_{FE} for Q_1 for reliable operation.

10.4 The TTL gate in Figure 10–2 is a 74LS00. How many U. L. can it drive in addition to the interface circuit?

10.5 Why can't a P-channel E-MOSFET be used in place of the JFET in Figure 10–3?

Section 10.4

10.6 Refer to the data sheet for the CMOS 4049 buffer. Determine how many standard TTL loads it can reliably drive at 25°C. Repeat for LS-TTL loads.

10.7 Refer to the circuit in Figure 10–5 and determine the following:

(a) The amount of current the CMOS output has to supply in its HIGH state.

(b) The approximate amount of current it has to sink in the LOW state.

(c) The amount of Q_2 base current when Q_2 is ON.

10.8 A technician tests the circuit of Figure 10–5 as an interface between a CMOS gate operating from V_{DD} = 15 V and a TTL input. The circuit works properly, but Q_1 heats up excessively when the CMOS output is at +15 V. The technician measures V_{BE} = +6 V for Q_1 when the CMOS output is at +15 V. What is causing Q_1 to overheat?

10.9 Consider the circuit of Figure 10–20 as a possible way to interface a high-voltage CMOS output to a TTL input. The diode is used to block the +15 V CMOS output from reaching the TTL input. The pull-up resistor brings the TTL input up to +5 V when the diode is OFF. Unfortunately, this circuit will not work in the LOW state. Explain why not, and then explain how the circuit can be modified to work.

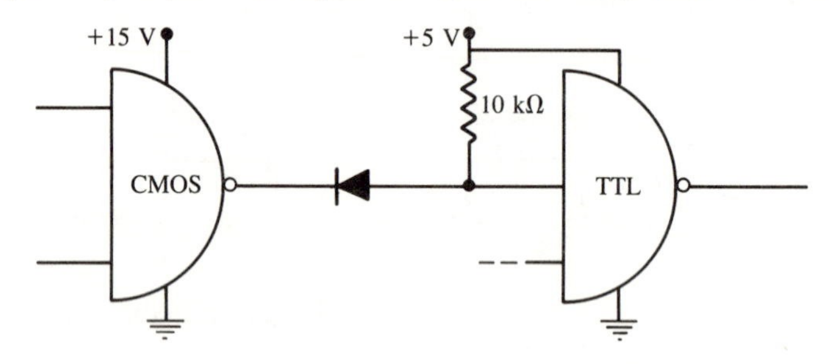

Figure 10–20

Sections 10.5–10.6

10.10 Consider the circuit of Figure 10–21. A and B are logic inputs that control the state of the two LEDs.

(a) Determine the status of each LED for each of the different logic input combinations.

(b) The red LED has $V_F \approx 1.7$ V and the green LED has $V_F \approx 2.4$ V. Determine the maximum current each gate has to sink.

10.11 Can 74LS38 NAND gates be used in the circuit of Figure 10–21?

10.12 Modify the circuit of Figure 10–21 to include a third logic input, C, which will prevent either LED from being turned ON whenever $C = 1$, regardless of the conditions on A and B.

10.13 The circuit of Figure 10–22 shows a CMOS output driving an LED through a buffer transistor. The LED has $V_F \approx 2.2$ V at $I_F = 20$ mA. Determine the required value for R_C if the LED current is to be 20 mA when the transistor is ON. Then determine an appropriate value for R_B if the transistor has $h_{FE} = 200$.

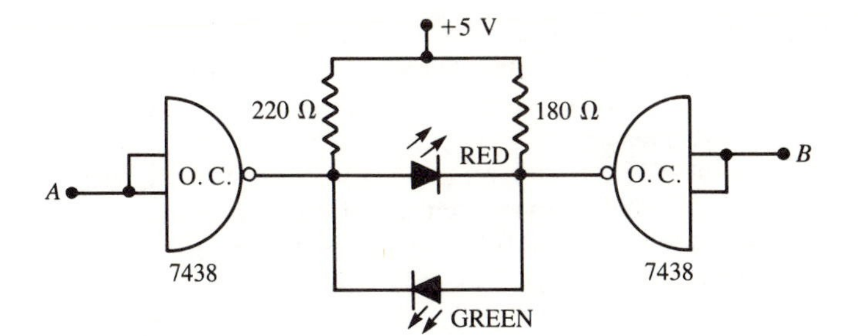

Figure 10–21

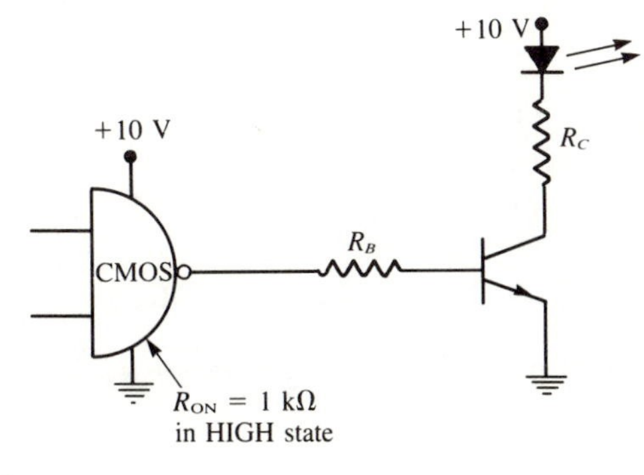

Figure 10–22

Section 10.7

10.14 Refer to the circuit in Figure 10–10(b). The AND-gate output is rated at 10 U. L. The transistor has $h_{FE} = 50$ and the collector supply voltage is $V_S = +36$ V.

(a) Determine the smallest permissible value for the resistor.

(b) How much load current will flow using this value?

10.15 The power transistor in Figure 10–10(b) is being operated as a switch so that it is either fully OFF or fully ON. This ensures that there will be minimal transistor power-dissipation. Verify this by calculating the collector dissipation in the OFF and ON states using the following values.

$$V_S = +40 \text{ V}$$
$$R_{\text{LOAD}} = 40 \ \Omega$$
$$h_{FE} = 120$$
$$V_{CE}(\text{SAT}) = 0.3 \text{ V @ } I_C = 1 \text{ A}$$
$$I_{CEO} \text{ (leakage)} = 0.1 \text{ mA}$$

10.16 Repeat Problem 10.15, but this time assume that the base current is *not* large enough to saturate the transistor, but instead produces only $I_C = 0.5$ A.

This should emphasize the importance of ensuring that sufficient base drive is available. Otherwise, the transistor will be dissipating a large amount of power when conducting and could become overheated and fail.

10.17 The TTL device in Figure 10–11 is a 7400 NAND gate. Determine the maximum number of other TTL loads the NAND gate can drive.

10.18 Repeat the preceding problem for a 74LS00 NAND gate.

10.19 What is the principal advantage of the V-MOSFET over a bipolar power transistor?

10.20 Why is it often necessary to have complete electrical isolation between the digital circuitry and the load?

10.21 In Figure 10–23, the relay controls two sets of normally open (N.O.) contacts. Contacts *a-b* switch AC power to the load. Contacts *c-d* connect the outputs of the two open-collector INVERTERs. Using the waveforms for logic inputs *A* and *B*, sketch the waveform for the load voltage. Ignore the time delay for relay switching.

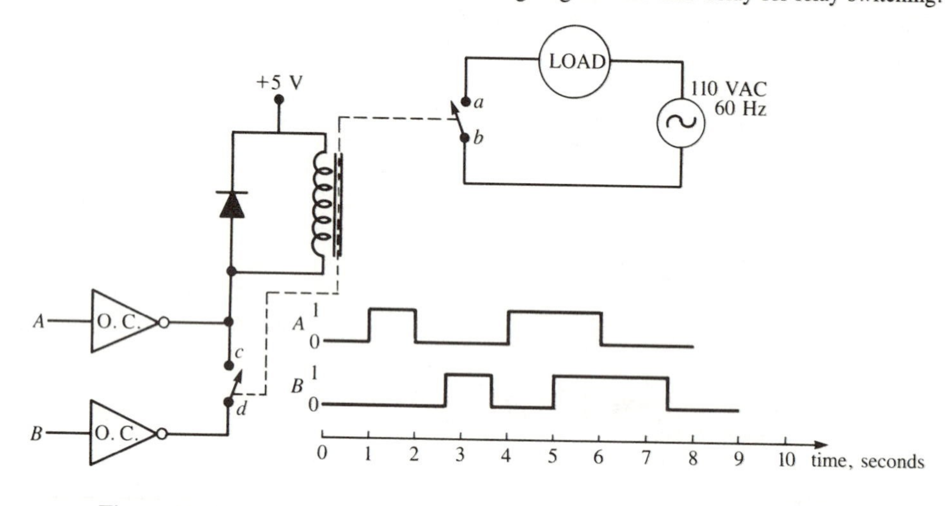

Figure 10–23

10.22 What are some of the advantages that opto-isolators have over mechanical relays?

10.23 One of the problems with switching large amounts of current using a TRIAC or an SCR is that a great deal of electromagnetic energy can be radiated when the switching occurs. For example, refer to the circuit of Figure 10–15 and the waveforms of 10–16. The load voltage and current waveforms have very fast transitions because of the relatively fast switching speed of the TRIAC in the opto-isolator. Because of the large current involved, this rapid switching produces a very large rate of change of current, *di/dt*. This *di/dt* will generate magnetic lines of flux that can induce spurious signals in nearby circuitry.

This problem can be eliminated by synchronizing the digital control signal to the *zero-crossings* of the AC source voltage so that the TRIAC is turned ON when the AC voltage is zero and there will be no sudden change in load current.

Figure 10–24 shows one method for implementing this zero-crossing switching. This circuit uses a high-speed, TTL-compatible voltage comparator (e.g., LM311) and a *D*-type FF to convert the original digital control signal to a new signal (*Q*) that is synced to the AC source voltage.

Study the circuit and write an explanation of its operation. Use waveform sketches to enhance your explanation.

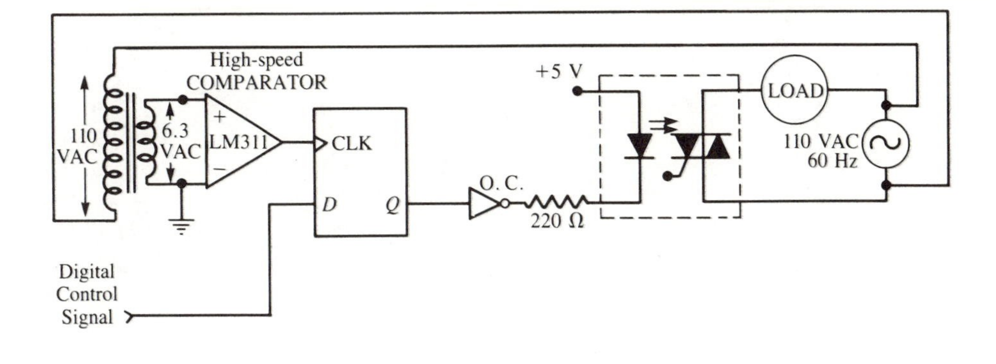

Figure 10–24

Section 10.8

10.24 Assume that the INVERTER in Figure 10–17 is a CMOS IC. A technician is testing the circuit and finds that the INVERTER output is staying at a constant LOW. When he places the oscilloscope probe at X to monitor the INVERTER input, he observes that the INVERTER output operates correctly (pulses). When he removes the scope from X, the output stays LOW. Can you explain his observations and conclude what is faulty in the circuit?

10.25 Determine the largest R that can be used in Figure 10–17 if the INVERTER is a 74LS14.

10.26 Change the photocell in Figure 10–19 so that it has R_{DARK} = 1 MΩ and R_{LIGHT} = 2.5 kΩ. Will this circuit still work?

10.27 Now change the photocell source to +5 V. Why won't this circuit still work? Can we get it to work by increasing the 820 Ω resistor?

10.28 Consider the circuit of Figure 10–25. The photocell has R_{DARK} = 1 MΩ and R_{LIGHT} = 1 kΩ. Why won't this circuit work? (HINT: the INVERTER output won't switch from HIGH to LOW when the photocell goes from light to dark.)

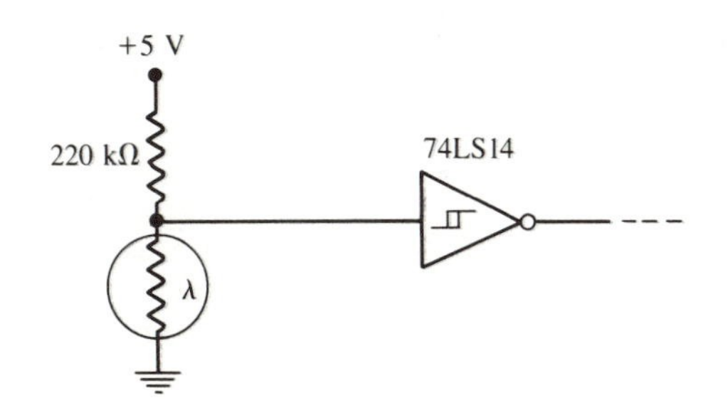

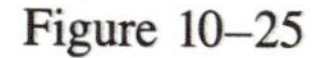
Figure 10–25

11 Pulse Generating Circuits

11.1 Introduction

We have used various types of pulse waveforms in all of our circuit analyses thus far, without being concerned with where they came from. In this chapter we will examine many of the kinds of circuits that are used to generate pulse waveforms. The waveforms will include rectangular pulses, square waves, spikes, and sweep (triangular) waveforms.

It would be impossible to present all of the hundreds of circuit arrangements that can generate pulses; the circuits that have been selected are representative of those most commonly used to generate pulse waveforms in modern circuits.

11.2 Unijunction Transistor Operation

Some common oscillator and timing circuits make use of a special semiconductor device, the *unijunction transistor* (UJT). Although it is assumed that the background of most students using this text includes a coverage of semiconductor devices such as the UJT, a brief description of UJT characteristics shall be given before discussing its circuit applications.

The UJT is a three-terminal device. It usually consists of a N-type silicon bar as illustrated in Figure 11–1(a). As shown, connections are made at each end of the bar forming the two bases, base 1 and base 2 (hereafter abbreviated B_1 and B_2). A PN junction is formed along one side of the N-type bar somewhere near the middle. A connection is made to the P-type material, called the *emitter* (E). The structure, then, contains only one PN junction (*uni*junction).

Examination of the UJT structure reveals that between B_1 and B_2 it is simply a pure resistor. The total resistance between B_1 and B_2 is called the *interbase resistance*, R_{BB}, and is typically between 4 kΩ and 10 kΩ. The portion of this resistance between B_2 and point x (cathode of the PN junction) is denoted R_{B_2}, and the portion between x and B_1 is denoted R_{B_1}. Clearly, $R_{BB} = R_{B_1} + R_{B_2}$. Typically, both R_{B_1} and R_{B_2} are in the range between 2 kΩ and 8 kΩ.

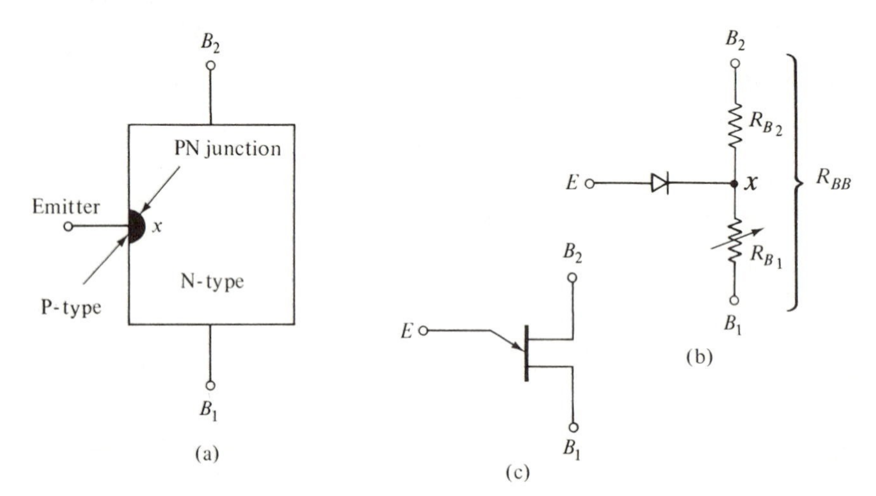

Figure 11–1 **UJT transistor (a) structure, (b) equivalent circuit, and (c) electronic symbol**

Between the emitter (E) and the bar we essentially have a PN diode. As such, we can draw an equivalent circuit for the UJT as shown in Figure 11–1 (b). Notice that R_{B_1} is shown as a variable resistance because during normal operation its resistance will change considerably. Part (c) of the figure shows the conventional electronic symbol for the UJT.

UJT operation To help understand UJT operation, consider the situation in Figure 11–2(a) and (b). Normally, B_2 is biased positively relative to B_1 through a voltage source V_{BB}. Because of this voltage source, point x will be at a voltage

$$V_x = \frac{R_{B_1}}{R_{B_1} + R_{B_2}} \times V_{BB} = \eta V_{BB}$$

where the parameter η (Greek letter eta) is called the *internal voltage divider ratio*. This parameter is one usually specified by the manufacturer, whereas R_{B_1} and R_{B_2} are not. Typically, η ranges from 0.4 to 0.9 for different UJTs.

The voltage V_x is at the cathode of the diode. Thus, as long as the emitter supply voltage V_{EE} is less than V_x, the diode will be reverse biased, and essentially no emitter current flows. Current will flow, however, between B_1 and B_2. In this state, the UJT is said to be OFF because the resistance between the emitter and the B_1 terminals is very high due to the OFF diode. In the OFF state the value of R_{B_1} is at its normal value between 2 kΩ and 8 kΩ.

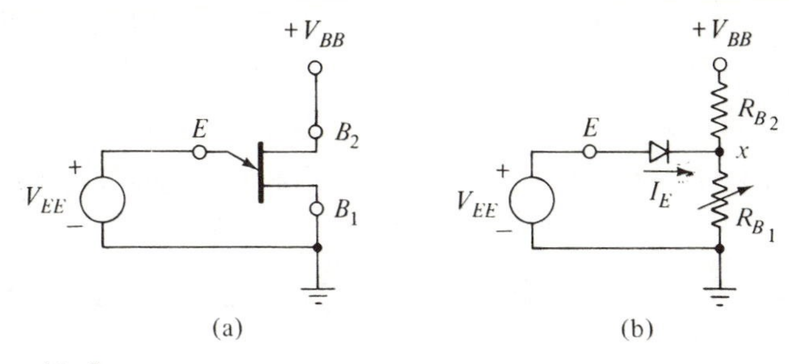

Figure 11–2

Turning ON the UJT When the emitter voltage is increased to the point where the diode begins to turn ON, emitter current will begin to flow. The value of V_{EE} needed to accomplish this is called the *peak-point voltage, V_P*. Since the diode is silicon, V_P is calculated as

$$\begin{aligned} V_P &= V_x + V_D \\ &= \eta V_{BB} + 0.5 \text{ V} \end{aligned} \qquad \textbf{(11–1)}$$

At the peak point, any slight increase in emitter voltage results in an increase in I_E. When I_E reaches a value called the *peak-point current, I_P* (a parameter of the UJT), the UJT immediately switches to its ON state within 1 μs. Values of I_P range from 2 μA to 50 μA.

In the ON state R_{B_1} drops to a low value typically less than 50 Ω. With the UJT ON and R_{B_1} at a low value, the impedance between E and B_1 drops to a low value, and I_E increases substantially. The following example should help to clarify this basic operation.

Example 11.1 Consider the circuit in Figure 11–2. Use V_{BB} = 12 V, R_{BB} = 6 kΩ, and η = 0.7, and determine V_P. Then, determine I_E when V_{EE} is at 10 V. Assume R_{B_1} (ON) = 50 Ω.

Solution: Using Equation (11–1), the peak-point voltage is

$$V_P = 0.7 \times 12 \text{ V} + 0.5 \text{ V} = 8.9 \text{ V}$$

When V_{EE} is made slightly higher than this value, the UJT turns ON. In the ON state R_{B_1} = 50 Ω. Figure 11–3 depicts the situation in which V_{EE} is at 10 V and the UJT is ON.

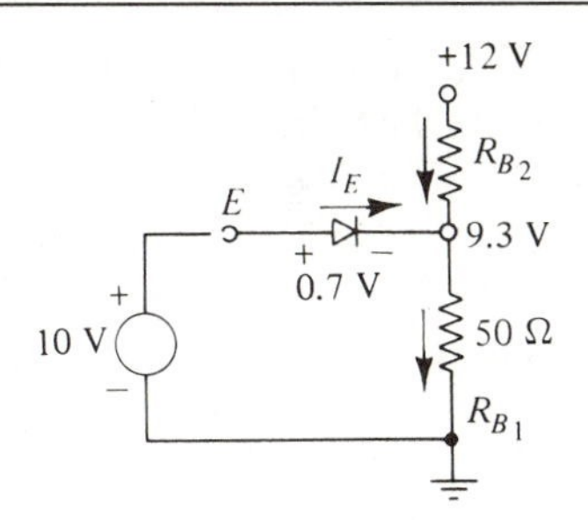

Figure 11–3

With V_{EE} = 10 V and the diode conducting at 0.7 V, the voltage at x is forced to be 9.3 V. The current through R_{B_2} is going to be very small compared with that through the much smaller R_{B_1}; therefore, we can assume that $I_E \approx I_{RB_1} = 9.3\ \text{V}/50\ \Omega =$ 186 mA. Thus, I_E switches from a very low value when $V_{EE} < V_P$ to a much larger value after the UJT has switched ON.

Turning OFF the UJT Once the UJT has turned ON, the value of emitter voltage can be reduced below V_P, and the UJT will remain ON. However, as V_{EE} is reduced, I_E will decrease. When I_E decreases below a value called the *valley current*, I_V, the UJT will revert back to its OFF state. Typically I_V values range between 1 mA and 10 mA. Once OFF, the UJT will stay OFF until V_{EE} is increased above V_P again.

Example 11.2 For the values of Example 11.1, determine the value of I_E after the UJT is ON and V_{EE} has been reduced to 2 V. Then, determine the value to which V_{EE} has to drop for the UJT to turn OFF if $I_V = 5$ mA.

Solution: With $V_{EE} = 2$ V, the voltage at x is 1.3 V, resulting in $I_E \approx 1.3\ \text{V}/50\ \Omega =$ 26 mA. To turn OFF the UJT, I_E has to drop below 5 mA. This requires that V_{EE} drop below

$$5\ \text{mA} \times 50\ \Omega + 0.7\ \text{V} = \mathbf{0.95\ V}$$

The calculations in the preceding examples contain many approximations; the intent is to give the student an understanding of the basic UJT operation. For example, it is very difficult to assume or estimate a value for R_{B_1} when the UJT is ON because it is not published on the manufacturers' specifications sheets, and it varies with circuit conditions.

11.3 Unijunction Oscillator Circuit

Figure 11–4(a) shows the basic UJT oscillator circuit. When the 20 V supply is initially applied to the circuit, the UJT is momentarily OFF because the capacitor holds the emitter voltage v_E at 0 V. If we assume a value of $\eta = 0.6$, the UJT will remain OFF until $v_E = V_P = 0.6 \times 20\ \text{V} + 0.5\ \text{V} = 12.5$ V. With the UJT in the OFF state, no emitter current will flow. As such, capacitor C will begin to charge toward 20 V through resistor R [see part (a) of Figure 11–4].

As C charges toward 20 V, the emitter voltage increases until it reaches a value equal to $V_P = 12.5$ V. At this point the UJT turns ON, and the resistance from emitter to base 1 becomes very small [see Figure 11–4(c)]. The capacitor begins to rapidly discharge toward 0 V through this low E-B_1 resistance. The capacitor discharge current and the current through R produce a relatively large emitter current that keeps the UJT ON as long as $i_E > I_V$. As the capacitor voltage rapidly discharges, i_E will rapidly decrease. If we assume that the current through R is less than I_V, then at some point the emitter current will drop below I_V and the UJT will turn OFF. When this occurs, the E-B_1 resistance becomes very large, and i_E is again zero.

With the UJT OFF, C again starts charging toward 20 V through R, and the same chain of events repeats itself. The capacitor voltage waveform [see Figure 11–4(c)] consists of repetitive charging and discharging intervals, with the charging time, t_{ch}, being much greater than the discharging time, t_{dch}. Note that the capacitor voltage is shown discharging to about 1 V. This is the UJT's *valley voltage*, V_V, which is the E-B_1 voltage at the point where $i_E = I_V$. The value of V_V is a parameter of the particular UJT and is difficult to predict accurately. Typically it is 1–2 V.

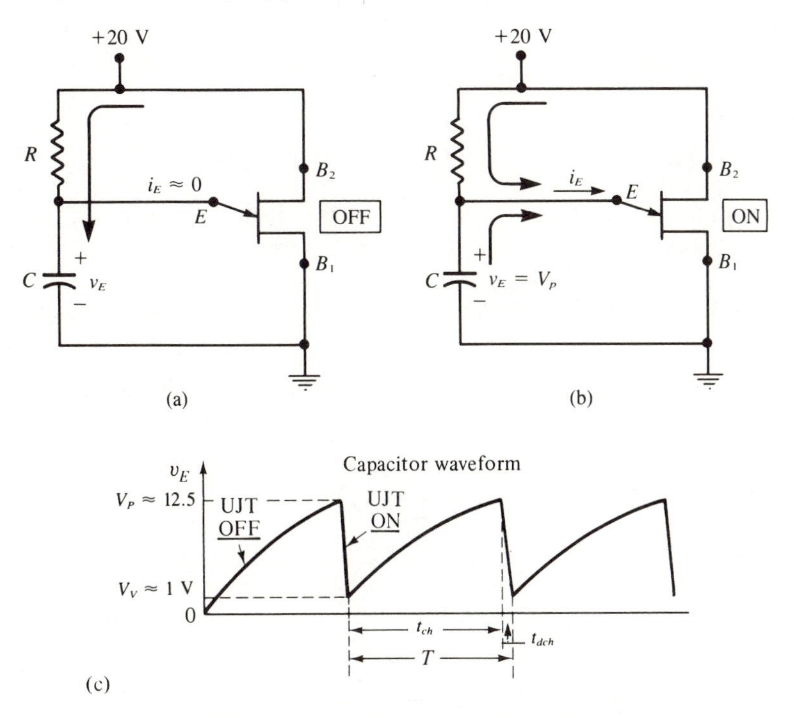

Figure 11–4 **Basic UJT oscillator (a) C charges through R while UJT is OFF ($i_E = 0$); (b) C discharges quickly through low E-B_1 resistance when UJT is ON; (c) Waveform across C**

Frequency of oscillations The period, T, of the v_E waveform is approximately equal to t_{ch} since t_{dch} is usually much smaller than t_{ch}. To find the value of T, we must set up the exponential expression for v_E and use it to determine how long it takes the capacitor to charge from V_V to V_P. The general expression is

$$v_E = V_F + (V_0 - V_F)\varepsilon^{-t/\tau}$$

Since $V_0 = V_V$ and $V_F = V_{BB}$ (C starts charging from V_V and heads toward V_{BB}), this becomes

$$v_E = V_{BB} + (V_V - V_{BB})\varepsilon^{-t/\tau}$$

We know that v_E will reach V_P at time $t = T$. Thus, we have

$$V_P = V_{BB} + (V_V - V_{BB})\varepsilon^{-T/\tau}$$

We want to solve for T. Rearranging, we have

$$\epsilon^{-T/\tau} = \frac{V_P - V_{BB}}{V_V - V_{BB}} = \frac{V_{BB} - V_P}{V_{BB} - V_V}$$

Taking the natural log (ln) of both sides

$$\frac{-T}{\tau} = \ln\left[\frac{V_{BB} - V_P}{V_{BB} - V_V}\right]$$

or

$$T = -\ln\left[\frac{V_{BB} - V_P}{V_{BB} - V_V}\right] \times \tau$$

Using the fact that $-\ln(x) = \ln(1/x)$ we have

$$T = \ln\left[\frac{V_{BB} - V_V}{V_{BB} - V_P}\right] \times \tau \qquad \textbf{(11–2)}$$

This expresses the period of the oscillations in terms of the circuit values (R, C, and V_{BB}) and the UJT parameters (V_V and V_P). Of course, the oscillation frequency is $1/T$.

Example 11.3 Determine the frequency of the oscillations for the circuit of Figure 11–4 in terms of R and C.

Solution: Using $V_P = 12.5$ V, $V_V = 1$ V, and $V_{BB} = 20$ V in Equation 11–2, we obtain

$$T = \ln\left[\frac{19}{7.5}\right] \times RC$$
$$\approx 0.93\ RC$$

so that

$$f = \frac{1}{T} \approx \frac{1.075}{RC}$$

Example 11.4 Repeat for $V_{BB} = 30$ V.

Solution: V_P will now be different.

$$V_P = \eta V_{BB} + 0.5 = 0.6(30) + 0.5$$
$$= 18.5 \text{ V}$$

Thus,

$$T = \ln\left[\frac{30 - 1}{30 - 18.5}\right] \times RC$$
$$\approx 0.925\ RC$$

so that

$$f = \frac{1}{T} \approx \frac{1.081}{RC}$$

These two examples show that the period and frequency are relatively independent of V_{BB}. We could have predicted this from Equation 11–2 by substituting the expression for V_P. That is,

$$T = \ln\left[\frac{V_{BB} - V_V}{V_{BB} - (\eta V_{BB} + 0.5)}\right] \times RC$$
$$= \ln\left[\frac{V_{BB} - V_V}{V_{BB} - \eta V_{BB} - 0.5}\right] \times RC$$

If we neglect V_V in the numerator, and the 0.5 V diode drop in the denominator since they are usually much smaller than V_{BB}, we have

$$T \approx \ln\left[\frac{V_{BB}}{V_{BB} - \eta V_{BB}}\right] \times RC$$

Dividing through by V_{BB}

$$T \approx \ln\left[\frac{1}{1 - \eta}\right] \times RC \qquad \textbf{(11–3)}$$

This approximate expression shows that T is independent of V_{BB} when we neglect V_V and V_D. This approximate expression is normally accurate enough for most purposes.

Example 11.5 Use this approximate expression to find T and f in Figure 11–4. Compare the results to the previous examples.

Solution:

$$T \approx \ln\left(\frac{1}{1 - 0.6}\right) \times RC$$

$$\approx 0.916\ RC$$

$$f = \frac{1}{T} \approx \frac{1.09}{RC}$$

These results are very close to those obtained using the more exact Equation 11–2.

Varying the frequency Since the parameter η is fixed for a given UJT, it is apparent from Equation 11–3 that the frequency of the oscillations can be varied only by varying the RC time constant. In variable frequency applications it is easiest to leave C fixed and use a variable R. In general, R can be varied over a relatively wide range. As R is varied, the waveform period varies proportionally. (The waveform amplitude is not affected since it depends only on V_P.) There are upper and lower limits on R which depend on the UJT parameters. The upper limit on R is dictated by the peak-point current (I_P) needed to turn ON the UJT. As the capacitor voltage approaches V_P, the emitter diode begins to conduct slightly. This small emitter current subtracts from the capacitor charging current, as shown in Figure 11–5(a). As such, if the current supplied through R is not large enough, it will be impossible to both charge C and supply the small I_P needed for the emitter current right at the turn-ON point. The value for R(MAX) can be determined by calculating the value of R that will supply a current equal to I_P when the capacitor voltage is right at the peak-point V_P. That is,

$$\frac{V_{BB} - V_P}{R(\text{MAX})} = I_P$$

or

$$R(\text{MAX}) = \frac{V_{BB} - V_P}{I_P} \qquad \textbf{(11–4)}$$

If R is increased above R(MAX), the UJT will never turn ON. The capacitor will charge up to a value close to V_P, but the UJT will not fire (turn ON) because the charging current will have been diverted to the UJT emitter, prohibiting any further capacitor charging. The capacitor waveform will appear as shown in Figure 11–5(b).

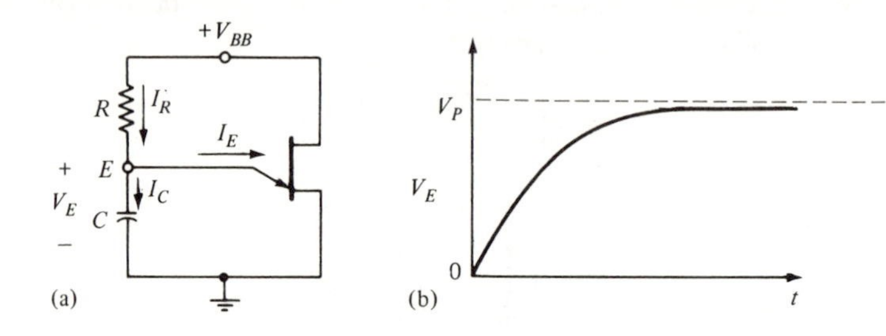

Figure 11–5 **Effect of using $R > R$(MAX)**

The lower limit on R is determined by the fact that the UJT's emitter current has to drop below I_V in order to cause turn-OFF. If R is too small, then when the capacitor has completed its discharge through the UJT, the current supplied through R will be larger than I_V, causing the UJT to remain ON. The value of R(MIN) can be determined by calculating the value of R that will supply a current equal to I_V when the UJT is ON. After the capacitor has discharged, the emitter voltage is very low, so almost all of V_{BB} is dropped across R. Thus,

$$R(\text{MIN}) \approx \frac{V_{BB}}{I_V} \tag{11–5}$$

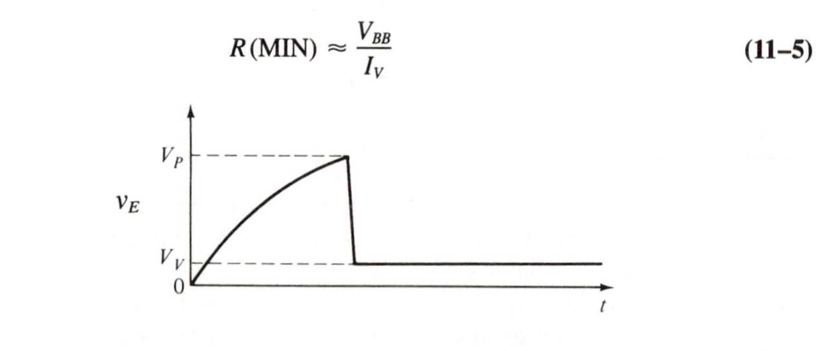

Figure 11–6 **Capacitor waveform when $R < R$(MIN)**

If a value of $R < R$(MIN) is used, the capacitor will charge to V_P the first time, firing the UJT, which will then remain ON, so no further oscillations will occur. In other words, only one cycle will occur, as illustrated in Figure 11–6.

Thus, for oscillations to occur, the value of R must be chosen between its two limits. The following examples will illustrate the range of frequencies obtainable by varying R.

Example 11.6 A UJT with $I_P = 5\ \mu$A, $I_V = 2$ mA, and $\eta = 0.63$ is used in a relaxation oscillator. Choose values for R, C, and V_{BB} to produce a sawtooth output with an amplitude of 10 V and a frequency that can be varied from a minimum of 10 Hz. Determine the possible range of frequencies that can be obtained by varying R.

Solution: Since V_P is required to be 10 V, then

$$V_P = \eta V_{BB} + 0.5\text{ V} = 10\text{ V}$$

resulting in

$$V_{BB} = \frac{9.5\text{ V}}{0.63} \approx 15\text{ V}$$

The period of the oscillations is approximately

$$T \approx RC \ln\left(\frac{1}{1-\eta}\right) \qquad \textbf{(11–3)}$$
$$\approx RC \ln(2.7)$$
$$\approx RC$$

The minimum oscillation frequency required is 10 Hz, which indicates a maximum period of 100 ms. Thus, the maximum value of RC must be 100 ms. To obtain this value, the largest possible value of R should be used so that the required value of C will be minimized, thereby minimizing discharge time.

$$R(\text{MAX}) = \frac{V_{BB} - V_P}{I_P} = \frac{15\text{ V} - 10\text{ V}}{5\ \mu\text{A}} \qquad \textbf{(11–4)}$$
$$= 1\text{ M}\Omega$$

Thus, R can be a 1 MΩ variable resistor. To find C,

$$1\text{ M}\Omega \times C = 100\text{ ms} = T(\text{MAX})$$

or

$$C = \frac{100 \times 10^{-3}}{10^6} = \mathbf{0.1\ \mu F}$$

The maximum frequency of oscillation will occur at the minimum value of R.

$$R(\text{MIN}) = \frac{V_{BB}}{I_V} = \frac{15\text{ V}}{2\text{ mA}} = 7.5\text{ k}\Omega \qquad \textbf{(11–5)}$$
$$T(\text{MIN}) = 7.5\text{ k}\Omega \times 0.1\ \mu\text{Fd} = 0.75\text{ ms}$$

Thus,

$$f(\text{MAX}) = \frac{1}{T(\text{MIN})} = \mathbf{1.33\ kHz}$$

resulting in a total frequency range of 10 Hz to 1.3 kHz by varying R between 7.5 kΩ and 1 MΩ.

A good choice would be to use a fixed 6.8 kΩ resistor in series with a 1 MΩ variable resistance. This would protect the UJT from excessive emitter current if the variable resistance is adjusted too low.

Example 11.7 Change C to 0.02 μF and determine the frequency range for the oscillator of the previous example.

Solution:

$$T(\text{MIN}) = 7.5\text{ k}\Omega \times 0.02\ \mu\text{F}$$
$$= 150\ \mu\text{s}$$

Therefore, $$f(\text{MAX}) = \frac{1}{150\ \mu\text{s}} = \mathbf{6.67\ kHz}$$

$$T(\text{MAX}) = 1\text{ M}\Omega \times 0.02\ \mu\text{F}$$
$$= 20\text{ ms}$$

Therefore, $$f(\text{MIN}) = \mathbf{50\ Hz}$$

Example 11.8 A technician is testing the UJT oscillator of Figure 11–7(a). The v_E waveform is supposed to have $V_P = 21.5$ V and a frequency of approximately 1.2 Hz, as shown. The oscilloscope display of v_E, however, shows a constant dc level of 18 V. The technician checks all of the circuit connections and replaces the UJT, but the result stays the same. What can be causing the malfunction?

Solution: One possibility is that R is too large.

$$R(\text{MAX}) = \frac{(30 - 21.5)\text{ V}}{4\ \mu\text{A}} = 2.125\ \text{M}\Omega \tag{11–4}$$

However, since R is only 680 kΩ, this cannot be the problem.

A more likely possibility is that the capacitor is "leaky"; that is, its dielectric insulation is allowing charges to flow from one plate to the other. This leakage can be represented by a parallel leakage resistance, R_L, across the capacitor, as shown in Figure 11–7(b). A good low-leakage capacitor will have $R_L \gg 10$ MΩ. Inexpensive electrolytic capacitors may have R_L values of only a few MΩ. Of course, a damaged capacitor can be more leaky than normal.

If R_L is too small, the capacitor will not be able to charge up to V_P. For example, with $R_L = 1$ MΩ in Figure 11–7(b), the capacitor can only charge to a maximum voltage determined by the voltage divider made up of R and R_L.

$$v_E(\text{MAX}) = \frac{R_L}{R + R_L} \times V_{BB}$$

$$= \mathbf{17.85\ V} < \boldsymbol{V_P}$$

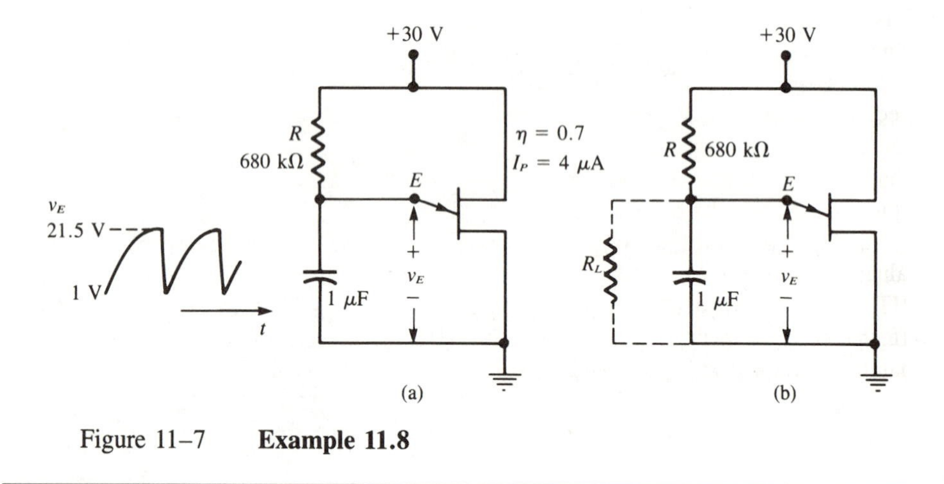

Figure 11–7 **Example 11.8**

11.4 UJT Oscillator-Spike Waveform

The basic UJT oscillator can be modified to produce a spike waveform across a resistor R_1 placed in series with B_1, as shown in Figure 11–8(a). The spike will appear across R_1 when C rapidly discharges through the ON UJT. R_1 is usually kept below 100 Ω to minimize the discharge time and, therefore, the duration of the spike.

Typical waveforms are shown in Figure 11–8(b) for a UJT with $V_P = 12.5$ V. The capacitor waveform, v_E, is the same as before except that its discharge time will be longer because of R_1. When v_E reaches 12.5 V, the UJT turns ON. At that instant the 12.5 V across C will divide between the UJT's E-B_1 resistance and R_1. Typically, 2 to 3 volts will be dropped across the UJT, so that the spike across R_1 will have an amplitude a couple of volts less than V_P.

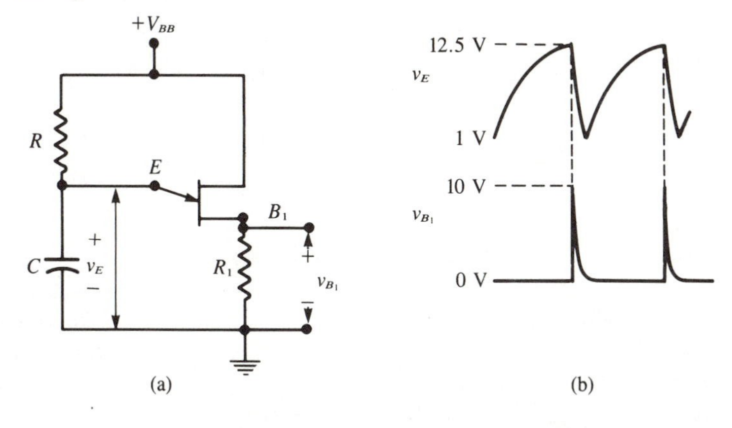

Figure 11–8 **(a) Addition of R_1 in series with B_1 produces spike outputs; (b) Typical waveforms when $V_P = 12.5$ V**

The width of the output spikes is determined mainly by the values of C and R_1. As either one is made smaller, the spikes become narrower. However, care must be taken that C is not made too low in value. Otherwise, as the UJT is turning ON and its resistance is going from a HIGH value to a LOW value, C will discharge significantly during the switching time of the UJT. As such, most of C's voltage will be dropped across the UJT (because its resistance is not LOW yet) and not across R_1. Then, when the UJT is fully ON there will be less capacitor voltage available (since it has already discharged somewhat), and the spike across R_1 will be diminished in amplitude. For this reason, C is typically kept at values of 0.01 μF or greater.

The repetition frequency of the spike waveform can be varied by varying the RC values, as discussed earlier. The amplitude of the spikes depends directly on the V_P of the UJT, which in turn depends on the supply voltage and η. The spike amplitude is also affected by C, as mentioned previously. In general, if R_1 is 50 Ω or greater and C is larger than 0.01 μF, the spike amplitude will be within 2 or 3 V of the V_P value.

11.5 Programmable UJT (PUT)

A programmable unijunction transistor (PUT) is a device that operates very similarly to the UJT although its physical structure is quite different. The PUT has a PNPN structure (Figure 11–9) with three terminals: anode, cathode, and gate. These correspond, respectively, to the emitter, base 1, and base 2 of the UJT. The major difference between the PUT and UJT operation is that *the peak-point voltage (V_P) of the PUT can be easily varied with an external voltage divider.*

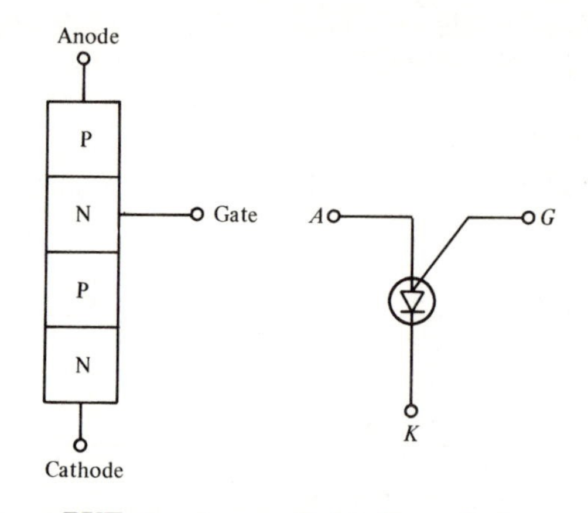

Figure 11–9 **PUT structure and circuit symbol**

This difference is illustrated in Figure 11–10, which shows the normal bias arrangement for the PUT and UJT. As we already know, the UJT will turn ON when its emitter voltage reaches V_P as determined by the UJT's *internal* voltage divider ratio, η. The PUT will turn ON when its anode-cathode voltage, v_{AK}, reaches V_P, where V_P is determined by the gate bias voltage, V_G, which in turn is determined by an *external* voltage divider made up of R_1 and R_2. In essence, then, the PUT behaves as a UJT whose η can be externally programmed to any desired value rather than being fixed at one value as in an ordinary UJT.

With $v_{AK} < V_P$, the PUT is OFF and acts as a very high resistance between anode and cathode. When v_{AK} exceeds V_P, the PUT turns ON, and its anode-cathode resistance drops to a very low value and acts like a constant voltage drop of around 1 V (that is, $v_{AK} \simeq 1$ V in the ON state). The PUT, like the UJT, is turned OFF when its anode current drops below I_V.

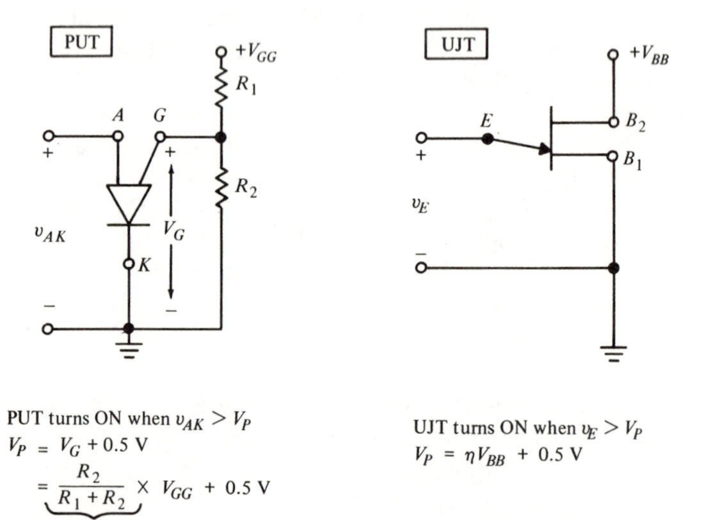

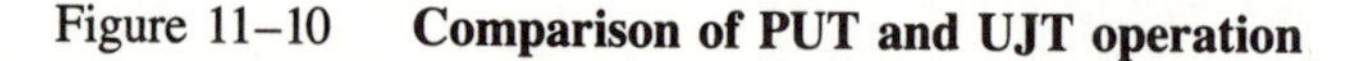

Figure 11–10 **Comparison of PUT and UJT operation**

PUT oscillator The PUT, since it operates like the UJT, can be used in any circuit that uses UJTs. For example, Figure 11–11 shows a PUT oscillator equivalent to the UJT oscillator studied earlier. The various circuit waveforms shown in the figure reveal the following important points:

1. The voltage divider biases the gate to +10 V, thereby establishing V_P as 10.5 V.
2. The capacitor charges toward +20 V (while the PUT is OFF) until $v_{AK} = V_P = 10.5$ V. At that point, the PUT is ON, and the capacitor rapidly discharges through the PUT. After C completes its discharge, the PUT turns OFF.
3. As C discharges, it produces a narrow spike of voltage across the 100 Ω cathode resistor. This spike amplitude is about 1 V less than V_P due to the 1 V anode-cathode voltage drop in the ON state.

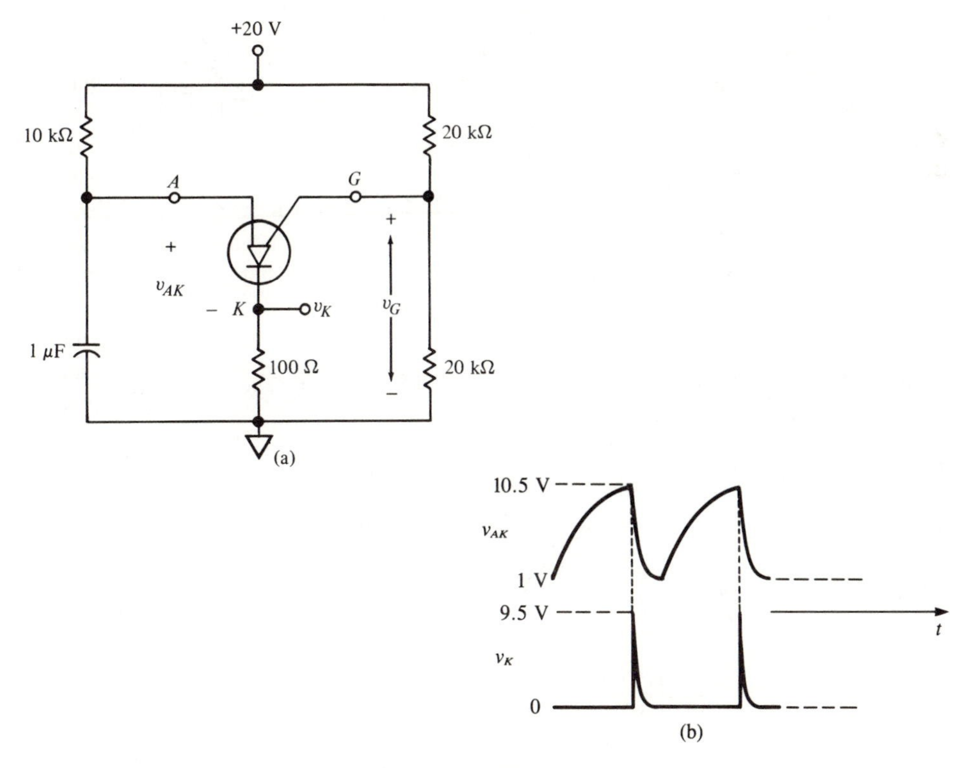

Figure 11–11 **PUT relaxation oscillator**

The period T can be calculated using the same basic formula that was used for the UJT, except that V_V is approximately 1 V.

$$T \approx \ln\left(\frac{V_{GG} - 1\text{ V}}{V_{GG} - V_P}\right) \times RC \tag{11–5}$$

The frequency of oscillations is simply $1/T$.

Advantages of the PUT In addition to its programmable feature, which allows the user to control V_P, the PUT also has several other desirable characteristics. It can operate over a wider range of frequencies than a UJT because it can be made to have much lower

I_P and much higher I_V values than a UJT. Recall that these parameters determine the frequency range of a UJT relaxation oscillator for a given value of C.

Even more importantly, the PUT can operate at lower supply voltages than a UJT—down to about 3 or 4 V. This characteristic makes them uniquely suited to low voltage applications such as with digital ICs that operate from a 5 V supply (TTL). For example, a PUT oscillator can be used to generate a clock waveform for a digital system. One method is shown in Figure 11–12. Here, a PUT oscillator operating from a 5 V supply generates a spike waveform that is connected to a clocked *JK* FF operating in the toggle mode ($J = K = 1$). The FF toggles on each spike, thereby producing a square wave output at *half* the frequencv of the PUT output waveform.

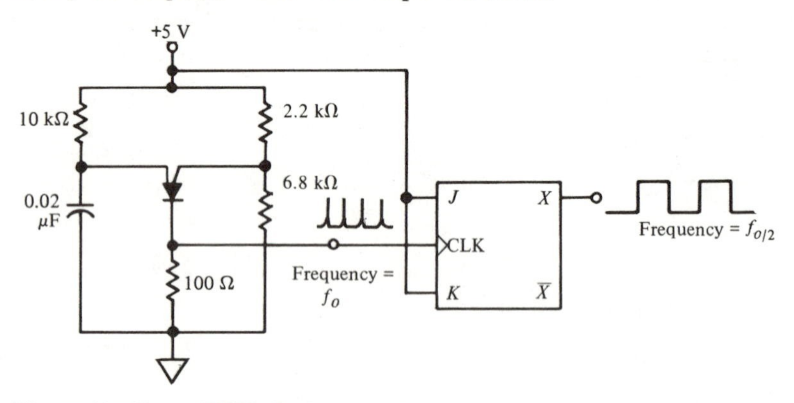

Figure 11–12 **PUT clock generator**

Example 11.9 The FF in Figure 11–12 is TTL. Determine the amplitude and frequency of the waveform at the FF's CLK input.

Solution:

$$V_P = \frac{6.8\text{ k}\Omega}{9\text{ k}\Omega} \times 5\text{ V} + 0.5\text{ V}$$
$$= 4.27\text{ V}$$

Thus, using Equation (11–5),

$$T = \ln\left[\frac{5\text{ V} - 1\text{ V}}{5\text{ V} - 4.27\text{ V}}\right] \times 10\text{ k}\Omega \times 0.02\ \mu\text{F}$$
$$\approx 0.34\text{ ms}$$
$$f = \frac{1}{T} \approx \mathbf{2.94\text{ kHz}}$$

The spike amplitude is $V_P - 1\text{ V} = \mathbf{3.27\text{ V}}$. This is larger than the $V_{IH}(\text{MIN}) = 2.4\text{ V}$ required by TTL.

11.6 Schmitt Trigger Oscillators

A Schmitt trigger circuit can produce oscillations if we use its output to charge a capacitor and then feed the capacitor voltage to the Schmitt trigger input. Any type of inverting

Schmitt trigger circuit can be used. It can be a digital Schmitt trigger such as the TTL 7414 inverter or the CMOS 40106 inverter, or it can be an op-amp Schmitt trigger.

CMOS circuit Figure 11–13(a) uses a CMOS Schmitt trigger inverter. The 40106 has typical threshold voltages of $V_{LT} = 3.9$ V and $V_{UT} = 5.9$ V when operated from $V_{DD} = 10$ V. Let's assume that the capacitor is initially at 0 V [see waveforms in Figure 11–13(b)]. This will produce 10 V at the inverter output and will begin to charge the capacitor toward +10 V through R. When v_{CAP} reaches $V_{UT} = 5.9$ V, the inverter output rapidly switches to 0 V. At this point, the capacitor begins to discharge toward 0 V through R and the low CMOS output. The capacitor will discharge until $v_{CAP} = V_{LT} = 3.9$ V, at which point the inverter output rapidly switches to +10 V and the capacitor begins charging toward +10 V again.

Clearly, v_{CAP} will consist of repetitively alternating charging and discharging portions, and v_{out} will be a rectangular pulse waveform whose period and pulse duration will depend on R, C, V_{UT}, and V_{LT}. The values of t_{ch} and t_{dch} can be determined by using the exponential expression for v_{CAP} during its charging and discharging interval. This is left as an exercise for the student. The results are

$$t_{ch} \approx 0.397\ RC$$

$$t_{dch} \approx 0.414\ RC$$

$$T = t_{ch} + t_{dch} \approx 0.811\ RC$$

$$f = \frac{1}{T}$$

The period and frequency of the oscillations can be easily changed by varying R over a wide range of values.

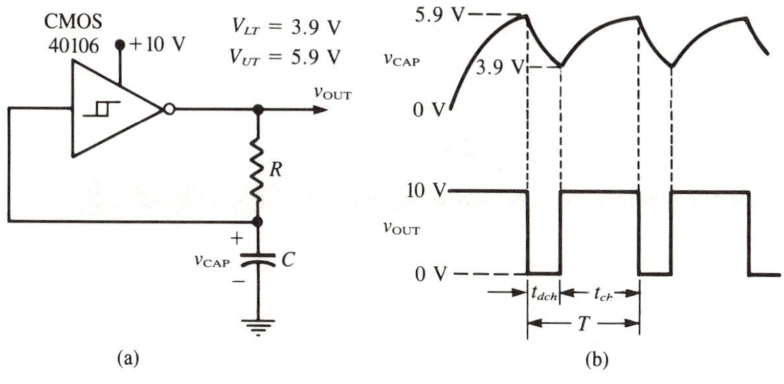

Figure 11–13 **CMOS Schmitt trigger oscillator and waveforms**

TTL circuit If a TTL Schmitt trigger is used, the basic operation is the same as previously described except that the LOW-state input current, I_{IL}, places a maximum limit on the value of R that can be used. As v_{CAP} is discharging toward V_{LT} (typically 0.8 V for TTL) through R and the inverter output, the I_{IL} current flows out of the inverter input terminal and tends to replenish the capacitor charge so that v_{CAP} might not discharge down to V_{LT}, and thus oscillations will cease. This can be avoided by making R small enough so that the discharge current it draws from C will exceed I_{IL}. The upper limit for R is 500 Ω for the 7414, and about 2 kΩ for the 74LS14.

The period and frequency of the TTL Schmitt trigger oscillator is difficult to predict accurately because of the effect of I_{IL} (which has a large tolerance and is temperature dependent). A crude approximation is $f \approx 1/RC$, which can be used to choose "ballpark" values of R and C for a desired frequency. R will generally have to be adjusted to bring the actual frequency acceptably close to the desired value.

Op-amp circuit Figure 11–14(a) shows how an op-amp Schmitt trigger can be used in an oscillator circuit. The op-amp and the 10 kΩ resistors form the Schmitt trigger. The op-amp output v_{OUT} feeds the RC circuit, and v_{CAP} provides the input to the Schmitt trigger.

For the values shown, the Schmitt trigger upper and lower threshold voltages are

$$V_{UT} = (+V_{SAT}) \times \frac{1\ \text{k}\Omega}{11\ \text{k}\Omega} \approx +1.23\ \text{V}$$

$$V_{LT} = (-V_{SAT}) \times \frac{1\ \text{k}\Omega}{11\ \text{k}\Omega} \approx -1.23\ \text{V}$$

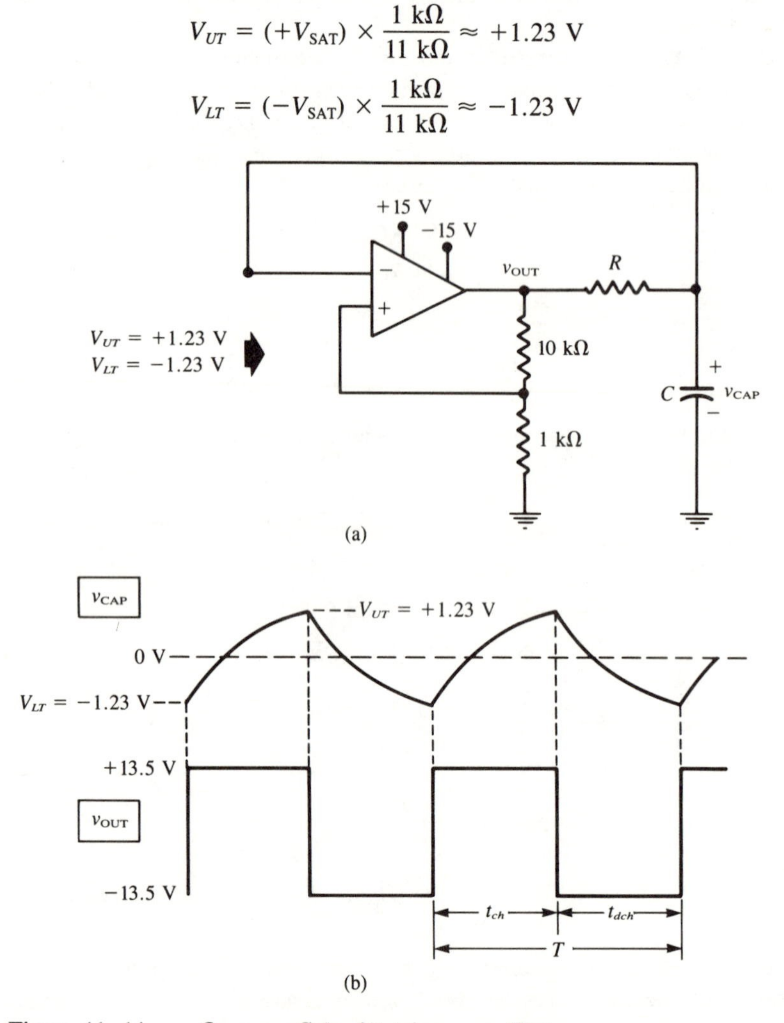

Figure 11–14 **Op-amp Schmitt trigger oscillator**

When v_{OUT} is at $+V_{SAT}$ = +13.5 V, v_{CAP} charges toward +13.5 V. However, when v_{CAP} reaches V_{UT} = +1.23 V, the op-amp output rapidly switches to $-V_{SAT}$ = −13.5 V. The capacitor then begins to discharge toward −13.5 V, until $v_{CAP} = V_{LT}$ = −1.23 V, at which point the op-amp output switches back to $+V_{SAT}$. The capacitor again starts charging toward +13.5 V, and the process repeats itself. The v_{CAP} waveform then consists of exponential excursions between V_{LT} and V_{UT}, while v_{OUT} switches between $\pm V_{SAT}$.

If the op-amp saturation voltages are approximately equal in magnitude, the v_{CAP} charging interval, t_{ch}, will be the same as the discharging interval, t_{dch}, so that v_{OUT} will be a *square wave*. Of course, the transition times of the v_{OUT} signal will be determined by the op-amp's slew rate limitation.

The period and frequency of the oscillations depend on V_{LT}, V_{UT}, V_{SAT}, R, and C. The formula for T can be derived by setting up the exponential equations for v_{CAP} during its charging and discharging intervals and solving for t_{ch} and t_{dch}. This is left as an end-of-chapter exercise, but here is the result.

$$\begin{aligned} T &= t_{ch} + t_{dch} \\ &= 2 \ln \left[\frac{V_{SAT} - V_{LT}}{V_{SAT} - V_{UT}} \right] \times RC \end{aligned} \qquad \textbf{(11–6)}$$

Example 11.10 Determine the frequency of v_{OUT} in Figure 11–14 for $R = 15\ \text{k}\Omega$ and $C = 0.01\ \mu\text{F}$.

Solution:

$$\begin{aligned} T &= 2 \ln \left[\frac{13.5\ \text{V} - (-1.23\ \text{V})}{13.5\ \text{V} - (1.23\ \text{V})} \right] \times 15\ \text{k}\Omega \times 0.01\ \mu\text{F} \\ &= 54.8\ \mu\text{s} \\ f &= \frac{1}{T} \approx \textbf{18.2 kHz} \end{aligned}$$

Example 11.11 Change the op-amp bias supplies to ±10 V and determine the frequency of v_{OUT}.

Solution: With ±10 V supplies, $\pm V_{SAT}$ will be about ±8.5 V. Thus,

$$V_{LT} = -\frac{1}{11} \times 8.5\ \text{V} = -0.77\ \text{V}$$

$$V_{UT} = \frac{1}{11} \times 8.5\ \text{V} = +0.77\ \text{V}$$

$$\begin{aligned} T &= 2 \ln \left[\frac{8.5\ \text{V} - (-0.77\ \text{V})}{8.5\ \text{V} - 0.77\ \text{V}} \right] \times 15\ \text{k}\Omega \times 0.01\ \mu\text{F} \\ &= 54.5\ \mu\text{s} \\ f &= \frac{1}{T} \approx \textbf{18.3 kHz} \end{aligned}$$

These examples show that the oscillation frequency is relatively independent of the op-amp bias voltages. We could have predicted this from Equation 11–6 because both V_{LT} and V_{UT} will change in direct proportion to any change in V_{SAT}. Thus, each term inside the parentheses in Equation 11–6 will make the same percentage change for a given change in the bias voltages.

The frequency can be readily changed by varying R or C. Of course, the maximum frequency is limited by the op-amp's switching speed and slew rate. Typically, this type of circuit can be useful up to 100 kHz if a fast op-amp is used.

11.7 The 555 IC Timer

The NE/SE 555 timer, a versatile IC, operates as a one-shot (*monostable* multivibrator) where it has one stable state, or as an oscillator (*astable* multivibrator) that has no stable states. The 555 IC has the following features:

(a) wide supply voltage range (5–15 V)
(b) can easily drive TTL when operated from a 5 V supply
(c) high output current capability (200 mA)
(d) low input signal current requirements (0.25 μA)
(e) excellent timing stability with variations in supply voltage and temperature

The internal circuitry of the 555 timer uses bipolar transistors in a multi-stage arrangement. The major stages are shown in the block diagram for the 555 timer in Figure 11–15. The circuit consists of two voltage comparators, a SET/CLEAR FF, a voltage-divider to provide the reference voltages for the comparators, a bipolar discharge transistor (Q_1), and a totem-pole buffer output stage. In addition to V_{CC} and ground, the IC has six other pins whose functions will now be described.

OUTPUT This pin is a totem-pole output capable of sourcing or sinking up to 200 mA. It can drive TTL directly when $V_{CC} = 5$ V is used. The OUTPUT level (HIGH or LOW) will be the same as the state of FF X.

DISCHARGE This pin is connected to an open-collector transistor that is driven by the FF $\overline{X}$ output. When $\overline{X}$ = LOW, transistor Q_1 will be OFF and the DISCHARGE pin can be assumed to be an open circuit. When $\overline{X}$ = HIGH, Q_1 will be ON, and the DISCHARGE pin can be considered to be a very low resistance to ground. As we shall see, this pin will be used as a discharge path for the capacitors used in the various 555 circuit applications.

THRESHOLD This pin drives the " + " input of comparator 1. The " − " input of comparator 1 is connected to a reference voltage set at $2/3\ V_{CC}$ via the voltage-divider formed by the three equal resistors. Whenever the THRESHOLD input voltage is $\geq 2/3\ V_{CC}$, the comparator's output will go HIGH to clear the FF so that $X = 0$ and $\overline{X} = 1$. Whenever THRESHOLD is $< 2/3\ V_{CC}$ the comparator's output is LOW and has no effect on the FF.

Thus, THRESHOLD can be used to produce OUTPUT = LOW and DISCHARGE = SHORT. Note that THRESHOLD cannot be used to set the FF. This latter function is performed by the TRIGGER input.

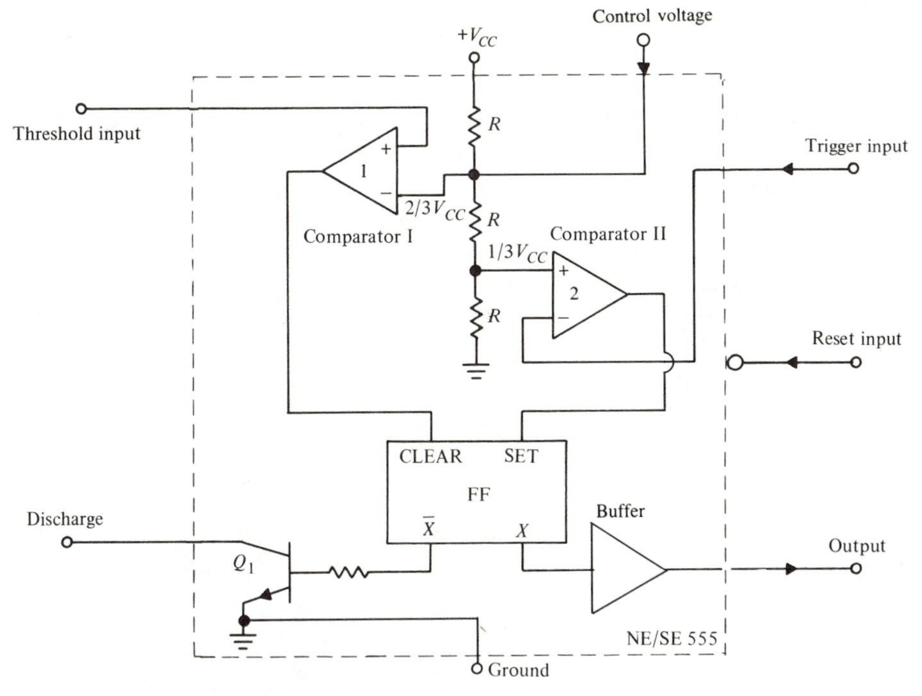

Figure 11–15 **Block diagram of 555 timer**

TRIGGER This input drives the " − " input of comparator 2. The " + " input of comparator 2 is tied to a reference voltage set at $1/3\ V_{CC}$. As long as the TRIGGER input is $\geq 1/3\ V_{CC}$, this comparator's output will be LOW and will have no effect on the FF. When TRIGGER is made $< 1/3\ V_{CC}$, comparator 2 goes HIGH and sets the FF to the $X = 1$, $\overline{X} = 0$ state. This produces OUTPUT = HIGH and DISCHARGE = OPEN. Note that TRIGGER cannot be used to clear the FF, only to set it.

Note that it is possible for both comparator outputs to be HIGH (i. e., when TRIGGER $< 1/3\ V_{CC}$ and THRESHOLD $> 2/3\ V_{CC}$). This puts HIGHs at both inputs to the FF and will cause both X and $\overline{X}$ to be HIGH. This condition is normally avoided.

RESET This is an override logic input that can be used to reset the FF to the $X = 0$, $\overline{X} = 1$ state at any time *independently* of the THRESHOLD and TRIGGER inputs. RESET is an active-LOW input, which, when not used, should be connected to V_{CC}.

CONTROL VOLTAGE This pin can be used to control the reference voltages connected to each comparator. In most applications this pin is not used, and the reference voltages are $2/3\ V_{CC}$ and $1/3\ V_{CC}$, as stated previously. The user can change these reference voltages by applying a voltage to the CONTROL VOLTAGE pin. For instance, with CONTROL VOLTAGE = 4 V, comparator 1 will have a reference of 4 V and comparator 2 will have a reference of 2 V. Of course, this changes the levels required for the THRESHOLD and TRIGGER inputs to perform their functions.

When the CONTROL VOLTAGE pin is not used, it can be left open. However, it is often connected to ground through a 0.01 μF capacitor to filter out any high frequency noise that may affect the comparators.

11.8 The 555 as a One-Shot

Figure 11–16 shows the 555 connected to operate as a one-shot. A timing capacitor, C_T, is connected between DISCHARGE and ground so that it is across transistor Q_1 (Figure 11–15). The capacitor is also tied to THRESHOLD so that its voltage, v_C, serves as the THRESHOLD input. The one-shot is triggered by applying a negative-going pulse at the TRIGGER input. This produces a positive-going pulse at the OUTPUT terminal, with a pulse duration determined solely by R_T and C_T.

Let's assume that initially OUTPUT = 0 V, DISCHARGE = SHORT (i. e., FFX is in $X = 0/\overline{X} = 1$ state), and TRIGGER is sitting at +4 V. With DISCHARGE = SHORT, C_T will be held at 0 V and will not be able to charge up through R_T. Thus, THRESHOLD = v_c = 0 V.

When TRIGGER is momentarily pulsed to below $V_{CC}/3 = 1.67$ V, the output switches to 5 V and DISCHARGE = OPEN (i. e., FFX is now in $X = 1/\overline{X} = 0$ state). With DISCHARGE now open, C_T can charge toward V_{CC} through R_T (see waveforms). When v_C has charged up to 2/3 V_{CC} = 3.33 V, the THRESHOLD input will cause OUTPUT to switch back to 0 V and DISCHARGE = SHORT. This will rapidly discharge C to approximately 0 V. The circuit is now back to its initial condition and will remain there until another TRIGGER pulse occurs.

The pulse at OUTPUT will have a duration t_p determined by the amount of time required for C_T to charge from 0 V to 3.3 V. This time can be readily determined using the exponential equation for v_C.

$$v_C = 5 - 5\epsilon^{-t/\tau}$$

Set $v_C = 3.33$ V at $t = t_p$ to obtain

$$3.33 \text{ V} = 5 - 5\epsilon^{-t_p/\tau}$$

Solving for t_p

$$\epsilon^{-t_p/\tau} = 0.333$$

$$\frac{-t_p}{\tau} = \ln 0.333 = -1.1$$

$$t_p = 1.1\tau = 1.1\, R_T\, C_T \quad \textbf{(11–7)}$$

Thus, each negative-going pulse at TRIGGER will produce a single OUTPUT pulse with this pulse duration. The value of t_p is independent of the duration of the TRIGGER pulse, *provided that* the TRIGGER input returns HIGH before C_T has reached 2/3 V_{CC}. Otherwise, the TRIGGER input will keep OUTPUT = HIGH past the point where THRESHOLD exceeds 3.3 V. For this reason, the TRIGGER input pulses are usually made much narrower than the desired OUTPUT t_p.

Example 11.12 What will happen to OUTPUT if a second TRIGGER pulse occurs during the t_p interval?

Solution: Nothing, as long as TRIGGER returns HIGH before the end of the t_p interval (see Figure 11–17). The first TRIGGER pulse sets FFX = 1 to produce OUTPUT = HIGH and DISCHARGE = OPEN. The second TRIGGER pulse has no effect since FFX is already set (HIGH).

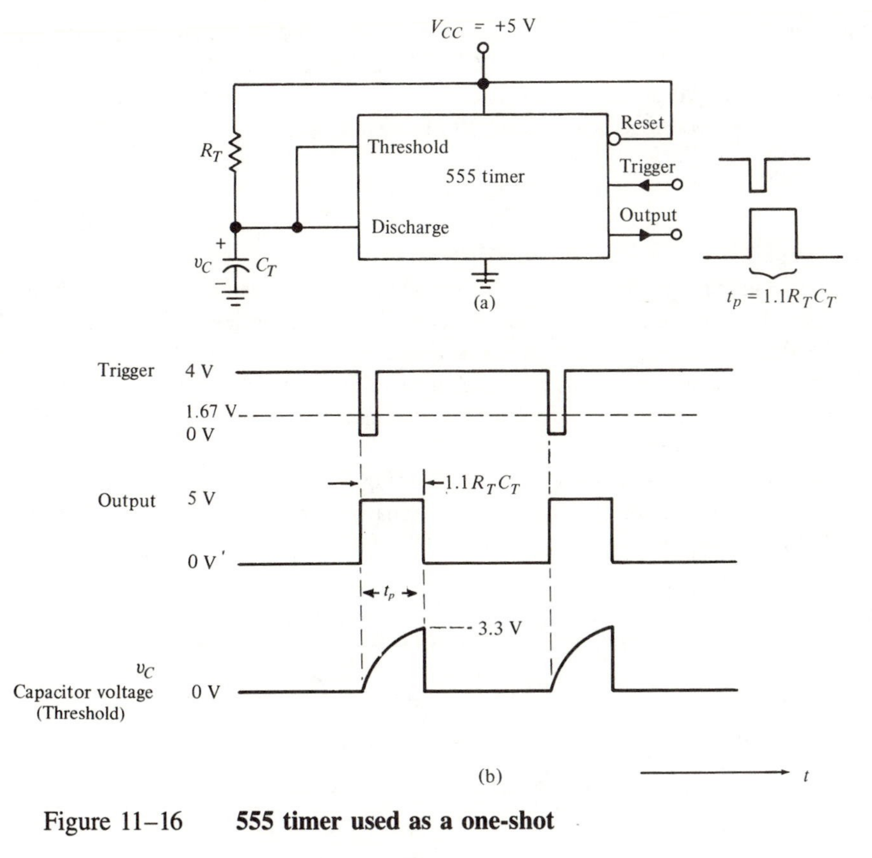

Figure 11–16 **555 timer used as a one-shot**

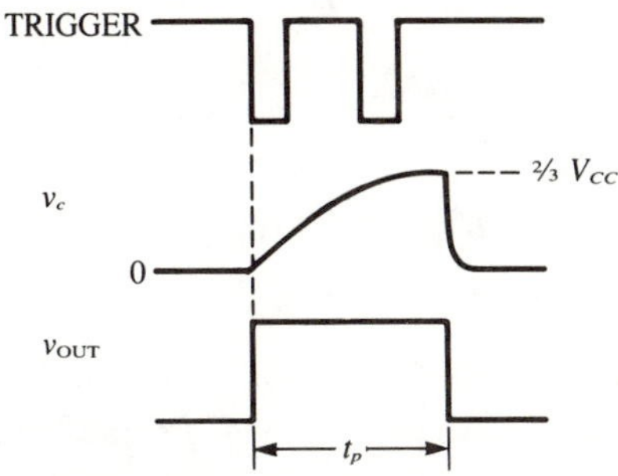

Figure 11–17 **TRIGGER pulses have no effect while C_T is still charging toward 2/3 V_{CC}**

Example 11.13 How is t_p affected by changes in V_{CC}?

Solution: t_p is the time it takes C_T to charge from 0 V to 2/3 V_{CC} as it heads toward V_{CC}. It will always take 1.1 time constants for C_T to charge to 2/3 of the final value it is charging toward, regardless of *what* this final value is. *Thus, V_{CC} will have no effect on t_p.*

To prove it mathematically, we can solve for t_p in the general case where $V_F = V_{CC}$. This is left as an end-of-chapter exercise.

Limits on R_T The minimum value of R_T is determined by the maximum current limit on Q_1 [see Figure 11–18(a)]. When DISCHARGE = SHORT, Q_1 is ON and holds the capacitor at 0 V. All of the current from V_{CC} through R_T is diverted through Q_1. For the 555 IC, Q_1's sink current is limited to 10 mA. Thus,

$$R_T(\text{MIN}) = \frac{V_{CC}}{10 \text{ mA}} \quad \textbf{(11–8)}$$

For example, at $V_{CC} = 5$ V, $R_T(\text{MIN}) = 500\ \Omega$.

The maximum limit on R_T is determined by the currents that will flow into the THRESHOLD and DISCHARGE terminals while C_T is charging [see Figure 11–18(b)]. The THRESHOLD input will draw a maximum of 250 nA. The DISCHARGE terminal will draw a maximum of 100 nA (Q_1's leakage current in the OFF state). The sum of these two currents will subtract from the current I_T being supplied by V_{CC}, thereby reducing the charging current available for C_T.

To assure there will be capacitor charging current, I_T must exceed 100 nA + 250 nA = 350 nA. Since I_T decreases as C_T charges, the smallest I_T occurs when C_T is at 2/3 V_{CC} and the voltage across R_T is 1/3 V_{CC}. Thus,

$$I_T(\text{MIN}) = \frac{1/3\ V_{CC}}{R_T} > 350 \text{ nA}$$

or

$$R_T < \frac{V_{CC}}{1{,}050 \text{ nA}} \approx \frac{V_{CC}}{1\ \mu\text{A}}$$

so that

$$R_T(\text{MAX}) \approx \frac{V_{CC}}{1\ \mu\text{A}} \quad \textbf{(11–9)}$$

For example, for $V_{CC} = 5$ V, $R_T(\text{MAX}) = 5$ MΩ, and for $V_{CC} = 15$ V, $R_T(\text{MAX}) = 15$ MΩ.

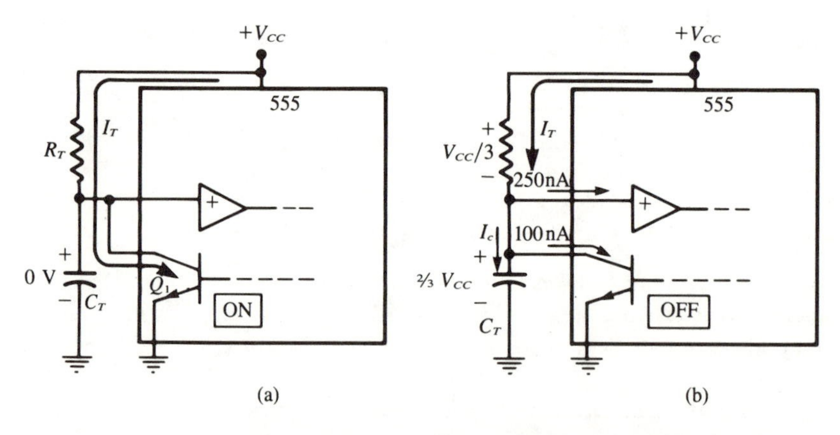

Figure 11–18 **(a) R_T(MIN) is determined by maximum collector current rating of Q_1; (b) R_T(MAX) is determined by small Q_1 leakage current and comparator input current**

Example 11.14 As stated earlier, the TRIGGER pulse duration has to be less than the OUTPUT t_p. How, then, can a wide pulse be used to trigger the 555 one-shot?

Solution: In order to use a wide pulse as the TRIGGER input, it must be converted to a narrow pulse or spike. This can be done using an *RC* differentiator (high-pass) circuit, as shown in Figure 11–19. R_1 and C_1 are used to convert the e_{IN} pulse to narrow spikes whose duration is much less than the OUTPUT t_p set by R_T and C_T.

Note that R_1 is connected to V_{CC} rather than to ground. This ensures that TRIGGER is biased normally at +10 V, its inactive state. The transitions of e_{IN} will produce 10 V spikes above and below this 10 V bias at TRIGGER. The negative-going spike causes TRIGGER to drop below 1/3 V_{CC}, thereby triggering the one-shot.

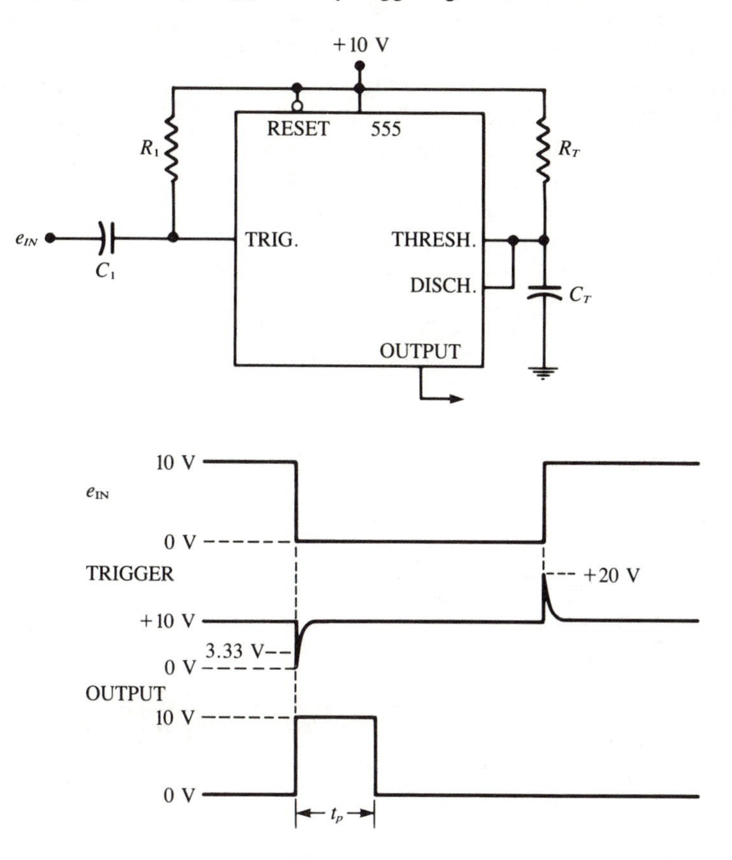

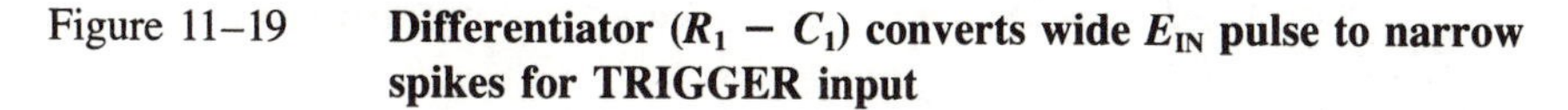

Figure 11–19 **Differentiator (R_1 − C_1) converts wide E_{IN} pulse to narrow spikes for TRIGGER input**

The $R_1 - C_1$ time constant is chosen to produce a negative-going spike whose duration is less than t_p. That is,

$$5\, R_1C_1 < t_p = 1.1\, R_TC_T$$

This ensures that TRIGGER will have risen back above $V_{CC}/3$ before the end of the t_p interval.

11.9 The 555 Oscillator

Figure 11–20(a) shows how the 555 can be connected to operate as an oscillator (astable multivibrator) that requires no trigger input signal. In this configuration, C_T is connected to both THRESHOLD and TRIGGER so that v_C is applied to both these inputs. Also note that the series resistors R_A and R_B have their common node connected to the DISCHARGE terminal.

The waveforms in Figure 11–20(b) show how the v_C and OUTPUT voltages vary with time. The capacitor waveform shows that C_T alternately charges up to 2/3 V_{CC} and discharges to 1/3 V_{CC}. While C_T is charging and before it reaches 2/3 V_{CC}, OUTPUT = HIGH and DISCHARGE = OPEN. With DISCHARGE open, C_T charges toward V_{CC} through R_A and R_B. When V_C = 2/3 V_{CC}, THRESHOLD is activated and produces OUTPUT = LOW and DISCHARGE = SHORT. With DISCHARGE = SHORT, C_T discharges toward ground only through R_B. When v_C discharges to 1/3 V_{CC}, TRIGGER is activated and produces OUTPUT = HIGH and DISCHARGE = OPEN. This causes C_T to start charging again, and the process keeps repeating itself.

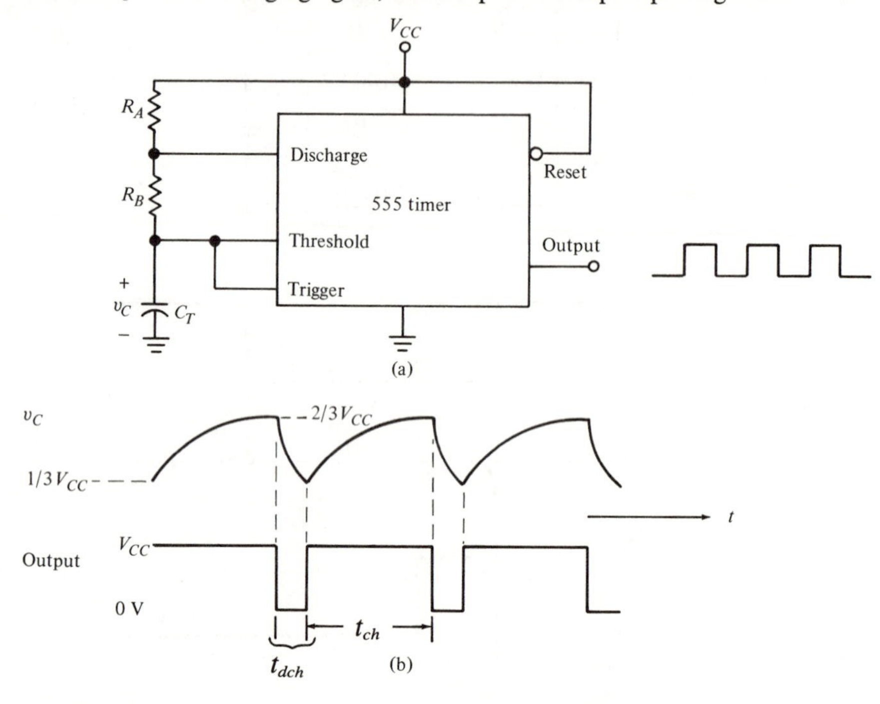

Figure 11–20 **555 timer used as an astable multivibrator**

Period and frequency To determine the period of the two waveforms, we need to find t_{ch} and t_{dch}. Using the exponential equation for v_C during its charging interval, it can be shown that

$$t_{ch} \approx 0.7\,(R_A + R_B)C_T$$

Likewise,

$$t_{dch} \approx 0.7\,R_B C_T$$

Thus,

$$T = t_{ch} + t_{dch} \approx 0.7\ (R_A + 2\ R_B)C_T \qquad \textbf{(11–10)}$$

and, of course, $f = 1/T$.

Note that t_{ch} and t_{dch} are different because the charging and discharging paths for C_T are different. Since $t_{ch} > t_{dch}$, the OUTPUT waveforms can never have a 50 percent duty cycle (square wave). However, in practice, t_{ch} and t_{dch} can be made reasonably close in value by choosing R_A to be much smaller than R_B. In other words, with $R_A \ll R_B$

$$t_{ch} \approx 0.7\ (R_A + R_B)C_T \approx 0.7\ R_BC_T \approx t_{dch}$$

Example 11.15 Select values for R_A, R_B, and C_T to produce an approximate square wave at 1 kHz using $V_{CC} = 5$ V. Choose the closest standard values for the R and C values.

Solution: We want $R_A \ll R_B$, so R_A should be made as small as possible. The minimum value for R_A, however, is determined by the same consideration we used for the 555 one-shot circuit. R_A limits the current flowing from V_{CC} to ground through the DISCHARGE terminal when DISCHARGE = SHORT. Thus,

$$R_A(\text{MIN}) = \frac{V_{CC}}{10\ \text{mA}} \qquad \textbf{(11–8)}$$

or $R_A(\text{MIN}) = 500\ \Omega$ for $V_{CC} = 5$ V. To be safe, we can choose $R_A =$ **680 Ω.**

Now we can choose R_B to be much larger than R_A. A factor of 100 is recommended. Thus, $R_B = 100\ R_A =$ **68 kΩ.**

The desired period is $T = 1/1$ kHz $= 1$ ms, which means that $t_{ch} = t_{dch} =$ 0.5 ms. Thus,

$$t_{dch} \approx 0.7\ R_BC_T = 0.5\ \text{ms}$$

or

$$C_T = \frac{0.5\ \text{ms}}{0.7 \times 68\ \text{k}\Omega}$$

$$\approx \mathbf{0.01\ \mu F}$$

Using $R_A = 680\ \Omega$, $R_B = 68\ \text{k}\Omega$ and $C_T = 0.01\ \mu$F, we can calculate the exact values as

$$t_{dch} = 0.7\ R_BC_T = 0.476\ \text{ms}$$

$$t_{ch} = 0.7(R_A + R_B)C_T = 0.481\ \text{ms}$$

$$T = t_{ch} + t_{dch} = 0.957\ \text{ms}$$

$$f = \frac{1}{T} = 1.045\ \text{kHz}$$

Note that t_{ch} and t_{dch} are almost equal, thereby producing an approximate square wave at OUTPUT.

If the frequency is to be adjusted more accurately, R_B can be made a variable resistance. Varying R_B will cause both t_{ch} and t_{dch} to change by the same amount (since R_A is so small), thereby allowing the period and frequency to be adjusted more accurately.

Example 11.16 Determine the frequency and duty cycle of the OUTPUT waveform when $R_A = 4.7\ \text{k}\Omega$, $R_B = 6.8\ \text{k}\Omega$, and $C_T = 0.033\ \mu\text{F}$.

Solution:

$$t_{ch} = 0.7(4.7\ \text{k}\Omega + 6.8\ \text{k}\Omega) \times (.033\ \mu\text{F})$$
$$= 265.7\ \mu\text{s}$$
$$t_{dch} = 0.7 \times 6.8\ \text{k}\Omega \times .033\ \mu\text{F} = 157.1\ \mu\text{s}$$
$$T = t_{ch} + t_{dch} = 422.8\ \mu\text{s}$$
$$f = \frac{1}{T} = \mathbf{2.37\ kHz}$$

The OUTPUT waveform is HIGH during the capacitor charging interval. Thus, the duty cycle is

$$\text{duty cycle} = \frac{t_{ch}}{T} \times 100\%$$
$$= \mathbf{62.8\%}$$

A variation of the basic 555 oscillator circuit will be introduced in Problems 11.27–11.28 at the end of the chapter.

11.10 Oscillators Made from Inverters

Figure 11–21(a) is a simple circuit that uses two TTL inverters to produce approximate square wave outputs at Q and $\overline{Q}$. The frequency of the oscillations is roughly equal to $1/(3\ RC)$. The value of R can be no larger than about 470 Ω for standard TTL, and no larger than about 1.8 kΩ for 74LS TTL. Thus, this oscillator is generally used only in fixed frequency applications.

Figure 11–21(b) is a modified version of the simple oscillator in Figure 11–21(a) which uses a quartz crystal element to control the frequency of the oscillations. The crystal will permit the circuit to oscillate only at the resonant frequency for which the crystal is cut. The value of coupling capacitor C is given by the formula next to the circuit diagram. A value of C is chosen such that X_C is LOW (<400 Ω) at the oscillation frequency. Also, for best operation the crystal impedance should be fairly LOW (500–600 Ω). This crystal-controlled oscillator provides much better frequency stability than the simpler circuit and should be used whenever this is a prime consideration. The simpler circuit, however, is more convenient for testing purposes or in applications where a tightly controlled frequency is not needed.

Logically controlled oscillators Either of the oscillators in Figure 11–21 can be modified so that the oscillator can be controlled by a separate logic control input. The control input, X, will determine whether or not oscillations will occur. Figure 11–22 shows how this is done for the simple oscillator.

The first inverter, I_1, has been replaced by a two-input NAND gate that has X as its other input. When X is HIGH the operation of the oscillator is unaffected; however, when X is held LOW the output of the NAND gate must go HIGH regardless of what is present at

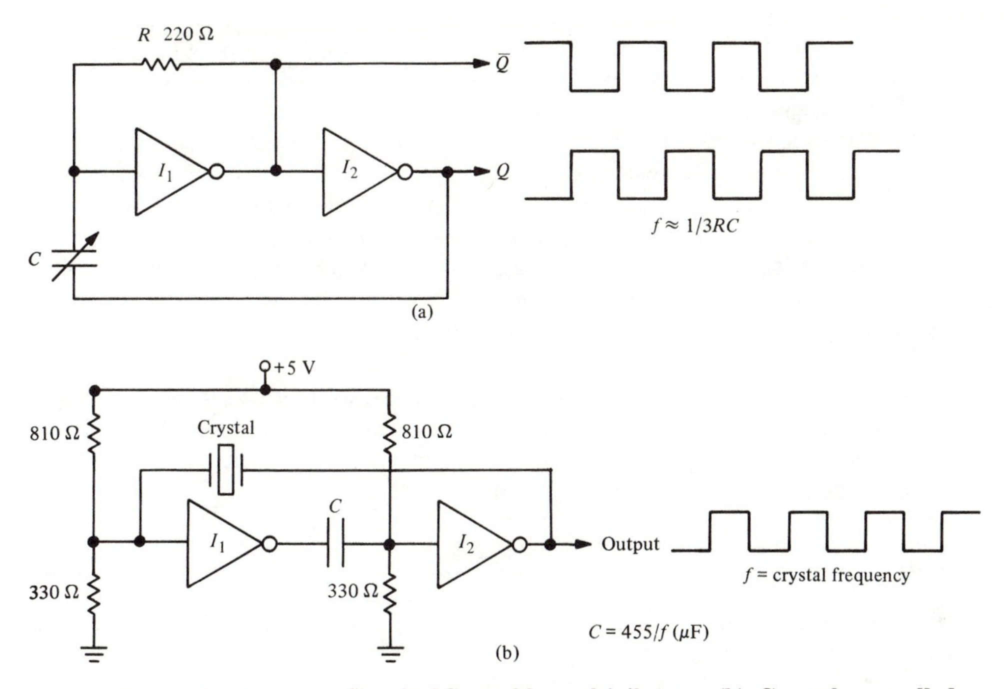

Figure 11–21 **(a) Simple IC astable multivibrator (b) Crystal-controlled astable multivibrator**

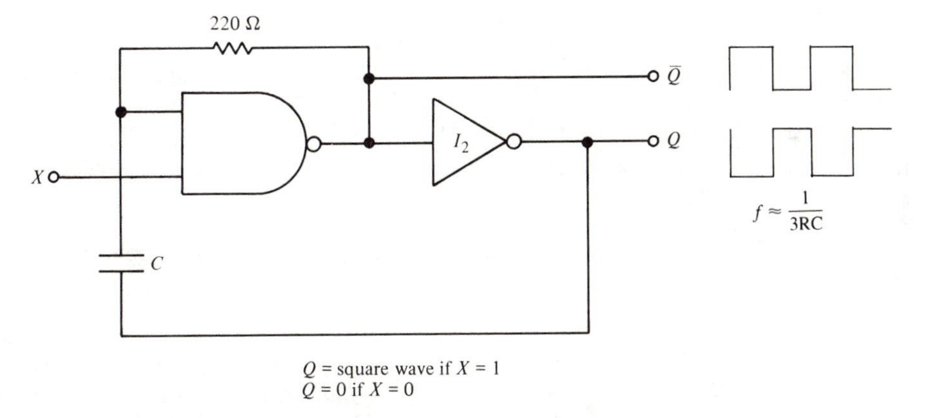

Figure 11–22 **Logic input *X* controls oscillator**

its other input. This in turn causes the I_2 output to stay LOW. Thus, there will be no oscillations at the outputs as long as $X = 0$. In this way, the oscillations can be controlled by any logic signal fed into input X.

11.11 Sweep-Voltage Waveform

A sweep-voltage waveform is one whose voltage starts from a baseline and increases at a uniform linear rate up to a peak amplitude, then rapidly returns to its baseline voltage (see Figure 11–23). A sweep waveform is also referred to as a *sawtooth,* or *time-base,* waveform. This latter term comes from the application of a sweep waveform in an oscilloscope to move the electron beam across the CRT at a uniform rate, thereby establishing the scope's time base.

The sweep waveform in Figure 11–23 has two basic parts: the sweep time, t_s, and the recovery time, t_{re}. The recovery time is also made up of two parts: the flyback time t_f (also called *retrace* time), and the OFF time, t_{OFF}. In some applications the OFF time is zero so that $t_{re} = t_f$. The total period of the sweep waveform is $T = t_s + t_{re}$.

The positive-going waveform in Figure 11–23 rests on a baseline voltage V_0 and makes a linear increase in the positive direction to a peak value V_1. A negative-going sweep waveform would decrease linearly from its baseline, V_0, to a more *negative* voltage, V_1.

Most methods for generating sweep voltages use the charging of a capacitor. However, the capacitor cannot be charged through a resistor as it was in the UJT or 555 oscillators, because this would produce an exponential waveform rather than a linear waveform. In order to produce a linear capacitor charging waveform, the capacitor charging current must be held constant. Since

$$\frac{dv_C}{dt} = \frac{i_C}{C} \tag{3–1}$$

it is clear that if i_C is held constant, the capacitor charging rate dv_C/dt will be constant and the capacitor voltage will change at a uniform rate.

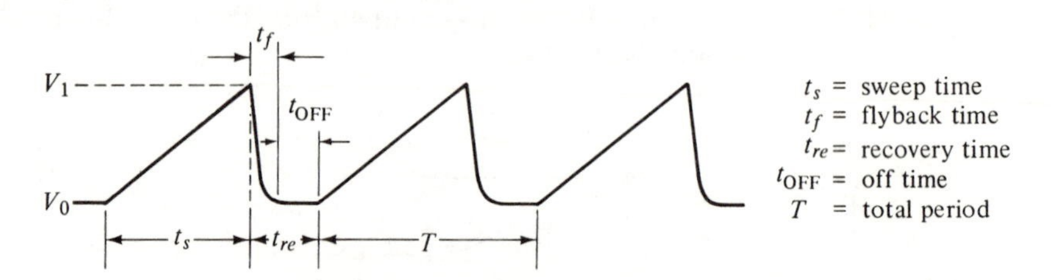

Figure 11–23 **General sweep waveform**

Example 11.17 Figure 11–24(a) shows a capacitor being charged from an initial voltage of 0 V. Determine the capacitor voltage after 5 ms and 15 ms if i_C is held constant at 10 mA.

Solution:

$$\frac{dv_C}{dt} = \frac{i_C}{c} = \frac{10 \text{ mA}}{5\ \mu\text{F}}$$
$$= 2{,}000 \text{ V/S} = 2 \text{ V/ms}$$

Since i_C is constant, the capacitor voltage will increase at the constant rate of 2 V/ms. Thus, after 5 ms, v_C will have increased from 0 V to

$$v_C\ (@t = 5 \text{ ms}) = \frac{2 \text{ V}}{\text{ms}} \times 5 \text{ ms}$$
$$= \mathbf{10\ V}$$

Likewise,

$$v_C\ (@t = 15 \text{ ms}) = \frac{2 \text{ V}}{\text{ms}} \times 15 \text{ ms}$$
$$= \mathbf{30\ V}$$

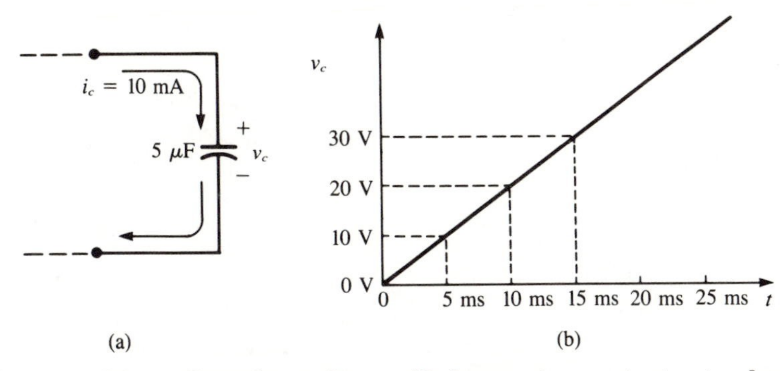

Figure 11–24 **Capacitor voltage will change at a constant rate whenever i_C is held constant**

The waveform for v_C will be a linear ramp, as shown in Figure 11–24(b). Of course, v_C will keep increasing at the rate of 2 V/ms only as long as i_C remains fixed at 10 mA.

11.12 Transistor Sweep Generator

A transistor can be used to provide a reasonably constant current to produce a linear capacitor charging waveform. The circuit in Figure 11–25(a) shows how this can be achieved using a PNP transistor connected in the common-base configuration. The transistor is biased to operate in the *active* region. Assume the capacitor is initially discharged so the collector is at ground potential. Since the base is biased at +10 V above ground, the *C-B* junction is reverse-biased with $V_{CB} = -10$ V. Clearly, the *E-B* junction is forward-biased, so the transistor is in the active region. The value of emitter current is determined as

$$I_E = \frac{20\text{ V} - 10.7\text{ V}}{4.7\text{ k}\Omega} \approx 2\text{ mA}$$

Figure 11–25(b) shows the initial operating point on the transistor common-base output curve corresponding to $I_E = 2$ mA. The collector current at the initial operation point is also approximately 2 mA. This current will begin to charge the capacitor positively at a uniform rate, given by

$$\frac{dv_C}{dt} = \frac{I_C}{C} = \frac{2\text{ mA}}{1\ \mu\text{F}} = 2\text{ V/ms}$$

As the capacitor charges, the transistor collector voltage becomes more positive, and the *C-B* reverse bias decreases (V_{CB} becomes less negative). For example, after 2.5 ms the capacitor will have charged to

$$v_C = 2.5\text{ ms} \times \frac{2\text{ V}}{\text{ms}} = 5\text{ V}$$

so $V_{CB} = -5$ V. As can be seen on the curve in Figure 11–25(b), the collector current is still essentially 2 mA at this point. In fact, even after 5 ms, when the capacitor voltage has reached 10 V and $V_{CB} = 0$, the collector current has barely dropped below 2 mA. Thus, between 0 and 10 V the capacitor voltage has increased linearly at a rate of 2 V/ms, as illustrated in Figure 11–25(c).

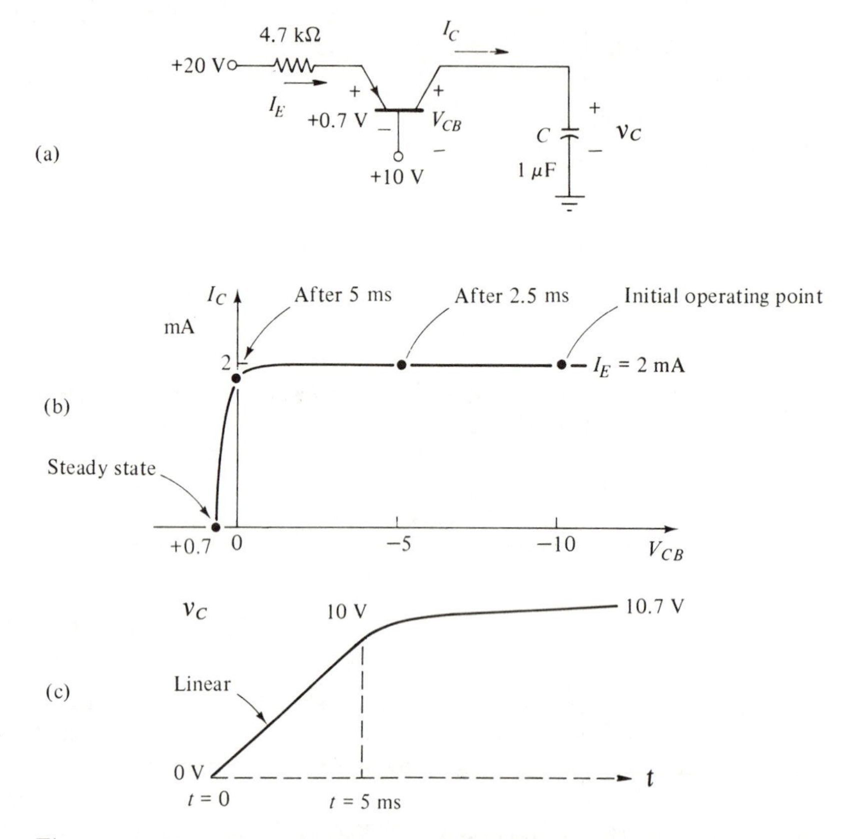

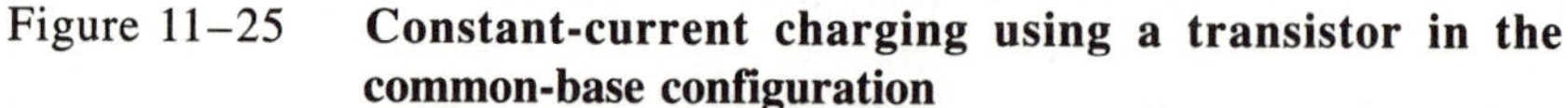
Figure 11–25 **Constant-current charging using a transistor in the common-base configuration**

At this point, as the capacitor charges further, the *C-B* junction becomes forward biased (V_{CB} goes positive), causing the collector current to decrease rapidly [see the curve in Figure 11–25(b)]. This decreasing collector current keeps charging the capacitor but at a slower and slower rate until eventually the collector current has dropped to zero. This occurs when v_C = 10.7 V and V_{CB} = +0.7 V. When I_C reaches zero, the capacitor stops charging and remains at 10.7 V.

It should be apparent from the output-voltage waveform in part (c) of Figure 11–25 that a linear sweep-voltage waveform can be obtained if the capacitor is discharged on or before the time it reaches 10 V. After 10 V the waveform becomes non-linear. In general, *the capacitor voltage will rise linearly until it reaches a value equal to the base bias voltage.*

Discharging the capacitor Now that we have a method for producing the linear part of a sweep waveform, consider how to quickly discharge the capacitor back to the baseline voltage. In practice, a short flyback time, t_f, is normally desirable, so the capacitor must be discharged rapidly. One common method is to use a transistor switch as shown in Figure 11–26.

Transistor Q_2 is connected across the capacitor and is turned ON and OFF by the input signal e_{IN}. This e_{IN} signal controls the sweep time, t_s, and the recovery time, t_{re}.

When e_{IN} is at 0 V, Q_2 is OFF and C is free to charge linearly at a rate determined by Q_1's collector current. When e_{IN} pulses positively, Q_2 turns ON and provides a very low resistance to discharge C to $\approx$ 0 V. v_C will remain at approximately 0 V until Q_2 is turned OFF when e_{IN} returns to 0 V. The resulting v_C signal is a periodic sweep waveform.

Example 11.18 The circuit of Figure 11–26 uses the following values: V_{EE} = 30 V, V_{BB} = 25 V, R_E = 22 kΩ, C = 0.05 μF, and R_B = 4.7 kΩ. The e_{IN} signal has a pulse duration of 1 ms, a frequency of 250 Hz and a +10 V amplitude. Determine the sweep waveform at v_C.

Solution: First we must determine the amount of time that e_{IN} is at 0 V during each cycle. Since its frequency is 250 Hz, e_{IN} has a period of 1/250 Hz = 4 ms. Thus, e_{IN} = 0 V for 3 ms of each period (see Figure 11–27). During this 3 ms, C will charge linearly at rate determined by Q_1's collector current I_{C_1}.

To find I_{C_1}, we must first find Q_1's emitter current.

$$I_{E_1} \approx \frac{30\text{ V} - 25\text{ V} - 0.7\text{ V}}{22\text{ k}\Omega} \approx 0.2\text{ mA}$$

While Q_1 is in the active region, we can assume that $I_{C_1} \approx I_{E_1}$. Therefore,

$$I_{C_1} \approx 0.2\text{ mA}$$

and the charging rate will be

$$\frac{dv_C}{dt} = \frac{0.2\text{ mA}}{0.05\ \mu\text{F}} = 4\text{ V/ms}$$

The v_C waveform will increase uniformly at this rate for 3 ms, from 0 V up to a peak voltage of

$$V_P = \frac{4\text{ V}}{\text{ms}} \times 3\text{ ms} = 12\text{ V}$$

After 3 ms, e_{IN} will turn ON Q_2 to discharge C and hold v_C at 0 V for the 1 ms pulse duration. The repetitive v_C waveform is shown in Figure 11–27(a). It has a sweep time of 3 ms, which is the e_{IN} = 0 V interval and a peak amplitude of 12 V.

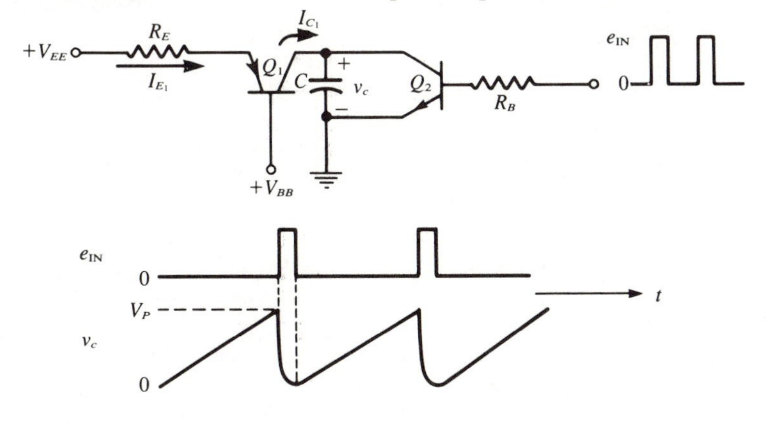

Figure 11–26 **Complete sweep generator circuit**

Example 11.19 If we keep e_{IN}'s pulse duration at 1 ms and decrease its frequency, what is the lowest frequency we can use and still obtain a linear sweep waveform?

Solution: As e_{IN}'s frequency is decreased, its period will increase; and since $t_p = 1$ ms, e_{IN} will be at 0 V for a longer time each cycle. This will allow v_C to charge further. However, v_C can charge linearly only up to a voltage equal to the base bias voltage, V_{BB}, which is set to 25 V. After that, it will level off, as shown in Figure 11–27(b), at about 25.7 V.

With $dv_C/dt = 4$ V/ms, we can determine how long it will take v_C to reach 25 V. We will call this the maximum allowable sweep time t_s(MAX).

$$t_s(\text{MAX}) = \frac{25\text{ V}}{4\text{ V/ms}} = 6.25\text{ ms}$$

Thus, e_{IN} can have a maximum period of 6.25 ms + 1 ms = 7.25 ms. This gives a minimum frequency of

$$f = \frac{1}{7.25\text{ ms}} \approx \mathbf{138\ Hz}$$

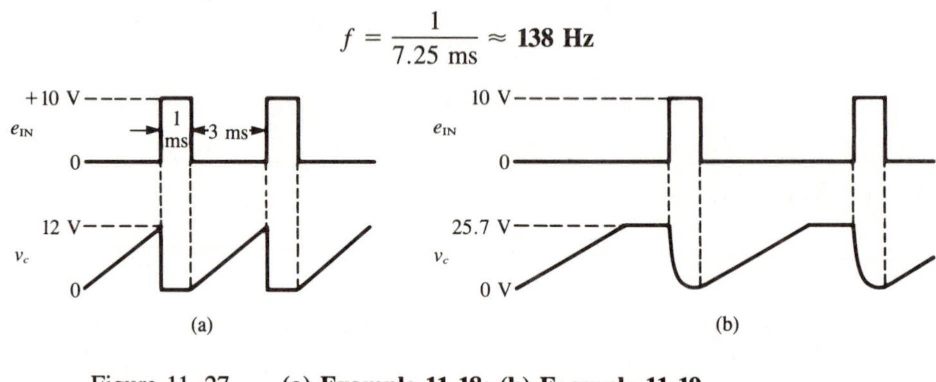

Figure 11–27 **(a) Example 11.18; (b) Example 11.19**

Determining flyback time, t_f When Q_2 is turned ON by e_{IN}, it provides a rapid discharge path for C. The amount of time required for C to discharge to 0 V will be determined by the ON resistance of Q_1, which is extremely difficult to predict. Thus, there is no simple way to determine t_f. We can, however, use the following approximate formula to get a "ballpark" value for t_f and to give us some insight as to how t_f can be controlled.

$$t_f \approx \frac{CV_P}{h_{FE_2} I_{B_2}} \qquad \textbf{(11–11)}$$

In this approximate equation, the $h_{FE_2} I_{B_2}$ term is the current gain and base current product for the discharge transistor Q_2. This $h_{FE_2} I_{B_2}$ is really Q_2's collector current, which flows as C discharges through Q_2. The CV_P term in the numerator is the actual charge (remember, $Q = CV$) on the capacitor when it is about to be discharged through Q_2. t_f is obtained by dividing this total charge, CV_P, by the rate at which this charge is being removed (recall that current = charge/time).

Thus, this relationship shows that t_f can be reduced if we increase the current gain of Q_2 or drive it harder by supplying more base current.

Example 11.20 Calculate t_f for the conditions given in Example 11.18. Assume $h_{FE} = 50$ for Q_2.

Solution: The base current of Q_2 when $e_{IN} = +10$ V will be

$$I_{B_2} = \frac{10\text{ V} - 0.7\text{ V}}{4.7\text{ k}\Omega} \approx 2\text{ mA}$$

We previously determined that $V_P = 12$ V, so we have

$$t_f \approx \frac{CV_P}{h_{FE}I_{B_2}} = \frac{0.05\ \mu\text{F} \times 12\text{ V}}{50 \times 2\text{ mA}}$$

$$\approx \mathbf{6\ \mu s}$$

Once the approximate value for t_f is known, the pulse duration for e_{IN} should be made greater than t_f to ensure enough time to discharge the capacitor all the way to 0 V. Since the t_f calculation is very approximate, the e_{IN} pulse duration should be made at least *twice* this value ($t_P \geqslant 2\ t_f$) to be on the safe side. Of course, t_f can be reduced by driving Q_2 harder, so that t_P can be made narrower if desired.

Effect of loading on sweep waveform Applications requiring a sweep waveform usually depend on it being as linear as possible. For example, a sweep waveform is used in an oscilloscope to produce the horizontal deflection of the electron beam. If the sweep waveform is perfectly linear, the beam will move across the screen at a uniform rate, say 1 cm/ms, which allows us to use the horizontal axis as a time base where each cm of horizontal deflection is a fixed unit of time.

The sweep generator circuit of Figure 11–26 is sufficiently linear for most applications since Q_1's collector current stays relatively constant as C charges. In fact, many industrial oscilloscopes use this circuit, such as Hewlett-Packard's 1220/1222 models.

To preserve its linearity, the sweep circuit output cannot be subjected to any significant resistive loading. Figure 11–28 shows a situation where the sweep waveform, v_c, is driving a load, R_L. The presence of R_L means that the capacitor will not be able to charge at a constant rate because its current will not be constant. Q_1's fixed collector current, I_{C_1}, will divide between the capacitor current i_c and the load current i_L. As v_C increases, R_L draws more and more current, thereby reducing the capacitor current and the charging rate. The overall result is an exponential, rather than linear, v_C waveform.

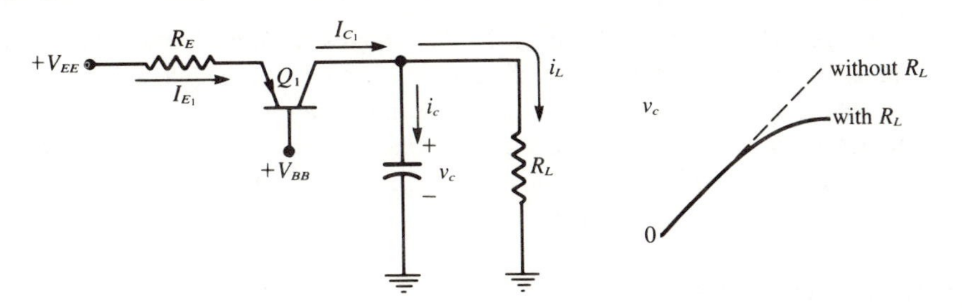

Figure 11–28 **Presence of load R_L can produce non-linear sweep waveform unless R_L is kept very large**

In order to minimize this effect, R_L should be kept large enough so that it will not draw more than *one percent* of the available current. For instance, in Example 11.18, the charging current from Q_1 was calculated to be 0.2 mA as the capacitor was charging to a peak voltage of 12 V. Thus, the current through R_L should be kept at or below

$$1\% \times 0.2 \text{ mA} = 2\ \mu\text{A}$$

so that

$$R_L(\text{MIN}) = \frac{12 \text{ V}}{2\ \mu\text{A}} = 6 \text{ M}\Omega$$

This is a fairly stringent restriction on R_L, and in many cases requires that a buffer be placed between v_c and the actual R_L so as to minimize the loading effect. The buffer is typically an emitter-follower or an op-amp voltage-follower.

Single-supply operation The sweep generator circuit we have been analyzing uses two dc supplies, V_{EE} and V_{BB}. In practice, only the V_{EE} supply is required, since V_{BB} is less than V_{EE} and can be derived from V_{EE} using a zener diode. This is illustrated in Figure 11–29 where the zener diode produces $V_{BB} = V_Z$.

The resistor R_z is chosen to provide enough current to keep the zener operating well above the zener knee current.

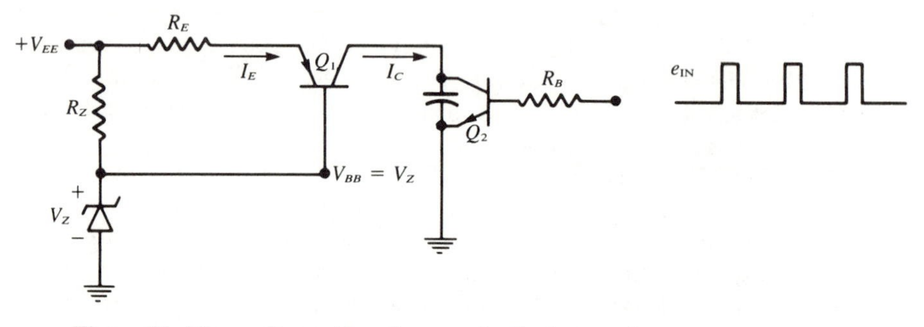

Figure 11–29 **Operation from a single dc supply**

11.13 Free-Running Sweep Generator

The sweep generator circuit of Figure 11–29 requires an input pulse signal, e_{IN}, to synchronize its operation. Without e_{IN}, the capacitor would charge up to a voltage slightly higher than V_{BB} and just stay there. A *free-running sweep generator* requires no input signal to periodically discharge C. Instead, the capacitor is automatically discharged when its voltage reaches a specific threshold value. Figure 11–30(a) shows how a UJT can be used to perform the discharge function.

The circuit operates as follows:

1. With C initially discharged, the UJT (Q_2) is held OFF since its emitter voltage is less than V_P, its peak-point voltage.
2. The collector current of Q_1, as determined by R_E, charges C at a constant rate. As C charges, it eventually reaches a voltage equal to V_P. At that instant, the UJT will turn ON.

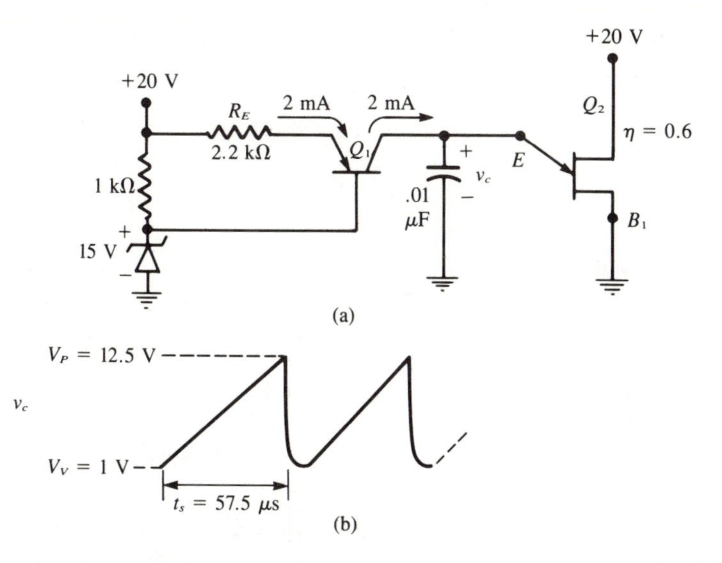

Figure 11–30 **(a) Free-running sweep generator using a UJT; (b) Sweep waveform for R_E = 2.2 kΩ**

3. The ON resistance of the UJT drops to a very low value and causes C to discharge rapidly through the UJT emitter and B_1 until the discharge current drops below I_V. At that instant, the UJT turns OFF, and the sequence is repeated beginning with step 1.
4. The sweep time (t_s), and therefore the frequency, can be most easily varied by varying R_E, which changes the capacitor charging current. As before, this charging current must be greater than the UJT's peak-point current (I_P) but lower than its valley current (I_V) for oscillations to occur.

Example 11.21 Determine the amplitude and sweep time of the v_C waveform in Figure 11–30.

Solution: The amplitude is determined by the UJT's peak-point voltage, V_P.

$$V_P = \eta V_{BB} + V_D = 0.6 \times 20 \text{ V} + 0.5 \text{ V}$$
$$= \mathbf{12.5 \text{ V}}$$

The capacitor will charge linearly from V_V = 1 V to V_P = 12.5 V, a total change of 11.5 V. The rate of change of v_C is

$$\frac{dv_C}{dt} = \frac{I_C}{C} \approx \frac{I_E}{C}$$

where

$$I_E \approx \frac{20 \text{ V} - 15 \text{ V} - 0.7 \text{ V}}{2.2 \text{ k}\Omega} \approx 2 \text{ mA}$$

Thus,

$$\frac{dv_C}{dt} \approx \frac{2 \text{ mA}}{0.01\ \mu\text{F}} = 200 \text{ V/ms}$$

The sweep time, t_s, is the time it takes C to charge by a total of 11.5 V when it is charging at a constant rate of 200 V/ms. That is,

$$t_s = \frac{\text{total voltage change}}{\text{rate of change}}$$
$$= \frac{11.5\ \text{V}}{200\ \text{V/ms}}$$
$$= 0.0575\ \text{ms} = \mathbf{57.5\ \mu s}$$

Example 11.22 Modify the circuit of Figure 11–30 so that its sweep time is 100 μs.

Solution: The easiest value to change is R_E. The value of R_E can be changed to vary the capacitor charging rate, and therefore the time required for v_C to reach V_P.

We want t_s=100 μs. Thus we have to find the required charging rate. That is,

$$t_s = \frac{\text{total voltage change}}{\text{rate of change}}$$
$$100\ \mu\text{s} = \frac{11.5\ \text{V}}{dv_C/dt}$$

so that

$$\frac{dv_C}{dt} = \frac{11.5\ \text{V}}{100\ \mu\text{s}} = 115\ \text{V/ms}$$

Therefore we have,

$$\frac{dv_C}{dt} = \frac{115\ \text{V}}{\text{ms}} = \frac{I_C}{0.01\ \mu\text{F}}$$

so that

$$I_C = 1.15\ \text{mA}$$

Since Q_1's emitter current is approximately equal to its collector current, we want

$$I_E \approx 1.15\ \text{mA} = \frac{20\ \text{V} - 15\ \text{V} - 0.7\ \text{V}}{R_E}$$

so that

$$R_E \approx \frac{4.3\ \text{V}}{1.15\ \text{mA}} = 3.74\ \text{k}\Omega$$

If a variable resistor is used for R_E, its value can be adjusted until t_s = 100 μs.

Example 11.23 Why won't the circuit of Figure 11–30 work properly if the UJT has I_V = 1 mA?

Solution: Q_1's collector current is 2 mA. When the UJT turns ON, this 2 mA will flow through the UJT's emitter in addition to the capacitor discharge current. Even if the capacitor's discharge current drops to zero, the 2 mA from Q_1 will exceed the UJT's I_V and will keep the UJT ON. This will prevent the sweep waveform at v_C from repeating its linear charge; instead, v_C will stay at 1 V after its first cycle.

11.14 Oscilloscope Triggered Sweep Circuit

Figure 11–31(a) is a simplified block diagram showing the principal components of an oscilloscope. The signal to be displayed on the CRT is the VERTICAL SIGNAL. This signal is applied to a VERTICAL AMPLIFIER block that drives the CRT's vertical deflection plates so that the vertical motion of the electron beam follows the variations of the VERTICAL SIGNAL. The VERTICAL SIGNAL is also applied to the TRIGGER GENERATOR circuit, which provides a trigger pulse for the SWEEP GENERATOR circuit. The sweep waveform is amplified by the HORIZONTAL AMPLIFIER block, which drives the CRT's horizontal deflection plates. Thus, the electron beam is gradually swept left to right across the CRT screen at a uniform rate as the sweep waveform linearly increases toward its peak, and then the beam is abruptly returned to the left side of the CRT screen on the flyback (or retrace) portion of the sweep waveform.

To produce a stable waveform display on the CRT, the sweep waveform has to be synchronized with the VERTICAL SIGNAL so that each sweep cycle starts at the same point on the VERTICAL SIGNAL. The waveforms in Figure 11–31(b) illustrate this point. The TRIGGER GENERATOR circuit produces a pulse each time the VERTICAL SIGNAL passes through point x in its cycle. This trigger pulse initiates the start of a sweep waveform that increases linearly for a time, t_s, determined by the SWEEP GENERATOR circuit. Note that the second trigger pulse occurs during the first t_s interval and has no effect on the sweep waveform. Once the sweep waveform returns to its baseline at flyback, it will stay there until the next trigger pulse initiates a new sweep cycle.

As these waveforms show, the VERTICAL SIGNAL goes through a variation from point x to point y during the sweep time. If we assume that the electron beam makes one complete trip across the CRT during t_s, then the beam will produce the trace shown in Figure 11–31(c). This same trace is produced over and over for each sweep waveform cycle so that it shows up on the CRT as a continuously displayed waveform. The CRT display does not show the electron beam retracing back to the left side of the screen during the sweep waveform flyback. This is because a special circuit turns off the electron beam during the flyback interval. This action is called *retrace blanking*.

TRIGGER GENERATOR circuit As previously stated, this circuit has to produce a pulse, or spike, corresponding to a particular point on the input VERTICAL SIGNAL, which we will call the *trigger point*. In Figure 11–31 the trigger point is point x. In practice, the trigger point can be adjusted to any voltage level on the VERTICAL SIGNAL; this will be called the TRIG LEVEL. In addition, the trigger point can be selected to occur on either the positive or negative sloping portion of the VERTICAL SIGNAL; this is called the TRIG SLOPE.

To illustrate, refer to Figure 11–32, which shows a typical VERTICAL SIGNAL. Several points on this waveform have been labelled as possible trigger points. The TRIG LEVEL and TRIG SLOPE for each of these points is given in the figure.

We will now look at one possible way to implement a TRIGGER GENERATOR circuit that can produce a trigger pulse corresponding to any selectable trigger point on an input waveform. The circuit diagram is shown in Figure 11–33(a). The VERTICAL SIGNAL is applied to a voltage comparator and compared to a dc reference voltage (TRIG LEVEL) that is adjustable between −15 V and +15 V. The comparator can be operated as either an *inverting* comparator or a *non-inverting* comparator, depending on the position of the TRIG SLOPE selector switch. Let's examine both cases.

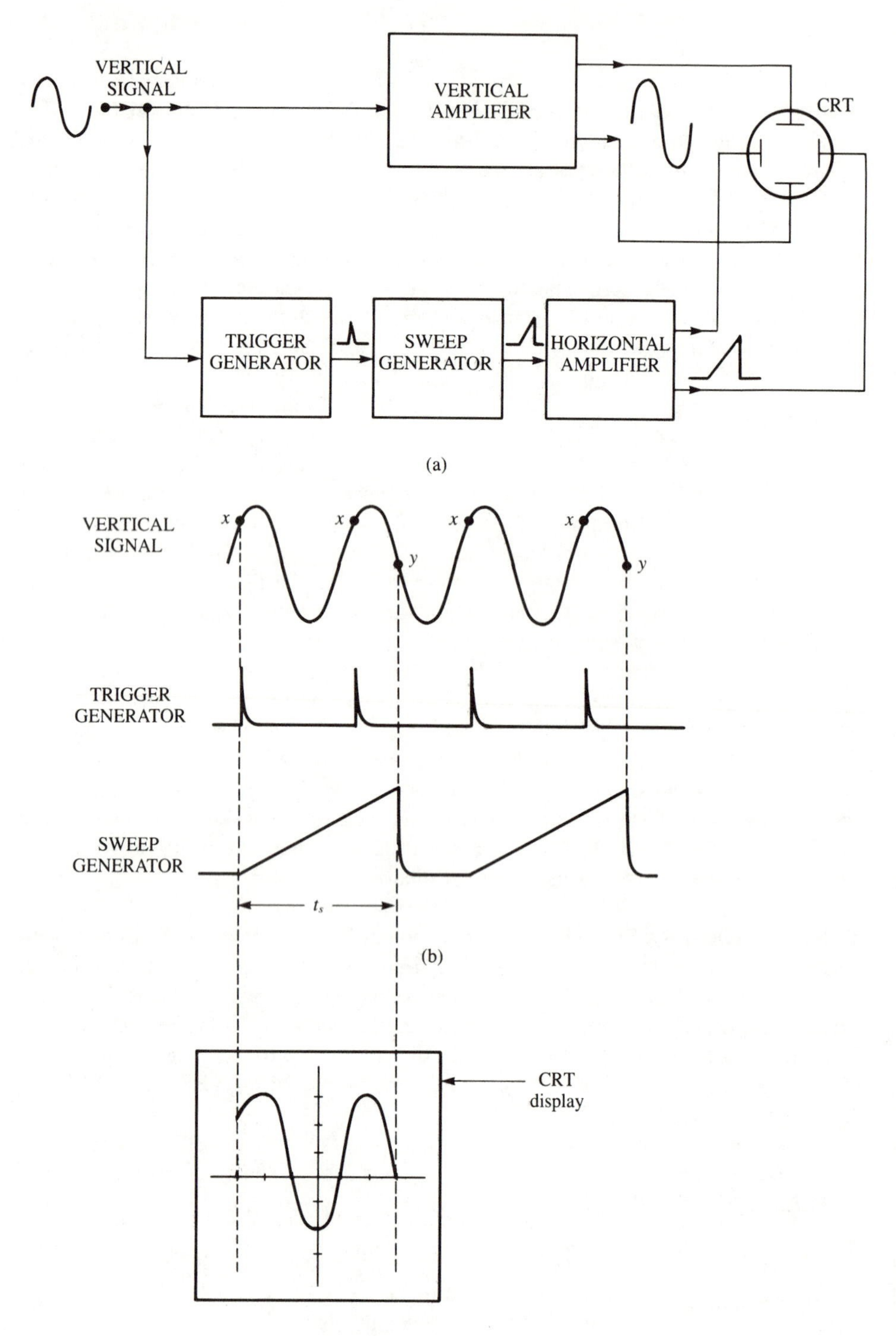

Figure 11–31 **(a) Simplified oscilloscope block diagram; (b) Typical waveforms; (c) CRT display**

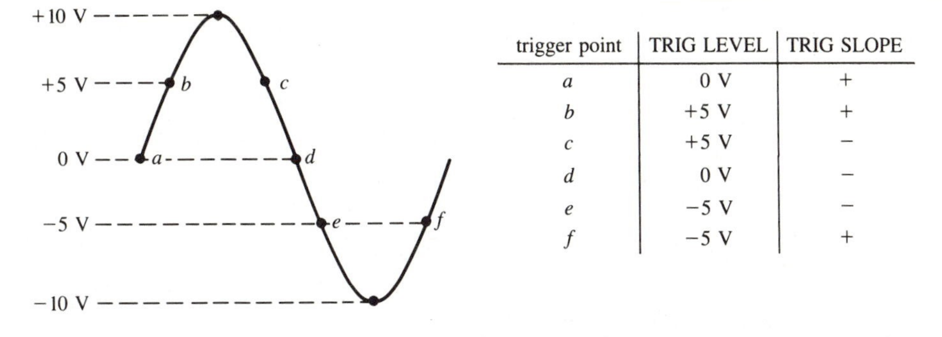

trigger point	TRIG LEVEL	TRIG SLOPE
a	0 V	+
b	+5 V	+
c	+5 V	−
d	0 V	−
e	−5 V	−
f	−5 V	+

Figure 11–32 **Various combinations of TRIG LEVEL and TRIG SLOPE settings**

Positive TRIG SLOPE With the TRIG SLOPE switch in the *up* position, VERTICAL SIGNAL is applied to the comparator's " + " input and TRIG LEVEL to the " − " input. If we assume TRIG LEVEL = +5 V, the circuit waveforms will appear as shown in Figure 11–33(b), where VERTICAL SIGNAL is a 20 V *p-p* sinewave. The comparator output, v_x, is a non-inverted, squared-up version of VERTICAL SIGNAL whose transitions occur when VERTICAL SIGNAL passes through +5 V (TRIG LEVEL). This signal is converted to narrow spikes by the *RC* differentiator. Thus, v_y consists of positive spikes that occur each time VERTICAL SIGNAL passes through +5 V in the positive-going direction. The TTL Schmitt trigger inverter converts the spikes to rectangular pulses. These will be the TRIGGER PULSES that we will feed to our triggered sweep circuit a little later.

Negative TRIG SLOPE With the TRIG SLOPE selector switch in the *down* position, the VERTICAL SIGNAL is connected to the comparator's " − " input and the TRIG LEVEL is connected to the " + " input, so that the comparator now operates as an inverting comparator with the resulting waveforms shown in Figure 11–33(c). The positive spikes at v_y now occur when VERTICAL SIGNAL passes through +5 V in the negative-going direction.

Example 11.24 What approximate amplitudes would you expect for the v_x and v_y waveforms in Figure 11–33(a)?

Solution: The comparator output, v_x, should switch between $-V_{SAT}$ and $+V_{SAT}$, which should be approximately −13.5 V and +13.5 V, respectively.

The positive spike at v_y will theoretically have an amplitude of +27 V since v_x makes a total transition of +27 V. However, the amplitude of the spike will depend on the relationship between the time constant of the differentiator circuit and the rise time, t_r, of v_x, which depends on the op-amp slew rate. Ordinarily, $\tau = RC$ will be kept small to produce narrow spikes, and t_r will be relatively slow due to the slew-rate limitation. This combination usually reduces the spikes at v_y to well below 27 V.

Example 11.25 What will happen if TRIG LEVEL is adjusted to +12 V in Figure 11–33(a)?

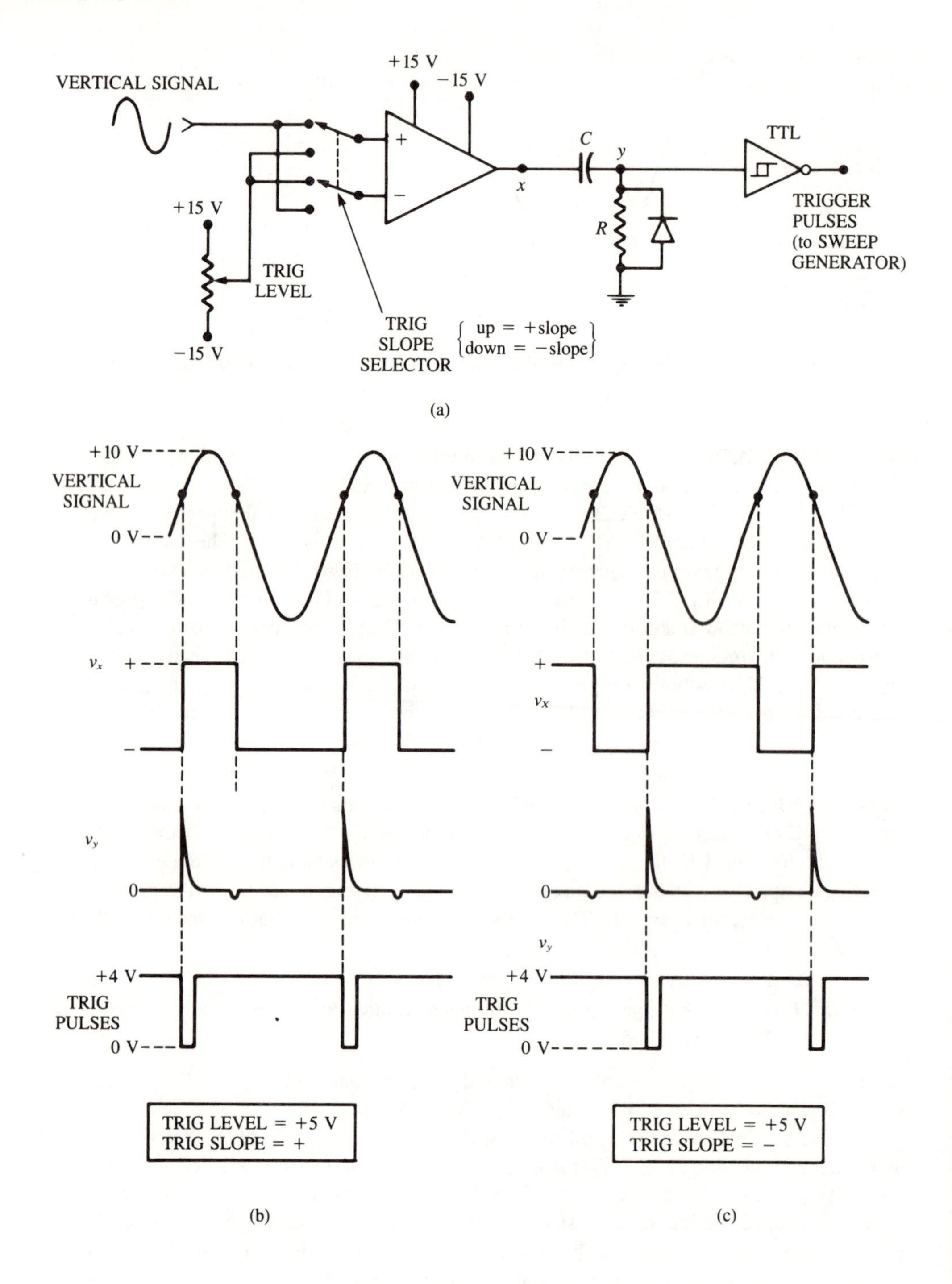

Figure 11–33 **(a) TRIGGER GENERATOR circuit with adjustable TRIG LEVEL and TRIG SLOPE; (b) Typical waveforms for positive TRIG SLOPE; (c) Negative TRIG SLOPE**

Solution: With the comparator reference voltage set at +12 V, its output will never switch states because VERTICAL SIGNAL will never reach the +12 V trigger point. Thus, v_x will be constant at either +13.5 V or −13.5 V, depending on the TRIG SLOPE setting. There will be no spikes at v_y, and the TRIGGER PULSE output will be a constant HIGH.

SWEEP GENERATOR circuit Now that we have a circuit that generates TRIGGER PULSES corresponding to a specific point on the repetitive VERTICAL SIGNAL, we will use these pulses to synchronize the generation of sweep waveforms. The circuit we will use is shown in Figure 11–34(a). The input signal to this circuit is the TRIGGER PULSE output from the TRIGGER GENERATOR circuit of Figure 11–33. All the various parts of this circuit are familiar to us. Transistor Q_1 produces a linear sweep waveform across C. Transistor Q_2 keeps v_c at approximately 0 V when Q_2 is ON. The UJT discharges C when v_C reaches V_P and produces a pulse at B_1 to clear the *JK* FF. This FF output $\overline{Q}$ turns Q_2 ON and OFF. The complete ciruit operates thusly:

1. Assume that initially the FF is in the $Q = 0/\overline{Q} = 1$ state, so that Q_2 is ON, $v_C \approx 0$ V, and the UJT is OFF.
2. The negative-going transition of the next TRIGGER PULSE will set the FF to the $Q = 1/\overline{Q} = 0$ state. This turns Q_2 OFF so that C will be allowed to charge linearly from Q_1's collector current.
3. When v_C reaches the value of the UJT's trigger voltage, V_P, the UJT turns ON and C begins to quickly discharge through the UJT and the 47 Ω resistor. The spike across the resistor is fed to the Schmitt trigger inverter, which produces a LOW pulse to clear the FF back to $Q = 0/\overline{Q} = 1$.
4. The HIGH level at $\overline{Q}$ causes Q_2 to turn ON. This action will allow C to discharge all the way to 0 V through Q_2. The UJT turns OFF and the circuit is back to its initial state (see step 1).

Example 11.26 Assume that VERTICAL SIGNAL is a 1 kHz, 20 V *p-p* sinewave. Sketch the various circuit waveforms for a TRIG LEVEL setting of −5 V and TRIG SLOPE = +.

Solution: The VERTICAL SIGNAL and TRIGGER PULSE waveforms are sketched in Figure 11–34(b) such that one TRIGGER PULSE occurs each time the sinewave passes through −5 V in the positive-going direction (point *x*).

The negative-going edge of the first TRIGGER PULSE will cause $\overline{Q}$ to go from HIGH to LOW. This action turns Q_2 OFF and allows C to begin charging, thereby initiating a sweep waveform cycle. To find the duration of the sweep waveform we first calculate the UJT's V_P as

$$V_P = 0.6 \text{ V} \times 10 \text{ V} + 0.5 \text{ V}$$
$$= 6.5 \text{ V}$$

Note that the V_{BB} = 10 V for the UJT can be derived from the 10 V zener diode.

Next, we calculate the charging rate.

$$I_{E_1} = \frac{15\text{ V} - 10\text{ V} - 0.7\text{ V}}{4.3\text{ k}\Omega} = 1\text{ mA}$$

$$I_{C_1} \approx I_{E_1} = 1\text{ mA}$$

$$\frac{dv_c}{dt} = \frac{1\text{ mA}}{0.33\ \mu\text{F}} = 3\text{ V/ms}$$

The capacitor will charge from 0 V to 6.5 V at this rate. Therefore,

$$t_s = \frac{6.5\text{ V}}{3\text{ V/ms}} = 2.17\text{ ms}$$

After 2.17 ms, $v_C = V_P$; the UJT turns ON, producing the clearing pulse for the FF bringing $\overline{Q}$ back HIGH. This action turns on Q_2 and holds v_c at 0 V.

v_C will stay at 0 V until the next TRIGGER PULSE sends $\overline{Q}$ back LOW and initiates a new sweep cycle. Note that the second and third TRIGGER PULSES had no effect because the FF was already in the $Q = 1/\overline{Q} = 0$ state.

Example 11.27 What portion of the VERTICAL SIGNAL will be displayed on the CRT screen?

Solution: The sweep waveform has $t_s = 2.17$ ms, which is a little more than two periods of the VERTICAL SIGNAL. The portion of the VERTICAL SIGNAL that occurs during the t_s interval (between points x and y) will be traced on the CRT.

Example 11.28 What happens to the electron beam in the CRT while the sweep waveform is resting at 0 V?

Solution: When the sweep waveform rapidly flies back to 0 V, the electron beam returns to the left side of the CRT screen and remains there until the next TRIGGER PULSE initiates the next sweep cycle. For the waveforms in Figure 11–33(b) this time duration is about 0.83 ms. In most modern scopes the electron beam would be turned off (blanked) during this time; otherwise, you would see a bright vertical line on the left side of the CRT, as the beam is deflected vertically by the VERTICAL SIGNAL variations but does not move horizontally.

The complete triggered sweep generator circuit (Figures 11–33 and 34) is a practical circuit that can be easily built and tested. A good test of its operation would be to feed the sweep waveform into an oscilloscope's horizontal amplifier input jack, and the VERTICAL SIGNAL to the scope's vertical input jack. Then, set the scope for *X-Y* operation (i. e., disengage the scope's time-base) and observe the resultant display. If your sweep circuit is working properly, it will provide the scope's horizontal time-base and should produce a stable display of the VERTICAL SIGNAL.

We will examine more aspects of this interesting circuit in the end-of-chapter problems.

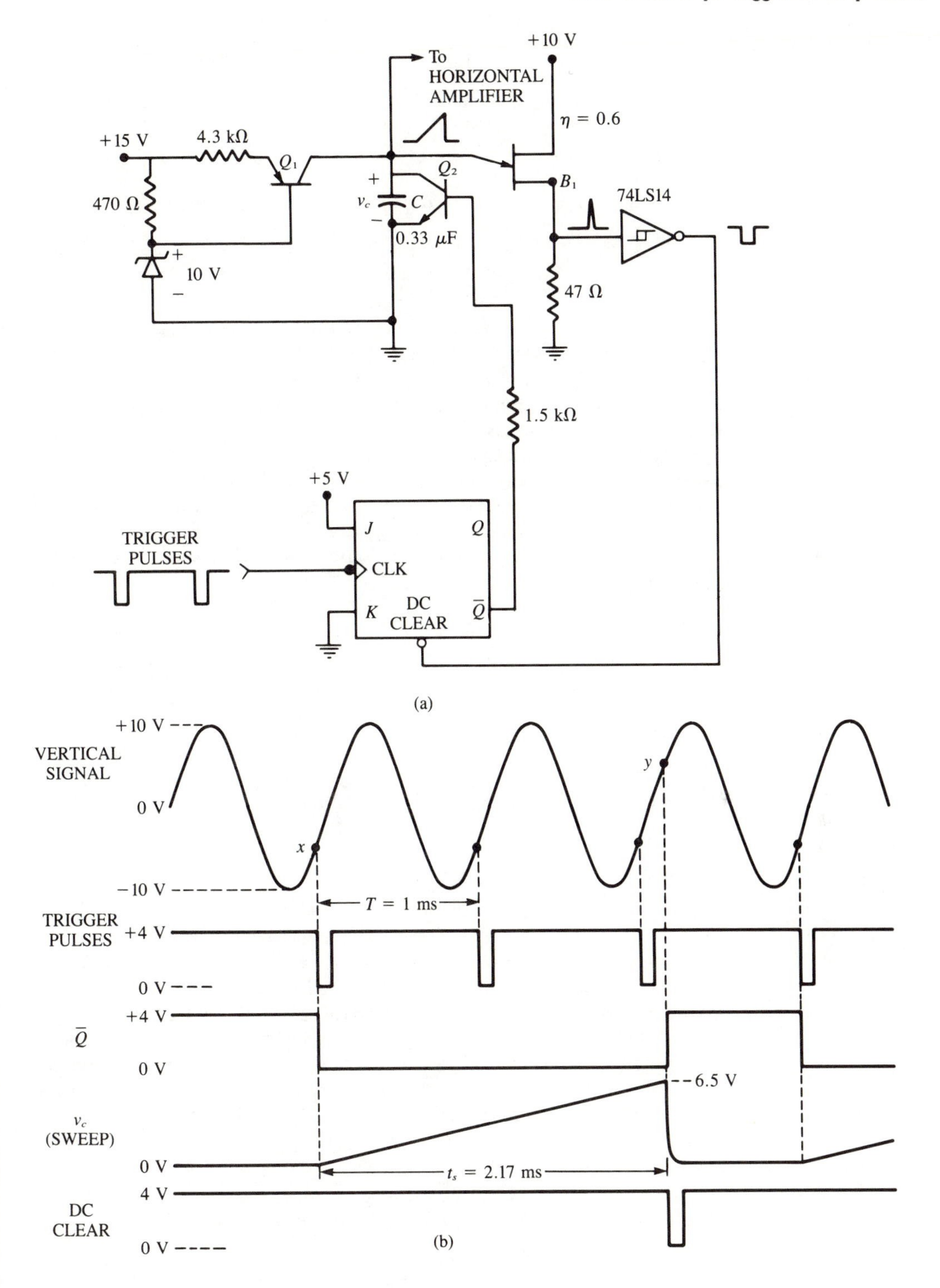

Figure 11–34 **(a) SWEEP GENERATOR synchronized by TRIGGER PULSES; (b) Waveforms for Example 11.26**

Questions

Sections 11.1–11.2

11.1 Consider the circuit in Figure 11–2. Use $V_{BB} = 20$ V, $R_{BB} = 8$ kΩ, $\eta = 0.65$, and $I_V = 2$ mA.

(a) Determine V_P.

(b) Determine I_E when $V_{EE} = 20$ V. [Assume R_{B_1}(ON) = 40 Ω].

(c) Determine I_E when V_{EE} is decreased from 20 V to 5 V.

(d) How low does V_{EE} have to be reduced in order to turn the UJT OFF?

Sections 11.3–11.4

11.2 Sketch the capacitor voltage waveform for a UJT oscillator with the following values: $V_{BB} = 20$ V, $R = 22$ kΩ, $C = 0.1$ μF, and $\eta = 0.5$.

11.3 Indicate the effect each of the following changes will have on the amplitude and frequency of the waveform of Problem 11.2: (a) decrease R; (b) increase η; (c) decrease V_{BB}; (d) increase C

11.4 The UJT of Problem 11.2 has $I_P = 10$ μA and $I_V = 2$ mA. Leaving $C = 0.1$ μF, determine the oscillator's total range of frequencies that can be obtained by varying R.

11.5 A technician uses a dual-trace oscilloscope to observe the v_E and v_{B_1} of the circuit in Figure 11–8. He finds that the circuit will not oscillate when the scope is connected to v_E. When the scope is disconnected from v_E, the correct waveform is observed at v_{B_1}. Explain his observations.

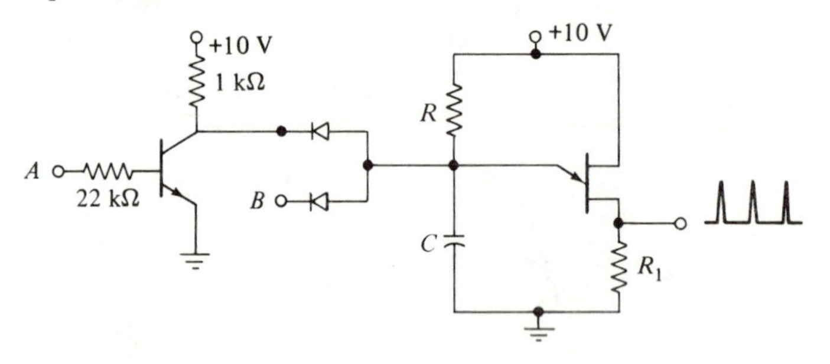

Figure 11–35

11.6 The circuit in Figure 11–35 is a *gated* UJT oscillator. Control inputs A and B determine whether the circuit will oscillate or not. Inputs A and B can be either 0 V or 10 V. Under what input conditions will oscillations occur?

11.7 Consider the circuit in Figure 11–36. The oscillator is supposed to produce pulses at a rate of 10 kHz. However, when power is applied to the circuit, the circuit fails to oscillate. The technician testing the circuit notices that when he turns the power off, the circuit temporarily oscillates as the power supply voltage drops to zero. Can you explain these observations and determine what is wrong with the circuit?

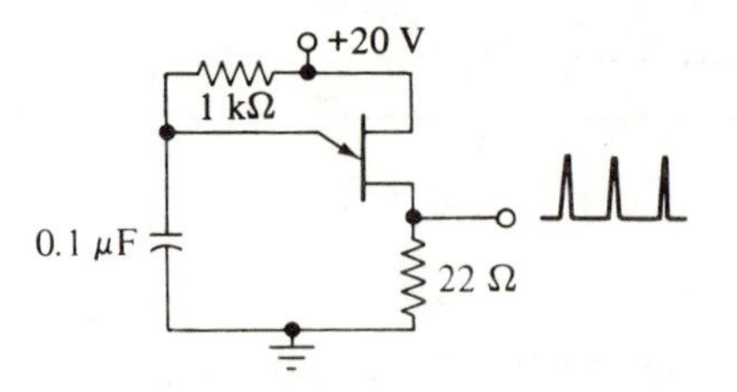

Figure 11–36

11.8 The circuit in Figure 11–37 is a *pulse-width detector*. Its function is to produce an output spike any time an input pulse occurs that is wider than a certain threshold value. The input pulses have randomly varying pulse widths. Describe the circuit operations, and calculate the threshold value of input pulse width needed to produce an output spike.

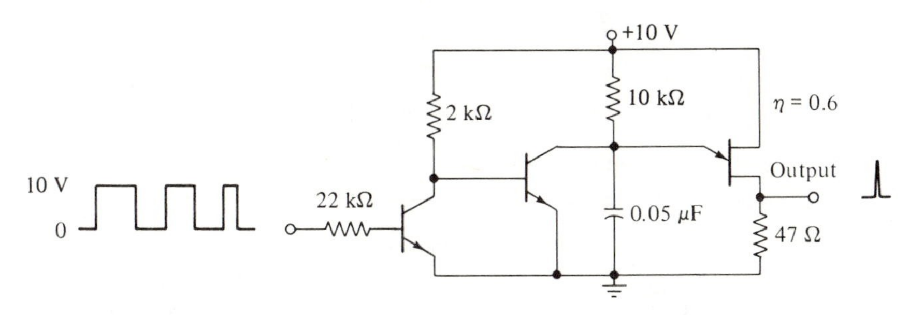

Figure 11–37

Section 11.5

11.9 Determine the frequency of the PUT oscillator in Figure 11–38. Sketch the waveforms of v_A and v_K showing approximate amplitudes.

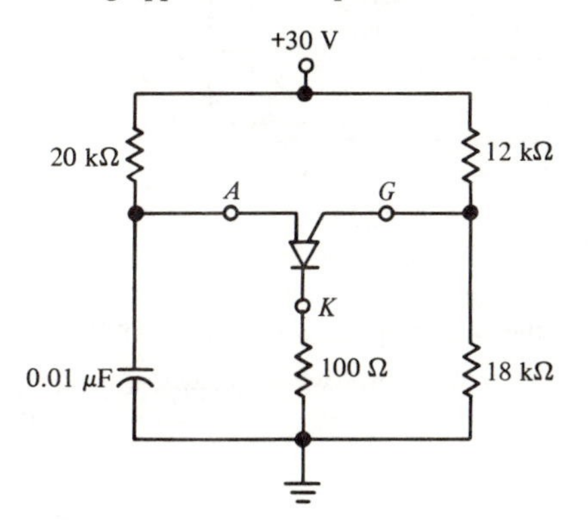

Figure 11–38

11.10 What are the advantages of a PUT over a UJT?

11.11 PUTs are particularly useful for providing long time delays because they can have I_P values as low as 0.1 μA, which allows very large values of R. Figure 11–39 shows how a PUT is used to produce a positive step-voltage output (X) after a long time delay following the opening of switch S_1. Determine values of R and C to produce a time delay of 3 min. The value of C is to be no larger than 50 μF. Assume $I_P = 0.1$ μA.

11.12 Show how a PUT oscillator can be used with a FF, a OS, and a logic gate to produce the waveforms of Figure 11–40. These waveforms could be control signals needed in a particular digital system.

Section 11.6

11.13 Derive the values of t_{ch} and t_{dch} (in terms of RC) for the circuit of Figure 11–13.

11.14 The circuit in Figure 11–41(a) uses a CMOS 4093 Schmitt trigger NAND gate in an oscillator configuration. The 4093 has the same V_{LT} and V_{UT} values as the 40106.

(a) Determine the frequency of the v_{OUT} waveform when CONTROL = 0.

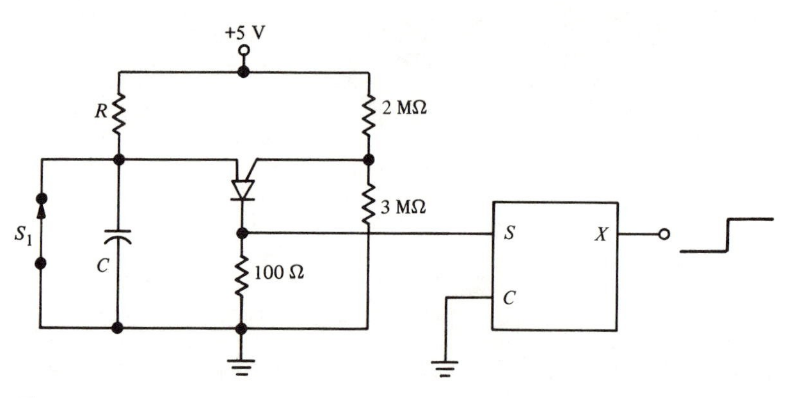

Figure 11–39

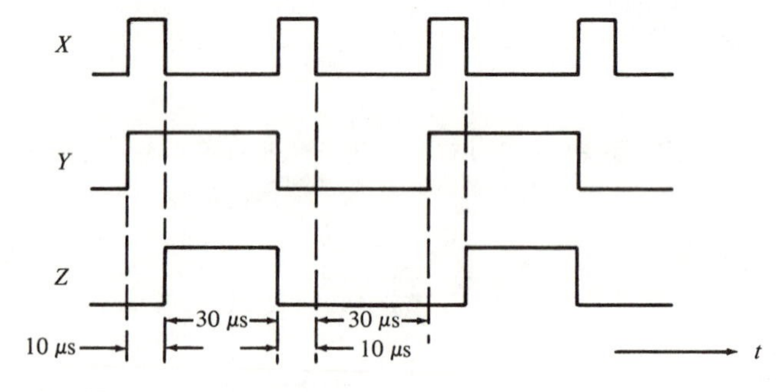

Figure 11–40

(b) Determine the v_{OUT} frequency when CONTROL = 1.

(c) Sketch the v_{OUT} waveform when the waveform of Figure 11–41(b) is used for CONTROL.

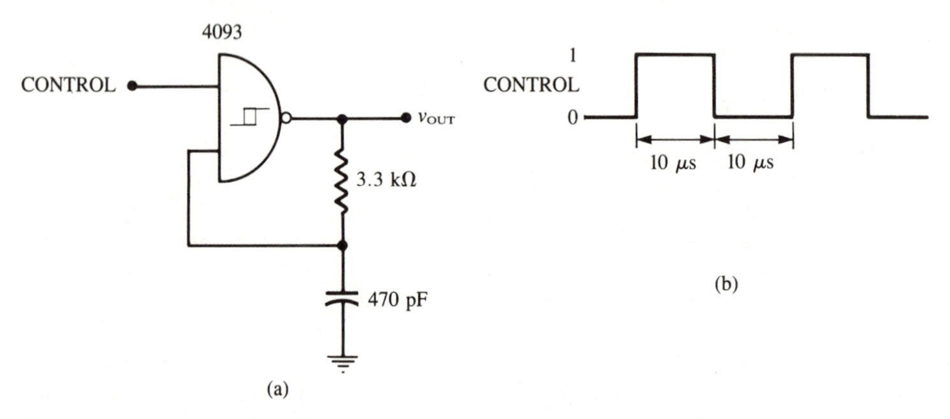

Figure 11–41

11.15 Derive the formula for the period of the op-amp oscillator (Equation 11–6).

11.16 The v_{OUT} waveform in Figure 11–14 has a duty cycle of approximately 50 percent because of the symmetry of the capacitor charge and discharge intervals. One way to change the duty cycle is to replace R by the parallel combination shown in Figure 11–42. Insert this in place of R in Figure 11–14, then determine the v_{CAP} and v_{OUT} waveforms for $R_1 = 100$ kΩ, $R_2 = 50$ kΩ, and $C = 0.02$ μF. The small diode forward-voltage drops can be neglected.

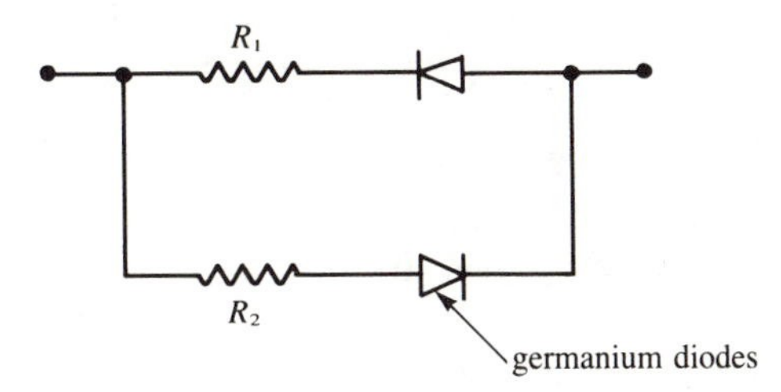

Figure 11–42

11.17 Modify the oscillator of Figure 11–14 so that the v_{OUT} waveform levels are suitable for driving a TTL device. Your modification should *not* affect the oscillator frequency.

Sections 11.7–11.8

11.18 Figure 11–43 shows how several 555 ICs can be connected to provide sequential timing pulses. Each 555 is connected as a OS whose OUTPUT pulse duration is controlled by its R_T and C_T values. Operation begins by applying a START pulse to the input of the first timer. At the end of its timing cycle it triggers the second timer, which in turn triggers the third timer. Draw the waveforms at the OUTPUTs and TRIGGER inputs of each timer. Assume R_{IN} at each TRIGGER input is very high, and R_{OUT} of OUTPUTs is very low.

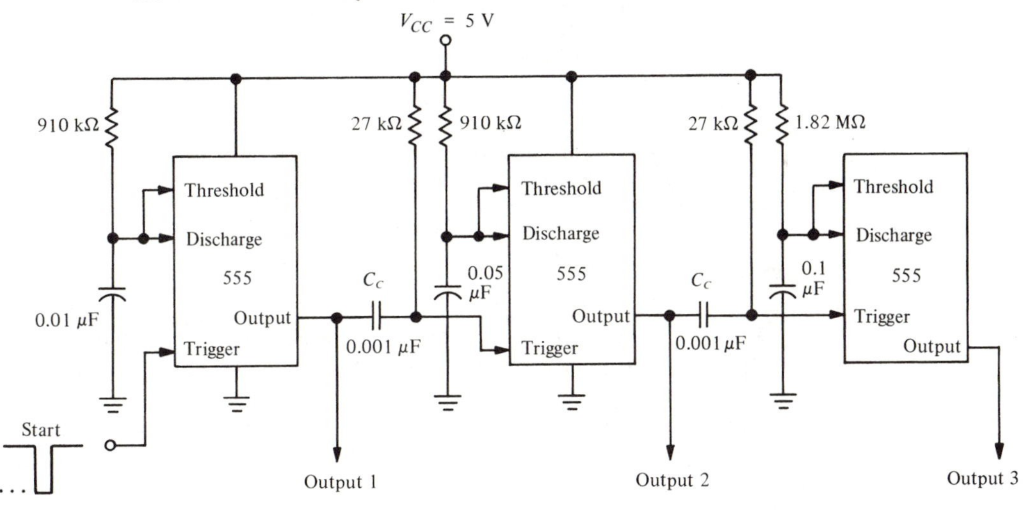

Figure 11–43

11.19 Why are the coupling capacitors (C_C) needed in the circuit of Figure 11–43?

11.20 Prove that $t_P = 1.1\ R_T C_T$ for the 555 one-shot, independent of the value of V_{CC}.

11.21 Determine the maximum t_P that can be obtained from a 555 one-shot operating at $V_{CC} = 10$ V and using $C_T = 1$ μF.

11.22 Refer to the circuit of Figure 11–19. The TRIGGER input waveform shows a positive-going spike that reaches a peak of +20 V. This large positive voltage might possibly damage the 555 comparator circuit. Where could you place a *conventional* diode to clip this positive spike? (HINT: why can't one end of the diode be connected to ground?)

11.23 The e_{IN} pulse in Figure 11–19 has a duration of 100 ms and $t_r = t_f = 10\ \mu s$. The one-shot uses $R_T = 4.7\ k\Omega$ and $C_T = 0.2\ \mu F$. Two of the following sets of values for R_1 and C_1 will *not* work properly. Explain why.

(a) $R_1 = 10\ k\Omega,\ C_1 = 0.5\ \mu F$

(b) $R_1 = 2.2\ k\Omega,\ C_1 = 220\ pF$

(c) $R_1 = 10\ k\Omega,\ C_1 = 2000\ pF$

11.24 Which case in Problem 11.23 will produce an output t_P that is longer than expected? Explain.

Section 11.9

11.25 Derive the formula for the period of the 555 oscillator (Equation 11–10).

11.26 Select values for R_A, R_B, and C_T to produce a 10 V, 5 kHz square wave from a 555 oscillator.

11.27 Figure 11–44 shows a modified 555 oscillator circuit where a diode has been placed across R_B. The diode is germanium, so its forward-voltage drop can be neglected. Derive an equation for the period of the oscillations $[T = 0.7(R_A + R_B)C_T]$. Then show that the output will be an approximate square wave when $R_A = R_B$.

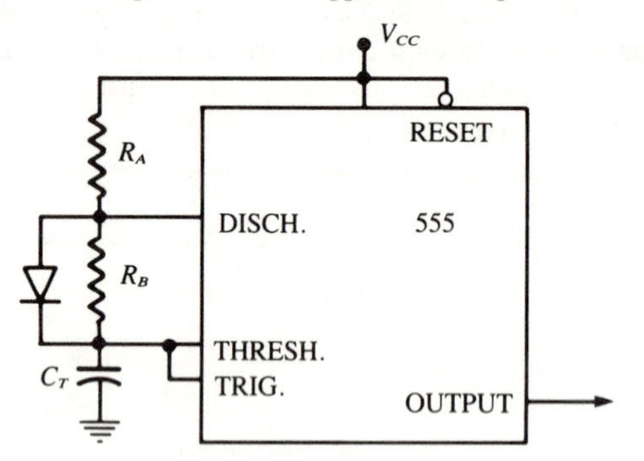

Figure 11–44

11.28 The circuit in Figure 11–44 can be used to produce duty cycles above or below 50 percent by choosing $R_A > R_B$ or $R_A < R_B$.

(a) Select values for R_A, R_B, and C_T to produce a 10 kHz output with a 25 percent duty cycle.

(b) Repeat for a 75 percent duty cycle.

11.29 A technician designs the circuit of Figure 11–44 to produce a 10 V, 1 kHz square wave using $R_A = R_B = 3.6\ k\Omega$ and $C_T = 0.2\ \mu F$. She constructs the circuit and observes the v_c and v_{OUT} waveforms shown in Figure 11–45. Which of the following could produce these erroneous waveforms?

(a) The diode is shorted.

(b) R_B is open.

(c) The diode is open.

(d) The diode is in backwards.

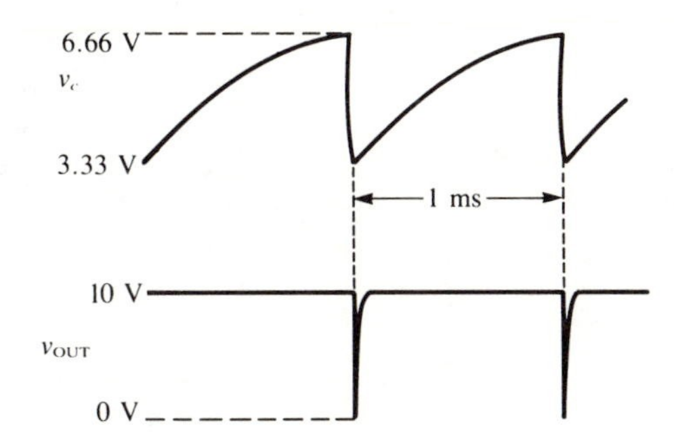

Figure 11–45

Section 11.10

11.30 Refer to Figure 11–22. Sketch the Q output waveform when $C = 1500$ pF and input X is a 50 kHz square wave.

Section 11.11

11.31 Sketch a sweep waveform with the following characteristics: baseline = 2 V; peak value = +18 V; sweep time = 4 ms; flyback time ≈ 0.

11.32 Determine the value of charging current required to produce the sweep waveform of Problem 11.31, using a 2 μF capacitor.

Section 11.12

11.33 The circuit of Figure 11–26 uses the following values: $V_{EE} = 24$ V, $V_{BB} = 15$ V, $R_E = 33$ kΩ, $C = 0.005$ μF, $R_B = 68$ kΩ, and $h_{FE} = 100$ for both transistors. The e_{IN} signal has a frequency of 4 kHz, a 20 percent duty cycle, and a +4 V amplitude. Determine the v_C waveform.

11.34 Repeat Problem 11.33 with $C = 0.002$ μF.

11.35 Change the value of R_E for the circuit of Problem 11.33 so that the sweep waveform has an amplitude of approximately 8 V.

11.36 (a) Determine the minimum e_{IN} pulse width that should be used for the circuit of Problem 11.33. Show what will happen to the v_C waveform if e_{IN} has a t_P smaller than this value.

(b) How can the circuit be changed so that a smaller t_P can be used without affecting the sweep waveform?

11.37 Determine the minimum R_L that the sweep output of Problem 11.33 can drive without significantly affecting the sweep linearity.

11.38 Describe the effects each of the following would have on the amplitude and linearity of the sweep waveform produced by the circuit of Figure 11–29. Sketch the waveform to support your answers.

(a) decrease of C

(b) increase of V_z

(c) a collector-emitter short in Q_1

(d) a leaky Q_2

(e) a collector-base open in Q_2

(f) zener diode in backwards

Section 11.13

11.39 The UJT of Figure 11–30 has $I_P = 5$ μA and $I_V = 5$ mA. Determine the range of sweep time values that can be obtained by varying R_E over its acceptable range.

11.40 What will v_C look like if R_E is made too large? Too small?

11.41 Use $\eta = 0.8$ and determine the v_C waveform for the circuit of Figure 11–30.

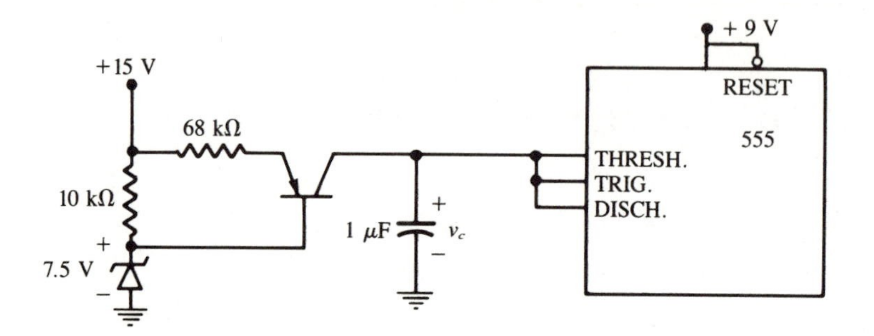

Figure 11–46

11.42 Figure 11–46 shows a free-running sweep generator that uses a 555 timer. Draw the complete v_C waveform assuming that C is initially at 0 V.

Section 11.14

11.43 Sketch a 12 V *p-p* sinewave. Indicate the following possible trigger points on the waveform:

TRIG LEVEL	TRIG SLOPE
0 V	+
−3 V	−
+5 V	−
+8 V	+
−2 V	+

11.44 Sketch the v_x, v_y, and TRIG PULSES for the circuit of Figure 11–33(a) for TRIG LEVEL = −5 V and TRIG SLOPE = + . Repeat for TRIG LEVEL = −5 V and TRIG SLOPE = − .

11.45 Why do you think a Schmitt trigger inverter is used in Figure 11–33 instead of a standard inverter?

11.46 The Schmitt trigger in Figure 11–33 has V_{LT} = 0.9 V and V_{UT} = 1.7 V. Under certain conditions the spikes at v_y could have insufficient amplitude to activate the Schmitt trigger. Describe some of these conditions; for each one, discuss what can be done to eliminate the problem.

11.47 Refer to Figure 11–34. Assume that TRIG LEVEL = +2 V and TRIG SLOPE = − . Change R_E to 22 kΩ and sketch the v_C waveform in relation to the VERTICAL SIGNAL. Then sketch how the display will appear on the CRT.

11.48 In the sweep circuit of Figure 11–34, the lower limit on R_E is *not* dependent on the I_V of the UJT. In other words, Q_1's constant collector current can be made larger than I_V and the circuit will still operate properly. Explain why this is so.

11.49 Assume that the triggered SWEEP GENERATOR of Figure 11–34 is used to provide the horizontal time-base in an oscilloscope. A technician is using the scope to display sinewaves of various frequencies, and he finds that the trace on the CRT never shows more than *one* cycle of the VERTICAL SIGNAL regardless of the frequency of this signal. (Illustrated in Figure 11–47.) Which of the following could cause this malfunction?

(a) Q_2 has a collector-base open.

(b) The K input of the FF has become disconnected from ground.

(c) The η of the UJT is much lower than 0.6.

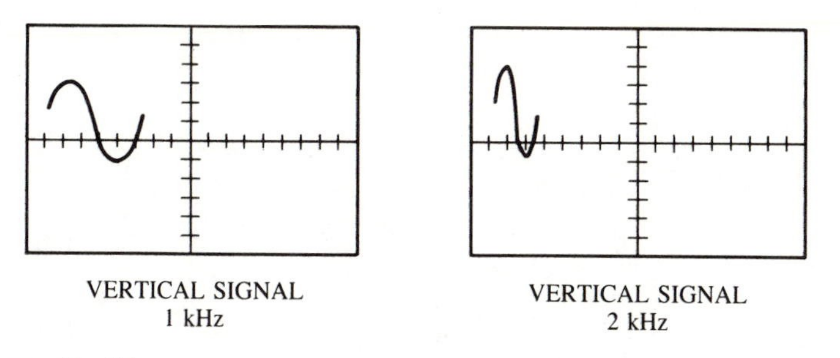

Figure 11–47

11.50 Show how each of the following malfunctions in the circuit of Figure 11–34 will effect the CRT display of the VERTICAL SIGNAL.

(a) Q_2 has an open collector.

(b) The 47 Ω resistor is open.

(c) Q_1 has an emitter-collector short.

Appendices

appendix A: binary number system

In digital technology there are many number systems in use. The most common systems are the *decimal, binary, octal,* and *hexadecimal*. The decimal system is clearly the most familiar to us since it is a tool we use every day. Examining some of its characteristics will help us to better understand the other systems.

The decimal system is composed of *ten* numerals or symbols which are commonly referred to as *digits*. These ten symbols are 0, 1, 2, 3, 4, 5, 6, 7, 8, and 9 and as we know can be used to express any quantity. The decimal system, also called the *base-10* system since it has ten digits, evolved naturally from the fact that people have ten fingers. In fact, the word *digit* is the Latin word for *finger*.

The decimal system is a *positional value system;* that is, the value of a digit depends on its position. For example, consider the decimal number 453. We know that the digit 4 actually represents 4 *hundreds,* that the 5 represents 5 *tens,* and that the 3 represents 3 *units*. In essence, the 4 carries the most weight of the three digits. For this reason, it is referred to as the *most significant digit (MSD)*. The 3 carries the least weight and is called the *least significant digit (LSD)*.

Consider as another example the number 27.35. This number is actually equal to 2 tens plus 7 units plus 3 tenths plus 5 hundredths, or in other words, $2 \times 10 + 7 \times 1 + 2 \times 0.1 + 5 \times 0.01$. The decimal point is used to separate the integer and the fractional parts of the number.

More rigorously, the various positions relative to the decimal point carry weights which can be expressed as powers of *ten*. This is illustrated in Figure A–1, where the number 2745.214 is represented. The decimal point separates the positive powers of ten from the negative powers of ten. The number 2745.214 is thus equal to

$$(2 \times 10^{+3}) + (7 \times 10^{+2}) + (4 \times 10^{1}) + (5 \times 10^{0}) + (2 \times 10^{-1}) + (1 \times 10^{-2}) + (4 \times 10^{-3})$$

In general, any number is simply the sum of the products of each digit value times its positional value.

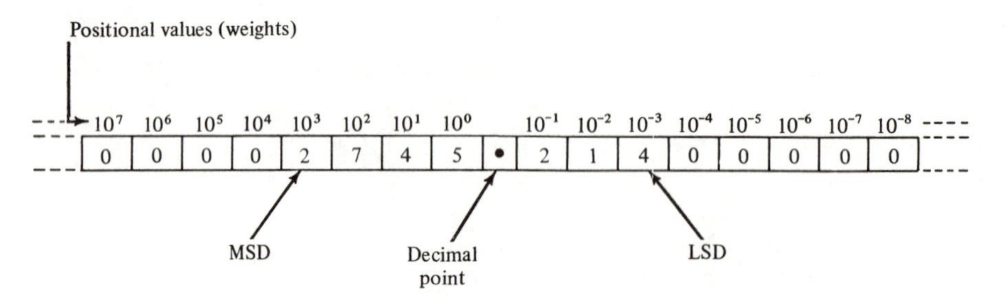

Figure A–1 **Decimal position values as powers of ten**

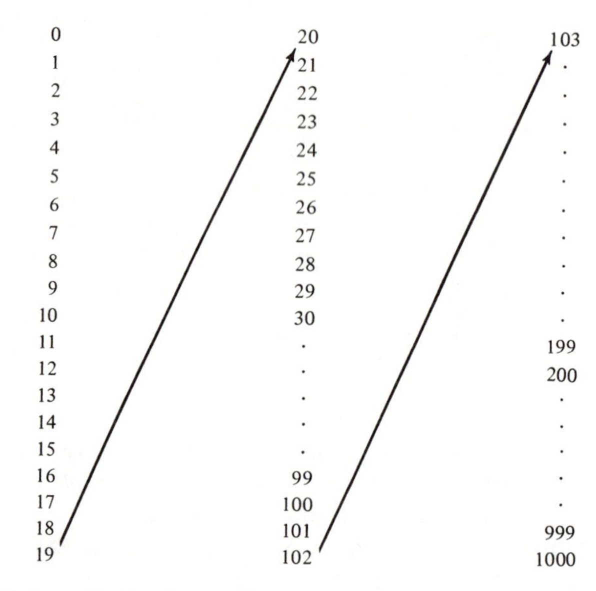

Figure A–2 **Decimal counting**

Decimal Counting

When counting in the decimal system, we start with 0 in the units position and take each symbol (digit) in progression until we reach 9. Then we add a 1 to the next higher position and start over with 0 in the first position (see Figure A–2). This process continues until the count of 99 is reached. Then, we add a 1 to the third position and start over with 0's in the first two positions. This same pattern is repeated as high as we wish to count.

It is important to note that in decimal counting the units position (LSD) changes upward with each step in the count, the tens position changes upward every 10 steps in the count, the hundreds position changes upward every 100 steps in the count, and so on.

Another characteristic of the decimal system is that using only *two* decimal places we can count through $10^2 = 100$ different numbers (0 through 99).* With *three* places we can count through 1000 numbers (0 through 999), and so on. In general, with N places or digits we can count through 10^N different numbers starting with and including zero. The largest number will always be $10^N - 1$.

Binary System

Unfortunately, the decimal number system does not lend itself to convenient implementation in digital systems. For example, it is very difficult to design electronic equipment so that it can work with ten different voltage levels (each one representing one of the decimal characters 0 through 9). On the other hand, it is very easy to design simple, accurate electronic circuits that operate with only *two* different voltage levels. For this reason the binary (base-2) number system is used in all digital systems.

In the binary system there are only two symbols or possible digit values—0 and 1. Even so, this base-2 system can be used to represent any quantity that can be represented by the decimal or other number systems. In general, it will take a greater number of binary digits to express a given quantity.

All of the statements made earlier concerning the decimal system are equally applicable to the binary system. The binary system is also a positional value system. Each binary digit has its own value or weight expressed as powers of *two*. This system is illustrated in Figure A–3. Here, places to the left of the *binary point* (counterpart of the decimal point) are positive powers of two, and places to the right are negative powers of two. The number 1011.101 is represented in the figure. To find its equivalent in the decimal system, we simply take the sum of the products of each digit value (0 or 1) times its positional value.

$$\begin{aligned} 1011.101_2 &= (1 \times 2^3) + (0 \times 2^2) + (1 \times 2^1) + (1 \times 2^0) \\ &\quad + (1 \times 2^{-1}) + (0 \times 2^{-2}) + (1 \times 2^{-3}) \\ &= 8 + 0 + 2 + 1 + 0.5 + 0 + 0.125 \\ 1011.101_2 &= 11.625_{10} \end{aligned}$$

Notice in this operation that subscripts (2 and 10) are used to indicate the base in which the particular number is expressed. This convention is used to avoid confusion whenever more than one number system is being employed.

In the binary system the term *binary digit* is often shortened to the common term *bit*. Thus, in the number expressed in Figure A–3 there are 4 bits to the left of the binary point representing the integer part of the number and 3 bits to the right of the binary point representing the fractional part. The *most significant bit (MSB)* is the leftmost bit, which is a 1. The *least significant bit (LSB)* is the rightmost bit, which is a 1. These are indicated in Figure A–3.

---

*Zero is counted as a number.

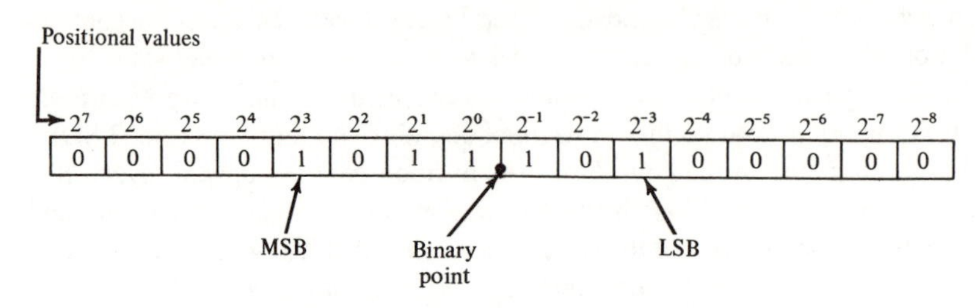

Figure A–3 **Binary position values as powers of two**

Binary Counting

When counting in the binary system we start with 0 in the first position (units position) and then proceed to 1 in the first position. Then, we add a 1 to the next-higher position (2^1, or twos, position) and start over with 0 in the first position. This procedure is illustrated in Figure A–4. When the count of 11_2 is reached, we add a 1 to the third position and start over with 0's in the first two positions. The process is followed continuously.

The binary counting sequence has an important characteristic, as shown in Figure A–4. The units bit (LSB) changes from 0 to 1 or 1 to 0 with *each* count. The second bit (twos position) stays at 0 for 2 counts, goes to 1 for 2 counts, then to 0 for 2 counts, and so on. The third bit (fours position) stays 0 for 4 counts, then 1 for 4 counts, and so on. The fourth bit (eights position) stays at 0 for 8 counts, then at 1 for 8 counts. If we wanted to count further, we would add more places, and this pattern would continue with 0's and 1's alternating in groups of 2^N. For example, using a fifth binary place, the fifth bit would alternate 16 0's, then 16 1's, and so on.

Binary				Decimal equivalent
Weights → $2^3 = 8$	$2^2 = 4$	$2^1 = 2$	$2^0 = 1$	
			0	0
			1	1
		1	0	2
		1	1	3
	1	0	0	4
	1	0	1	5
	1	1	0	6
	1	1	1	7
1	0	0	0	8
1	0	0	1	9
1	0	1	0	10
1	0	1	1	11
1	1	0	0	12
1	1	0	1	13
1	1	1	0	14
1	1	1	1	15
			LSB	

Figure A–4 **Binary counting sequence**

As we saw for the decimal system, it is also true of the binary system that using N bits or places we can go through 2^N counts. For example, with 2 bits we can go through $2^2 = 4$ counts (00_2 through 11_2), with 4 bits we can go through $2^4 = 16$ counts (0000_2 through 1111_2), and so on. The last count will always be all 1's and is equal to $2^N - 1$ in the decimal system. For example, using 4 bits, the last count is $1111_2 = 2^4 - 1 = 15_{10}$.

appendix B: digital IC data sheets

The data sheets on the following pages appear through the courtesy of Fairchild Semiconductors, Inc., and RCA Corporation.

FAIRCHILD TTL/SSI • 9N00/5400, 7400

QUAD 2-INPUT NAND GATE

LOGIC AND CONNECTION DIAGRAM

DIP (TOP VIEW)

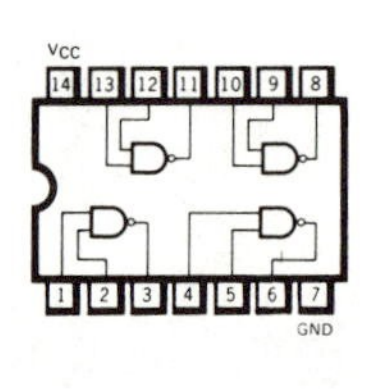

FLATPAK (TOP VIEW)

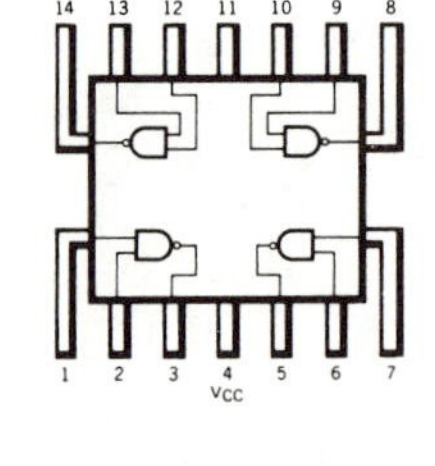

Positive logic: $Y = \overline{AB}$

SCHEMATIC DIAGRAM
(EACH GATE)

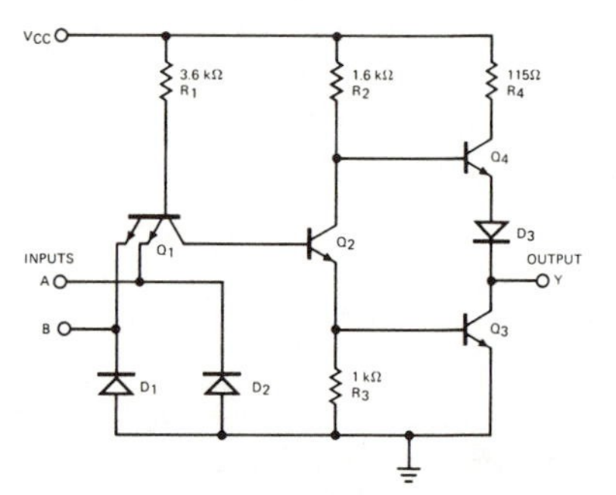

Component values shown are typical

RECOMMENDED OPERATING CONDITIONS

PARAMETER	9N00XM/5400XM			9N00XC/7400XC			UNITS
	MIN.	TYP.	MAX.	MIN.	TYP.	MAX.	
Supply Voltage V_{CC}	4.5	5.0	5.5	4.75	5.0	5.25	Volts
Operating Free-Air Temperature Range	−55	25	125	0	25	70	°C
Normalized Fan-Out from Each Output, N			10			10	U.L.

X = package type; F for Flatpak, D for Ceramic Dip, P for Plastic Dip. See Packaging Information Section for packages available on this product.

ELECTRICAL CHARACTERISTICS OVER OPERATING TEMPERATURE RANGE (Unless Otherwise Noted)

SYMBOL	PARAMETER	LIMITS			UNITS	TEST CONDITIONS (Note 1)		TEST FIGURE
		MIN.	TYP. (Note 2)	MAX.				
V_{IH}	Input HIGH Voltage	2.0			Volts	Guaranteed Input HIGH Voltage		1
V_{IL}	Input LOW Voltage			0.8	Volts	Guaranteed Input LOW Voltage		2
V_{OH}	Output HIGH Voltage	2.4	3.3		Volts	V_{CC} = MIN., I_{OH} = 0.4 mA, V_{IN} = 0.8 V		2
V_{OL}	Output LOW Voltage		0.22	0.4	Volts	V_{CC} = MIN., I_{OL} = 16 mA, V_{IN} = 2.0 V		1
I_{IH}	Input HIGH Current			40	μA	V_{CC} = MAX., V_{IN} = 2.4 V	Each Input	4
				1.0	mA	V_{CC} = MAX., V_{IN} = 5.5 V		
I_{IL}	Input LOW Current			−1.6	mA	V_{CC} = MAX., V_{IN} = 0.4 V, Each Input		3
I_{OS}	Output Short Circuit Current	−20		−55	mA	9N00/5400	V_{CC} = MAX.	5
	(Note 3)	−18		−55	mA	9N00/7400		
I_{CCH}	Supply Current HIGH		4.0	8.0	mA	V_{CC} = MAX., V_{IN} = 0V		6
I_{CCL}	Supply Current LOW		12	22	mA	V_{CC} = MAX., V_{IN} = 5.0 V		6

SWITCHING CHARACTERISTICS (T_A = 25°C)

SYMBOL	PARAMETER	LIMITS			UNITS	TEST CONDITIONS	TEST FIGURE
		MIN.	TYP.	MAX.			
t_{PLH}	Turn Off Delay Input to Output		11	22	ns	V_{CC} = 5.0 V C_L = 15 pF R_L = 400Ω	A
t_{PHL}	Turn On Delay Input to Output		7.0	15	ns		

NOTES:
(1) For conditions shown as MIN. or MAX., use the appropriate value specified under recommended operating conditions for the applicable device type.
(2) Typical limits are at V_{CC} = 5.0 V, 25°C.
(3) Not more than one output should be shorted at a time.

FAIRCHILD SUPER HIGH SPEED TTL/SSI • 9S00/54S00, 74S00

QUAD 2-INPUT NAND GATE

LOGIC AND CONNECTION DIAGRAM

DIP (TOP VIEW)

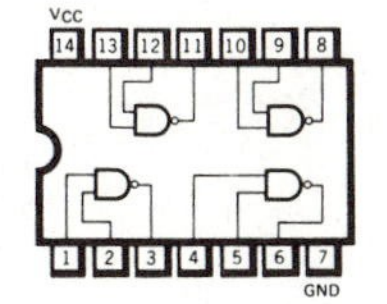

Positive logic: $Y = \overline{AB}$

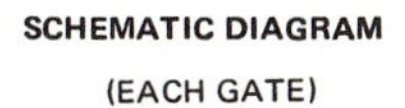

SCHEMATIC DIAGRAM

(EACH GATE)

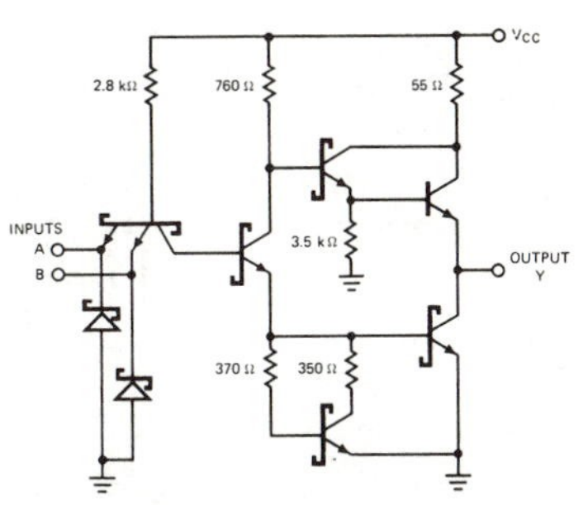

Component values shown are typical.

RECOMMENDED OPERATING CONDITIONS

PARAMETER	9S00XM/54S00XM			9S00XC/74S00XC			UNITS
	MIN.	TYP.	MAX.	MIN.	TYP.	MAX.	
Supply Voltage V_{CC}	4.5	5.0	5.5	4.75	5.0	5.25	Volts
Operating Free-Air Temperature Range	−55	25	125	0	25	75	°C
Input Loading for Each Input			1.25			1.25	U.L.

X = package type; F for Flatpak, D for Ceramic Dip, P for Plastic Dip. See Packaging Information Section for packages available on this product.

ELECTRICAL CHARACTERISTICS OVER OPERATING TEMPERATURE RANGE (Unless Otherwise Noted)

SYMBOL	PARAMETER		LIMITS MIN.	TYP. (Note 2)	MAX.	UNITS	TEST CONDITIONS (Note 1)	
V_{IH}	Input HIGH Voltage		2.0			Volts	Guaranteed Input HIGH Voltage	
V_{IL}	Input LOW Voltage				0.8	Volts	Guaranteed Input LOW Voltage	
V_{CD}	Input Clamp Diode Voltage			−0.65	−1.2	Volts	V_{CC} = MIN., I_{IN} = -18mA	
V_{OH}	Output HIGH Voltage	XM	2.5	3.4		Volts	V_{CC} = MIN., I_{OH} = -1.0mA, V_{IN} = 0.8V	
		XC	2.7	3.4				
V_{OL}	Output LOW Voltage			0.35	0.5	Volts	V_{CC} = MIN., I_{OL} = 20mA, V_{IN} = 2.0V	
I_{IH}	Input HIGH Current			1.0	50	μA	V_{CC} = MAX., V_{IN} = 2.7V	Each Input
					1.0	mA	V_{CC} = MAX., V_{IN} = 5.5V	
I_{IL}	Input LOW Current			−1.4	−2.0	mA	V_{CC} = MAX., V_{IN} = 0.4V	Each Input
I_{OS}	Output Short Circuit Current (Note 3)		−40	−65	−100	mA	V_{CC} = MAX., V_{OUT} = 0V	
I_{CCH}	Supply Current HIGH			10.8	16.0	mA	V_{CC} = MAX., V_{IN} = 0V	
I_{CCL}	Supply Current LOW			25.2	36.0	mA	V_{CC} = MAX., Inputs Open	

SWITCHING CHARACTERISTICS (T_A = 25°C)

SYMBOL	PARAMETER	LIMITS MIN.	TYP.	MAX.	UNITS	TEST CONDITIONS	TEST FIGURES
t_{PLH}	Turn Off Delay Input to Output	2.0	3.0	4.5	ns	V_{CC} = 5.0V	DD
t_{PHL}	Turn On Delay Input to Output	2.0	3.0	5.0	ns	C_L = 15pF	

NOTES:
(1) For conditions shown as MIN. or MAX., use the appropriate value specified under recommended operating conditions for the applicable device type.
(2) Typical limits are at V_{CC} = 5.0V, 25°C.
(3) Not more than one output should be shorted at a time.

FAIRCHILD TTL/SSI • 9N01/5401, 7401

QUAD 2-INPUT NAND GATE (WITH OPEN-COLLECTOR OUTPUT)

LOGIC AND CONNECTION DIAGRAM

SCHEMATIC DIAGRAM
(EACH GATE)

DIP (TOP VIEW) **FLATPAK** (TOP VIEW)

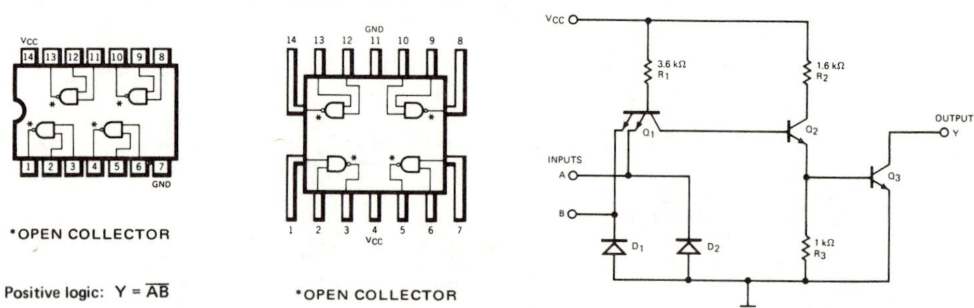

*OPEN COLLECTOR

Positive logic: $Y = \overline{AB}$

*OPEN COLLECTOR

Component values shown are typical.

RECOMMENDED OPERATING CONDITIONS

PARAMETER	9N01XM/5401XM			9N01XC/7401XC			UNITS
	MIN.	TYP.	MAX.	MIN.	TYP.	MAX.	
Supply Voltage V_{CC}	4.5	5.0	5.5	4.75	5.0	5.25	Volts
Operating Free-Air Temperature Range	–55	25	125	0	25	70	°C
Normalized Fan-Out from Each Output, N			10			10	U.L.

X = package type; F for Flatpak, D for Ceramic Dip, P for Plastic Dip. See Packaging Information Section for packages available on this product.

ELECTRICAL CHARACTERISTICS OVER OPERATING TEMPERATURE RANGE (Unless Otherwise Noted)

SYMBOL	PARAMETER	LIMITS			UNITS	TEST CONDITIONS (Note 1)		TEST FIGURE
		MIN.	TYP. (Note 2)	MAX.				
V_{IH}	Input HIGH Voltage	2.0			Volts	Guaranteed Input HIGH Voltage		1
V_{IL}	Input LOW Voltage			0.8	Volts	Guaranteed Input LOW Voltage		7
I_{OH}	Output HIGH Current			0.25	mA	V_{CC} = MIN., V_{OH} = 5.5 V, V_{IL} = 0.8 V		7
V_{OL}	Output LOW Voltage			0.4	Volts	V_{CC} = MIN., I_{OL} = 16 mA, V_{IN} = 2.0 V (On Level)		1
I_{IH}	Input HIGH Current			40	μA	V_{CC} = MAX., V_{IN} = 2.4 V	Each Input	4
				1.0	mA	V_{CC} = MAX., V_{IN} = 5.5 V		
I_{IL}	Input LOW Current			–1.6	mA	V_{CC} = MAX., V_{IN} = 0.4 V,	Each Input	3
I_{CCH}	Supply Current HIGH		4.0	8.0	mA	V_{CC} = MAX., V_{IN} = 0 V		6
I_{CCL}	Supply Current LOW		12	22	mA	V_{CC} = MAX., V_{IN} = 5.0 V		6

SWITCHING CHARACTERISTICS (T_A = 25°C)

SYMBOL	PARAMETER	LIMITS			UNITS	TEST CONDITIONS		TEST FIGURE
		MIN.	TYP.	MAX.				
t_{PLH}	Turn Off Delay Input to Output		35	45	ns	R_L = 4.0 kΩ	V_{CC} = 5.0 V C_L = 15 pF	A
t_{PHL}	Turn On Delay Input to Output		8.0	15	ns	R_L = 400Ω		

NOTES:
(1) For conditions shown as MIN. or MAX., use the appropriate value specified under recommended operating conditions for the applicable device type.
(2) Typical limits are at V_{CC} = 5.0 V, 25°C.

54/7404
54H/74H04
54S/74S04
54S/74S04A
54LS/74LS04

HEX INVERTER

CONNECTION DIAGRAMS

PINOUT A

PINOUT B

ORDERING CODE: See Section 9

PKGS	PIN OUT	COMMERCIAL GRADE	MILITARY GRADE	PKG TYPE
		V_{CC} = +5.0 V ±5%, T_A = 0° C to +70° C	V_{CC} = +5.0 V ±10%, T_A = -55° C to +125° C	
Plastic DIP (P)	A	7404PC, 74H04PC 74S04PC, 74S04APC 74LS04PC		9A
Ceramic DIP (D)	A	7404DC, 74H04DC 74S04DC, 74S04ADC 74LS04DC	5404DM, 54H04DM 54S04DM, 54S04ADM 54LS04DM	6A
Flatpak (F)	A	74S04FC, 74S04AFC 74LS04FC	54S04FM, 54S04AFM 54LS04FM	3I
	B	7404FC, 74H04FC	5404FM, 54H04FM	

INPUT LOADING/FAN-OUT: See Section 3 for U.L. definitions

PINS	54/74 (U.L.) HIGH/LOW	54/74H (U.L.) HIGH/LOW	54/74S (U.L.) HIGH/LOW	54/74LS (U.L.) HIGH/LOW
Inputs	1.0/1.0	1.25/1.25	1.25/1.25	0.5/0.25
Outputs	20/10	12.5/12.5	25/12.5	10/5.0

DC AND AC CHARACTERISTICS: See Section 3*

SYMBOL	PARAMETER	54/74 Min	54/74 Max	54/74H Min	54/74H Max	54/74S Min	54/74S Max	54/74LS Min	54/74LS Max	UNITS	CONDITIONS	
I_{CCH}	Power Supply		12		26		24		2.4	mA	V_{IN} = Gnd	V_{CC} = Max
I_{CCL}	Current		33		58		54		6.6		V_{IN} = Open	
t_{PLH} t_{PHL}	Propagation Delay		22 15		10 10	2.0 2.0	4.5 5.0		10 10	ns	Fig. 3-1, 3-4	
t_{PLH} t_{PHL}	Propagation Delay (54/74S04A only)					1.0 1.0	3.5 4.0			ns	Fig. 3-1, 3-4	

*DC limits apply over operating temperature range; AC limits apply at T_A = +25° C and V_{CC} = +5.0 V.

FAIRCHILD TTL/SSI • 9N05/5405, 7405

HEX INVERTER
(WITH OPEN-COLLECTOR OUTPUT)

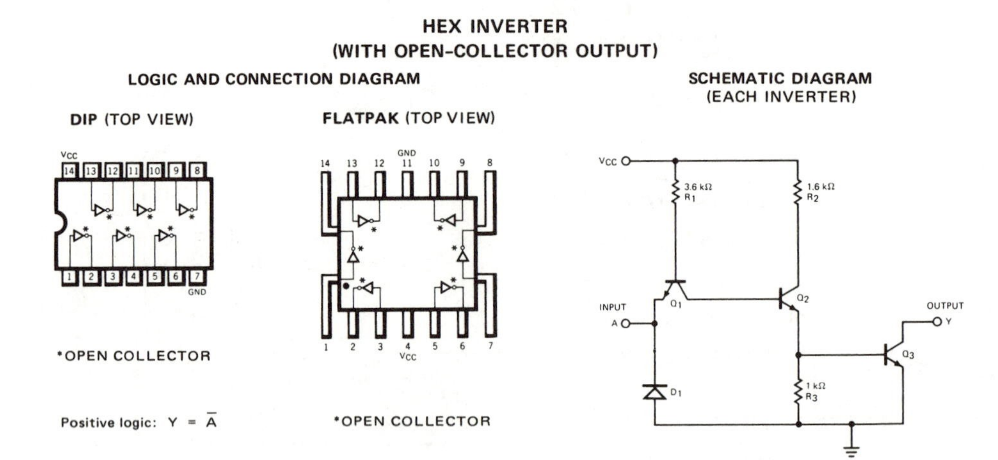

Component values shown are typical.

RECOMMENDED OPERATING CONDITIONS

PARAMETER	9N05XM/5405XM			9N05XC/7405XC			UNITS
	MIN.	TYP.	MAX.	MIN.	TYP.	MAX.	
Supply Voltage V_{CC}	4.5	5.0	5.5	4.75	5.0	5.25	Volts
Operating Free-Air Temperature Range	−55	25	125	0	25	70	°C
Normalized Fan-Out from Each Output, N			10			10	U.L.

X = package type; F for Flatpak, D for Ceramic Dip, P for Plastic Dip. See Packaging Information Section for packages available on this product.

ELECTRICAL CHARACTERISTICS OVER OPERATING TEMPERATURE RANGE (Unless Otherwise Noted)

SYMBOL	PARAMETER	LIMITS			UNITS	TEST CONDITIONS (Note 1)		TEST FIGURE
		MIN.	TYP. (Note 2)	MAX.				
V_{IH}	Input HIGH Voltage	2.0			Volts	Guaranteed Input HIGH Voltage		15
V_{IL}	Input LOW Voltage			0.8	Volts	Guaranteed Input LOW Voltage		17
I_{OH}	Output HIGH Current			0.25	mA	V_{CC} = MIN., V_{OH} = 5.5 V, V_{IN} = 0.8 V		17
V_{OL}	Output LOW Voltage			0.4	Volts	V_{CC} = MIN., I_{OL} = 16 mA, V_{IN} = 2.0 V (On Level)		15
I_{IH}	Input HIGH Current			40	μA	V_{CC} = MAX., V_{IN} = 2.4 V		18
				1.0	mA	V_{CC} = MAX., V_{IN} = 5.5 V		
I_{IL}	Input LOW Current			−1.6	mA	V_{CC} = MAX., V_{IN} = 0.4 V		18
I_{CCH}	Supply Current HIGH		6	12	mA	V_{IN} = 0 V	V_{CC} = 5.0 V, T_A = 25°C	20
I_{CCL}	Supply Current LOW		18	33	mA	V_{IN} = 5 V		20

SWITCHING CHARACTERISTICS (T_A = 25°)

SYMBOL	PARAMETER	LIMITS			UNITS	TEST CONDITIONS		TEST FIGURE
		MIN.	TYP.	MAX.				
t_{PLH}	Turn Off Delay Input to Output		40	55	ns	R_L = 4kΩ	V_{CC} = 5.0 V	A
t_{PHL}	Turn On Delay Input to Output		8.0	15	ns	R_L = 400Ω	C_L = 15 pF	

NOTES:
(1) For conditions shown as MIN. or MAX., use the appropriate value specified under recommended operating conditions for the applicable device type.
(2) Typical limits are at V_{CC} = 5.0 V, 25°C.

54/7414
54LS/74LS14
HEX SCHMITT TRIGGER INVERTER

CONNECTION DIAGRAM
PINOUT A

ORDERING CODE: See Section 9

PKGS	PIN OUT	COMMERCIAL GRADE	MILITARY GRADE	PKG TYPE
		V_{CC} = +5.0 V ±5%, T_A = 0°C to +70°C	V_{CC} = +5.0 V ±10%, T_A = -55°C to +125°C	
Plastic DIP (P)	A	7414PC, 74LS14PC		9A
Ceramic DIP (D)	A	7414DC, 74LS14DC	5414DM, 54LS14DM	6A
Flatpak (F)	A	7414FC, 74LS14FC	5414FM, 54LS14FM	3I

INPUT LOADING/FAN-OUT: See Section 3 for U.L. definitions

PINS	54/74 (U.L.) HIGH/LOW	54/74LS (U.L.) HIGH/LOW
Inputs	1.0/1.0	0.5/0.25
Outputs	20/10	10/5.0

DC AND AC CHARACTERISTICS: See Section 3*

SYMBOL	PARAMETER	54/74 Min	54/74 Max	54/74LS Min	54/74LS Max	UNITS	CONDITIONS	
V_{T+}	Positive-going Threshold Voltage	1.5	2.0	1.5	2.0	V	V_{CC} = +5.0 V	
V_{T-}	Negative-going Threshold Voltage	0.6	1.1	0.6	1.1	V	V_{CC} = +5.0 V	
$V_{T+} - V_{T-}$	Hysteresis Voltage	0.4		0.4		V	V_{CC} = +5.0 V	
I_{T+}	Input Current at Positive-going Threshold	-0.43**		-0.14**		mA	V_{CC} = +5.0 V, V_{IN} = V_{T+}	
I_{T-}	Input Current at Negative-going Threshold	-0.56**		-0.18**		mA	V_{CC} = +5.0 V, V_{IN} = V_{T-}	
I_{IL}	Input LOW Current		-1.2		-0.4	mA	V_{CC} = Max, V_{IN} = 0.4 V	
I_{OS}	Output Short Circuit Current	-18	-55	-20	-100	mA	V_{CC} = Max, V_{OUT} = 0 V	
I_{CCH}	Power Supply Current		36		16	mA	V_{IN} = Gnd	V_{CC} = Max
I_{CCL}			60		21		V_{IN} = Open	
t_{PLH}	Propagation Delay		22		22	ns	Figs. 3-1, 3-15	
t_{PHL}			22		22			

*DC limits apply over operating temperature range; AC limits apply at T_A = +25°C and V_{CC} = +5.0 V. **Typical Value

FAIRCHILD TTL/SSI • 9N20/5420, 7420

DUAL 4-INPUT NAND GATE

LOGIC AND CONNECTION DIAGRAM

DIP (TOP VIEW) **FLATPAK** (TOP VIEW)

Positive logic: $Y = \overline{ABCD}$

NC — No internal connection.

SCHEMATIC DIAGRAM
(EACH GATE)

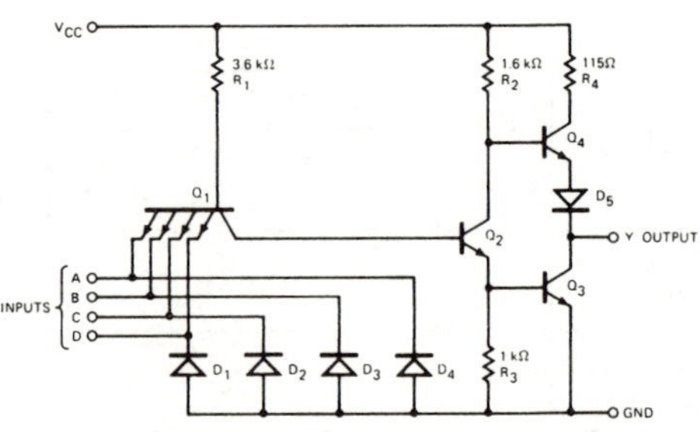

Component values shown are typical.

RECOMMENDED OPERATING CONDITIONS

PARAMETER	9N20XM/5420XM			9N20XC/7420XC			UNITS
	MIN.	TYP.	MAX.	MIN.	TYP.	MAX.	
Supply Voltage V_{CC}	4.5	5.0	5.5	4.75	5.0	5.25	Volts
Operating Free-Air Temperature Range	−55	25	125	0	25	70	°C
Normalized Fan Out From Each Output, N			10			10	U.L.

X = package type; F for Flatpak, D for Ceramic Dip, P for Plastic Dip. See Packaging Information Section for packages available on this product.

ELECTRICAL CHARACTERISTICS OVER OPERATING TEMPERATURE RANGE (Unless Otherwise Noted)

SYMBOL	PARAMETER	LIMITS			UNITS	TEST CONDITIONS (Note 1)		TEST FIGURE
		MIN.	TYP. (Note 2)	MAX.				
V_{IH}	Input HIGH Voltage	2.0			Volts	Guaranteed Input HIGH Voltage		1
V_{IL}	Input LOW Voltage			0.8	Volts	Guaranteed Input LOW Voltage		2
V_{OH}	Output HIGH Voltage	2.4	3.3		Volts	V_{CC} = MIN., I_{OH} = −0.4 mA, V_{IN} = 0.8 V		2
V_{OL}	Output LOW Voltage		0.22	0.4	Volts	V_{CC} = MIN., I_{OL} = 16 mA, V_{IN} = 2.0 V		1
I_{IH}	Input HIGH Current			40	μA	V_{CC} = MAX., V_{IN} = 2.4 V	Each Input	4
				1.0	mA	V_{CC} = MAX., V_{IN} = **5.5** V		
I_{IL}	Input LOW Current			−1.6	mA	V_{CC} = MAX., V_{IN} = 0.4 V, Each Input		3
I_{OS}	Output Short Circuit Current	−20		−55	mA	9N20/5420	V_{CC} = MAX.	5
	(Note 3)	−18		−55	mA	9N20/7420		
I_{CCH}	Supply Current HIGH		2.0	4.0	mA	V_{CC} = MAX., V_{IN} = 0 V		6
I_{CCL}	Supply Current LOW		6.0	11	mA	V_{CC} = MAX., V_{IN} = 5.0 V		6

SWITCHING CHARACTERISTICS (T_A = 25°C)

SYMBOL	PARAMETER	LIMITS			UNITS	TEST CONDITIONS	TEST FIGURE
		MIN.	TYP.	MAX.			
t_{PLH}	Turn Off Delay Input to Output		12	22	ns	V_{CC} = 5.0 V C_L = 15 pF R_L = 400Ω	A
t_{PHL}	Turn On Delay Input to Output		8.0	15	ns		

NOTES:
(1) For conditions shown as MIN. or MAX., use the appropriate value specified under recommended operating conditions for the applicable device type.
(2) Typical limits are at V_{CC} = 5.0 V, 25°C.
(3) Not more than one output should be shorted at a time.

FAIRCHILD TTL/SSI • 9N37/5437, 7437 • 9N38/5438, 7438

QUAD 2-INPUT NAND BUFFER

LOGIC AND CONNECTION DIAGRAM

DIP (TOP VIEW) **FLATPAK** (TOP VIEW)

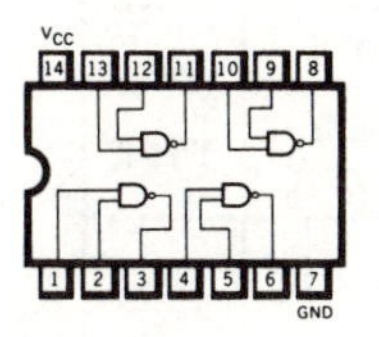

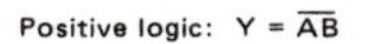

Positive logic: $Y = \overline{AB}$

SCHEMATIC DIAGRAM
(EACH BUFFER)

9N37/5437, 7437
(TOTEM-POLE OUTPUT)

9N38/5438, 7438
(OPEN-COLLECTOR OUTPUT)

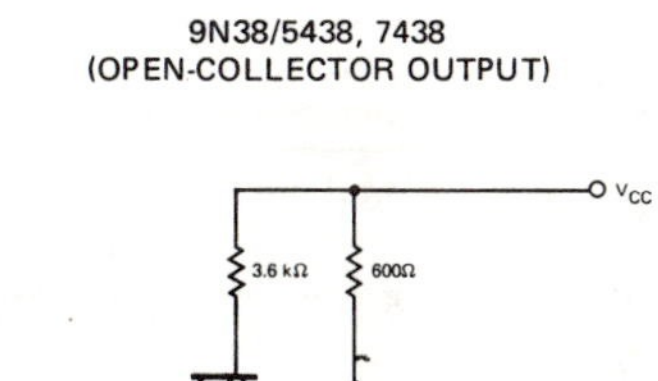

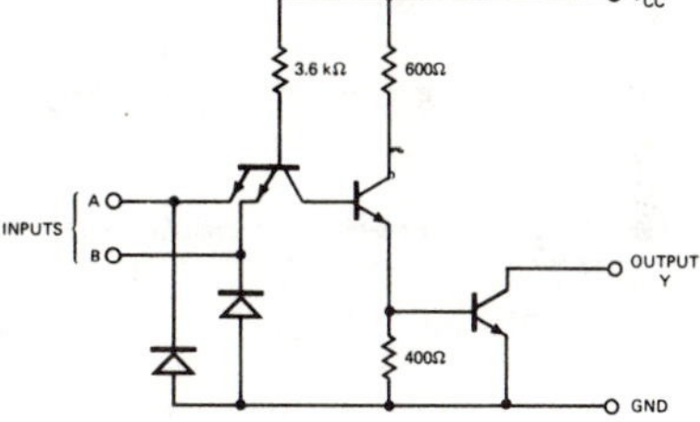

RECOMMENDED OPERATING CONDITIONS

PARAMETER	9N37XM/5437XM 9N38XM/5438XM			9N37XC/7437XC 9N38XC/7438XC			UNITS
	MIN.	TYP.	MAX.	MIN.	TYP.	MAX.	
Supply Voltage V_{CC}	4.5	5.0	5.5	4.75	5.0	5.25	Volts
Operating Free-Air Temperature Range	–55	25	125	0	25	70	°C
Normalized Fan Out from Each Output, N			30			30	U.L.

X = package type; F for Flatpak, D for Ceramic Dip, P for Plastic Dip. See Packaging Information Section for packages available on this product.

54/7438
54LS/74LS38

QUAD 2-INPUT NAND BUFFER
(With Open-Collector Output)

CONNECTION DIAGRAM
PINOUT A

Vcc 14, 13, 12, 11, 10, 9, 8; 1, 2, 3, 4, 5, 6, GND 7

ORDERING CODE: See Section 9

PKGS	PIN OUT	COMMERCIAL GRADE	MILITARY GRADE	PKG TYPE
		V_{CC} = +5.0 V ±5%, T_A = 0° C to +70° C	V_{CC} = +5.0 V ±10%, T_A = -55° C to +125° C	
Plastic DIP (P)	A	7438PC, 74LS38PC		9A
Ceramic DIP (D)	A	7438DC, 74LS38DC	5438DM, 54LS38DM	6A
Flatpak (F)	A	7438FC, 74LS38FC	5438FM, 54LS38FM	3I

INPUT LOADING/FAN-OUT: See Section 3 for U.L. definitions

PINS	54/74 (U.L.) HIGH/LOW	54/74LS (U.L.) HIGH/LOW
Inputs	1.0/1.0	0.5/0.25
Outputs	OC**/30	OC**/15 (7.5)

DC AND AC CHARACTERISTICS: See Section 3*

SYMBOL	PARAMETER	54/74 Min	54/74 Max	54/74LS Min	54/74LS Max	UNITS	CONDITIONS	
V_{OL}	Output LOW Voltage		0.4			V	V_{IN} = 2.0 V, V_{CC} = Min, I_{OL} = 48 mA	
I_{OH}	Output HIGH Current				250	μA	V_{OH} = 5.5 V, V_{CC} = Min, V_{IN} = V_{IL}	
I_{CCH}	Power Supply Current		8.5		2.0	mA	V_{IN} = Gnd	V_{CC} = Max
I_{CCL}			54		12		V_{IN} = Open	
t_{PLH} t_{PHL}	Propagation Delay		22 18		22 22	ns	Figs. 3-2, 3-4	

*DC limits apply over operating temperature range; AC limits apply at T_A = +25° C and V_{CC} = +5.0 V.
**OC—Open Collector

FAIRCHILD TTL/SSI • 9N73/5473, 7473 • 9N107/54107, 74107

DUAL JK MASTER/SLAVE FLIP-FLOP WITH SEPARATE CLEARS AND CLOCKS

DESCRIPTION – The TTL/SSI 9N73/5473, 7473 and 9N107/54107, 74107 are Dual JK Master/Slave flip-flops with a separate clear and a separate clock for each flip-flop. Inputs to the master section are controlled by the clock pulse. The clock pulse also regulates the state of the coupling transistors which connect the master and slave sections. The sequence of operation is as follows: 1) Isolate slave from master. 2) Enter information from J and K inputs to master. 3) Disable J and K inputs. 4) Transfer information from master to slave.

LOGIC AND CONNECTION DIAGRAM

9N73/5473, 7473

DIP (TOP VIEW)

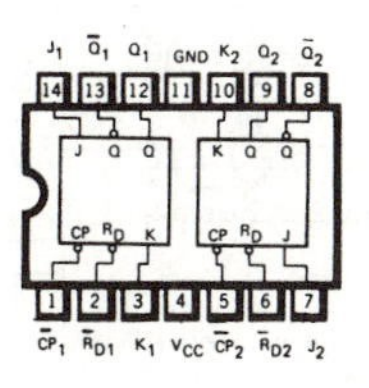

FLATPAK (TOP VIEW)

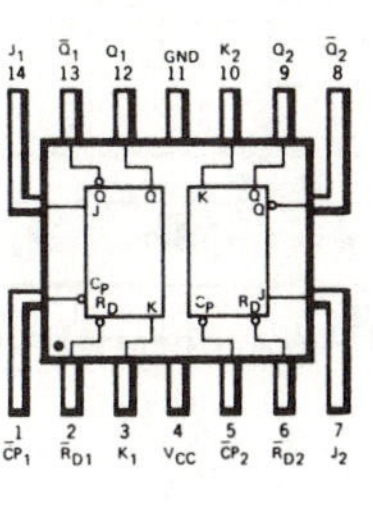

9N107/54107, 74107

DIP (TOP VIEW)

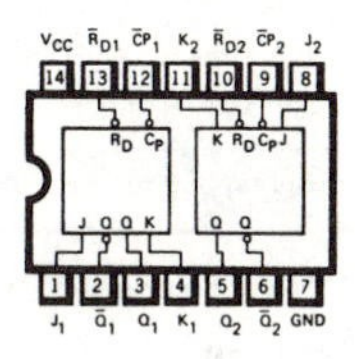

Positive logic:
LOW input to clear sets Q to LOW level
Clear is independent of clock

TRUTH TABLE

t_n		t_{n+1}
J	K	Q
L	L	Q_n
L	H	L
H	L	H
H	H	$\bar{Q}_n$

NOTES:
t_n = Bit time before clock pulse.
t_{n+1} = Bit time after clock pulse.

CLOCK WAVEFORM

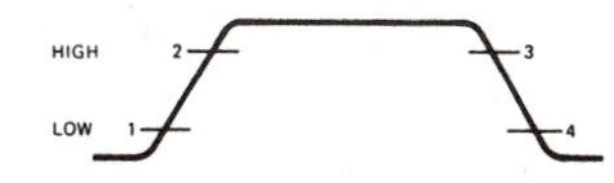

LOGIC DIAGRAM
(EACH FLIP-FLOP)

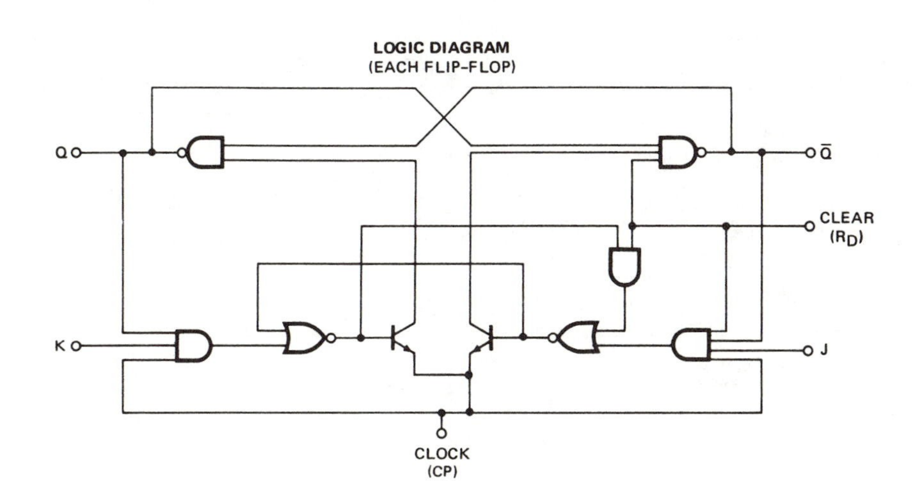

FAIRCHILD TTL/SSI • 9N73/5473, 7473 • 9N107/54107, 74107

RECOMMENDED OPERATING CONDITIONS

PARAMETER	9N73XM/5473XM 9N107XM/54107XM			9N73XC/7473XC 9N107XC/74107XC			UNITS
	MIN.	TYP.	MAX.	MIN.	TYP.	MAX.	
Supply Voltage V_{CC}	4.5	5.0	5.5	4.75	5.0	5.25	Volts
Operating Free-Air Temperature Range	−55	25	125	0	25	70	°C
Normalized Fan Out from Each Output, N			10			10	U.L.
Width of Clock Pulse, $t_{p(clock)}$ (See Fig. E)	20			20			ns
Width of Clear Pulse, $t_{p(clear)}$ (See Fig. F)	25			25			ns
Input Setup Time, t_{setup} (See Fig. E)	$\geqslant t_{p(clock)}$			$\geqslant t_{p(clock)}$			
Input Hold Time, t_{hold}	0			0			

X = package type; F for Flatpak, D for Ceramic Dip, P for Plastic Dip. See Packaging Information Section for packages available on this product.

ELECTRICAL CHARACTERISTICS OVER OPERATING TEMPERATURE RANGE (Unless Otherwise Noted)

SYMBOL	PARAMETER	LIMITS			UNITS	TEST CONDITIONS (Note 1)		TEST FIGURE
		MIN.	TYP. (Note 2)	MAX.				
V_{IH}	Input HIGH Voltage	2.0			Volts	Guaranteed Input HIGH		46 & 47
V_{IL}	Input LOW Voltage			0.8	Volts	Guaranteed Input LOW		46 & 47
V_{OH}	Output HIGH Voltage	2.4	3.5		Volts	V_{CC} = MIN., I_{OH} = −0.4 mA		46
V_{OL}	Output LOW Voltage		0.22	0.4	Volts	V_{CC} = MIN., I_{OL} = 16 mA		47
I_{IH}	Input HIGH Current at J or K			40	μA	V_{CC} = MAX., V_{IN} = 2.4 V		49
				1.0	mA	V_{CC} = MAX., V_{IN} = 5.5 V		
	Input HIGH Current at Clock or Clear			80	μA	V_{CC} = MAX., V_{IN} = 2.4 V		49
				1.0	mA	V_{CC} = MAX., V_{IN} = 5.5 V		
I_{IL}	Input LOW Current at J or K			−1.6	mA	V_{CC} = MAX., V_{IN} = 0.4 V		48
	Input LOW Current at Clear or Clock			−3.2	mA	V_{CC} = MAX., V_{IN} = 0.4 V		48
I_{OS}	Output Short Circuit Current (Note 3)	−20		−57	mA	9N73/5473; 9N107/54107	V_{CC} = MAX.	50
		−18		−57	mA	9N73/7473; 9N107/74107	V_{IN} = 0 V	
I_{CC}	Supply Current		20	40	mA	V_{CC} = MAX.		49

SWITCHING CHARACTERISTICS (T_A = 25°C)

SYMBOL	PARAMETER	LIMITS			UNITS	TEST CONDITIONS	TEST FIGURE
		MIN.	TYP.	MAX.			
f_{max}	Maximum Clock Frequency	15	20		MHz		E
t_{PLH}	Turn Off Delay Clear to Output		16	25	ns	V_{CC} = 5.0 V	F
t_{PHL}	Turn On Delay Clear to Output		25	40	ns	C_L = 15 pF	F
t_{PLH}	Turn Off Delay Clock to Output	10	16	25	ns	R_L = 400Ω	E
t_{PHL}	Turn On Delay Clock to Output	10	25	40	ns		E

NOTES:
(1) For conditions shown as MIN. or MAX., use the appropriate value specified under recommended operating conditions for the applicable device type.
(2) Typical limits are at V_{CC} = 5.0 V, 25°C.
(3) Not more than one output should be shorted at a time.

FAIRCHILD TTL/SSI • 9N86/5486, 7486

QUAD 2-INPUT EXCLUSIVE OR GATE

DESCRIPTION – The TTL/SSI 9N86/5486, 7486 is a Quad 2-input Exclusive OR gate designed to perform the function: $Y = A\overline{B} + \overline{A}B$. When the input states are complementary, the output goes to the HIGH level.

Input clamping diodes are provided to minimize transmission line effects. On chip input buffers are also provided to lower the fan in requirement to only 1 U.L. (unit load). The 9N86/5486, 7486 is fully compatible with all members of the Fairchild TTL family.

LOGIC AND CONNECTION DIAGRAM

DIP (TOP VIEW)

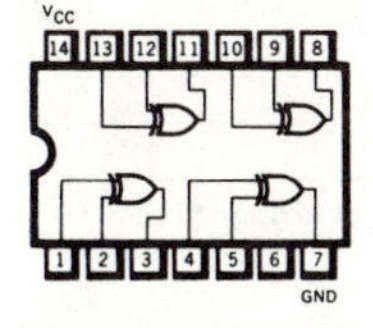

FLATPAK (TOP VIEW)

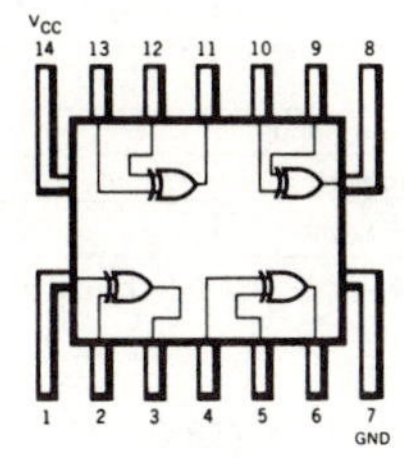

TRUTH TABLE

INPUTS		OUTPUT
A	B	Y
L	L	L
L	H	H
H	L	H
H	H	L

Positive logic: $Y = A \oplus B$

SCHEMATIC DIAGRAM

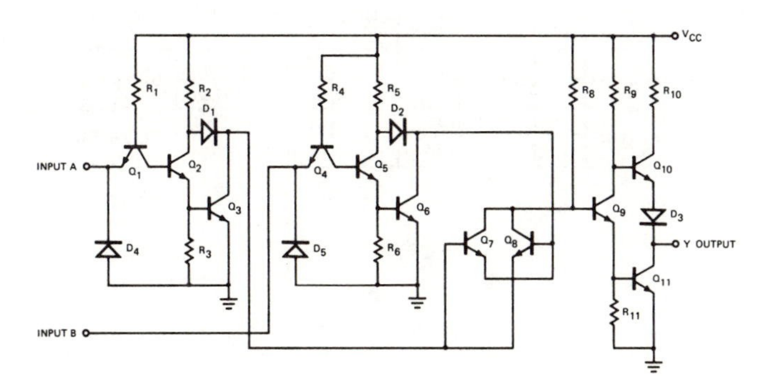

1/4 of Circuit shown.

FAIRCHILD TTL/SSI • 9N86/5486, 7486

RECOMMENDED OPERATING CONDITIONS

PARAMETER		9N86XM/5486XM			9N86XC/7486XC			UNITS
		MIN.	TYP.	MAX.	MIN.	TYP.	MAX.	
Supply Voltage V_{CC}		4.5	5.0	5.5	4.75	5.0	5.25	Volts
Operating Free-Air Temperature Range		–55	25	125	0	25	70	°C
Normalized Fan Out	LOW Level			10			10	U.L.
from Each Output N	HIGH Level			20			20	U.L.

X = package type; F for Flatpak, D for Ceramic Dip, P for Plastic Dip. See Packaging Information Section for packages available on this product.

ELECTRICAL CHARACTERISTICS OVER OPERATING TEMPERATURE RANGE (Unless Otherwise Noted)

SYMBOL	PARAMETER	LIMITS MIN.	TYP. (Note 2)	MAX.	UNITS	TEST CONDITIONS (Note 1)		TEST FIGURE
V_{IH}	Input HIGH Voltage	2.0			Volts	Guaranteed Input HIGH Voltage		98
V_{IL}	Input LOW Voltage			0.8	Volts	Guaranteed Input LOW Voltage		98
V_{OH}	Output HIGH Voltage	2.4			Volts	V_{CC} = MIN., I_{OH} = –800 µA, V_{IH} = 2.0 V, V_{IL} = 0.8 V		98
V_{OL}	Output LOW Voltage			0.4	Volts	V_{CC} = MIN., I_{OL} = 16 mA, V_{IH} = 2.0 V, V_{IL} = 0.8 V		99
I_{IH}	Input HIGH Current			40	µA	V_{CC} = MAX., V_{IN} = 2.4 V	(Each Input)	100
				1.0	mA	V_{CC} = MAX., V_{IN} = 5.5 V		
I_{IL}	Input LOW Current			–1.6	mA	V_{CC} = MAX., V_{IN} = 0.4 V (Each Input)		101
I_{OS}	Output Short Circuit Current	–20		–55	mA	9N86/5486	V_{CC} = MAX., V_{IH} = 4.5 V	102
	(Note 3)	–18		–55	mA	9N86/7486	V_{IL} = 0 V	
I_{CC}	Supply Current		30	43	mA	9N86/5486	V_{CC} = MAX., V_{IN} = 4.5 V	103
			30	50	mA	9N86/7486		

SWITCHING CHARACTERISTICS (T_A = 25°C)

SYMBOL	PARAMETER	LIMITS MIN.	TYP.	MAX.	UNITS	TEST CONDITIONS		TEST FIGURE
t_{PLH}	Turn Off Delay Input to Output		15	23	ns	Other Input Low	V_{CC} = 5.0 V	S
			18	30		Other Input High	C_L = 15 pF	
t_{PHL}	Turn On Delay Input to Output		11	17		Other Input Low	R_L = 400Ω	
			13	22		Other Input High		

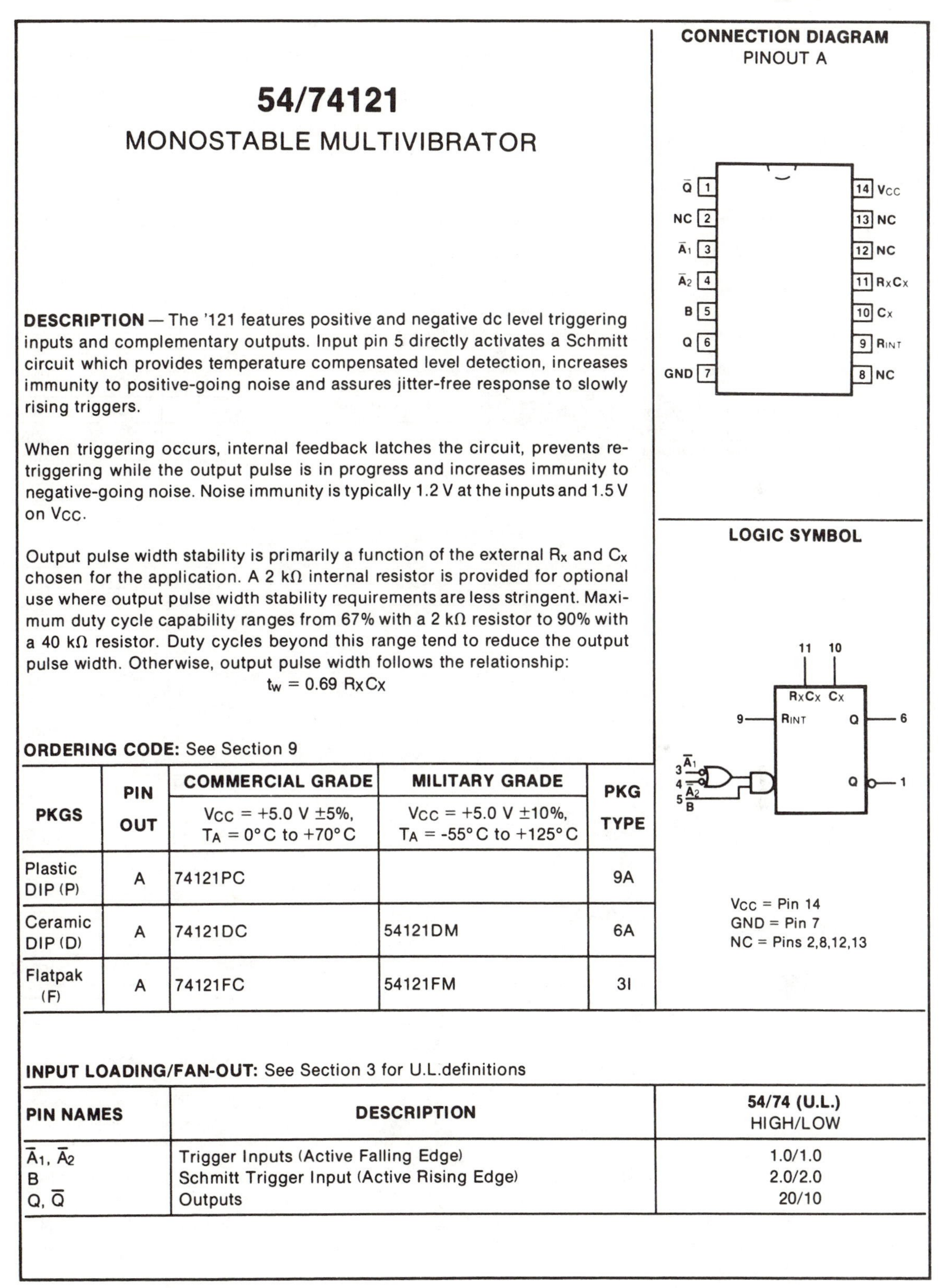

54/74121

MONOSTABLE MULTIVIBRATOR

DESCRIPTION — The '121 features positive and negative dc level triggering inputs and complementary outputs. Input pin 5 directly activates a Schmitt circuit which provides temperature compensated level detection, increases immunity to positive-going noise and assures jitter-free response to slowly rising triggers.

When triggering occurs, internal feedback latches the circuit, prevents re-triggering while the output pulse is in progress and increases immunity to negative-going noise. Noise immunity is typically 1.2 V at the inputs and 1.5 V on V_{CC}.

Output pulse width stability is primarily a function of the external R_x and C_x chosen for the application. A 2 kΩ internal resistor is provided for optional use where output pulse width stability requirements are less stringent. Maximum duty cycle capability ranges from 67% with a 2 kΩ resistor to 90% with a 40 kΩ resistor. Duty cycles beyond this range tend to reduce the output pulse width. Otherwise, output pulse width follows the relationship:

$$t_w = 0.69\, R_X C_X$$

CONNECTION DIAGRAM
PINOUT A

LOGIC SYMBOL

V_{CC} = Pin 14
GND = Pin 7
NC = Pins 2,8,12,13

ORDERING CODE: See Section 9

PKGS	PIN OUT	COMMERCIAL GRADE	MILITARY GRADE	PKG TYPE
		V_{CC} = +5.0 V ±5%, T_A = 0°C to +70°C	V_{CC} = +5.0 V ±10%, T_A = -55°C to +125°C	
Plastic DIP (P)	A	74121PC		9A
Ceramic DIP (D)	A	74121DC	54121DM	6A
Flatpak (F)	A	74121FC	54121FM	3I

INPUT LOADING/FAN-OUT: See Section 3 for U.L. definitions

PIN NAMES	DESCRIPTION	54/74 (U.L.) HIGH/LOW
$\overline{A}_1$, $\overline{A}_2$	Trigger Inputs (Active Falling Edge)	1.0/1.0
B	Schmitt Trigger Input (Active Rising Edge)	2.0/2.0
Q, $\overline{Q}$	Outputs	20/10

CD4011UB, CD4012UB, CD4023UB Types

COS/MOS NAND Gates

High-Voltage Types (20-Volt Rating)

Quad 2 Input – CD4011UB
Dual 4 Input – CD4012UB
Triple 3 Input – CD4023UB

The RCA-CD4011UB, CD4012UB, and CD4023UB NAND gates provide the system designer with direct implementation of the NAND function and supplement the existing family of COS/MOS gates.

The CD4011UB, CD4012UB, and CD4023UB types are supplied in 14-lead hermetic dual-in-line ceramic packages (D and F suffixes), 14-lead dual-in-line plastic packages (E suffix), and in chip form (H suffix).

Features:

- **Propagation delay time = 30 ns (typ). at C_L = 50 pF, V_{DD} = 10 V**
- **Standardized symmetrical output characteristics**
- **100% tested for quiescent current at 20 V**
- **Maximum input current of 1 μA at 18 V over full package temperature range; 100 nA at 18 V and 25°C**
- **5-V, 10-V, and 15-V parametric ratings**
- **Meets all requirements of JEDEC Tentative Standard No. 13A, "Standard Specifications for Description of 'B' Series CMOS Devices"**

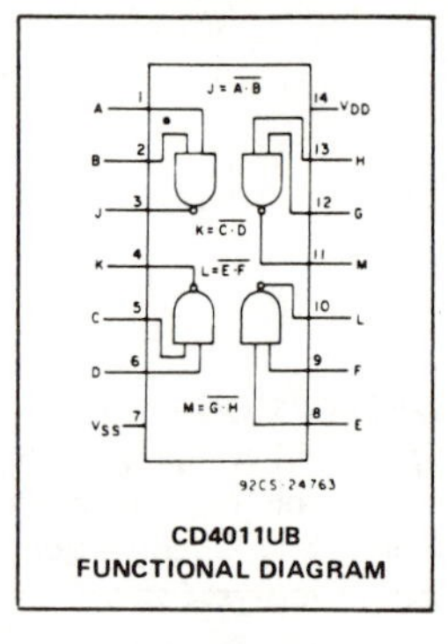

CD4011UB
FUNCTIONAL DIAGRAM

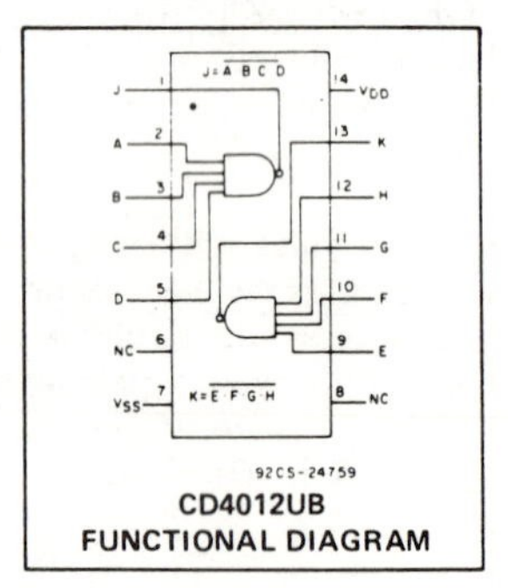

CD4012UB
FUNCTIONAL DIAGRAM

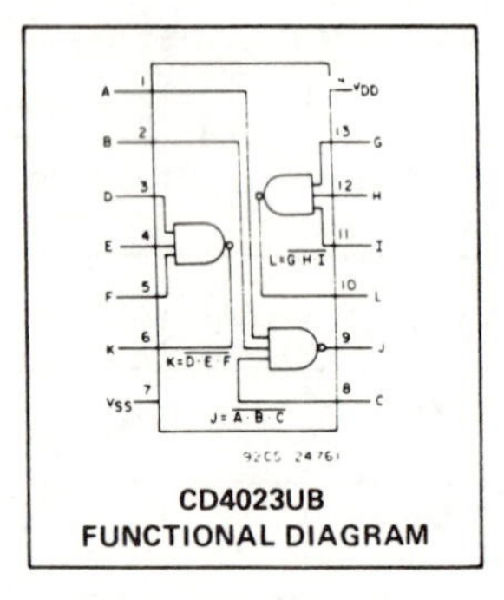

CD4023UB
FUNCTIONAL DIAGRAM

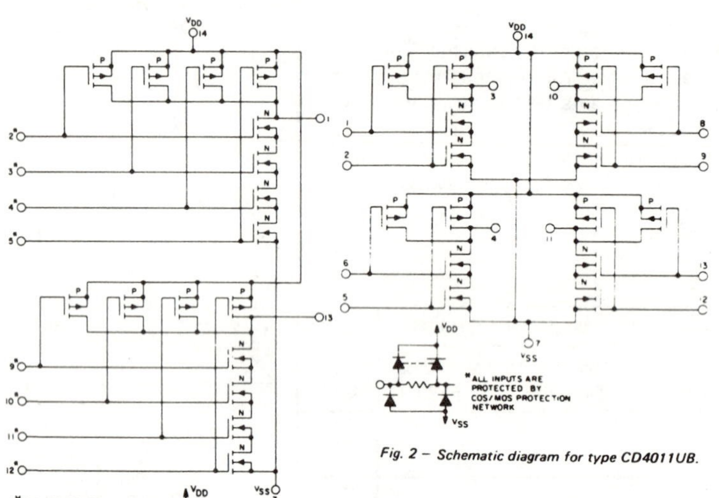

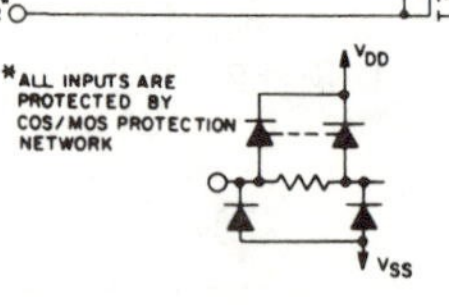

Fig. 1 – Schematic diagram for type CD4012UB.

Fig. 2 – Schematic diagram for type CD4011UB.

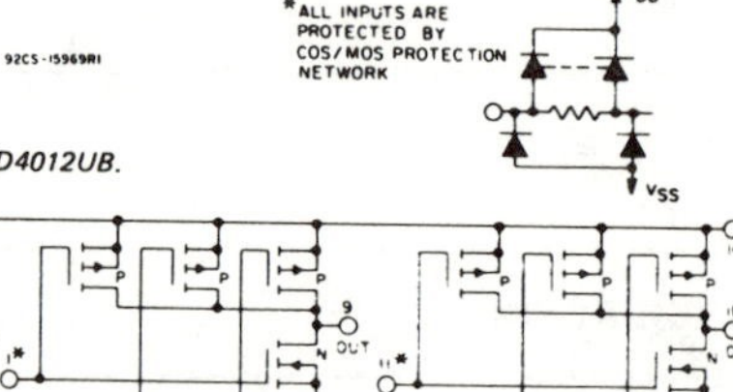

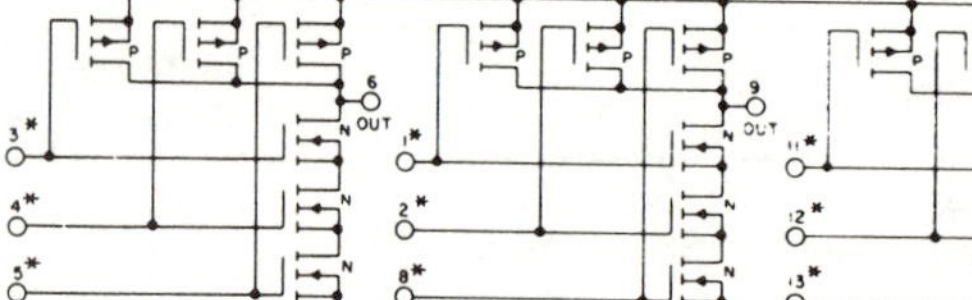

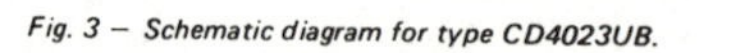

Fig. 3 – Schematic diagram for type CD4023UB.

RECOMMENDED OPERATING CONDITIONS

For maximum reliability, nominal operating conditions should be selected so that operation is always within the following ranges.

CHARACTERISTIC	MIN.	MAX.	UNITS
Supply Voltage Range (For T_A= Full Package Temperature Range)	3	18	V

CD4011UB, CD4012UB, CD4023UB Types

DYNAMIC ELECTRICAL CHARACTERISTICS
At $T_A = 25°C$, Input t_r, $t_f = 20$ ns, and $C_L = 50$ pF, $R_L = 200k\,\Omega$

CHARACTERISTIC	TEST CONDITIONS	V_{DD} VOLTS	ALL TYPES LIMITS TYP.	ALL TYPES LIMITS MAX	UNITS
Propagation Delay Time, t_{PHL}, t_{PLH}		5 10 15	60 30 25	120 60 50	ns
Transition Time, t_{THL}, t_{TLH}		5 10 15	100 50 40	200 100 80	ns
Input Capacitance, C_{IN}	Any Input		10	15	pF

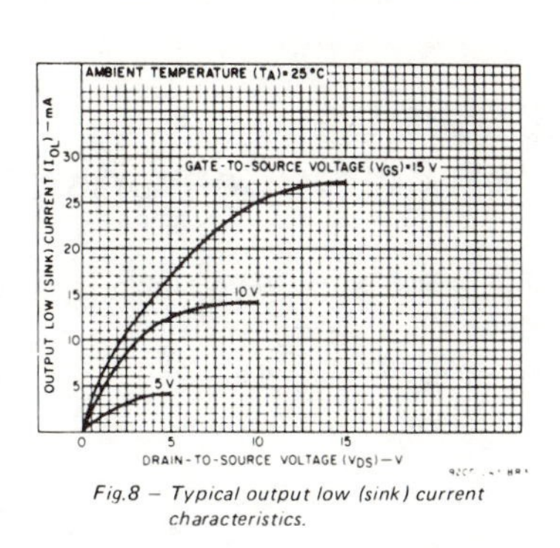

Fig.8 – Typical output low (sink) current characteristics.

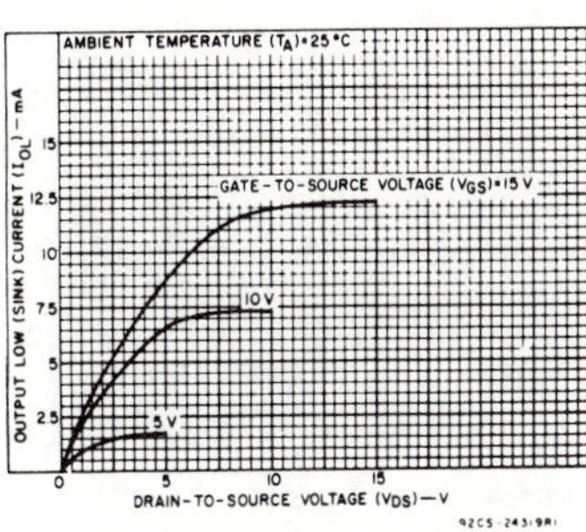

Fig. 9 – Minimum output low (sink) current characteristics.

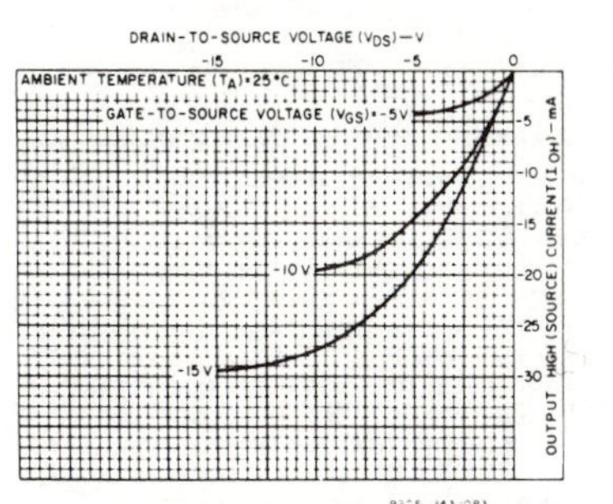

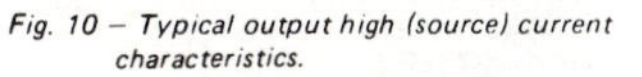

Fig. 10 – Typical output high (source) current characteristics.

Fig. 11 – Minimum output high (source) current characteristics.

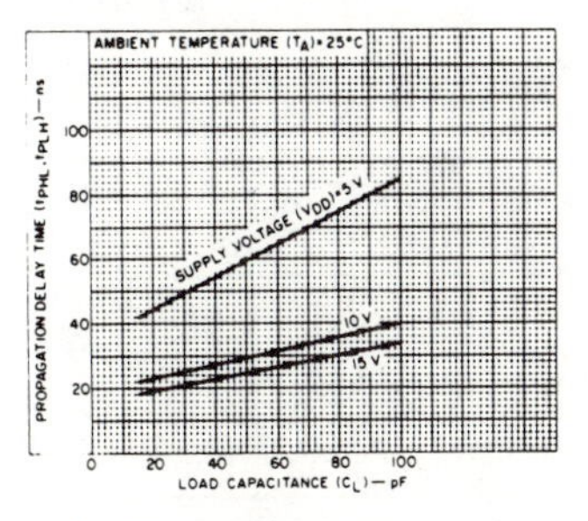

Fig.12 – Typical propagation delay time vs. load capacitance.

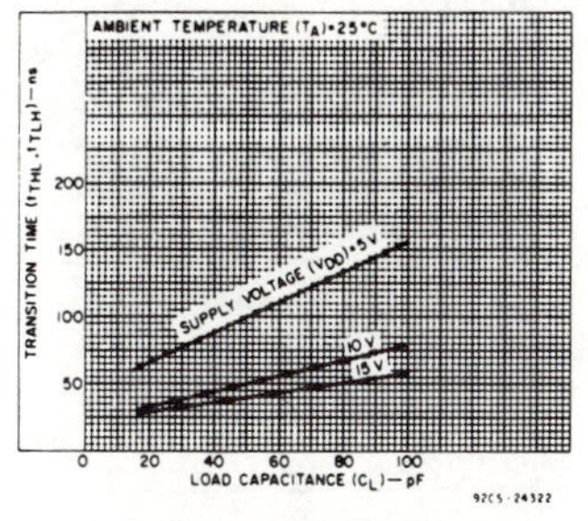

Fig.13 – Typical transition time vs. load capacitance.

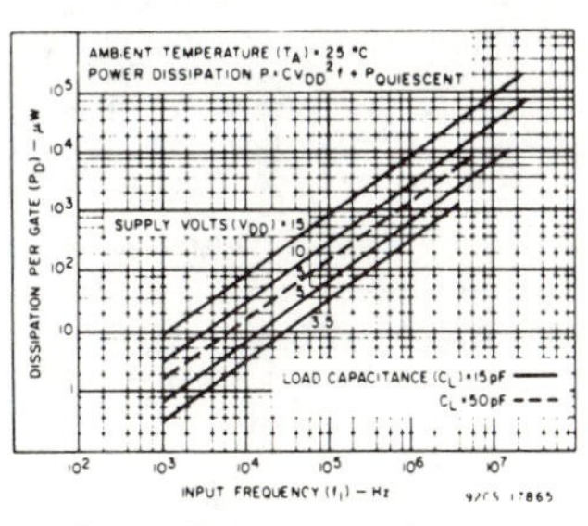

Fig.14 – Typical power dissipation vs. frequency characteristics.

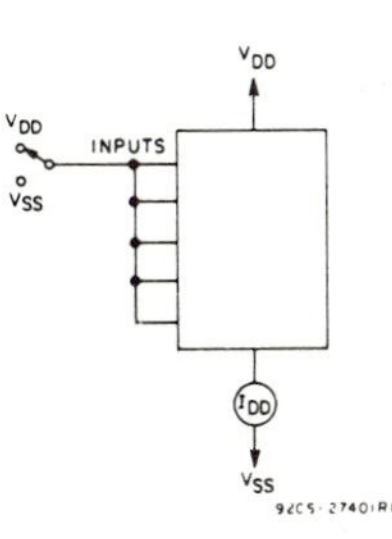

Fig.15 – Quiescent device current test circuit.

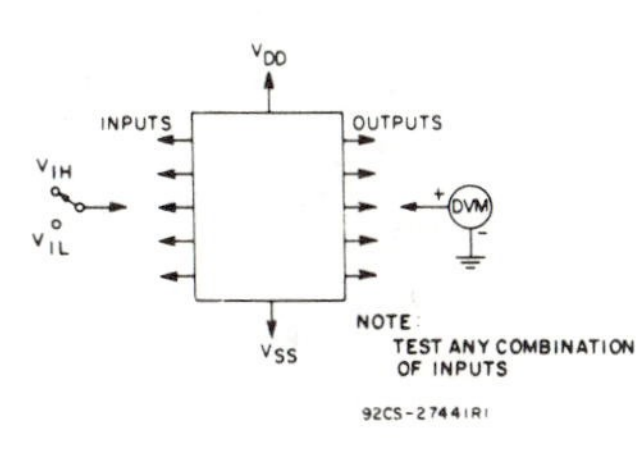

Fig.16 – Input voltage test circuit.

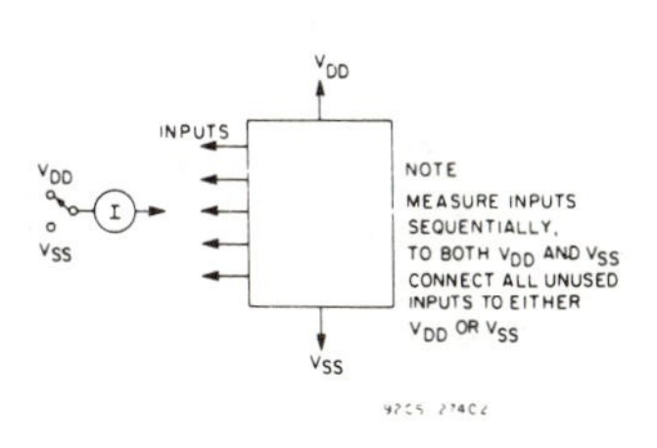

Fig.17 – Input current test circuit.

CD4049UB, CD4050B Types

COS/MOS Hex Buffer/Converters

High-Voltage Types (20-Volt Rating)

CD4049UB—Inverting Type

CD4050B—Non-Inverting Type

Features:

- High sink current for driving 2 TTL loads
- High-to-low level logic conversion
- 100% tested for quiescent current at 20 V
- Maximum input current of 1 μA at 18 V over full package-temperature range; 100 nA at 18 V and 25°C
- 5-, 10-, and 15-volt parametric ratings

Applications:

- COS/MOS to DTL/TTL hex converter
- COS/MOS current "sink" or "source" driver
- COS/MOS high-to-low logic-level converter

A 3 2 G=Ā
B 5 4 H=B̄
C 7 6 I=C̄
D 9 10 J=D̄
E 11 12 K=Ē
F 14 15 L=F̄
V_{CC} 1
V_{SS} 8
NC = 13
NC = 16
92CS-27506

CD4049UB FUNCTIONAL DIAGRAM

The RCA-CD4049UB and CD4050B are inverting and non-inverting hex buffers, respectively, and feature logic-level conversion using only one supply voltage (V_{CC}). The input-signal high level (V_{IH}) can exceed the V_{CC} supply voltage when these devices are used for logic-level conversions. These devices are intended for use as COS/MOS to DTL/TTL converters and can drive directly two DTL/TTL loads. (V_{CC}=5 V, $V_{OL} \leq$ 0.4 V, and $I_{OL} \geq$ 3.2 mA.)

The CD4049UB and CD4050B are designated as replacements for CD4009UB and CD4010B, respectively. Because the CD4049UB and CD4050B require only one power supply, they are preferred over the CD4009UB and CD4010B and should be used in place of the CD4009UB and CD4010B in all inverter, current driver, or logic-level conversion applications. In these applications the CD4049UB and CD4050B are pin compatible with the CD4009UB and CD4010B respectively, and can be substituted for these devices in existing as well as in new designs. Terminal No. 16 is not connected internallyon the CD4049UB or CD4050B, therefore, connection to this terminal is of no consequence to circuit operation. For applications not requiring high sink-current or voltage conversion, the CD4069UB Hex Inverter is recommended.

The CD4049UB and CD4050B types are supplied in 16-lead hermetic dual-in-line ceramic packages (D and F suffixes), 16-lead dual-in-line plastic package (E suffix), and in chip form (H suffix).

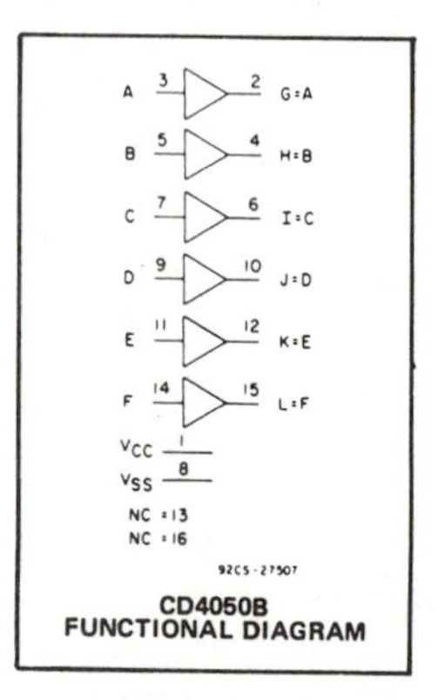

CD4050B FUNCTIONAL DIAGRAM

MAXIMUM RATINGS, *Absolute-Maximum Values:*

DC SUPPLY-VOLTAGE RANGE, (V_{CC}) (Voltages referenced to V_{SS} Terminal)	−0.5 to +20 V
INPUT VOLTAGE RANGE, ALL INPUTS	−0.5 to +20.5 V
DC INPUT CURRENT, ANY ONE INPUT	±10 mA
POWER DISSIPATION PER PACKAGE (P_D):	
For T_A = −40 to +60°C (PACKAGE TYPE E)	500 mW
For T_A = +60 to +85°C (PACKAGE TYPE E)	Derate Linearly at 12 mW/°C to 200 mW
For T_A = −55 to +100°C (PACKAGE TYPES D, F)	500 mW
For T_A = +100 to +125°C (PACKAGE TYPES D, F)	Derate Linearly at 12 mW/°C to 200 mW
DEVICE DISSIPATION PER OUTPUT TRANSISTOR	
FOR T_A = FULL PACKAGE-TEMPERATURE RANGE (All Package Types)	100 mW
OPERATING-TEMPERATURE RANGE (T_A):	
PACKAGE TYPES D, F, H	−55 to +125°C
PACKAGE TYPE E	−40 to +85°C
STORAGE TEMPERATURE RANGE (T_{stg})	−65 to +150°C
LEAD TEMPERATURE (DURING SOLDERING):	
At distance 1/16 ± 1/32 inch (1.59 ± 0.79 mm) from case for 10 s max.	+265°C

RECOMMENDED OPERATING CONDITIONS at T_A=25°C, Except as Noted.
For maximum reliability, nominal operating conditions should be selected so that operation is always within the following ranges:

CHARACTERISTIC	LIMITS		UNITS
	Min.	Max.	
Supply-Voltage Range (V_{CC}) (For T_A=Full Package-Temperature Range)	3	18	V
Input Voltage Range (V_{IN})	V_{CC}*	18	V

*The CD4049 and CD4050 have high-to-low-level voltage conversion capability but not low-to-high-level; therefore it is recommended that $V_{IN} \geq V_{CC}$.

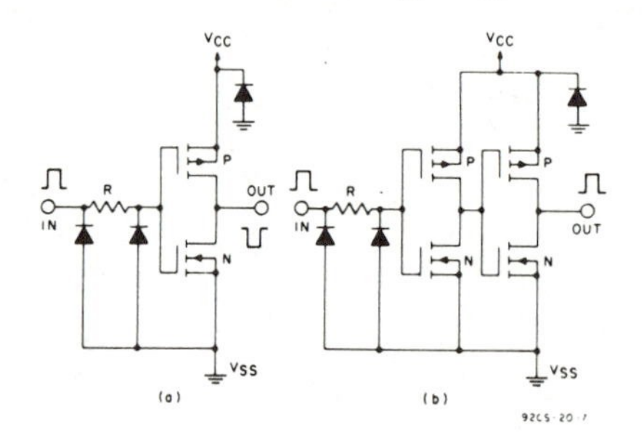

Fig. 1—a) Schematic diagram of CD4049UB, 1 of 6 identical units;
b) Schematic diagram of CD4050B, 1 of 6 identical units.

CD4049UB, CD4050B Types

DYNAMIC ELECTRICAL CHARACTERISTICS at T_A=25°C; Input t_r,t_f=20 ns, C_L=50 pF, R_L=200 kΩ

CHARACTERISTIC	CONDITIONS		LIMITS ALL PKGS.		UNITS
	V_{IN}	V_{CC}	Typ.	Max.	
Propagation Delay Time: Low-to-High, t_{PLH} CD4049UB	5	5	60	120	ns
	10	10	32	65	
	10	5	45	90	
	15	15	25	50	
	15	5	45	90	
CD4050B	5	5	70	140	
	10	10	40	80	
	10	5	45	90	
	15	15	30	60	
	15	5	40	80	
High-to-Low, t_{PHL} CD4049UB	5	5	32	65	ns
	10	10	20	40	
	10	5	15	30	
	15	15	15	30	
	15	5	10	20	
CD4050B	5	5	55	110	
	10	10	22	55	
	10	5	50	100	
	15	15	15	30	
	15	5	50	100	
Transition Time: Low-to-High, t_{TLH}	5	5	80	160	ns
	10	10	40	80	
	15	15	30	60	
High-to-Low, t_{THL}	5	5	30	60	
	10	10	20	40	
	15	15	15	30	
Input Capacitance, C_{IN} CD4049UB	–	–	15	22.5	pF
CD4050B	–	–	5	7.5	

appendix C: values of ϵ^x and ϵ^{-x}

x	Function	0.00	0.01	0.02	0.03	0.04	0.05	0.06	0.07	0.08	0.09
0.0	ϵ^x	1.0000	1.0101	1.0202	1.0305	1.0408	1.0513	1.0618	1.0725	1.0833	1.0942
	ϵ^{-x}	1.0000	0.9900	0.9802	0.9704	0.9608	0.9512	0.9418	0.9324	0.9231	0.9139
0.1	ϵ^x	1.1052	1.1163	1.1275	1.1388	1.1503	1.1618	1.1735	1.1853	1.1972	1.2093
	ϵ^{-x}	0.9048	0.8958	0.8869	0.8781	0.8694	0.8607	0.8521	0.8437	0.8353	0.8270
0.2	ϵ^x	1.2214	1.2337	1.2461	1.2546	1.2712	1.2840	1.2969	1.3100	1.3231	1.3364
	ϵ^{-x}	0.8187	0.8106	0.8025	0.7945	0.7856	0.7788	0.7711	0.7634	0.7558	0.7483
0.3	ϵ^x	1.3499	1.3634	1.3771	1.3910	1.4049	1.4191	1.4333	1.4477	1.4623	1.4770
	ϵ^{-x}	0.7408	0.7334	0.7261	0.7189	0.7118	0.7047	0.6977	0.6907	0.6839	0.6771
0.4	ϵ^x	1.4918	1.5068	1.5220	1.5373	1.5527	1.5683	1.5841	1.6000	1.6161	1.6323
	ϵ^{-x}	0.6703	0.6637	0.6570	0.6505	0.6440	0.6376	0.6313	0.6250	0.6188	0.6126
0.5	ϵ^x	1.6487	1.6653	1.6820	1.6989	1.7160	1.7333	1.7507	1.7683	1.7860	1.8040
	ϵ^{-x}	0.6065	0.6005	0.5945	0.5886	0.5827	0.5769	0.5712	0.5655	0.5599	0.5543
0.6	ϵ^x	1.8221	1.8404	1.8589	1.8776	1.8965	1.9155	1.9348	1.9542	1.9739	1.9939
	ϵ^{-x}	0.5488	0.5434	0.5379	0.5326	0.5273	0.5220	0.5169	0.5117	0.5066	0.5017
0.7	ϵ^x	2.0138	2.0340	2.0544	2.0751	2.0959	2.1170	2.1383	2.1598	2.1815	2.2034
	ϵ^{-x}	0.4966	0.4916	0.4868	0.4819	0.4771	0.4724	0.4677	0.4630	0.4584	0.4538
0.8	ϵ^x	2.2255	2.2479	2.2705	2.2933	2.3164	2.3396	2.3632	2.3869	2.4109	2.4351
	ϵ^{-x}	0.4493	0.4449	0.4404	0.4360	0.4317	0.4274	0.4232	0.4190	0.4148	0.4107
0.9	ϵ^x	2.4596	2.4843	2.5093	2.5345	2.5600	2.5857	2.6117	2.6379	2.6645	2.6912
	ϵ^{-x}	0.4066	0.4025	0.3985	0.3946	0.3906	0.3867	0.3829	0.3791	0.3753	0.3716
1.0	ϵ^x	2.7183	2.7456	2.7732	2.8011	2.8292	2.8577	2.8864	2.9154	2.9447	2.9743
	ϵ^{-x}	0.3679	0.3642	0.3606	0.3570	0.3535	0.3499	0.3465	0.3430	0.3396	0.3362
1.1	ϵ^x	3.0042	3.0344	3.0649	3.0957	3.1268	3.1582	3.1899	3.2220	3.2544	3.2871
	ϵ^{-x}	0.3329	0.3296	0.3263	0.3230	0.3198	0.3166	0.3135	0.3104	0.3073	0.3042
1.2	ϵ^x	3.3201	3.3535	3.3872	3.4212	3.4556	3.4903	3.5254	3.5609	3.5966	3.6328
	ϵ^{-x}	0.3012	0.2982	0.2952	0.2923	0.2894	0.2865	0.2837	0.2808	0.2780	0.2753
1.3	ϵ^x	3.6693	3.7062	3.7434	3.7810	3.8190	3.8574	3.8962	3.9354	3.9749	4.0149
	ϵ^{-x}	0.2725	0.2698	0.2671	0.2645	0.2618	0.2592	0.2567	0.2541	0.2516	0.2491
1.4	ϵ^x	4.0552	4.0960	4.1371	4.1787	4.2207	4.2631	4.3060	4.3492	4.3929	4.4371
	ϵ^{-x}	0.2466	0.2441	0.2417	0.2393	0.2369	0.2346	0.2322	0.2299	0.2276	0.2254
1.5	ϵ^x	4.4817	4.5267	4.5722	4.6182	4.6646	4.7115	4.7588	4.8066	4.8550	4.9037
	ϵ^{-x}	0.2231	0.2209	0.2187	0.2165	0.2144	0.2122	0.2101	0.2080	0.2060	0.2039
1.6	ϵ^x	4.9530	5.0028	5.0531	5.1039	5.1552	5.2070	5.2593	5.3122	5.3656	5.4195
	ϵ^{-x}	0.2019	0.1999	0.1979	0.1959	0.1940	0.1920	0.1901	0.1882	0.1864	0.1845
1.7	ϵ^x	5.4739	5.5290	5.5845	5.6407	5.6973	5.7546	5.8124	5.8709	5.9299	5.9895
	ϵ^{-x}	0.1827	0.1809	0.1791	0.1773	0.1755	0.1738	0.1720	0.1703	0.1686	0.1670
1.8	ϵ^x	6.0496	6.1104	6.1719	6.2339	6.2965	6.3598	6.4237	6.4883	6.5535	6.6194
	ϵ^{-x}	0.1653	0.1637	0.1620	0.1604	0.1588	0.1572	0.1557	0.1541	0.1526	0.1511
1.9	ϵ^x	6.6859	6.7531	6.8210	6.8895	6.9588	7.0287	7.0993	7.1707	7.2427	7.3155
	ϵ^{-x}	0.1496	0.1481	0.1466	0.1451	0.1437	0.1423	0.1409	0.1395	0.1381	0.1367

Tables courtesy of John Wiley & Sons, Inc., from Donald P. Leach, *Basic Electric Circuits*.

x	Function	0.00	0.01	0.02	0.03	0.04	0.05	0.06	0.07	0.08	0.09
2.0	ϵ^x	7.3891	7.4633	7.5383	7.6141	7.6906	7.7679	7.8460	7.9248	8.0045	8.8049
	ϵ^{-x}	0.1353	0.1340	0.1327	0.1313	0.1300	0.1287	0.1275	0.1262	0.1249	0.1237
2.1	ϵ^x	8.1662	8.2482	8.3311	8.4149	8.4994	8.5849	8.6711	8.7583	8.8463	8.9352
	ϵ^{-x}	0.1225	0.1212	0.1200	0.1188	0.1177	0.1165	0.1153	0.1142	0.1130	0.1119
2.2	ϵ^x	9.0250	9.1157	9.2073	9.2999	9.3933	9.4877	9.5831	9.6794	9.7767	9.8749
	ϵ^{-x}	0.1108	0.1097	0.1086	0.1075	0.1065	0.1054	0.1044	0.1033	0.1023	0.1013
2.3	ϵ^x	9.9742	10.074	10.716	10.278	10.381	10.486	10.591	10.697	10.805	10.913
	ϵ^{-x}	0.1003	0.0993	0.0983	0.0973	0.0963	0.0954	0.0944	0.0935	0.0926	0.0916
2.4	ϵ^x	11.023	11.134	11.246	11.359	11.473	11.588	11.705	11.822	11.941	12.061
	ϵ^{-x}	0.0907	0.0898	0.0889	0.0880	0.0872	0.0863	0.0854	0.0846	0.0837	0.0829
2.5	ϵ^x	12.182	12.305	12.429	12.553	12.680	12.807	12.936	13.066	13.197	13.330
	ϵ^{-x}	0.0821	0.0813	0.0805	0.0797	0.0789	0.0781	0.0773	0.0765	0.0758	0.0750
2.6	ϵ^x	13.464	13.599	13.736	13.874	14.013	14.154	14.296	14.440	14.585	14.732
	ϵ^{-x}	0.0743	0.0735	0.0728	0.0721	0.0714	0.0707	0.0699	0.0693	0.0686	0.0679
2.7	ϵ^x	14.880	15.029	15.180	15.333	15.487	15.643	15.800	15.959	16.119	16.281
	ϵ^{-x}	0.0672	0.0665	0.0659	0.0652	0.0646	0.0639	0.0633	0.0627	0.0620	0.0614
2.8	ϵ^x	16.445	16.610	16.777	16.945	17.116	17.288	17.462	17.637	17.814	17.993
	ϵ^{-x}	0.0608	0.0602	0.0596	0.0590	0.0584	0.0578	0.0573	0.0567	0.0561	0.0556
2.9	ϵ^x	18.174	18.357	18.541	18.728	18.916	19.106	19.298	19.492	19.688	19.886
	ϵ^{-x}	0.0550	0.0545	0.0539	0.0534	0.0529	0.0523	0.0518	0.0513	0.0508	0.0503
3.0	ϵ^x	20.086	20.287	20.491	20.697	20.905	21.115	21.328	21.542	21.758	21.977
	ϵ^{-x}	0.0498	0.0493	0.0488	0.0483	0.0478	0.0474	0.0469	0.0464	0.0460	0.0455
3.1	ϵ^x	22.198	22.421	22.646	22.874	23.104	23.336	23.571	23.807	24.047	24.288
	ϵ^{-x}	0.0450	0.0446	0.0442	0.0437	0.0433	0.0429	0.0424	0.0420	0.0416	0.0412
3.2	ϵ^x	24.533	24.779	25.028	25.280	25.534	25.790	26.050	26.311	26.576	26.843
	ϵ^{-x}	0.0408	0.0404	0.0400	0.0396	0.0392	0.0388	0.0384	0.0380	0.0376	0.0373
3.3	ϵ^x	27.113	26.385	27.660	27.938	28.219	28.503	28.789	29.079	29.371	29.666
	ϵ^{-x}	0.0369	0.0365	0.0362	0.0358	0.0354	0.0351	0.0347	0.0344	0.0340	0.0337
3.4	ϵ^x	29.964	30.265	30.569	30.877	31.187	31.500	31.817	32.137	32.460	32.786
	ϵ^{-x}	0.0334	0.0330	0.0327	0.0324	0.0321	0.0317	0.0314	0.0311	0.0308	0.0305
3.5	ϵ^x	33.115	33.448	33.784	34.124	34.467	34.813	35.163	35.517	35.874	36.234
	ϵ^{-x}	0.0302	0.0299	0.0296	0.0293	0.0290	0.0287	0.0284	0.0282	0.0279	0.0276
3.6	ϵ^x	36.598	36.966	37.338	37.713	38.092	38.475	38.861	39.252	39.646	40.045
	ϵ^{-x}	0.0273	0.0271	0.0268	0.0265	0.0263	0.0260	0.0257	0.0255	0.0252	0.0250
3.7	ϵ^x	40.447	40.854	41.264	41.679	42.098	42.521	42.948	43.380	43.816	44.256
	ϵ^{-x}	0.0247	0.0245	0.0242	0.0240	0.0238	0.0235	0.0233	0.0231	0.0228	0.0226
3.8	ϵ^x	44.701	45.150	45.604	46.063	46.525	46.993	47.465	47.942	48.424	48.911
	ϵ^{-x}	0.0224	0.0221	0.0219	0.0217	0.0215	0.0213	0.0211	0.0209	0.0207	0.0204
3.9	ϵ^x	49.402	49.899	50.400	50.907	51.419	51.935	52.457	52.985	53.517	54.055
	ϵ^{-x}	0.0202	0.0200	0.0198	0.0196	0.0195	0.0193	0.0191	0.0189	0.0187	0.0185

x	Function	0.00	0.01	0.02	0.03	0.04	0.05	0.06	0.07	0.08	0.09
4.0	ϵ^x	54.598	55.147	55.701	56.261	56.826	57.397	57.974	58.557	59.145	59.740
	ϵ^{-x}	0.0183	0.0181	0.0180	0.0178	0.0176	0.0174	0.0172	0.0171	0.0169	0.0167
4.1	ϵ^x	60.340	60.947	61.559	62.178	62.803	63.434	64.072	64.714	65.366	66.023
	ϵ^{-x}	0.0166	0.0164	0.0162	0.0161	0.0159	0.0158	0.0156	0.0155	0.0153	0.0151
4.2	ϵ^x	66.686	67.357	68.033	68.717	69.408	70.105	70.810	71.522	72.240	72.966
	ϵ^{-x}	0.0150	0.0148	0.0147	0.0146	0.0144	0.0143	0.0141	0.0140	0.0138	0.0137
4.3	ϵ^x	73.700	74.440	75.189	75.944	76.708	77.478	78.257	79.044	79.838	80.640
	ϵ^{-x}	0.0136	0.0134	0.0133	0.0132	0.0130	0.0129	0.0128	0.0127	0.0125	0.0124
4.4	ϵ^x	81.451	82.269	83.096	83.931	84.775	85.627	86.488	87.357	88.235	89.121
	ϵ^{-x}	0.0123	0.0122	0.0120	0.0119	0.0118	0.0117	0.0116	0.0114	0.0113	0.0112
4.5	ϵ^x	90.017	90.922	91.836	92.759	93.691	94.632	95.583	96.544	97.514	98.494
	ϵ^{-x}	0.0111	0.0110	0.0109	0.0108	0.0107	0.0106	0.0105	0.0104	0.0103	0.0102
4.6	ϵ_x	99.484	100.48	101.49	102.51	103.54	104.58	105.64	106.70	107.77	108.85
	ϵ^{-x}	0.0101	0.0100	0.0099	0.0098	0.0097	0.0096	0.0095	0.0094	0.0093	0.0092
4.7	ϵ^x	109.95	111.05	112.17	113.30	114.43	115.58	116.75	117.92	119.10	120.30
	ϵ^{-x}	0.0091	0.0090	0.0089	0.0088	0.0087	0.0087	0.0086	0.0085	0.0084	0.0083
4.8	ϵ^x	121.51	122.73	123.97	125.21	126.47	127.74	129.02	130.32	131.63	132.95
	ϵ^{-x}	0.0082	0.0081	0.0081	0.0080	0.0079	0.0078	0.0078	0.0077	0.0076	0.0075
4.9	ϵ^x	134.29	135.64	137.00	138.38	139.77	141.17	142.59	144.03	145.47	146.94
	ϵ^{-x}	0.0074	0.0074	0.0073	0.0072	0.0072	0.0071	0.0070	0.0069	0.0069	0.0068
5.0	ϵ^x	148.41	149.90	151.41	152.93	154.47	156.02	157.59	159.17	160.77	162.39
	ϵ^{-x}	0.0067	0.0067	0.0066	0.0065	0.0065	0.0064	0.0063	0.0063	0.0062	0.0062
5.1	ϵ^x	164.02	165.67	167.34	169.02	170.72	172.43	174.16	175.91	177.68	179.47
	ϵ^{-x}	0.0061	0.0060	0.0060	0.0059	0.0059	0.0058	0.0057	0.0057	0.0056	0.0056
5.2	ϵ^x	181.27	183.09	184.93	186.79	188.67	190.57	192.48	194.42	196.37	198.34
	ϵ^{-x}	0.0055	0.0055	0.0054	0.0054	0.0053	0.0052	0.0052	0.0051	0.0051	0.0050
5.3	ϵ^x	200.34	202.35	204.38	206.44	208.51	210.61	212.72	214.86	217.02	219.20
	ϵ^{-x}	0.0050	0.0049	0.0049	0.0048	0.0048	0.0047	0.0047	0.0047	0.0046	0.0046
5.4	ϵ^x	221.41	223.63	225.88	228.15	230.44	232.76	235.10	237.46	239.85	242.26
	ϵ^{-x}	0.0045	0.0045	0.0044	0.0044	0.0043	0.0043	0.0043	0.0042	0.0042	0.0041
5.5	ϵ^x	244.69	247.15	249.64	252.14	254.68	257.24	259.82	262.43	265.07	267.74
	ϵ^{-x}	0.0041	0.0040	0.0040	0.0040	0.0039	0.0039	0.0038	0.0038	0.0038	0.0037
5.6	ϵ^x	270.43	273.14	275.89	278.66	281.46	284.29	287.15	290.03	292.95	295.89
	ϵ^{-x}	0.0037	0.0037	0.0036	0.0036	0.0036	0.0035	0.0035	0.0034	0.0034	0.0034
5.7	ϵ^x	298.87	301.87	304.90	307.97	311.06	314.19	317.35	320.54	323.76	327.01
	ϵ^{-x}	0.0033	0.0033	0.0033	0.0032	0.0032	0.0032	0.0032	0.0031	0.0031	0.0031
5.8	ϵ_x	330.30	333.62	336.97	340.36	343.78	347.23	350.72	354.25	357.81	361.41
	ϵ^{-x}	0.0030	0.0030	0.0030	0.0029	0.0029	0.0029	0.0029	0.0028	0.0028	0.0028
5.9	ϵ^x	365.04	368.71	372.41	376.15	379.93	383.75	387.61	391.51	395.44	399.41
	ϵ^{-x}	0.0027	0.0027	0.0027	0.0027	0.0026	0.0026	0.0026	0.0026	0.0025	0.0025

Answers to Selected Problems

Chapter 1

1.6 False.
1.7 *b*
1.8 None.
1.11 18 V
1.12 −3 V
1.17 $V_{OUT} = 0$ V only when $V_{IN_1} = V_{IN_2} = 10$ V
1.18 Clockwise (CW).
1.19 (a) −10 V, +10 V, −16 V, −6 V;
(b) 2 V, −2 V, 12 V, 10 V
1.20 72 V, CW
1.21 9 V
1.22 10 V, −4 V, 14 V

Chapter 2

2.1 120 ns, 50 ns, 245 ns
2.3

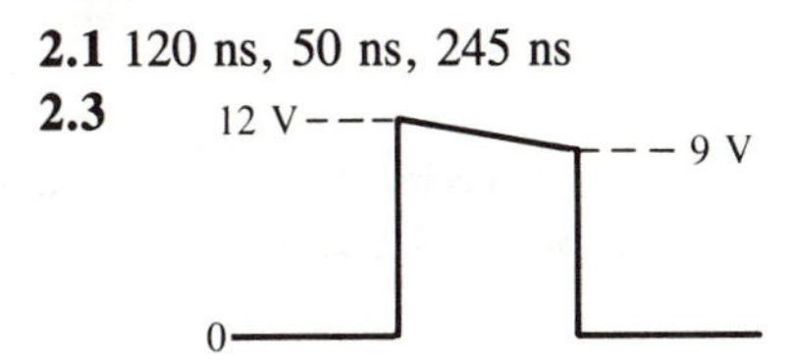

2.6 83.3 Hz, 25%, 3.25 V
2.9 (b) 50 μs; (c) −2.8 V

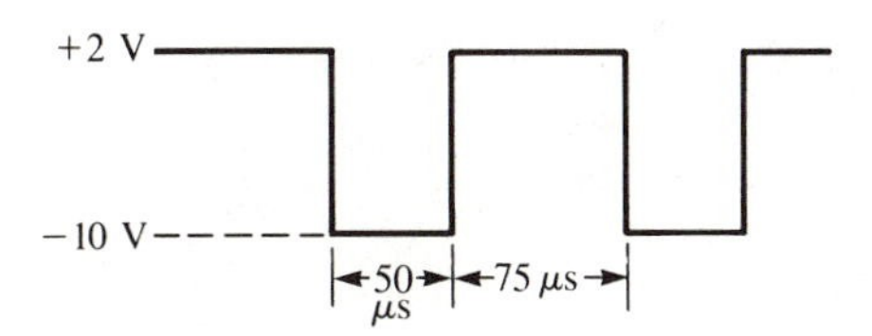

2.11 3.5 V
2.14 8 kpps
2.16 83.3, 166.6, 250, 333.3, 416.6, and 500 kHz
2.17 20 ns
2.19 No.
2.20 Yes.
2.21 4.8 V DC
2.23 2.1 kHz, 143 MHz
2.25 500 MHz, 525 MHz
2.27 2.33 MHz

Chapter 3

3.1 1.2 ms
3.2 −4.44 V, −7.56 V
3.5 0.2725, 0.779, 0.0012, 0.98
3.6 1.39
3.7 $-12 + 12\,\epsilon^{-t/1.2\text{ ms}}$
3.8 −6.78 V, −12 V
3.9 1.23 ms
3.10 $40\,\epsilon^{-t/15\,\mu\text{s}}$
3.11 $-20\,\epsilon^{-t/.22\text{ ms}}$
3.13 $13.6 + 2.2\,\epsilon^{-t/\tau}$; $4.4 - 2.2\,\epsilon^{-t/\tau}$
3.14 $f_{lc} \approx 100$ Hz
3.15 15.9 pF
3.17 220 μs
3.18 Nine.
3.21 5 percent.
3.23 *c* and *d*
3.28 False.
3.30 4; 8; 4
3.31 4 V; 4 V
3.35 0.5 μF
3.36 (a) 1 V; (b) 2 V; 5 V; (c) 5 s
3.41 5 pF
3.44 50 MΩ, 0.66 pF
3.45 1.56 inches
3.46 1000 ohms
3.47 250 ohms

Chapter 4

4.1 (a) 24 V, 0 mA; (b) 0.7 V, 5.7 mA; (c) 50 V, 7.57 mA; (d) 44 V, 0 mA; (e) 0.36 V, 0 mA;
4.8 0.53 mA, 40 mA, 0 V
4.10 ≈0 V
4.11 2.28 kΩ
4.12 *a, b, d*
4.14 (a), (b), (d), (e) decreases; (c) increases; (f) no effect on t_{ON}
4.15 (a), (b), (d) increases; (c), (e), (f) decreases
4.17 50 mV; 11.4 V
4.20 +20 V; 0 V
4.21 420 kΩ
4.22 *b*

Chapter 5

5.1 (a) 2.5 V; (b) −2.5 V; (c) −10.5 V; (d) +10.5 V
5.2 1.9500744 V
5.3 0.21 mV
5.5 −30 μV, 600 pA
5.12 (a) 9.9975 mV p-p; (b) 9.9965 mV p-p; (c) 9.9964 mV p-p
5.13 9.9964 mV p-p
5.14 (a) 3.91 V; (b) 3.45 V
5.15 200
5.17 200 mV p-p
5.18 −10 V
5.21 15 kΩ, 750 kΩ
5.24 (a) 71.4 kHz; (b) 2.5 kHz; (c) 2.5 MHz
5.25 12.8 μs
5.26 54 V/μs
5.28 1.92 kΩ
5.29 8.33 kΩ

Chapter 6

6.1 100 V p-p
6.2 0 V
6.3 796 ohms
6.4 200 mV p-p
6.8 106 ohms
6.9 $v_{OUT} = e_i$; 10 V
6.10 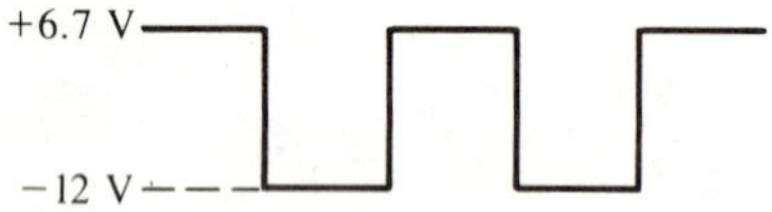
+6.7 V
−12 V
6.13 35 μV
6.14 8.2 V peak
6.15 6.66 mA
6.16 −2 V negative spikes
6.18 23.5 V
6.20 *b*
6.24 (a) Increase; (b) Decrease; (c) Increase; (d) Increase.
6.26 124
6.27 20.66 pF
6.30 *a*
6.31 *a*
6.32 0.7 V
6.33 (a) 7.5 milliohms; (b) 440 ohms
6.34 3.00135 V, 2.99865 V, 2.7 mV
6.37 False.
6.38 $t_r = t_f = 4.3$ μs; One possible set of values is $R = 3.9$ kΩ and C = 270 pF.
6.39 333.3 μs; 0.5 ms

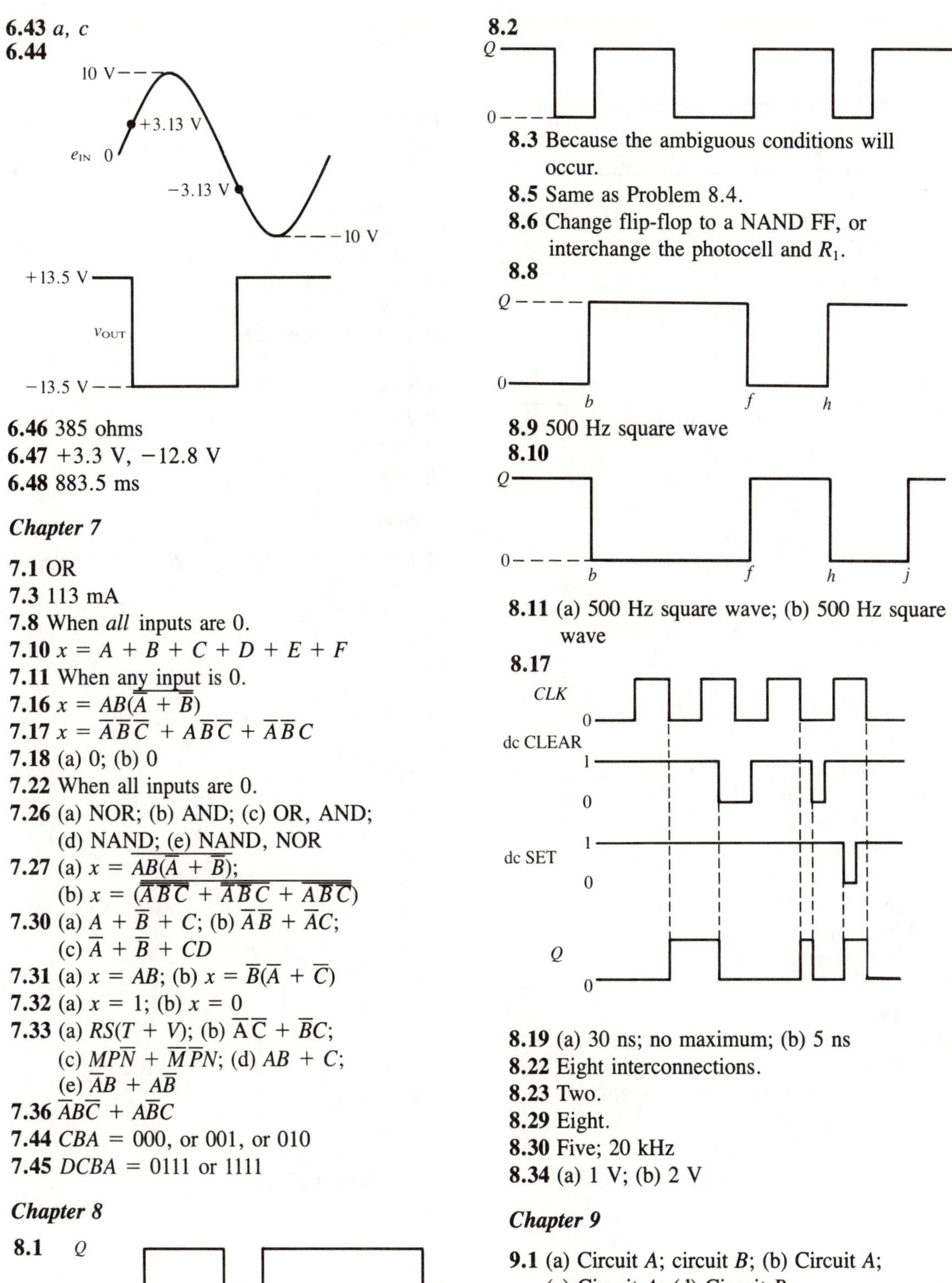

6.43 *a, c*

6.44

6.46 385 ohms

6.47 +3.3 V, −12.8 V

6.48 883.5 ms

Chapter 7

7.1 OR

7.3 113 mA

7.8 When *all* inputs are 0.

7.10 $x = A + B + C + D + E + F$

7.11 When any input is 0.

7.16 $x = AB(\overline{\overline{A} + \overline{B}})$

7.17 $x = \overline{A}\,\overline{B}\,\overline{C} + A\overline{B}\,\overline{C} + \overline{A}\,\overline{B}C$

7.18 (a) 0; (b) 0

7.22 When all inputs are 0.

7.26 (a) NOR; (b) AND; (c) OR, AND; (d) NAND; (e) NAND, NOR

7.27 (a) $x = \overline{AB(\overline{\overline{A} + \overline{B}})}$;
(b) $x = \overline{(\overline{\overline{A}\,\overline{B}\,\overline{C}} + \overline{\overline{A}\,\overline{B}C} + \overline{A\overline{B}\,\overline{C}})}$

7.30 (a) $A + \overline{B} + C$; (b) $\overline{A}\,\overline{B} + \overline{A}C$; (c) $\overline{A} + \overline{B} + CD$

7.31 (a) $x = AB$; (b) $x = \overline{B}(\overline{A} + \overline{C})$

7.32 (a) $x = 1$; (b) $x = 0$

7.33 (a) $RS(T + V)$; (b) $\overline{A}\,\overline{C} + \overline{B}C$; (c) $MP\overline{N} + \overline{M}\,\overline{P}N$; (d) $AB + C$; (e) $\overline{A}B + A\overline{B}$

7.36 $\overline{A}B\overline{C} + A\overline{B}C$

7.44 CBA = 000, or 001, or 010

7.45 $DCBA$ = 0111 or 1111

Chapter 8

8.1

8.2

8.3 Because the ambiguous conditions will occur.

8.5 Same as Problem 8.4.

8.6 Change flip-flop to a NAND FF, or interchange the photocell and R_1.

8.8

8.9 500 Hz square wave

8.10

8.11 (a) 500 Hz square wave; (b) 500 Hz square wave

8.17

8.19 (a) 30 ns; no maximum; (b) 5 ns

8.22 Eight interconnections.

8.23 Two.

8.29 Eight.

8.30 Five; 20 kHz

8.34 (a) 1 V; (b) 2 V

Chapter 9

9.1 (a) Circuit *A*; circuit *B*; (b) Circuit *A*; (c) Circuit *A*; (d) Circuit *B*.

9.3 1.075 mA

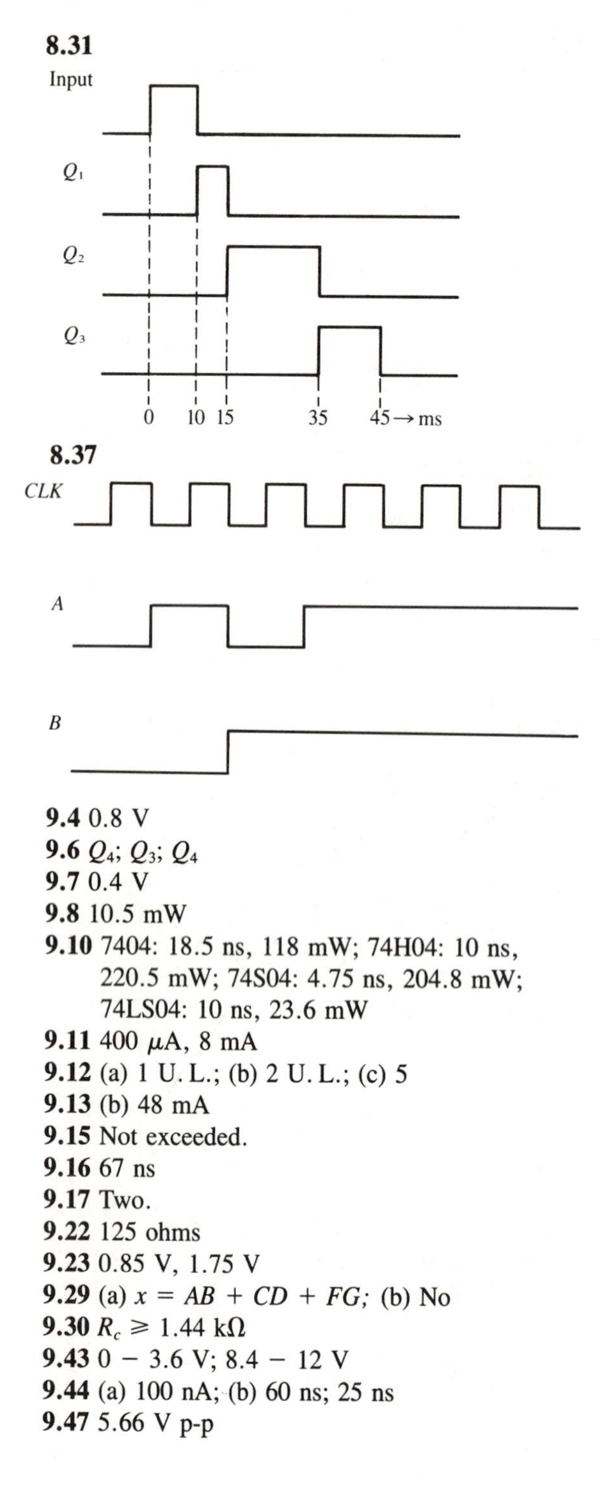

9.4 0.8 V
9.6 Q_4; Q_3; Q_4
9.7 0.4 V
9.8 10.5 mW
9.10 7404: 18.5 ns, 118 mW; 74H04: 10 ns, 220.5 mW; 74S04: 4.75 ns, 204.8 mW; 74LS04: 10 ns, 23.6 mW
9.11 400 μA, 8 mA
9.12 (a) 1 U.L.; (b) 2 U.L.; (c) 5
9.13 (b) 48 mA
9.15 Not exceeded.
9.16 67 ns
9.17 Two.
9.22 125 ohms
9.23 0.85 V, 1.75 V
9.29 (a) $x = AB + CD + FG$; (b) No
9.30 $R_c \geqslant 1.44$ kΩ
9.43 0 − 3.6 V; 8.4 − 12 V
9.44 (a) 100 nA; (b) 60 ns; 25 ns
9.47 5.66 V p-p
9.48 3 V, 4.5 V, 4.5 V, 6 V

Chapter 10

10.2 479 Ω to 9.37 kΩ
10.3 25
10.4 Four.
10.6 Two; seven.
10.7 (a) zero; (b) 91.5 μA; (c) 0.43 mA
10.10 (b) 37.1 mA; 42.8 mA
10.13 390 Ω; 92 kΩ
10.14 (a) 288 Ω; (b) 625 mA
10.15 4 mW; 300 mW
10.16 10 W
10.17 Seven.
10.25 1 kΩ
10.26 Yes.
10.27 No.

Chapter 11

11.1 (a) 13.5 V; (b) 482.5 mA; (c) 107.5 mA; (d) 0.78 V
11.4 15 Hz to 1.44 kHz
11.6 $A = 0$, $B = 10$ V
11.8 0.525 ms
11.9 $T = 0.185$ ms
11.14 (a) 0; (b) 795 kHz
11.16 1.79 kHz
11.21 11 s
11.24 *a*
11.26 1.5 K, 150 K, 950 pF
11.29 *d*
11.32 8 mA
11.33 $V_P = 10$ V

10 V
0
0
200 250
450 500 → μs

11.35 41.5 K
11.36 (a) 20.6 μs
11.37 4 MΩ
11.39 23 μs to 23 ms
11.49 *b*

index